D1569126

Linear Systems, Fourier Transforms, and Optics

WILEY SERIES IN PURE AND APPLIED OPTICS

Advisory Editor

Stanley S. Ballard, University of Florida

ALLEN AND EBERLY • *Optical Resonance and Two-Level Atoms*
BABCOCK • *Silicate Glass Technology Methods*
BOND • *Crystal Technology*
CATHEY • *Optical Information Processing and Holography*
CAULFIELD AND LU • *The Applications of Holography*
EBERLY AND LAMBROPOULOS • *Multiphoton Processes*
GASKILL • *Linear Systems, Fourier Transforms, and Optics*
GERRARD AND BURCH • *Introduction of Matrix Methods in Optics*
HUDSON • *Infrared System Engineering*
JUDD AND WYSZECKI • *Color in Business, Science, and Industry,* Third Edition
KNITTL • *Optics of Thin Films*
LENGYEL • *Lasers,* Second Edition
LEVI • *Applied Optics, A Guide to Optical System Design,* Volume I
LOUISELL • *Quantum Statistical Properties of Radiation*
MALACARA • *Optical Shop Testing*
McCARTNEY • *Optics of the Atmosphere; Scattering by Molecules and Particles*
MOLLER AND ROTHSCHILD • *Far-Infrared Spectroscopy*
PRATT • *Laser Communications Systems*
ROGERS • *Noncoherent Optical Processing*
SHULMAN • *Optical Data Processing*
WILLIAMS AND BECKLUND • *Optics*
ZERNIKE AND MIDWINTER • *Applied Nonlinear Optics*

Linear Systems, Fourier Transforms, and Optics

JACK D. GASKILL
Professor of Optical Sciences
Optical Sciences Center
University of Arizona

John Wiley & Sons, New York / Chichester / Brisbane / Toronto

Library of Congress Cataloging in Publication Data:

Gaskill, Jack D.
Linear systems, Fourier transforms, and optics.

(Wiley series in pure and applied optics)
Includes bibliographical references and index.
1. Optics. 2. Fourier transformations.
3. System analysis. I. Title.

QC355.2.G37 535 78-1118
ISBN 0-471-29288-5

Printed in the United States of America

10 9 8 7 6 5 4 3

To my students—past and future

Preface

Since the introduction of the laser in 1960, the application of communication theory to the analysis and synthesis of optical systems has become extremely popular. Central to the theory of communication is that part of mathematics developed by Jacques Fourier, who first undertook a systematic study of the series and integral expansions that now bear his name. Also important to communication theory are the concepts associated with linear systems and the characterization of such systems by mathematical operators. Although there are a number of books available that provide excellent treatments of these topics individually, in my opinion there has not been a single book that adequately combines all of them in a complete and orderly fashion. To illustrate, most of the good books on Fourier analysis contain very little material about optics, and most of those devoted to optical applications of communication theory assume that the reader has prior familiarity with Fourier analysis and linear systems.

In writing this book I have attempted to remedy the situation just described by including complete treatments of such important topics as general harmonic analysis, linear systems, convolution, and Fourier transformation, first for one-dimensional signals and then for two-dimensional signals. The importance attached to these topics becomes apparent with the observation that they comprise over 60% of the material in the book. Following the development of this strong mathematical foundation, the phenomenon of diffraction is investigated in considerable depth. Included in this study are Fresnel and Fraunhofer diffraction, the effects of lenses on diffraction, and the propagation of Gaussian beams, with particularly close attention being paid to the conditions required for validity of the theory. Finally, the concepts of linear systems and Fourier analysis are combined with the theory of diffraction to describe the image-forming process in terms of a linear filtering operation for both coherent and incoherent imaging. With this background in Fourier optics the reader should be prepared to undertake more advanced studies of such topics as holography and optical data processing, for which there already exist several good books and innumerable technical papers.

The book evolved from a set of course notes developed for a one-semester course at the University of Arizona. This course, which is basically an applied mathematics course presented from the viewpoint of an engineer-turned-opticist, is intended primarily for students in the first year of a graduate program in optical sciences. The only absolute prerequisite for the course is a solid foundation in differential and integral calculus; a background in optics, although helpful, is not required. (To aid those with no previous training in optics, a section on geometrical optics is included as Appendix 2.) Consequently, the book should be suitable for courses in disciplines other than optical sciences (e.g., physics and electrical engineering). In addition, by reducing the amount of material covered, by altering the time allotted to various topics, and/or by revising the performance standards for the course, the book could be used for an undergraduate-level course. For example, the constraints of an undergraduate course might dictate the omission of those parts of the book concerned with descriptions of two-dimensional functions in polar coordinate systems (Sec. 3-4), convolution in polar coordinates (Sec. 9-2), and Hankel transforms (Sec. 9-5). The subjects of diffraction and image formation might still be investigated in some detail, but the student would be required to solve only those problems that can be described in rectangular coordinates. On the other hand, the book might be adapted for a one-quarter course in linear systems and Fourier analysis by omitting the chapters on diffraction theory and image formation altogether.

A carefully designed set of problems is provided at the end of each chapter to help guide the reader through the learning process in an orderly manner. Some of these problems have parts that are entirely independent of one another, whereas other problems have closely related parts. By careful selection of exercises (or combinations of exercises), an instructor can emphasize a particular topic to any desired degree. For example, if the student is required only to be familiar with a certain operation, a single part of an appropriate problem might be assigned. On the other hand, if the student is required to be highly proficient in performing that operation, all parts of the problem might be assigned. Many of the problems request that sketches of various functions be provided, and students often complain that such a task is not only tedious but of questionable value. However, a simple sketch can be a very important ingredient of the problem-solving process as illustrated by two famous sayings: you don't understand it if you can't sketch it, and a word is only worth a millisketch. Since there are many more exercises than will normally be required for a single course offering, different sets of exercises can be assigned each time the course is given—at least for a few times. As a final comment about the problems, individuals who can work all of them may feel confident that they have mastered the material superbly.

Because this book deals with applied mathematics, I did not feel it necessary to emphasize such topics as convergence and existence to the extent a pure mathematician might have. In addition, my engineering treatment of certain other topics (e.g., delta functions) is likely to produce some minor discomfort within the graves of a number of deceased mathematicians. Nevertheless, I have attempted to be as mathematically rigorous and precise as possible without losing sight of the objectives of the book. Wherever practical I have attempted to relate the physics of a process to the mathematics describing it and to present examples that illustrate these relationships. Although the book was written as a textbook, it should also serve as a useful reference for those already well versed in the areas of Fourier analysis, diffraction theory, and image formation. The following items should be of particular interest to these individuals: the extensive tables of properties and pairs of Fourier transforms and Hankel transforms; the completely general formulation of the effects of lenses on the diffraction phenomenon; the presentation of some surprising aspects (which are well known, but not widely known) of Gaussian beam propagation; and the completely general formulation of coherent and incoherent image formation.

I gratefully acknowledge the contributions of the many individuals who have played a part in the development of this book. Although an attempt to list all of their names would be impractical, I would like to single out a few to whom I am particularly indebted. Listed more or less according to the chronology of their contributions, they are Jim Omura, who, as a graduate teaching assistant at Stanford University, first kindled my interest in the theory of communication; Joe Goodman, who made me aware of the benefits to be gained by applying communication theory to the field of optics; Roland Shack, who patiently tried to teach me something about the field of optics so that I might apply communication theory to it; Howard Morrow, whose many probing questions in the classroom contributed to my education and encouraged me to spend more time on the preparation of my lectures; Mary Cox and Roland Payne, who read the initial portions of the manuscript and made many helpful suggestions regarding organization and terminology; Vini Mahajan and John Greivenkamp, who carefully read portions of the original draft and prevented many substantive errors from reaching the final draft; Janet Rowe and Martha Stockton, who typed the manuscript and frequently kept me from being dashed upon the shoals of bad grammar; Don Cowen, who prepared the illustrations so beautifully; and my wife, Marjorie, who proofread the final typescript with painstaking care. Finally, I wish to acknowledge the contributions of all those individuals whose names do not appear above, but whose efforts in bringing the book to fruition are appreciated no less.

JACK D. GASKILL

Tucson, Arizona
January 1978

Contents

Linear Systems, Fourier Transforms, and Optics

CHAPTER 1

INTRODUCTION

This book was written primarily for use as a textbook and was designed specifically to help the reader master the fundamental concepts of linear systems, Fourier analysis, diffraction theory, and image formation. It should not be regarded as an advanced treatise on communication theory and Fourier optics, nor as a compilation of recently reported results in these fields. For those interested in such treatments, a number of excellent books have already been written (see, for example, Refs. 1-1 through 1-3). However, as a word of caution, many of these books presume a good understanding of linear systems and Fourier analysis, without which they are of little value. Once a good understanding of the prerequisite topics has been acquired, these books can be more readily understood.

The tutorial nature of the present book was adopted with the average student in mind; those individuals who find every concept to be trivial and every result immediately obvious will no doubt consider it to be somewhat elementary and tedious (but, then, it wasn't intended for them anyway). The philosophy employed is basically the following: it is best to learn how to walk before attempting to run. With this in mind, each section or chapter was designed to provide background material for later sections or chapters. In addition, the problems provided at the end of each chapter were devised to supplement the discussions of those chapters and to reinforce the important results thereof.

Idealizations were employed from time to time to simplify the development of the material and, whenever a physically untenable situation resulted from such an idealization, an attempt was made to explain the reasons for—and consequences of—the idealization. If all problems had been attacked without the use of simplifying approximations or assumptions, but with brute force alone, it would frequently have been difficult to arrive at any worthwhile conclusions. By first seeing how an idealized problem is handled, students may often obtain the solution to the non-idealized problem more readily. In fact, the effects of nonideal conditions may sometimes be regarded simply as perturbations of the ideal solution.

1-1 ORGANIZATION OF THE BOOK

The task of organizing a book can be quite difficult, and frequently there seems to be no optimum way of arranging the material. For example, consider the convolution theorem of Fourier transforms. This theorem greatly simplifies the evaluation of certain convolution integrals, but to understand it the reader must know not only what a Fourier transform is but also what a convolution integral is. However, because convolution integrals are often difficult to evaluate without employing the convolution theorem, it might seem fruitless to explore the former in any detail until the latter is understood. The latter, on the other hand, cannot be understood until..., etc. As a result, it is not clear whether the study of Fourier analysis should precede or follow the study of convolution. One way out of this predicament would be to omit both topics, but the use of such a tactic here would not serve the purpose of the book.

The determination of the final arrangement of material was based on a number of factors, including the classroom experience of the author (both as a student and as an instructor), suggestions of students familiar with the material, plus a number of educated guesses. The solution chosen for the dilemma mentioned above was to give a brief introduction to the Fourier transform in Chapter 4 followed by a detailed discussion of the convolution operation in Chapter 6. Finally, an intensive study of the Fourier transform, which included the convolution theorem, was undertaken in Chapter 7. Thus, the desired result was obtained by alternating between the two main subjects involved.

A similar question arose regarding the best order of presentation for the Fourier series and the Fourier integral. Because the former is merely a special case of the latter, it might appear that the more general Fourier integral should be discussed first. However, it seems to be easier for the beginning student to visualize the decomposition process when the basis set is a discrete set of harmonically related sine and cosine functions rather than when it is a continuous set of nonharmonically related sines and cosines. Consequently, it was deemed desirable to begin with the Fourier series. This type of rationale was also employed to determine the most suitable arrangement of other material.

1-2 CONTENTS OF THE BOOK

In Chapter 2 we present an elementary review of various properties and classes of mathematical functions, and we describe the manner in which these functions may be used to represent physical quantities. It is anticipated that Chapter 2 will require only a cursory reading by most individuals, but those with weak backgrounds in mathematics may want to devote a bit more time to it. In Chapter 3 we introduce a number of special functions that will prove to be of great utility in later chapters. In particular, we will find the rectangle function, the sinc function, the delta function, and the comb function to be extremely useful. Also, several special functions of two variables are described. As a suggestion, the section on coordinate transformations (Sec. 3-5) might be omitted—or given only minimal attention—until the two-dimensional operations of Chapter 9 are encountered.

Next, in Chapter 4, we explore the fundamentals of harmonic analysis and learn how various arbitrary functions may be represented by linear combinations of other, more elementary, functions. We then investigate, in Chapter 5, the description of physical systems in terms of mathematical operators, and we introduce the notions of linearity and shift invariance. Following this, the impulse response function, the transfer function, and the eigenfunctions associated with linear shift-invariant systems are discussed. Chapter 6 is devoted to studies of the convolution, cross-correlation, and autocorrelation operations, and the properties of these operations are explored in considerable depth. In addition, we derive the following fundamental result for linear shift-invariant systems: the output is given by the convolution of the input with the impulse response of the system.

In Chapter 7 we investigate the properties of the Fourier transformation and learn of its importance in the analysis of linear shift-invariant systems. In this regard the output spectrum of such a system is found to be given by the product of the input spectrum and the transfer function of the system, a consequence of the convolution theorem of Fourier transformation. Then, in Chapter 8, we describe the characteristics of various types of linear filters and discuss their applications in signal processing and recovery. We also consider the so-called matched filter problem and study various interpretations of the sampling theorem.

The material in Chapter 9 is designed to extend the student's previous knowledge of one-dimensional systems and signals to two dimensions. In particular, an investigation of convolution and Fourier transformation in two dimensions is conducted, and the Hankel transform and its properties are studied. Additionally, the line response and edge response functions are introduced. In Chapter 10 we explore the propagation and diffraction of optical wave fields, in both the Fresnel and Fraunhofer regions, and we study the effects of lenses on the diffraction process. Particular attention is devoted to the curious properties of Gaussian beams in the last section of this chapter.

Finally, in Chapter 11, the concepts of linear systems and Fourier analysis are combined with the theory of diffraction to describe the process of image formation in terms of a linear filtering operation. This is done for both coherent and incoherent imaging, and the corresponding impulse response functions and transfer functions are discussed in detail.

Several special functions are tabulated in Appendix 1 and, for those with little or no previous training in optics, the fundamentals of geometrical image formation and aberrations are presented in Appendix 2.

REFERENCES

1-1. J. W. Goodman, *Introduction to Fourier Optics*, McGraw-Hill, New York, 1968.

1-2. R. J. Collier, C. B. Burckhardt, and L. H. Lin, *Optical Holography*, Academic Press, New York, 1971.

1-3. W. T. Cathey, *Optical Information Processing and Holography*, Wiley, New York, 1974.

CHAPTER 2

REPRESENTATION OF PHYSICAL QUANTITIES BY MATHEMATICAL FUNCTIONS

To simplify the analysis of various scientific and engineering problems, it is almost always necessary to represent the physical quantities encountered by mathematical functions of one type or another. There are many types of functions, and the choice of an appropriate one for a specific situation depends largely on the nature of the problem at hand and the characteristics of the quantity to be represented. In this chapter we discuss several of the important properties and classes of functions, and the manner in which these functions are used to represent physical quantities. In addition, a review of complex numbers and phasors is presented. Some of the topics included may seem too elementary for this book, but we emphasize that you, the reader, should have a good understanding of them prior to tackling the more difficult topics of the chapters to follow. Also, it was felt that you would benefit by having all of these diverse topics available in one place for easy reference.

2-1 CLASSES AND PROPERTIES OF FUNCTIONS

There are many ways in which functions may be classified, and for the sake of brevity we shall restrict our attention to only those classes that will be of interest to us in the later chapters, namely, those important in our study of linear systems and optics.

General

Perhaps one of the most basic distinctions to be made when discussing the mathematical representation of physical phenomena is that which separates *deterministic phenomena* from *random phenomena*. The behavior of a deterministic phenomenon is completely predictable, whereas the behavior of a random phenomenon has some degree of uncertainty associated with it. To make this distinction a little clearer, let us consider the observation, or measurement, of some time-varying quantity associated with a particular phenomenon. Let us assume that the quantity of interest has been observed for a very long time and that we have a very good record of its past behavior. If, by knowing its past behavior, we were able to predict its future behavior exactly, we would say that this quantity is deterministic. On the other hand, if we were unable to predict its future behavior exactly, we would say that it is random.

Actually it is not entirely fair to draw a sharp dividing line between these two type of phenomena as we have just done. No phenomena are truly deterministic and none are completely random—it is a matter of degree. It is also a matter of ignorance on the part of the observer. The more we know about the factors governing the behavior of a particular phenomenon, the more likely we are to think of it as being deterministic. Conversely, the less we know about these factors, the more likely we are to say that it is random. We might conclude then that the use of the descriptors "deterministic" and "random" is not entirely proper for the classification of physical phenomena. Nevertheless, such usage is widespread among engineers and scientists and we shall adopt these terms for our work here.

Mathematical functions are used to represent various physical quantities associated with the phenomenon under investigation. When we deal with deterministic phenomena, these functions can often be expressed in terms of explicit mathematical formulas. For example, the motion of a simple pendulum is highly deterministic and can be described as a function of time according to an explicit formula. In contrast, the motion of the waves on the ocean is quite random in nature and cannot be so-described.

There is another point to consider regarding the distinction between deterministic and random quantities, and to make this point let us consider the transmission of a typical telephone message. Before the caller actually transmits the message, the individual on the receiving end does not know exactly what the message will be. (If he did, there would be no reason for making the call.) Consequently, as far as the listener is concerned, the message has some degree of uncertainty associated with it prior to transmission, and he is therefore required to treat it as a random message. However, once the message is received, there is no longer any uncertainty about it and the listener may now consider it to be deterministic. Thus the distinction between determinism and randomness must take into account the epoch for which the distinction is being made.

Because the treatment of random phenomena is beyond the scope of this book, we shall deal only with those phenomena that can be treated as if they are deterministic.

Many physical quantities can be represented by *scalar functions*, whereas others must be described by *vector functions*. The pressure P of an enclosed gas, for example, is a scalar quantity that depends on the gas temperature T, another scalar quantity. Hence P may be described by a scalar function of the scalar variable T. On the other hand, the electric field $\mathbf{E}$ associated with a propagating electromagnetic wave is a vector quantity that depends on position $\mathbf{r}$ and time t; it therefore must be represented by a vector function of the vector variable $\mathbf{r}$ and the scalar variable t. We are concerned primarily with scalar functions of scalar variables in our studies of linear systems and optics.

A function may be thought of as the "rule" relating the dependent variable to one or more independent variables. Suppose we are given the function

$$y=f(x), \tag{2.1}$$

where y is the dependent variable, x is the independent variable, and $f(\cdot)$ is the "rule" relating these two variables. If there exists only one value of y for each value of x, then y is called a *single-valued function* of x; if there exist more than one value, y is known as a *multiple-valued function* of x. To illustrate, let us consider the functions $y=x^2$ and $y=\pm\sqrt{x}$, with the restriction that x be real and nonnegative. The first of these is a single-valued function, whereas the second is double valued, and their graphs are shown in Fig. 2-1. In either of these cases the range of y includes only real numbers; therefore y is known as a *real-valued function*. If we had allowed

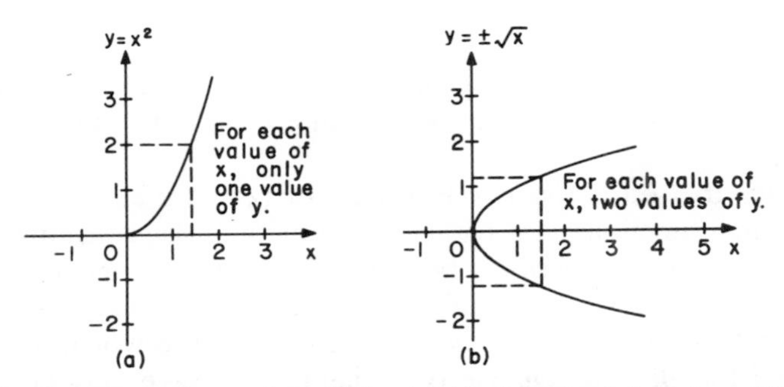

Figure 2-1 Functions of a real variable. (*a*) Single-valued function. (*b*) Double-valued function.

x to take on both positive and negative values, the latter of these functions would have been *complex-valued*. We shall deal only with real-valued functions in the remainder of this section, and will take up complex-valued functions in Sec. 2-2.

Another very important distinction to be made in the classification of functions has to do with whether they are *periodic* or *nonperiodic*. A periodic function $f(x)$ has the property that, for all x,

$$f(x)=f(x+nX), \tag{2.2}$$

where n is an integer and X is a real positive constant known as the period of $f(x)$. [Here it is assumed that X is the smallest number for which Eq. (2.2) is satisfied.] From this expression, it is clear that a periodic function repeats itself exactly after fixed intervals of nX. The reciprocal of the period is called the *fundamental frequency* of the function. When the independent variable is a spatial coordinate, such that x and X have dimensions of length, the fundamental frequency has dimensions of inverse length. In this case we use the symbol ξ_0 to represent the fundamental *spatial frequency* of the periodic function, i.e.,

$$\xi_0=\frac{1}{X}\cdot \tag{2.3}$$

The fundamental spatial frequency of a periodic function describes how many repetitions the function makes per unit length and is measured in units of cycles/meter. If the independent variable is time, as in

$$g(t)=g(t+nT), \tag{2.4}$$

where T is now the period, we shall use the symbol ν_0 to represent the fundamental *temporal frequency* of the function. Thus,

$$\nu_0=\frac{1}{T}, \tag{2.5}$$

and the units for ν_0 are cycles/second or Hertz. The fundamental temporal frequency specifies the number of repetitions made by the function per unit time.

Functions that do not satisfy Eqs. (2.2) or (2.4) are called nonperiodic, or aperiodic, functions. In reality, all functions representing physical quantities must be nonperiodic because, for such functions to be strictly periodic, certain physical principles would have to be violated. For example, a radio wave cannot be truly periodic because this would imply that it has existed for all time. On the other hand, this wave may be so nearly periodic that to consider it otherwise might be imprudent. In deciding

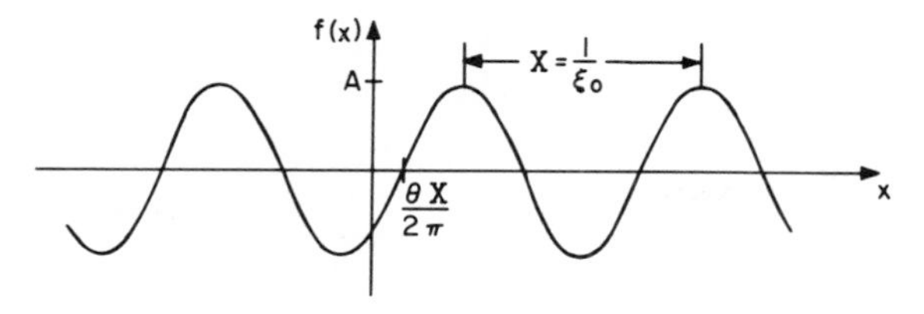

Figure 2-2 Sinusoidal function of amplitude A, frequency ξ_0, and phase shift θ.

whether some physical quantity may be treated as a periodic or nonperiodic quantity, a judgement of some sort will be necessary.

Perhaps the most common and most important periodic function is the sinusoidal function

$$f(x)=A\sin(2\pi\xi_0 x-\theta), \tag{2.6}$$

the graph of which is shown in Fig. 2-2. It is easy to see that this function repeats itself exactly after intervals of $nX=n/\xi_0$. The quantity θ, sometimes called the *phase shift*, is an arbitrary constant that determines the position of the function along the x-axis. The sine function is zero whenever its argument is equal to an integral multiple of π; thus the zeroes of the function given by Eq. (2.6) occur at the points $x=(n\pi+\theta)X/2\pi$. Another periodic function $g(t)=g(t+nT)$, this time of arbitrary form, is shown in Fig. 2-3.

An example of a nonperiodic function is the Gaussian function

$$h(x)=Ae^{-\pi(x/b)^2} \tag{2.7}$$

where b is a real positive number. The graph of this function is shown in Fig. 2-4.

There is another class of functions called *almost-periodic* functions. These functions are composed of the sum of two or more periodic functions whose periods are incommensurate. To illustrate, consider the two periodic functions $f_1(x)$ and $f_2(x)$ shown in Fig. 2-5, with periods X_1

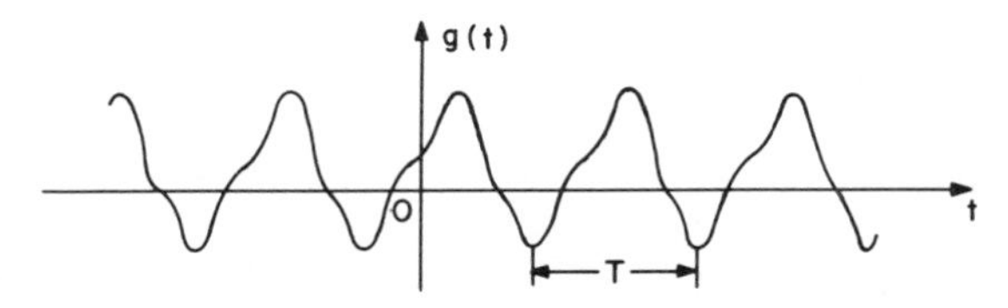

Figure 2-3 Periodic function of period T.

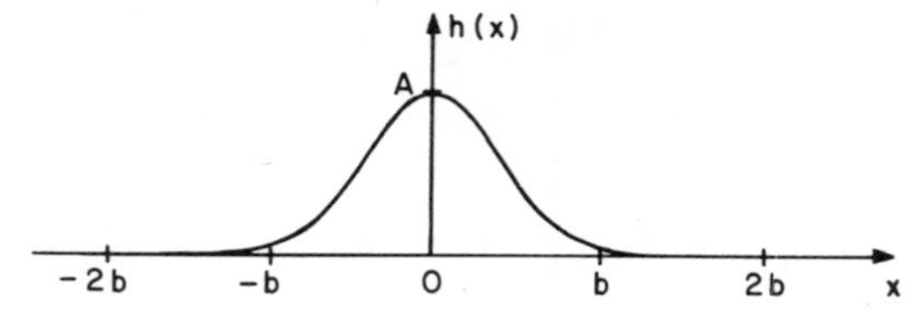

Figure 2-4 The Gaussian function is a nonperiodic function.

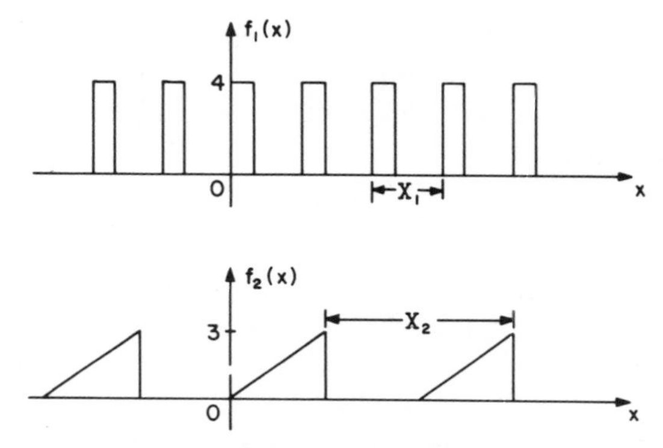

Figure 2-5 Periodic functions of period X_1 and X_2.

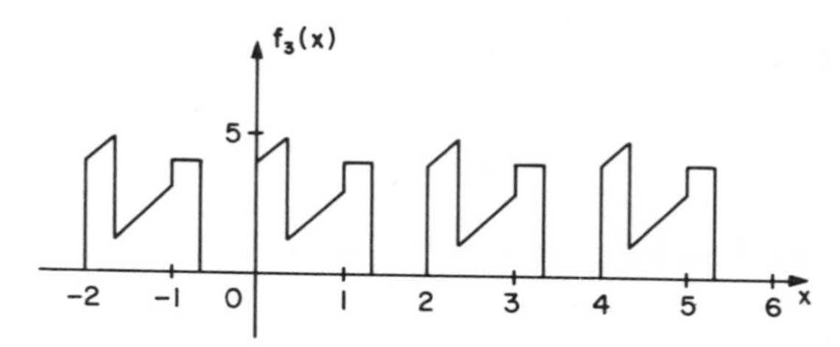

Figure 2-6 Sum of $f_1(x)$ and $f_2(x)$ of Fig. 2-5 when $X_2/X_1=2$.

and X_2, respectively. If we choose X_1 and X_2 such that their ratio is a rational number, then the sum $f_3(x)=f_1(x)+f_2(x)$ will be periodic. For example, suppose that $X_1=1$ and $X_2=2$; the ratio $X_2/X_1=2$ is rational and $f_3(x)$ is periodic (with period $X_3=2$) as shown in Fig. 2-6.

If, however, we had chosen $X_1=1$ and $X_2=\sqrt{3}$, $f_3(x)$ would no longer be periodic because there is no rational number R such that $RX_1=X_3$. As a result, $f_3(x)$ will be "almost periodic" as shown in Fig. 2-7. This function will very nearly repeat itself as x increases, but it will never repeat itself exactly because there is no number X_3 such that $f_3(x)=f_3(x+X_3)$.

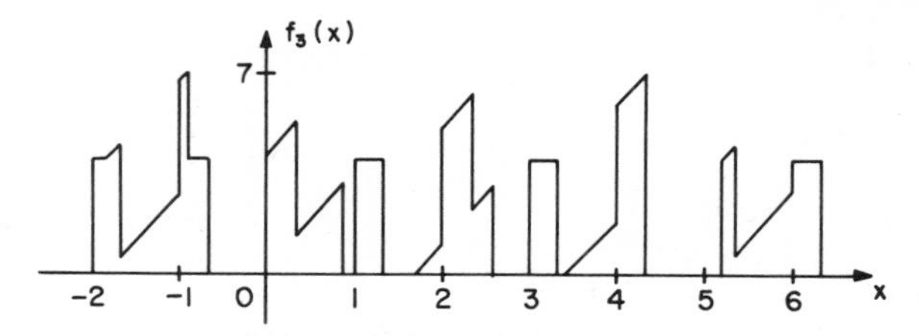

Figure 2-7 An almost-periodic function.

As will be shown in Chapter 4, many nonperiodic functions can be thought of as being composed of a linear combination (sum) of periodic functions.

Symmetry Properties

In the study of linear systems, problems may often be simplified by taking advantage of certain symmetry properties of functions. For example, a function $e(x)$ with the property

$$e(x)=e(-x) \tag{2.8}$$

is called an *even function* of x, whereas a function $o(x)$ satisfying the equality

$$o(x)=-o(-x) \tag{2.9}$$

is said to be an *odd function* of x. To see what Eqs. (2.8) and (2.9) mean, let us look at the graphs of two such functions (see Fig. 2-8). For the even function, the curve to the left of the origin is simply the reflection about the vertical axis of the curve to the right of the origin. For the odd function, the curve to the left of the origin is obtained by first reflecting the curve to the right of the origin about the vertical axis, and then reflecting

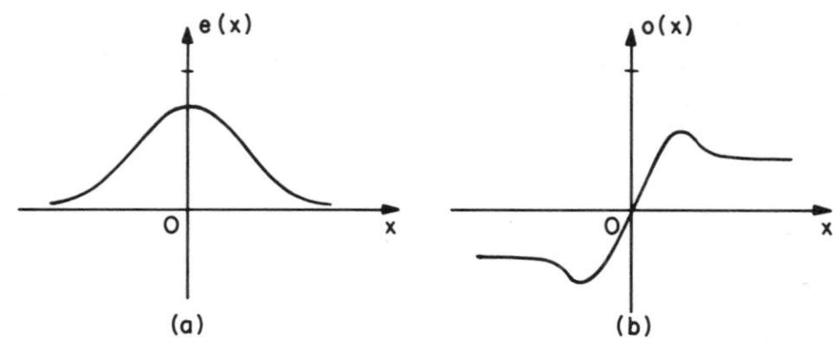

Figure 2-8 Symmetrical functions. (*a*) Even function. (*b*) Odd function.

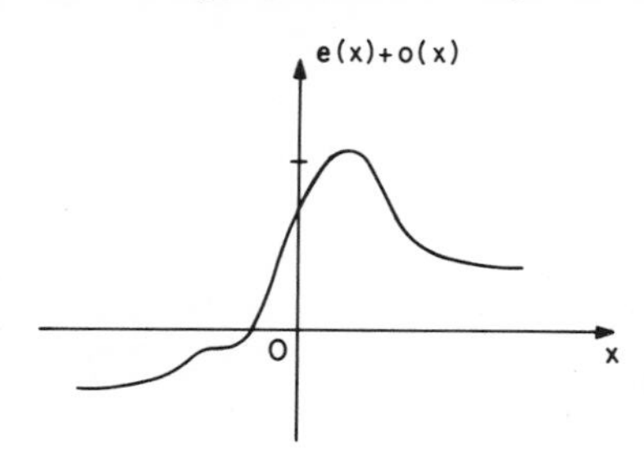

Figure 2-9 A function that is neither even nor odd.

this "reflection" about the horizontal axis. (Here "left" and "right" may be interchanged without altering the results, and the order in which the reflections are made is immaterial.) It is interesting to demonstrate this physically with a pair of small mirrors.

The sum of an even function and an odd function will be neither even nor odd; this obvious result is illustrated in Fig. 2-9, in which is graphed the sum of the functions shown in Fig. 2-8.

It is easy to show that any arbitrary function $f(x)$ may be expressed as the sum of an even part $f_e(x)$ and an odd part $f_o(x)$, i.e.,

$$f(x) = f_e(x) + f_o(x), \tag{2.10}$$

where

$$f_e(x) = \tfrac{1}{2}[f(x) + f(-x)] \tag{2.11}$$

and

$$f_o(x) = \tfrac{1}{2}[f(x) - f(-x)]. \tag{2.12}$$

Other interesting results pertaining to the symmetry properties of functions are: the product of two even functions is an even function, the product of two odd functions is an even function, and the product of an even function and an odd function is odd. These results may be useful in simplifying certain integral operations, as we shall now discuss.

Suppose we wish to evaluate the definite integral of the even function $e(x)$ on the interval $(-a, a)$. With α as the dummy variable of integration, we may write

$$\int_{-a}^{a} e(\alpha)\,d\alpha = \int_{-a}^{0} e(\alpha)\,d\alpha + \int_{0}^{a} e(\alpha)\,d\alpha. \tag{2.13}$$

Using the property $e(\alpha) = e(-\alpha)$, it may be shown that the integrals to the right of the equality sign in Eq. (2.13) are equal, with the result

$$\int_{-a}^{a} e(\alpha)\,d\alpha = 2\int_{0}^{a} e(\alpha)\,d\alpha. \tag{2.14}$$

Thus, the definite integral of an even function, evaluated between the limits $-a$ and a, is just twice the integral of the function evaluated from zero to a.

In a similar fashion, for the odd function $o(x)$, it is easy to show that

$$\int_{-a}^{a} o(\alpha)\,d\alpha \equiv 0. \tag{2.15}$$

Hence the definite integral of an odd function on the interval $(-a,a)$ is identically zero!

A geometrical interpretation of these results is given below. The operation of evaluating the definite integral of a function may be thought of as finding the area lying "under" the graph of the function between the limits of integration as illustrated in Fig. 2-10. Where the function is positive, its area is positive, and where the function is negative, its area is negative. Now let us look at Fig. 2-11(a), where the shaded region depicts the area

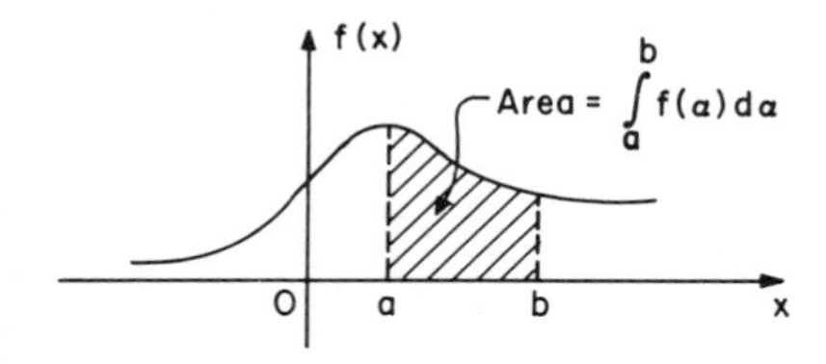

Figure 2-10 The integral of a function as an area.

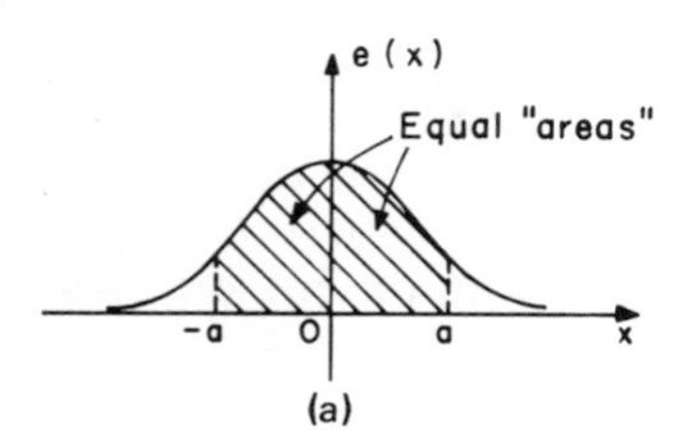

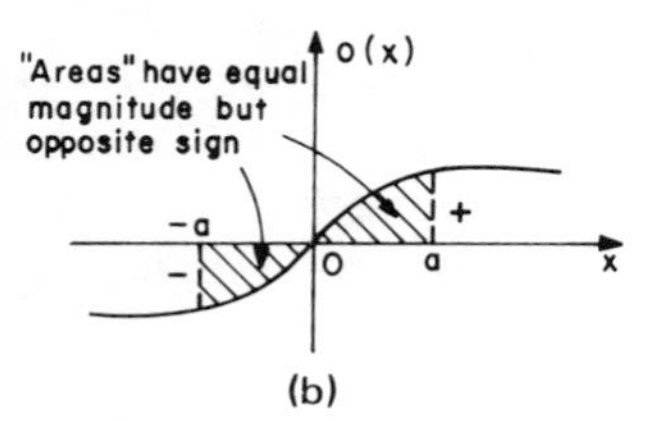

Figure 2-11 The area on the interval $(-a,a)$. (*a*) Even function. (*b*) Odd function.

under an even function in the interval $-a$ to a. It is clear that the portions of the shaded region to the left and to the right of the origin have equal areas, and thus the total area under the curve is just equal to twice the area of either of these portions. Similarly, for the odd function shown in Fig. 2-11(b), the portions of the shaded region to the left and to the right of the origin have areas of equal magnitude by opposite sign. Hence, the total area of the odd function is zero.

Using the above results, we find that for any even function $e(x)$ and any odd function $o(x)$,

$$\int_{-a}^{a} e(\alpha)o(\alpha)\,d\alpha \equiv 0. \tag{2.16}$$

Also, for any two even functions $e_1(x)$ and $e_2(x)$,

$$\int_{-a}^{a} e_1(\alpha)e_2(\alpha)\,d\alpha = 2\int_{0}^{a} e_1(\alpha)e_2(\alpha)\,d\alpha, \tag{2.17}$$

and for any two odd functions $o_1(x)$ and $o_2(x)$

$$\int_{-a}^{a} o_1(\alpha)o_2(\alpha)\,d\alpha = 2\int_{0}^{a} o_1(\alpha)o_2(\alpha)\,d\alpha. \tag{2.18}$$

Finally, from Eqs. (2.10), (2.14), and (2.15), we have for any function $f(x)$

$$\int_{-a}^{a} f(\alpha)\,d\alpha = \int_{-a}^{a} [f_e(\alpha) + f_0(\alpha)]\,d\alpha = 2\int_{0}^{a} f_e(\alpha)\,d\alpha. \tag{2.19}$$

Two-Dimensional Functions

So far, most of the discussion in this chapter has been about functions of the form $y = f(x)$, where the dependent variable y is related to the *single* independent variable x by the "rule" $f(\cdot)$. This kind of function is referred to as a *one-dimensional function* because it depends only on one independent variable. A *two-dimensional function*, on the other hand, is a "rule" relating a single dependent variable to two independent variables. Such functions are used extensively in the analysis of optical systems, and for this reason we include them here. As an illustration, suppose we wish to specify the transmittance of a photographic plate; in general, the transmittance will vary from point to point on the plate, and therefore it must be represented by a two-dimensional function of the spatial coordinates. In the rectangular coordinates we might write this as

$$t = g(x, y), \tag{2.20}$$

where t denotes the transmittance, x and y are the independent variables, and $g(\cdot,\cdot)$ is the "rule."

A one-dimensional function is usually represented graphically by a curve, as illustrated by the previous figures in this chapter. When a rectangular coordinate system is used for a graph, as in all of these figures, the value of the function for any value of the independent variable is given by the "height" of the curve above the horizontal axis at that point. Note that this height may be either positive or negative. Similarly, the graph of a two-dimensional function may be associated with a surface in space. For example, consider the two-dimensional Gaussian function.

$$f(x,y)=A\exp\left[-\pi\left(\frac{x^2+y^2}{d^2}\right)\right], \tag{2.21}$$

whose graph is shown plotted in rectangular coordinates in Fig. 2-12. The value of this function at any point (x,y) is just the "height" of the surface above the $x-y$ plane. In general, this height may be either positive or negative, depending on whether or not the surface lies above or below the $x-y$ plane. We might wish to express the the function of Eq. (2.21) in polar coordinates; to do so we let

$$r=+\sqrt{x^2+y^2} \tag{2.22}$$

$$\theta=\tan^{-1}\left(\frac{y}{x}\right), \tag{2.23}$$

and thus we obtain

$$g(r,\theta)=Ae^{-\pi(r/d)^2}. \tag{2.24}$$

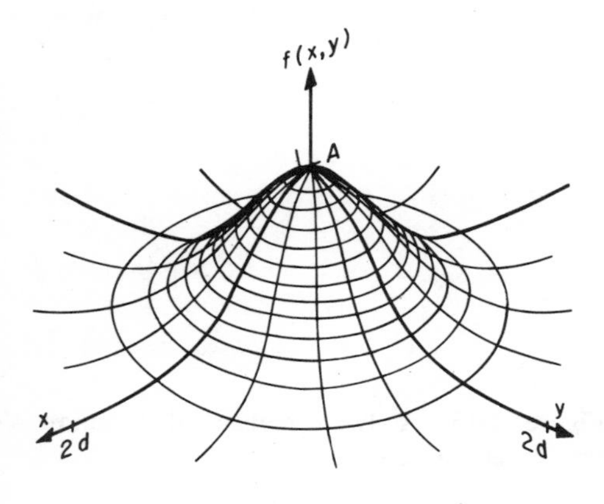

Figure 2-12 The two-dimensional Gaussian function.

Note that there is no θ-dependence for this particular function.

A two-dimensional function is said to be *separable* in a particular coordinate system if it can be written as the product of two one-dimensional functions, each of which depends only on one of the coordinates. Thus, the function $f(x, y)$ is separable if

$$f(x,y)=g(x)h(y), \tag{2.25}$$

where $g(x)$ is a function only of x and $h(y)$ is a function only of y.

A given two-dimensional function may be separable in one coordinate system and not in another. To illustrate, let us consider the function $f(x, y)$ shown in Fig. 2-13. This function is described by

$$\begin{aligned} f(x,y)&=1, \quad |x|\leqslant b, \quad |y|\leqslant d \\ &=0, \quad \text{elsewhere,} \end{aligned} \tag{2.26}$$

where b and d are real positive constants. It is not hard to see that it may also be written as

$$f(x,y)=g(x)h(y), \tag{2.27}$$

where

$$\begin{aligned} g(x)&=1, \quad |x|\leqslant b \\ &=0, \quad \text{elsewhere} \end{aligned} \tag{2.28}$$

and

$$\begin{aligned} h(y)&=1, \quad |y|\leqslant d \\ &=0, \quad \text{elsewhere.} \end{aligned} \tag{2.29}$$

Thus this function is separable in x and y. It would not be separable,

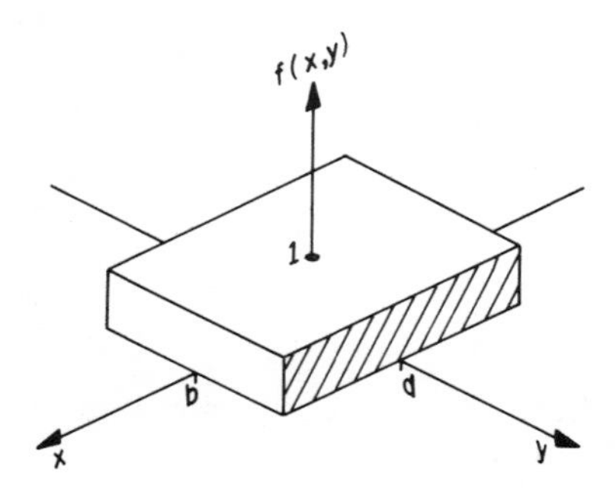

Figure 2-13 A two-dimensional function that is separable in x and y.

however, in polar coordinates. Nor would it be separable for a rectangular coordinate system that had been rotated with respect to the $x-y$ coordinate axes of Fig. 2-13 (other than by multiples of 90°). It is interesting to note that the two-dimensional Gaussian function of Eq. (2.21) is separable in both rectangular and polar coordinates, a property most functions do not possess.

To gain a better understanding of the behavior of separable functions, it may be helpful to consider the following development. Suppose we are given the separable function $f(x, y)=g(x)h(y)$. For some particular value of x, say $x=x_1$, we have

$$f(x_1, y)=g(x_1)h(y). \tag{2.30}$$

But $g(x_1)$ is just a constant, so we might think of the "height" of the function along the line $x=x_1$, as being determined by $g(x_1)$ and its "shape," or functional form, as being determined by $h(y)$. For a different value of x, say $x=x_2$, the function $f(x_2, y)$ will have the same shape as $f(x_1, y)$, but it may possibly have a different height, now determined by $g(x_2)$. Thus, from this point of view, as x varies over its entire range, $f(x, y)$ will be a function whose shape is that of $h(y)$ and whose height varies as $g(x)$. It is obvious that the roles of the height and shape of $g(x)$ and $h(y)$ may be interchanged without altering the behavior of the function. To illustrate this point of view, consider the functions $g(x)$ and $h(y)$, shown in Fig. 2-14(a) and (b), from which we form the separable function $f(x, y)=g(x)h(y)$. The graph of $f(x, y)$ is shown in Fig. 2-14(c), which shows the behavior described above.

The reason we are spending so much time on separable functions is that we shall make extensive use of their properties later when we discuss two-dimensional convolution and two-dimensional Fourier transforms. Also, these functions occur quite frequently in the study of optical systems.

We shall discuss one last topic before going on to complex numbers; that topic concerns the integration of two-dimensional functions. We may consider the operation of evaluating the definite integral of such a function to be that of finding the volume lying "under" the associated surface within the region specified by the limits of integration. This volume, of course, may be either positive or negative, depending on the behavior of the function. If the function is separable, then we obtain the result

$$\int_a^b \int_c^d f(\alpha,\beta)\,d\alpha\, d\beta=\left[\int_a^b g(\alpha)\,d\alpha\right]\left[\int_c^d h(\beta)\,d\beta\right], \tag{2.31}$$

which is just the product of the one-dimensional integrals of $g(x)$ and $h(y)$. Thus the volume of the separable function $f(x, y)=g(x)h(y)$ is simply the product of the areas of $g(x)$ and $h(y)$.

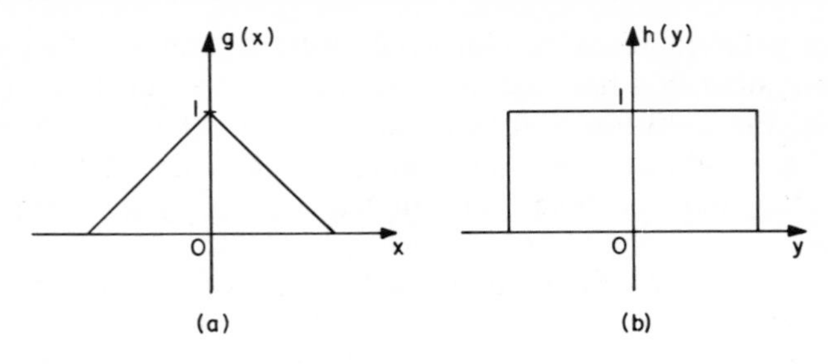

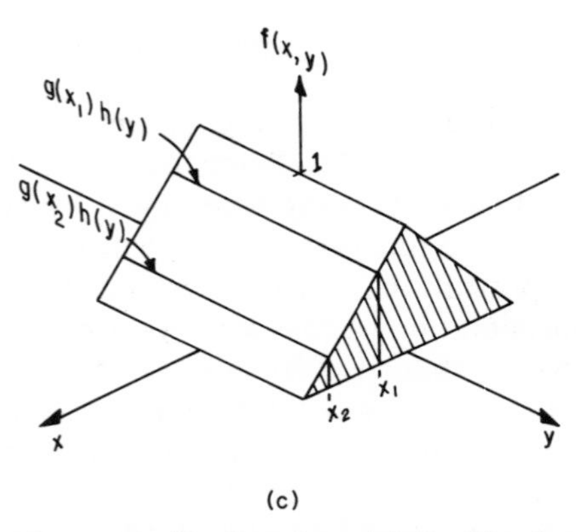

Figure 2-14 Nature of a separable function. (*a*) Profile along *x*-axis. (*b*) Profile along *y*-axis. (*c*) The entire function.

2-2 COMPLEX NUMBERS AND PHASORS

In this section we shall review several of the properties and uses of complex numbers and phasors. Those with backgrounds in electrical engineering should find the material presented to be quite elementary, and may wish to skim through it rather quickly. However, for those who are not familiar with this material, we point out that it is quite important to have a good understanding of these techniques and concepts, as we shall make extensive use of them in progressing through the book.

Complex Numbers

Given any two real numbers v and w, we can form the sum

$$u = v + jw, \tag{2.32}$$

where $j = \sqrt{-1}$. In general we say that u is a *complex number*; however, if

$w=0$, then u is simply a real number, and if $v=0$, u is an *imaginary number*. The number v is called the *real part* (or real component) of u, and w is called the *imaginary part* (or imaginary component). The real and imaginary parts of u are denoted by

$$v=\text{Re}\{u\}, \tag{2.33}$$

$$w=\text{Im}\{u\}. \tag{2.34}$$

Thus, from Eq. (2.32),

$$u=\text{Re}\{u\}+j\,\text{Im}\{u\}. \tag{2.35}$$

We stress that both the real and imaginary parts are themselves real numbers, and the symbol j merely *precedes* the imaginary part. The representation given in Eq. (2.32) is known as the rectangular form of the complex number u.

Any complex number is determined by two real numbers. In fact, there is a one-to-one correspondence between complex numbers and ordered pairs of real numbers. Since there is also a one-to-one correspondence between ordered pairs of real numbers and points in the Cartesian plane, it follows that a complex number $u=v+jw$ may be represented by the point (v,w) in this plane. For example, suppose we are given the complex number $u=3+j2$. We may represent this number by the point in the Cartesian plane whose abscissa is 3 and whose ordinate is 2, as shown in Fig. 2-15.

When a complex number is depicted in this fashion, the Cartesian plane is often referred to as the *complex plane*. The horizontal axis is called the *axis of reals*, and the vertical axis is called the *axis of imaginaries*. Sometimes the terms "real axis" and "imaginary axis" are used in place of the above, but this seems to imply that the former axis exists and the latter does not, which of course does not make much sense. Therefore, to avoid confusion, we shall refrain from such usage.

We now mention another way of thinking about complex numbers. There is a one-to-one correspondence between ordered pairs of real num-

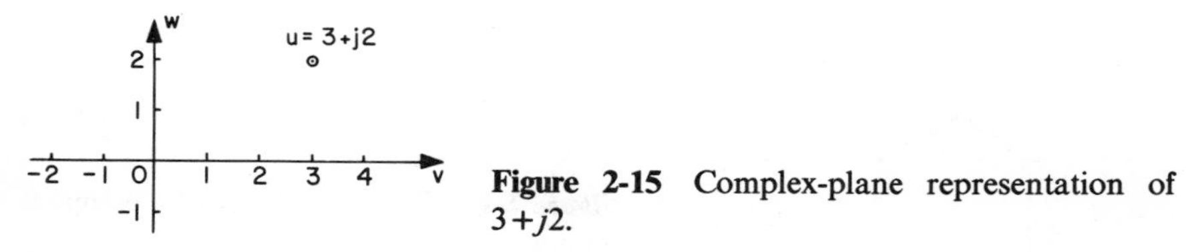

Figure 2-15 Complex-plane representation of $3+j2$.

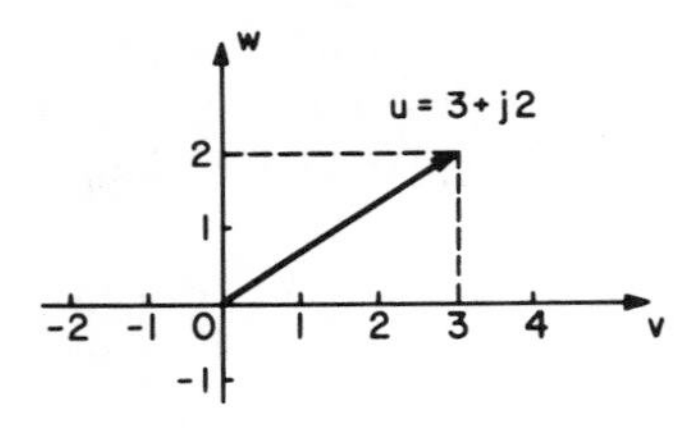

Figure 2-16 Vector representation of $3+j2$.

bers and vectors. Thus the complex number $u=3+j2$ may be thought of as a vector whose components are 3 and 2, respectively, as shown in Fig. 2-16.

Perhaps the most useful way of representing complex numbers for our purposes is in their polar form. The polar form of the complex number $u=v+jw$ is given by

$$u=re^{j\phi}, \tag{2.36}$$

where

$$r=|u|=\sqrt{v^2+w^2} \tag{2.37}$$

and

$$\phi=\arg(u)=\tan^{-1}\left(\frac{w}{v}\right). \tag{2.38}$$

The quantity r is known as the *absolute value*, or *modulus*, of u, whereas ϕ is its *argument*, or *phase*. Both r and ϕ are real numbers, and in addition r is nonnegative. The complex-plane representation of this number is shown in Fig. 2-17, where u is depicted as a vector whose magnitude is given by r and whose direction angle with respect to the v-axis is given by ϕ. We shall take the positive sense for ϕ to be in the counterclockwise direction, as is normally done.

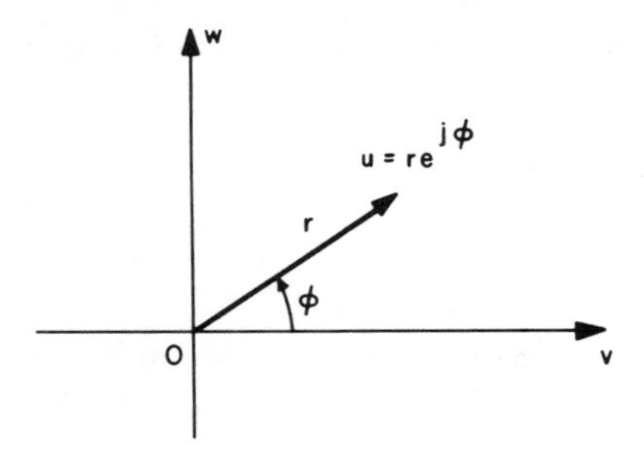

Figure 2-17 The polar form of a complex number.

In the previous example, for which $u=3+j2$, we find that

$$r=\sqrt{(3)^2+(2)^2}=\sqrt{13}=3.61, \tag{2.39}$$

$$\phi=\tan^{-1}\left(\tfrac{2}{3}\right)=33.7^\circ=0.591\,\text{rad}. \tag{2.40}$$

Thus, in polar form, this number is written

$$u=3.61\,e^{j0.591} \tag{2.41}$$

When a complex number is given in polar form, its real and imaginary components may be determined by using Euler's formula, i.e.,

$$\begin{aligned} u&=re^{j\phi}\\ &=r(\cos\phi+j\sin\phi), \end{aligned} \tag{2.42}$$

so that

$$\text{Re}\{u\}=r\cos\phi \tag{2.43}$$

and

$$\text{Im}\{u\}=r\sin\phi. \tag{2.44}$$

The *complex conjugate* of the complex number u is denoted by u^* and is obtained by replacing each j with $-j$; thus the complex conjugate of $u=v+jw$ is given by

$$u^*=v-jw. \tag{2.45}$$

In polar form, the complex conjugate of $u=r\exp\{j\phi\}$ becomes

$$u^*=re^{-j\phi}. \tag{2.46}$$

Complex Algebra

We now list several rules of complex algebra that are frequently needed when working with complex numbers. This list is intended to serve only as a quick reference, and therefore few details are included. For the various operations listed, we shall use the two complex numbers

$$u_1=v_1+jw_1=r_1e^{j\phi_1}, \tag{2.47}$$

$$u_2=v_2+jw_2=r_2e^{j\phi_2}. \tag{2.48}$$

The rules for addition and multiplication are

$$u_1+u_2=(v_1+v_2)+j(w_1+w_2), \tag{2.49}$$

$$u_1u_2=(v_1v_2-w_1w_2)+j(v_1w_2+v_2w_1) \tag{2.50}$$

$$=r_1r_2e^{j(\phi_1+\phi_2)}. \tag{2.51}$$

Some results of these rules are

$$u_1+u_1^*=2v_1, \tag{2.52}$$

$$u_1-u_1^*=j2w_1, \tag{2.53}$$

$$u_1u_1^*=|u_1|^2=v_1^2+w_1^2=r_1^2. \tag{2.54}$$

The operation of division is a little more involved than those of addition and multiplication unless the dividend and divisor are given in polar form. If they are in rectangular form, the divisor must be rationalized before the quotient can be expressed in rectangular form, i.e.,

$$\frac{u_1}{u_2}=\frac{u_1u_2^*}{u_2u_2^*}=\frac{u_1u_2^*}{|u_2|^2}$$

$$=\left(\frac{v_1v_2+w_1w_2}{v_2^2+w_2^2}\right)+j\left(\frac{v_2w_1-v_1w_2}{v_2^2+w_2^2}\right). \tag{2.55}$$

This operation is much more easily performed in polar form, as shown below.

$$\frac{u_1}{u_2}=\frac{r_1e^{j\phi_1}}{r_2e^{j\phi_2}}=\frac{r_1}{r_2}e^{j(\phi_1-\phi_2)}. \tag{2.56}$$

Other useful rules are

$$|u_1u_2|=|u_1|\cdot|u_2|, \tag{2.57}$$

$$|u_1u_1^*|=|u_1|^2, \tag{2.58}$$

$$(u_1+u_2)^*=u_1^*+u_2^*, \tag{2.59}$$

$$(u_1u_2)^*=u_1^*u_2^*, \tag{2.60}$$

$$\left(\frac{u_1}{u_2}\right)^*=\frac{u_1^*}{u_2^*}, \tag{2.61}$$

$$\frac{1}{u_1}=\frac{u_1^*}{|u_1|^2}. \tag{2.62}$$

It should be pointed out that

$$\mathrm{Re}\{u_1+u_2\}=\mathrm{Re}\{u_1\}+\mathrm{Re}\{u_2\}, \tag{2.63}$$

$$\mathrm{Im}\{u_1+u_2\}=\mathrm{Im}\{u_1\}+\mathrm{Im}\{u_2\}, \tag{2.64}$$

but in general

$$\mathrm{Re}\{u_1u_2\}\neq\mathrm{Re}\{u_1\}\cdot\mathrm{Re}\{u_2\}, \tag{2.65}$$

$$\mathrm{Im}\{u_1u_2\}\neq\mathrm{Im}\{u_1\}\cdot\mathrm{Im}\{u_2\}. \tag{2.66}$$

We now list the rules for finding integral powers and roots of complex numbers. For n a positive integer,

$$(u_1)^n=r_1^n e^{jn\phi_1}, \tag{2.67}$$

and

$$(u_1)^{1/n}=(r_1)^{1/n}\exp\left[j\left(\frac{\phi_1+k2\pi}{n}\right)\right], \tag{2.68}$$

where $k=0,1,2,3,\ldots,n-1$. It is interesting to note that there are exactly n distinct values of $(u_1)^{1/n}$.

Some of these operations have very helpful geometrical interpretations. For example, the addition of complex numbers may be treated as a problem in the addition of vectors. Since every complex number may be associated with a vector, as indicated in Fig. 2-16, the sum of two complex numbers may be associated with a vector that is the sum of the vectors corresponding to the individual terms. This is illustrated in Fig. 2-18, which shows the sum of the two complex numbers $u_1=1+j3$ and $u_2=-3-j1$. This sum may also be obtained from Eq. (2.49) and is equal to $-2+j2$.

The operation of multiplication is another that has a useful geometrical interpretation. To see this, consider the rule given by Eq. (2.51) for finding the product of the complex numbers u_1 and u_2. This product may be

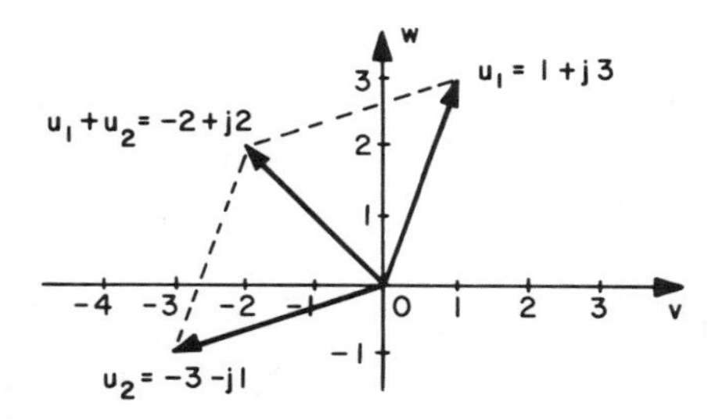

Figure 2-18 The sum of two complex numbers.

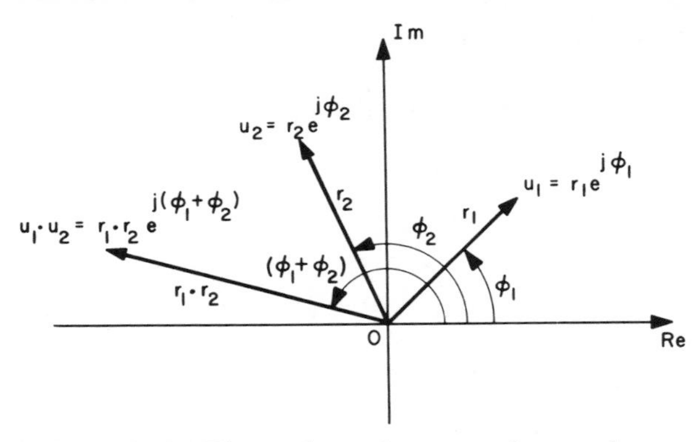

Figure 2-19 The product of two complex numbers.

associated with a vector whose magnitude is equal to the product of the magnitudes of the vectors corresponding to u_1 and u_2, and whose direction angle is the sum of the direction angles of these vectors. This interpretation is illustrated in Fig. 2-19.

There are other geometrical interpretations of complex algebra, but we shall not discuss them here because they are not relevant to our work.

Phasors

The word phasor is often used by mathematicians to mean any complex number. In engineering it is frequently used to denote a complex exponential function of constant modulus and linear phase, that is, a function with purely harmonic behavior. Here, however, we shall use it in a much more general way; we shall use phasor to mean a *complex-valued function* of one or more real variables. Suppose, for example, we are given the two real-valued functions $v(x)$ and $w(x)$. We form the complex-valued function

$$u(x) = v(x) + jw(x), \tag{2.69}$$

which we shall call a phasor. In polar form this function is written

$$u(x) = a(x)e^{j\phi(x)}, \tag{2.70}$$

where

$$a(x) = |u(x)| = \left[v^2(x) + w^2(x)\right]^{1/2} \tag{2.71}$$

and

$$\phi(x) = \arg\left[u(x)\right] = \tan^{-1}\left[\frac{w(x)}{v(x)}\right] \tag{2.72}$$

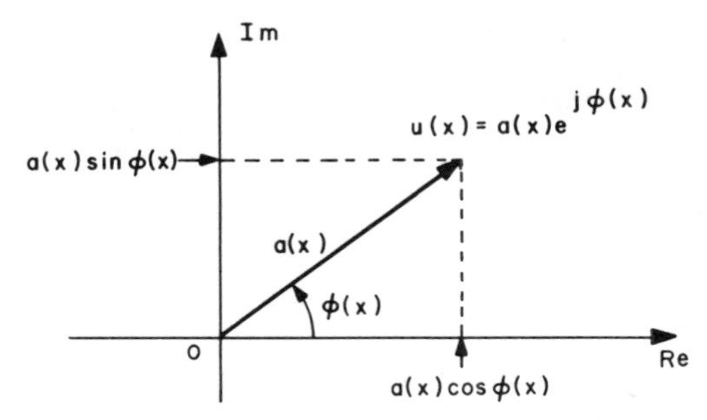

Figure 2-20 Complex-plane representation of a phasor.

are the modulus and phase, respectively, of $u(x)$. (We have elected to denote the modulus by $a(x)$ rather than by $r(x)$ in order to avoid certain problems in notation later on.) Both $a(x)$ and $\phi(x)$ are real-valued functions, and in addition $a(x)$ is nonnegative.

Phasors may be represented in the complex plane in a fashion similar to that used for representing complex numbers, as illustrated in Fig. 2-20. Here the phasor $u(x)=a(x)\exp\{j\phi(x)\}$ is associated with a vector whose magnitude is given by $a(x)$ and whose direction angle is given by $\phi(x)$. As $a(x)$ varies with x, the length of the vector varies, and as $\phi(x)$ varies with x, the direction of the vector varies. Following Eqs. (2.43) and (2.44), we find that the real and imaginary components of this phasor are given by

$$\mathrm{Re}\{u(x)\}=a(x)\cos\phi(x), \tag{2.73}$$

$$\mathrm{Im}\{u(x)\}=a(x)\sin\phi(x). \tag{2.74}$$

We note that, if both $a(x)$ and $\phi(x)$ are nonvarying, $u(x)$ simply becomes a complex constant.

The phasor diagram of Fig. 2-20 provides a very useful geometrical interpretation of complex-valued functions, and such diagrams are frequently used when dealing with phasors. However, they are by no means the only way of representing these functions. Other methods include individually graphing the real and imaginary components, individually graphing the modulus and phase, and graphing the function in three dimensions. We shall now demonstrate these various methods by using each of them to represent the phasor

$$u(x)=Ae^{j2\pi\xi_0 x}, \tag{2.75}$$

which has a constant modulus and linearly varying phase. The complex-plane representation of this phasor is the same as that shown in Fig. 2-20,

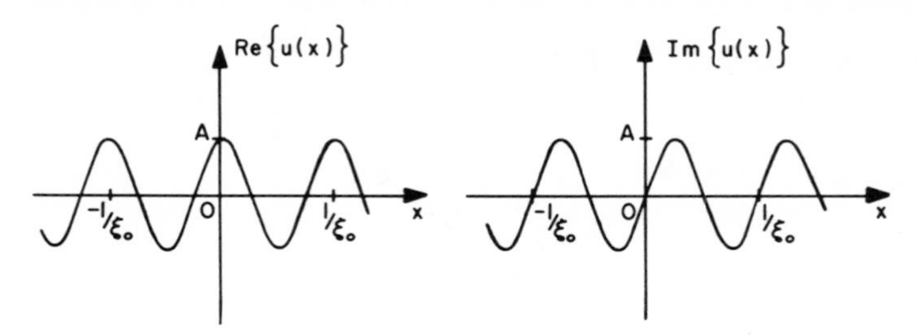

Figure 2-21 Real and imaginary components of $A \exp\{j2\pi\xi_0 x\}$.

with $a(x)=A$ and $\phi(x)=2\pi\xi_0 x$, so we shall not include a new figure. In the second representation mentioned above, we individually graph the real and imaginary components of $u(x)$, which are given by

$$\mathrm{Re}\{u(x)\}=A\cos(2\pi\xi_0 x) \tag{2.76}$$

and

$$\mathrm{Im}\{u(x)\}=a\sin(2\pi\xi_0 x). \tag{2.77}$$

The graphs of these functions are shown in Fig. 2-21. The next method involves the individual graphing of the modulus and phase of $u(x)$, which were specified above to be

$$|u(x)|=a(x)=A \tag{2.78}$$

and

$$\arg[u(x)]=\phi(x)=2\pi\xi_0 x. \tag{2.79}$$

These graphs are shown in Fig. 2-22. Finally, for the last method mentioned, we erect a three-dimensional rectangular coordinate system. The variable x is plotted along one axis, and the real and imaginary components of the phasor are plotted along the other two. The result is the three-dimensional curve, a helix in this case, that is shown in Fig. 2-23. This curve is generated by the tip of a constant-magnitude vector, oriented perpendicular to the x-axis, which rotates at a constant rate in a counterclockwise direction as it moves in the positive x-direction. It is interesting to observe that if we were to look "back down the x-axis," we would merely see the complex-plane representation of this phasor. Also, the vertical and horizontal projections of this curve orthogonal to the x-axis are just the real and imaginary components of the phasor, respectively, as shown in the figure.

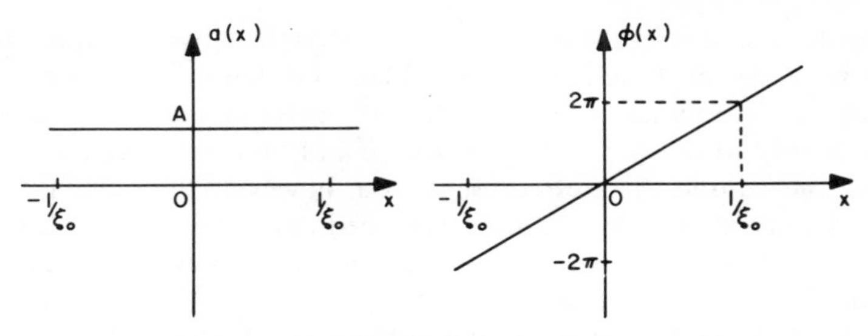

Figure 2-22 Modulus and phase of $A\exp\{j2\pi\xi_0 x\}$.

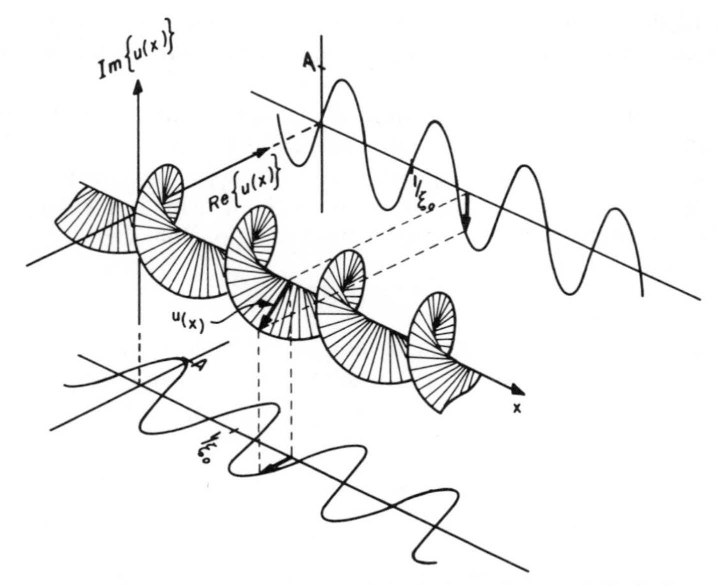

Figure 2-23 Three-dimensional depiction of $A\exp\{j2\pi\xi_0 x\}$.

We now briefly mention a subject that will come up again when we discuss Fourier transforms. A complex-valued function, or phasor, whose real part is an even function and whose imaginary part is odd is said to be *hermetian*, while a phasor whose real part is odd and whose imaginary part is even is called *antihermetian*. The Fourier transforms of such functions possess certain special properties, as we shall see in Chapter 7.

As discussed previously, the complex-plane representation is very useful in visualizing the behavior of phasors. It is also particularly helpful in finding the sum of two or more phasors. Suppose, for example, we wish to find the sum of the two functions $2\exp\{j2\pi\nu_0 t\}$ and $\exp\{j4\pi\nu_0 t\}$. We might convert these functions to rectangular form and use trigonometric identities to find the result, but this could get rather involved (particularly if there were several functions being added). Or, we might graph the real and imaginary components of these phasors in rectangular coordinates and add the graphs at every point. This, too, could be a very tedious process. By using the complex-plane representation, however, the behavior of the sum can be visualized quite readily. We need only treat these two phasors as vectors and find their vector sum for various values of t. Figure 2-24 shows the history of this resultant vector at intervals of $1/8\nu_0$, one-eighth the period of the more slowly varying phasor. The dashed vectors represent the phasors being added at the various times, whereas the solid vectors depict their sum. The lightly dashed curve shows how the modulus of the resulting phasor varies with time, and the behavior of its phase can be visualized by noting how the direction angle of the solid vectors varies with time.

So far in this section, we have limited our discussion to phasors that are determined by functions of a single independent variable. Later on, however, we will be dealing frequently with phasors that depend on two independent variables, and so we mention them here. For example, consider the two-dimensional complex-valued function (phasor)

$$u(x,y)=a(x,y)e^{j\phi(x,y)}. \tag{2.80}$$

Although many of the concepts developed for one-dimensional complex-valued functions are still useful in thinking about such a function, it is in general much more difficult to visualize and to represent graphically. For

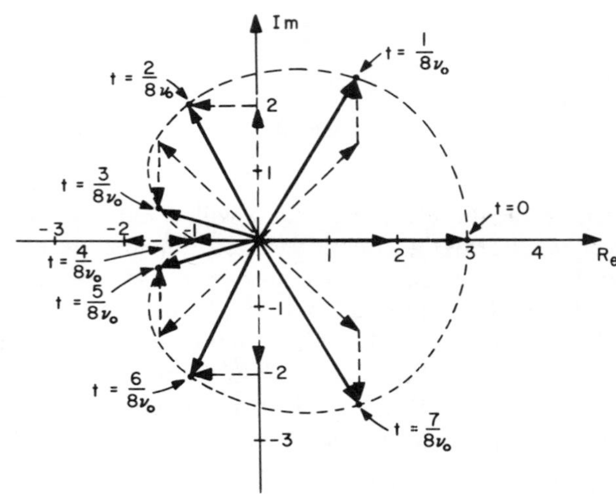

Figure 2-24 Sum of two phasors of different amplitudes and frequencies.

example, the modulus and phase must be graphed separately as surfaces rather than simple curves. Even if we put Eq. (2.80) in rectangular form, we still must graph the real and imaginary parts as surfaces. Finally, we cannot construct the equivalent of Fig. 2-23 because to do so would require more than three dimensions.

Graphical representations of these two-dimensional functions are somewhat simplified, however, if the functions are separable. In that event, graphs corresponding to those of Figs. 2-21, 2-22, or 2-23 can be drawn separately for each of the independent variables.

2-3 REPRESENTATION OF PHYSICAL QUANTITIES

There are many physical quantities that could be used as examples in this section, but we shall restrict our attention to just a few important ones. To begin with, consider the representation of phenomena whose behavior is sinusoidal; then we proceed to the representation of amplitude and phase-modulated waves, and finally to the description of a monochromatic light wave. Not only do these quantities allow many important concepts to be demonstrated nicely, but they will also be of interest to us in later chapters.

As will be seen, it is often advantageous to represent physical quantities by phasors, which are complex-valued functions. However, unless this representation is properly formulated, a considerable amount of confusion can result. For example, a time-varying voltage is often represented by the phasor $a(t)\exp\{j\phi(t)\}$, which implies that the voltage consists of a real part and an imaginary part, and this does not make sense physically. Such a voltage is more properly represented as either the real part or the imaginary part of the appropriate phasor, which are both real-valued functions, and not by the phasor itself. In most cases where a physical quantity is represented by a phasor, there is an implicit understanding that it is the real or imaginary part of this phasor that is of interest. As long as this is realized and as long as care is taken not to violate any of the rules of complex algebra [see, for example, Eqs. (2.65) and (2.66)], such a representation should pose no major problems.

At this point we introduce the term *signal*, which we shall use loosely to mean a function, representing a specific physical quantity, that possesses information in which we are interested. We do not restrict its use to the more familiar electrical, audio, or visual signals, but will find it helpful to include virtually any quantity of interest. For example, it is often useful to consider the transmittance function of a photographic transparency to be a two-dimensional "signal" for a coherent optical system. Such usage is quite common in the engineering world and should not cause any serious misunderstandings.

Sinusoidal Signals

As previously indicated, sinusoidal functions are of great importance in dealing with various engineering problems, partly because they accurately describe the behavior of many phenomena, but primarily because so many other functions can be decomposed into a linear combination of these sinusoids (a process called harmonic analysis, which is the principal topic of Chapter 4). In addition, sinusoids are eigenfunctions of linear, shift-invariant systems, a characteristic that makes them particularly useful for our work here. The significance of this property will be discussed more fully in Chapter 5.

Let us now consider the real-valued function

$$v(t)=A\cos(2\pi\nu_0 t+\theta), \tag{2.81}$$

which might be used to represent any physical quantity whose behavior is sinusoidal, e.g., the line voltage of a power distribution system, the oscillations of a pendulum, etc. From Eq. (2.73) we see that $v(t)$ may be written as

$$v(t)=\mathrm{Re}\{u(t)\}, \tag{2.82}$$

where

$$u(t)=A\exp[j(2\pi\nu_0 t+\theta)]. \tag{2.83}$$

Geometrically, using our convention for phasor diagrams, $v(t)$ is simply the projection of the vector $u(t)$ on the axis of reals, as can be seen with reference to Fig. 2-20. Thus any cosine function can be written as the real part of the appropriate phasor, and in a similar fashion any sine function can be described as the imaginary part of a phasor.

The utility of using phasor notation becomes apparent when attempting to perform certain operations on sinusoidal signals. For example, suppose we wished to find the sum

$$s(t)=\sum_{i=1}^{n} v_i(t) \tag{2.84}$$

of n cosinusoidal signals of the form

$$v_i(t)=A_i\cos(2\pi\nu_i t+\theta_i), \tag{2.85}$$

where $i=1,2,\ldots,n$ and A_i, ν_i, and θ_i are arbitrary real constants. For a large number of terms, it might be very difficult to calculate this sum, but by defining the phasor

$$u_i(t) = A_i \exp[j(2\pi\nu_i t + \theta_i)], \tag{2.86}$$

we may use Eq. (2.83) to write

$$v_i(t) = \mathrm{Re}\{u_i(t)\}. \tag{2.87}$$

Finally, from Eq. (2.63), we obtain

$$\sum_{i=1}^{n} v_i(t) = \mathrm{Re}\left\{ \sum_{i=1}^{n} u_i(t) \right\}. \tag{2.88}$$

Thus the sum of the n cosinusoidal signals is just equal to the real part of the sum of the n corresponding phasors, or, in a phasor diagram, it is just the projection on the horizontal axis of the sum of the n corresponding vectors. We will not give a specific example of this, because it is simply an extension of the one illustrated in Fig. 2-24. (Also, see Figs. 4-11 and 4-14.)

Another useful concept concerning the representation of sinusoidal signals is the following. Suppose we are given the signal $A\cos(2\pi\nu_0 t)$. We may use Euler's formula to write

$$A\cos(2\pi\nu_0 t) = \frac{A}{2}\left[e^{j2\pi\nu_0 t} + e^{-j2\pi\nu_0 t}\right], \tag{2.89}$$

which is just the sum of two phasors of constant modulus $A/2$ and linear phase $\pm 2\pi\nu_0 t$. In a phasor diagram, the first of these phasors rotates counterclockwise with time at a rate of $2\pi\nu_0$ rad/sec, whereas the second rotates clockwise at the same rate. The imaginary parts of each of these phasors are always of opposite sign, thus canceling one another, but their real parts always have the same sign, thus adding to produce the real-valued cosine function $A\cos(2\pi\nu_0 t)$. This is illustrated in Fig. 2-25. It is also instructive to associate the negative sign in the exponent of the second phasor with the fundamental frequency of that phasor rather than with the entire exponent, that is,

$$A\cos(2\pi\nu_0 t) = \frac{A}{2}\left[e^{j2\pi\nu_0 t} + e^{j2\pi(-\nu_0)t}\right]. \tag{2.90}$$

Thus we may consider the signal to be composed of a "positive-frequency" component and a "negative-frequency" component, the former a phasor rotating counterclockwise because its fundamental frequency is positive, and the latter a phasor rotating clockwise due to its negative fundamental frequency. It may be rather difficult at first to grasp the meaning of a

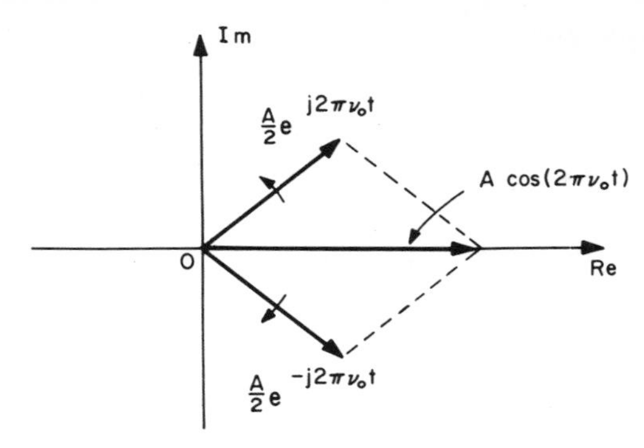

Figure 2-25 The cosine function as the sum of positive- and negative-frequency phasors.

"negative-frequency" phasor, but the concept just mentioned can be quite helpful. It should be pointed out that the notion of negative frequency will be dealt with regularly in the chapter on Fourier transforms.

Modulated Waves

Modulation, a process in which a *modulating signal* is used to control some property of a *carrier wave* in a prescribed fashion, is an important part of all communications systems, including radio, television, telephone, etc. Although there are many types of modulation, we shall discuss only those for which the carrier wave exhibits sinusoidal behavior. In addition, we shall restrict our attention in this section to temporal modulation—the modulation of time-varying waves—although there is no necessity to do so. We use this approach simply because most of the readers are probably more familiar with temporally modulated waves, such as those used in radio and television broadcasting, than they are with spatially modulated waves. Later on, when optical applications are discussed, the concepts developed here for temporal modulation will be applied directly to spatial modulation.

The general expression for a modulated wave with a sinusoidal carrier is given by

$$v(t)=a(t)\cos[\phi(t)], \tag{2.91}$$

where $a(t)$ may be thought of as the "instantaneous amplitude" of the carrier wave and $\phi(t)$ its "instantaneous phase." For *amplitude modulation* (AM), $a(t)$ is linearly related to the modulating signal $m(t)$, whereas the phase is independent of this signal and usually has the form

$$\phi(t)=2\pi\nu_c t-\theta, \tag{2.92}$$

where ν_c is the fundamental frequency of the carrier wave and θ is an arbitrary real constant. Often $a(t)$ is written as

$$a(t)=A[1+m(t)] \tag{2.93}$$

where A is a real positive constant and $m(t)$ is a real-valued function. Thus for AM, Eq. (2.91) becomes

$$v(t)=A[1+m(t)]\cos(2\pi\nu_c t-\theta). \tag{2.94}$$

When $a(t)$ is put in the form of Eq. (2.93), the condition

$$m(t)\geqslant -1 \tag{2.95}$$

is usually assumed so that $a(t)$ will be nonnegative. If $m(t)$ is a slowly varying function with respect to the oscillations of the carrier wave, as is generally the case, $a(t)$ is called the *envelope* of the modulated wave. In Fig. 2-26 we show an arbitrary modulating signal, beginning at time t_0, and the resulting modulated wave.

It is also instructive to consider the process of modulation from a phasor point of view. To illustrate, let us write Eq. (2.94) as

$$\begin{aligned} v(t)&=A[1+m(t)]\,\mathrm{Re}\{e^{j2\pi\nu_c t}\}\\ &=\mathrm{Re}\{A[1+m(t)]e^{j2\pi\nu_c t}\}, \end{aligned} \tag{2.96}$$

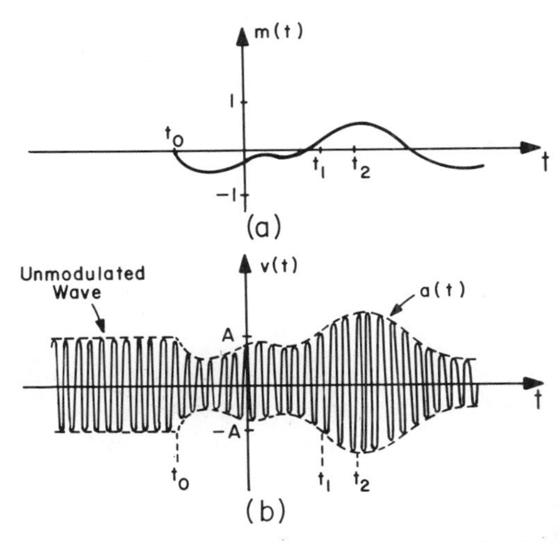

Figure 2-26 Amplitude modulation. (*a*) Modulating signal. (*b*) Modulated carrier wave.

where we have now omitted the constant θ to simplify the notation. Thus $v(t)$ is simply the real part of a phasor whose modulus is $a(t)=A[1+m(t)]$ and whose phase varies linearly with time. In a phasor diagram, we may think of $v(t)$ as being the horizontal projection of a vector $u(t)$ that is rotating in a counterclockwise direction at a constant rate of $2\pi\nu_c$ rad/sec, and whose length is varying as $a(t)$. In Fig. 2-27 we show two such diagrams, the first representing this vector at the time t_1 and the second representing it at the time t_2. These times correspond to the t_1 and t_2 of Fig. 2-26 and are separated by an interval equal to three periods of the carrier wave. The behavior of $v(t)$ is readily visualized from such diagrams.

Again let us refer to Eq. (2.91). For *phase modulation* (PM), the instantaneous amplitude $a(t)$ is constant and the instantaneous phase $\phi(t)$ depends linearly on the modulating signal. This dependence is often expressed as

$$\phi(t)=2\pi\nu_c t+\Delta\phi m(t), \tag{2.97}$$

where the quantity $\Delta\phi m(t)$ is the instantaneous phase deviation, and again ν_c is the carrier frequency. Thus for the PM case, Eq. (2.91) becomes

$$v(t)=A\cos\left[2\pi\nu_c t+\Delta\phi m(t)\right]. \tag{2.98}$$

Graphs of $\Delta\phi m(t)$, $\phi(t)$, and the resulting PM wave are shown in Fig. 2-28 for an arbitrary modulating signal $m(t)$. From Fig. 2-28(c) it may be seen that the amplitude of the oscillations is constant, but their position, or spacing, varies with $m(t)$.

The change of position of these oscillations may also be thought of as resulting from a change in the "instantaneous frequency" ν_{in} of the carrier, which is equal to the slope of the $\phi(t)$ curve of Fig. 2-28(b) divided by 2π, i.e.,

$$\begin{aligned}\nu_{\text{in}}&=\frac{1}{2\pi}\frac{d\phi(t)}{dt}\\&=\nu_c+\frac{\Delta\phi}{2\pi}\frac{dm(t)}{dt}.\end{aligned}$$

Thus the instantaneous frequency of a PM wave is linearly related to the derivative of the modulating signal. (In a third type of modulation, known as *frequency modulation* (FM), it is the instantaneous frequency that depends linearly on $m(t)$. Actually, PM and FM are simply different types of *angle modulation*, and because of their similarities we shall limit our discussion to PM.)

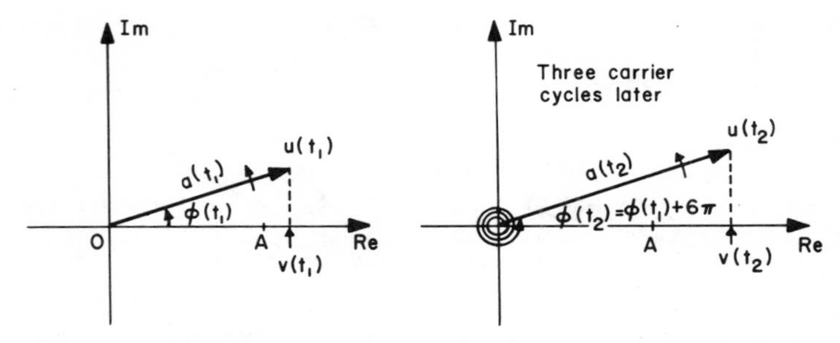

Figure 2-27 Phasor representation of amplitude modulation.

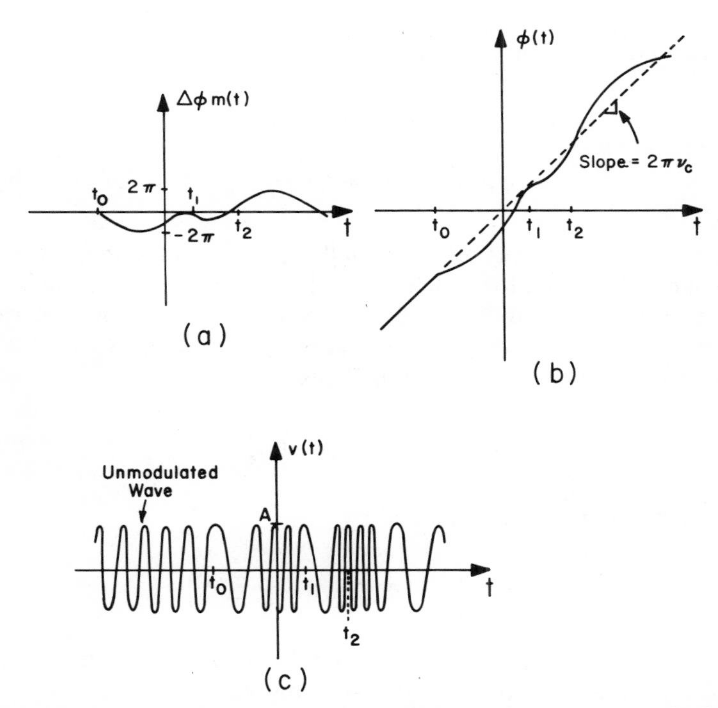

Figure 2-28 Phase modulation. (*a*) Instantaneous phase deviation. (*b*) Instantaneous phase. (*c*) Modulated carrier wave.

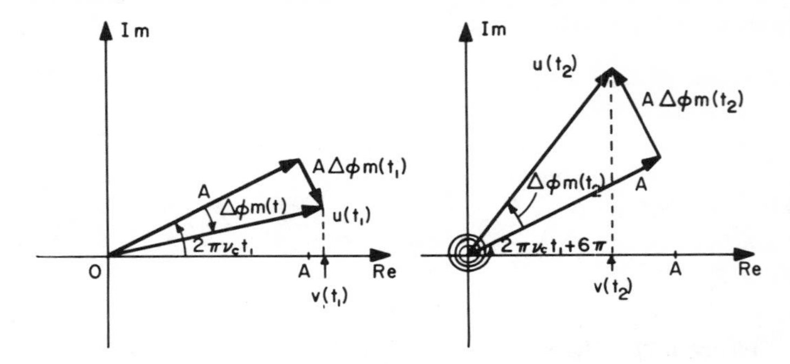

Figure 2-29 Phasor representation of phase modulation.

In phasor notation, Eq. (2.98) becomes

$$v(t) = A\,\text{Re}\{\exp(j[2\pi\nu_c t + \Delta\phi m(t)])\}$$

$$= A\,\text{Re}\{e^{j\Delta\phi m(t)} e^{j2\pi\nu_c t}\}. \tag{2.99}$$

To simplify the development, we shall now restrict ourselves to the "narrowband" PM case, for which the peak phase deviation is small. Once the narrowband case is understood, an extension to the wideband case is readily made. If $\Delta\phi m(t) \ll 1$, we may approximate $\exp\{j\Delta\phi m(t)\}$ by the first two terms of its series expansion. Thus

$$e^{j\Delta\phi m(t)} \cong 1 + j\Delta\phi m(t), \tag{2.100}$$

and Eq. (2.99) may be approximated by

$$v(t) \cong \text{Re}\{A[1 + j\Delta\phi m(t)]e^{j2\pi\nu_c t}\}, \tag{2.101}$$

which is nearly the same as the expression for an AM wave given by Eq. (2.96). The difference is that the time-varying part of the modulus is no longer in phase with the constant part. This can best be seen with reference to phasor diagrams, and in Fig. 2-29 we show two such diagrams as we did for the AM case. From these diagrams it can be seen that $v(t)$ is the horizontal projection of a vector $u(t)$ whose length is nearly constant (this is a result of our approximation—it actually has a constant length), but which no longer rotates at a constant rate. As this vector rotates, it "wobbles" back and forth about the position it would have if it represented a pure sinusoid, this wobbling being governed by the modulating signal.

Monochromatic Light Waves

The last topic of this section is concerned with the mathematical representation of light waves and is included to familiarize the reader with some of the concepts that will be needed later on in the study of optical applications. Here, of course, we will be able to do no more than scratch the surface of this exceedingly complex subject, and for those who wish a more detailed treatment, there are a number of excellent books available (e.g., Refs. 2-1, 2-2, and 2-3).

In general, a light wave must be represented by a vector function of position and time. There are special cases, however, for which such a wave may be adequately described by a scalar function, and we limit our discussion here to those cases for which a scalar representation is valid. As our first restriction, we shall consider only monochromatic light waves —waves consisting of a single temporal-frequency component. (In reality no such waves can exist, but this idealization is an extremely useful one even so because there are sources that can be made to emit very nearly monochromatic light; e.g., lasers.) Furthermore, we shall assume that these monochromatic light waves are linearly polarized, and we shall associate the scalar function describing them with the magnitude of either the electric-field vector or the magnetic-field vector. We shall use the real-valued scalar function

$$u(\mathbf{r},t)=a(\mathbf{r})\cos\left[2\pi\nu_0 t-\phi(\mathbf{r})\right] \tag{2.102}$$

to represent such a linearly polarized, monochromatic light wave, where $\mathbf{r}$ is a position vector and ν_0 is the temporal frequency of the wave. The function $a(\mathbf{r})$ is known as the amplitude of the wave and the argument of the cosine function is called its phase. The surfaces in space defined by the equation $\phi(\mathbf{r})=\text{constant}$ are called *co-phasal surfaces*, or more commonly, *wavefronts*. Both $a(\mathbf{r})$ and $\phi(\mathbf{r})$ are real-valued scalar functions of position.

The function $u(\mathbf{r},t)$ is a solution of the scalar wave equation

$$\nabla^2 u(\mathbf{r},t)-\frac{n^2}{c^2}\frac{\partial^2 u(\mathbf{r},t)}{\partial t^2}=0, \tag{2.103}$$

where n is the refractive index of the medium in which the wave is propagating, c is the speed of light in vacuum, and ∇^2 is the Laplacian operator (for rectangular coordinates)

$$\nabla^2=\frac{\partial^2}{\partial x^2}+\frac{\partial^2}{\partial y^2}+\frac{\partial^2}{\partial z^2}. \tag{2.104}$$

We shall assume that this medium is homogeneous and that its properties are time-independent; thus, the refractive index is constant (e.g., $n=1$ in vacuum).

At this point it will be advantageous for us to define the phasor

$$u(\mathbf{r})=a(\mathbf{r})e^{j\phi(\mathbf{r})}, \tag{2.105}$$

which is known in optics as the *complex amplitude* of the wave $u(\mathbf{r},t)$, so that this wave may be expressed as

$$u(\mathbf{r},t)=\mathrm{Re}\{u^*(\mathbf{r})e^{j2\pi\nu_0 t}\}. \tag{2.106}$$

The advantage gained is the following: it can be shown that not only must $u(\mathbf{r},t)$ be a solution of the scalar wave equation, but also that the complex amplitude $u(\mathbf{r})$ must satisfy the time-independent Helmholz equation

$$\nabla^2 u(\mathbf{r})+n^2k_0^2u(\mathbf{r})=0, \tag{2.107}$$

where $k_0=2\pi\nu_0/c=2\pi/\lambda_0$ is the wave number associated with $u(\mathbf{r},t)$, λ_0 being the wavelength in vacuum. As a result many problems of interest in optics may be solved, and many intermediate operations simplified, by dropping the $\mathrm{Re}\{\cdot\}$ operator and working directly with the function $u(\mathbf{r})$, which depends only on the spatial coordinates and not on the time. However, if these operations are not linear, care must be taken to insure that none of the rules of complex algebra are violated in obtaining the final solution. We also point out that even though the complex amplitude is very useful in solving problems, it is not a true representation of the actual wave because a complex-valued function cannot be used to describe a real physical quantity directly. In addition, as mentioned above, the complex amplitude does not exhibit the time dependence of the actual wave. When the complex amplitude alone is used to represent this wave, the time-harmonic behavior is understood, and Eq. (2.106) must be used to obtain the physical solution for the wave.

It is interesting to note that the expression for the complex amplitude, Eq. (2.105), is quite similar to the expression for a general modulated wave given by Eq. (2.91). Of course $u(\mathbf{r})$ is complex-valued and depends on the three independent variables (x,y,z), whereas the modulated wave of Eq. (2.91) was a one-dimensional real-valued function, but the similarities readily become apparent with a little thought. At a *fixed instant* in time, the modulus $a(\mathbf{r})$ of the complex amplitude describes how the amplitude of the light wave $u(\mathbf{r},t)$ varies with position (similar to AM), whereas its argument $\phi(\mathbf{r})$ describes how the phase of the wave varies with position (similar to PM). Thus, the concepts developed for temporal modulation will be of use to us in our study of optical waves. We shall defer any further discussion of light waves until Chapter 10.

REFERENCES

2-1 M. Born and E. Wolf, *Principles of Optics*, 3rd ed., Pergamon Press, New York, 1965.

2-2 J. M. Stone, *Radiation and Optics: An Introduction to the Classical Theory*, McGraw-Hill, New York, 1963.

2-3 J. W. Goodman, *Introduction to Fourier Optics*, McGraw-Hill, New York, 1968.

PROBLEMS

2-1. Given an arbitrary function $f(x)=f_e(x)+f_0(x)$, where $f_e(x)$ represents the even part and $f_0(x)$ the odd part, show that:

a. $f_e(x)=\frac{1}{2}[f(x)+f(-x)]$.

b. $f_0(x)=\frac{1}{2}[f(x)-f(-x)]$.

2-2. Given the complex number $u=v+jw$, show that:

a. $\text{Re}\{u\}=\frac{1}{2}(u+u^*)$.

b. $\text{Im}\{u\}=(1/2j)(u-u^*)$.

2-3. Given the complex numbers $u_1=v_1+jw_1$ and $u_2=v_2+jw_2$, show that:

a. $\text{Re}\{u_1+u_2\}=\frac{1}{2}(u_1+u_2)+\frac{1}{2}(u_1+u_2)^*$.

b. $(u_1+u_2)+(u_1-u_2)^*=2\,\text{Re}\{u_1\}+j2\,\text{Im}\{u_2\}$.

c. $\text{Re}\{u_1u_2^*\}\neq\text{Re}\{u_1\}\text{Re}\{u_2^*\}$.

d. $u_1u_2^*+u_1^*u_2$ is real valued.

e. $|u_1+u_2|^2=|u_1|^2+|u_2|^2+u_1u_2^*+u_1^*u_2$.

2-4. Find all of the roots of the following equations, and show the locations of these roots in the complex plane.

a. $x^3=1$.

b. $x^3=8e^{j\pi}$.

c. $x^4=4e^{j(\pi/3)}$.

2-5. Calculate the following complex sums and sketch a complex-plane representation of each.

a. $u=e^{j(\pi/6)}+e^{j(5\pi/6)}$.

b. $u=2e^{j(\pi/3)}-2e^{j(2\pi/3)}$.

c. $u=1+\sqrt{2}\,e^{j(\pi/4)}+e^{j(\pi/2)}$.

2-6. Let $u(x)=A\exp\{j2\pi\xi_0x\}$, where A and ξ_0 are real positive constants. Find, and sketch as functions of x, the following:

a. $|u(x)|^2$.

b. $u(x)+u^*(x)$.

c. $|u(x)+u^*(x)|^2$.

CHAPTER 3

SPECIAL FUNCTIONS

In solving scientific and engineering problems, it is often helpful to employ the use of functions that cannot be described by single algebraic expressions over their entire domain, i.e., functions that must be described in a piecewise fashion. To simplify much of our work later on, we now define several such piecewise functions and assign special notation to them. In addition, for the sake of compactness, we introduce special notation for a few functions that do not require a piecewise representation.

It should be stressed that there is nothing sacred about the choice of notation to be used here. With each choice there are advantages and disadvantages, and the argument as to which is the best will never be settled. Many authors in the past have developed their own special notation, with varying degrees of success, and an attempt has been made here to utilize the best points of this previous work and to eliminate the bad points. If a particular symbol or abbreviation has achieved popularity and has no serious shortcomings, it was adopted for our use. In many cases, however, entirely new expressions have been introduced because previously used notation was felt to be inferior in some respect.

Some authors have denoted special functions by single letters or symbols, whereas others have used abbreviations for their special functions. The former method has the advantage that a minimum of writing is required, but it also has several disadvantages. For example, one author may use a certain symbol to represent one function and another author may use that symbol to describe a different function, with the result that keeping track of the various quantities of interest can become quite a chore. Also, because of the limited supply of letters and symbols, individual authors themselves are often forced to use a particular symbol to represent more than one function. This requires the introduction of new fonts, overbars, tildes, primes, subscripts, etc., and things are bad enough in that respect without any additional complications. One last disadvantage has to do with the utility of this method in the classroom; it is quite often difficult for the instructor to quickly and unambiguously reproduce the various symbols at the chalkboard, or for the student to do the same in his notebook.

Disadvantages associated with the use of abbreviations are that more writing and more space are required to represent the desired function, and again different authors may use the same abbreviation to describe different functions, although here the disagreement is usually in the normalization and scaling. On the positive side, this method does have the advantage that the abbreviations can be quite descriptive in nature, which can significantly reduce the time required to become well acquainted with the definition.

The above discussion provides the rationale for adopting the present set of notation, which is a combination of symbols and descriptive abbreviations. We trust that no one will be offended by any of our choices.

3-1 ONE-DIMENSIONAL FUNCTIONS

The functions defined and discussed in this section are all real-valued, one-dimensional functions of the real independent variable x. In our notation, x_0 is a real constant that essentially determines the "position" of the function along the x-axis, and the real constant b is a scaling factor that regulates the orientation of the function about the point $x=x_0$ and is usually proportional to its "width." (The latter is not true in the case of the step function and the sign function.) In general, x_0 may be zero, positive, or negative, and b may be either positive or negative. When $x_0=0$, we may consider the function to be unshifted; if $x_0>0$, the function is shifted by an amount x_0 to the right, and if $x_0<0$, it is shifted by an amount $|x_0|$ to the left. In addition, if b is changed from positive to negative, the function is reflected about the line $x=x_0$. Another point of interest is that with our choice of definitions and when the concept is meaningful, the area of the various functions is just equal to $|b|$.

For the arguments of our special functions to be dimensionless, x_0 and b must have the same units as the independent variable. This condition is implicit in all of the following definitions and figures, and, unless otherwise specified, the figures have been drawn for both x_0 and b positive.

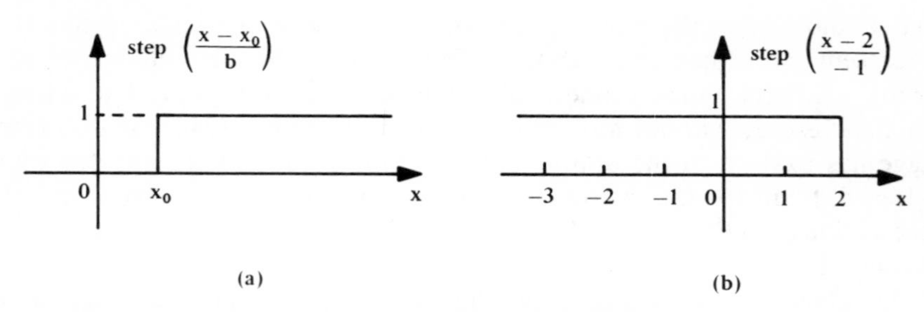

Figure 3-1 The step function. (*a*) For positive *b*. (*b*) For negative *b*.

The Step Function

We define the step function to be

$$\operatorname{step}\left(\frac{x-x_0}{b}\right)=\begin{cases}0, & \dfrac{x}{b}<\dfrac{x_0}{b}\\ \frac{1}{2}, & \dfrac{x}{b}=\dfrac{x_0}{b}\\ 1, & \dfrac{x}{b}>\dfrac{x_0}{b}\end{cases}. \tag{3.1}$$

This function is illustrated in Fig. 3-1(a), and we see that it has a discontinuity at the point $x=x_0$. Note that in this case it is not meaningful to talk about the width or area of the function, and the only purpose of the constant b is to allow the function to be reflected about the line $x=x_0$, as illustrated in Fig. 3-1(b).

The utility of the step function is that it can be used as a "switch" to turn another function on or off at some point. For example, the product given by $\operatorname{step}(x-1)\cos(2\pi x)$ is identically zero for $x<1$ and simply $\cos(2\pi x)$ for $x>1$, as shown in Fig. 3-2.

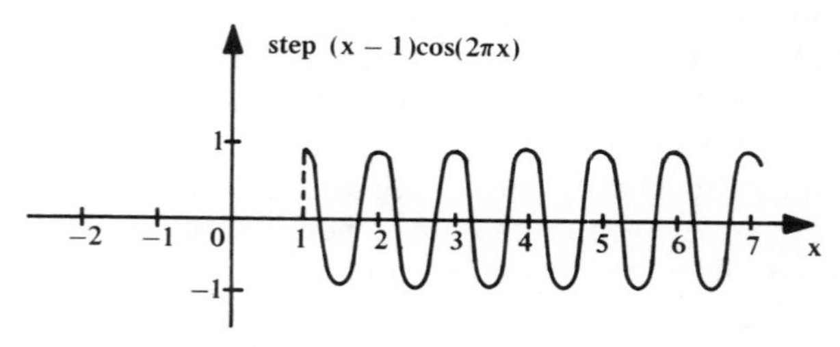

Figure 3-2 The step function as a switch.

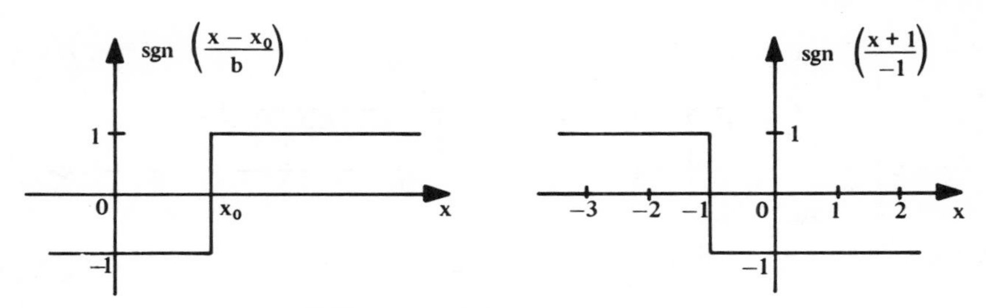

Figure 3-3 The sign function.

The Sign Function (Pronounced "Signum")

The sign function, which is similar to the step function, is given by

$$\operatorname{sgn}\left(\frac{x-x_0}{b}\right)=\begin{cases}-1, & \dfrac{x}{b}<\dfrac{x_0}{b}\\ 0, & \dfrac{x}{b}=\dfrac{x_0}{b}\\ 1, & \dfrac{x}{b}>\dfrac{x_0}{b}\end{cases}. \tag{3.2}$$

With reference to Fig. 3-3, it is easy to see that this function and the step function are related by the expression

$$\operatorname{sgn}\left(\frac{x-x_0}{b}\right)=2\operatorname{step}\left(\frac{x-x_0}{b}\right)-1. \tag{3.3}$$

As with the step function the concepts of width and area are meaningless and the polarity of the constant b merely determines the orientation of the function. The sign function may be used to reverse the polarity of another function at some point.

The Rectangle Function

One of the most useful functions to be defined here is the rectangle function, which is given by

$$\operatorname{rect}\left(\frac{x-x_0}{b}\right)=\begin{cases}0, & \left|\dfrac{x-x_0}{b}\right|>\frac{1}{2}\\ \frac{1}{2}, & \left|\dfrac{x-x_0}{b}\right|=\frac{1}{2}\\ 1, & \left|\dfrac{x-x_0}{b}\right|<\frac{1}{2}\end{cases}. \tag{3.4}$$

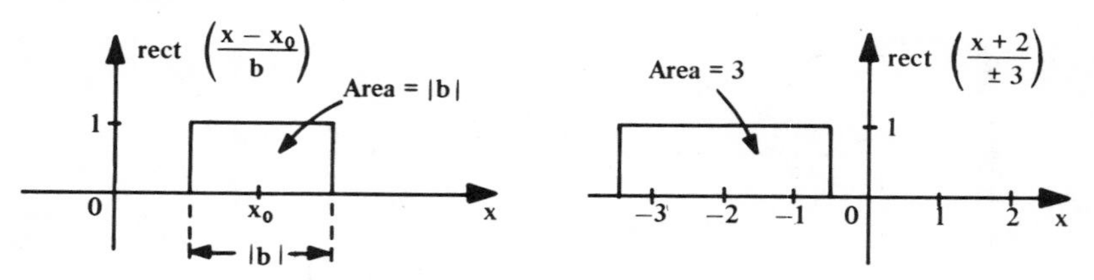

Figure 3-4 The rectangle function.

It has a height of unity, it is centered at the point $x = x_0$, and both its width and area are equal to $|b|$, as may be seen in Fig. 3-4. In the time domain the rectangle function can be used to represent a gating operation in an electrical circuit, a shutter operation in a camera, etc., whereas in the space domain it is often used to describe such quantities as the transmittance of a slit aperture. We shall make a great deal of use of this function.

The Ramp Function

We define the ramp function to be

$$\operatorname{ramp}\left(\frac{x-x_0}{b}\right)=\begin{cases} 0, & \dfrac{x}{b}\leqslant\dfrac{x_0}{b} \\ \left|\dfrac{x-x_0}{b}\right|, & \dfrac{x}{b}>\dfrac{x_0}{b} \end{cases}, \tag{3.5}$$

and we illustrate this function in Fig. 3-5. In this definition the concepts of width and area are again meaningless, but the slope of the nonzero portion of the function is equal to b^{-1}. It is easy to show that the ramp function is related to the integral of the step function by the expression

$$\operatorname{ramp}\left(\frac{x-x_0}{b}\right)=\frac{1}{b}\int_{x_0}^{x}\operatorname{step}\left(\frac{\alpha-x_0}{b}\right)d\alpha. \tag{3.6}$$

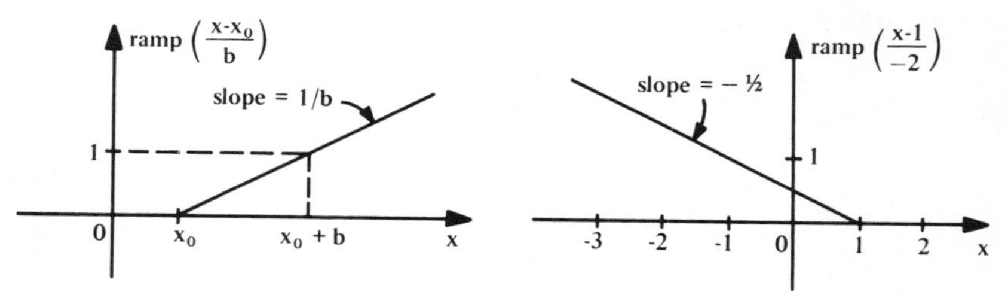

Figure 3-5 The ramp function.

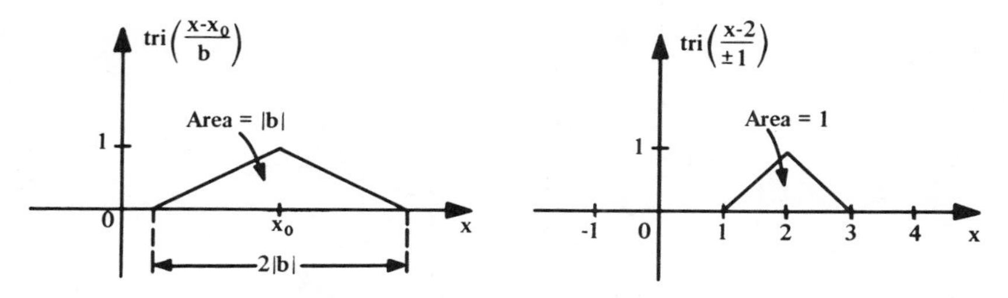

Figure 3-6 The triangle function.

The Triangle Function

The triangle function is another function that will be of considerable importance to us later on, and it is defined by

$$\operatorname{tri}\left(\frac{x-x_0}{b}\right)=\begin{cases} 0, & \left|\dfrac{x-x_0}{b}\right| \geqslant 1 \\ 1-\left|\dfrac{x-x_0}{b}\right|, & \left|\dfrac{x-x_0}{b}\right|<1 \end{cases}. \tag{3.7}$$

The width of the base of the triangle function is now $2|b|$, as can be seen in Fig. 3-6, but its area is still $|b|$.

There are various ways in which the triangle function may be formed from the previously defined functions. For example, the function $\operatorname{tri}[(x-1)/2]$ may be written as

$$\operatorname{tri}\left(\frac{x-1}{2}\right)=\operatorname{ramp}\left(\frac{x+1}{2}\right)\operatorname{step}\left(\frac{x-1}{-1}\right)+\operatorname{ramp}\left(\frac{x-3}{-2}\right)\operatorname{step}\left(\frac{x-1}{1}\right). \tag{3.8}$$

In this expression you will note that we used scaling factors of ± 1 in the step functions; however, it is only the signs of these scaling factors that are of importance here and not their magnitude. The values of ± 1 were chosen simply because they seemed to be the most convenient. It is interesting to investigate other combinations of functions that will yield the triangle function of Eq. (3.8).

The Sinc Function (Pronounced "Sink")

We use Bracewell's (Ref. 3-1) definition for this function, i.e.,

$$\operatorname{sinc}\left(\frac{x-x_0}{b}\right)=\frac{\sin\pi\left(\dfrac{x-x_0}{b}\right)}{\pi\left(\dfrac{x-x_0}{b}\right)}. \tag{3.9}$$

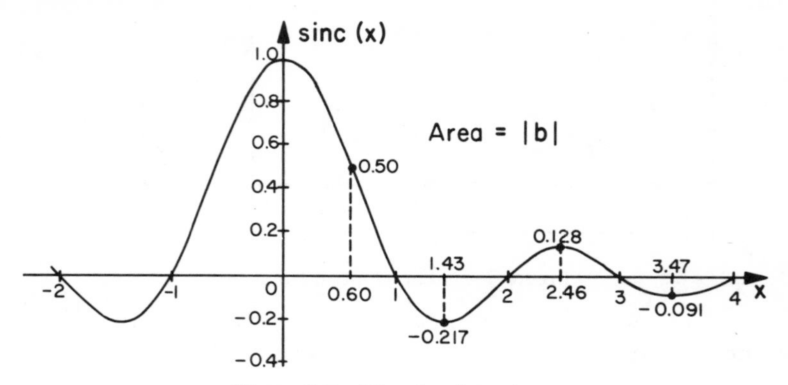

Figure 3-7 The sinc function.

It is often asked why this function is defined with the factor of π included. The answer is very simple: when the function is defined as in Eq. (3.9), its zeros occur at $x_0 \pm nb$, and when it is defined without the factor of π, the zeros occur at $x_0 \pm n\pi b$, where $n = 1, 2, 3, \ldots$. Thus, with Bracewell's definition, the zeros are much more conveniently located than for the other definition. In addition, the height of this function is unity, its width between the first two zeros is $2|b|$, and its area, including positive and negative lobes, is just equal to $|b|$. These "nice" properties make Bracewell's definition the most attractive for our purposes, but as we pointed out earlier there is nothing sacred about it. The graph of this function is shown in Fig. 3-7, where we have let $x_0 = 0$ and $b = 1$ to simplify the drawing. In addition, it is tabulated in the Appendix.

The sinc function will show up quite often in our studies of linear systems, and it is interesting to note that there is a close relationship between this function and the rectangle function; appropriately scaled and shifted, each is the Fourier transform of the other. But let's not worry about that just yet.

The Sinc² Function

The square of the sinc function will also be encountered quite frequently, and therefore we include it as one of our special functions. This function can be used to describe the one-dimensional irradiance profile in the Fraunhofer diffraction pattern of a slit, and there exists the same Fourier transform relationship between it and the triangle function that exists between the sinc function and the rectangle function. The graph of $\text{sinc}^2(x)$ is shown in Fig. 3-8, where again we have let $x_0 = 0$ and $b = 1$, and a table of numerical values is given in the Appendix.

It is obvious that the zeros of the sinc^2 function and those of the sinc function occur at the same points, but it is not so obvious that the areas of

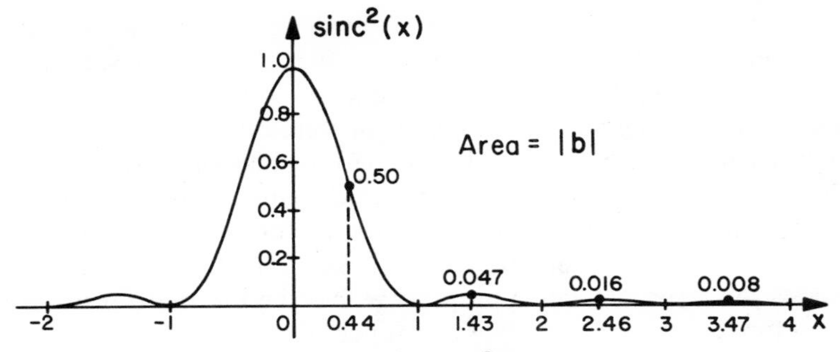

Figure 3-8 The sinc² function.

these functions are equal. At first glance it looks as if sinc(x) has the larger area, but when the negative lobes are taken into account, its total area is exactly the same as that of sinc²(x) (which has only positive lobes). This curious fact can of course be shown by direct integration, but later we shall see how it can be determined very easily by using a theorem involving the Fourier transform.

The Gaussian Function

We define the Gaussian function to be

$$\mathrm{Gaus}\left(\frac{x-x_0}{b}\right)=\exp\left[-\pi\left(\frac{x-x_0}{b}\right)^2\right], \tag{3.10}$$

where we have followed Bracewell (Ref. 3-2) by including the factor of π in the exponent. When defined in this fashion, the Gaussian function has a height of unity and its area is equal to $|b|$, the same "nice" properties we

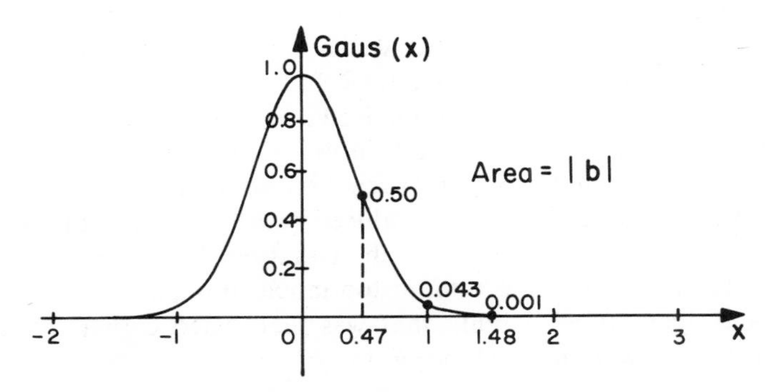

Figure 3-9 The Gaussian function.

observed for our sinc function. The graph of this function is shown in Fig. 3-9, and once more we have chosen $x_0=0$ and $b=1$ to simplify the drawing. It is also tabulated in the Appendix.

The Gaussian function is frequently encountered in the field of statistics, although usually in a somewhat different form than that given here. We shall find, however, that due to its many unique and interesting properties, it is a very useful function for our study of linear systems. One such property is that it is a smooth function, i.e., all of its derivatives are continuous. Another interesting property is that the Fourier transform of a Gaussian function is another Gaussian function.

Discussion

Now let us stop momentarily for a brief review. Note once again that in the definitions for the rectangle, triangle, sinc, sinc^2, and Gaussian functions, the central ordinate is always unity and the area is always equal to the magnitude of the scaling factor, $|b|$. If we wish to obtain a version of one of these functions whose height is different from unity, we need only multiply that function, as defined above, by the appropriate constant. In addition, we can obtain any width or area desired by making the proper choice of the constant b. Again we point out that our definitions are not hallowed, but their advantages will become more apparent and appreciated as we get further in the book.

It is interesting to note the similarities between several of our special functions. For example, the main lobe of the sinc^2 function very closely resembles the Gaussian function. For that matter, both of these are quite similar to the main lobe of the sinc function, which itself is not too different in shape from the simple triangle function. To illustrate these similarities, the functions $\text{sinc}(x)$, $\text{sinc}^2(x)$, $\text{Gaus}(x)$, and $\text{tri}(x)$ are all graphed together in Fig. 3-10.

In our definitions for the step and rectangle functions, we chose the value at the point of discontinuity to be $\frac{1}{2}$ rather than zero or unity. This is just the average of the values of the function as the discontinuity is approached from above and from below. There is no compelling reason for this choice, but as Bracewell points out (Ref. 3-3) certain consistencies are then likely to be observed. We know, of course, that real physical quantities will not exhibit such behavior, but in many cases these quantities can be represented quite satisfactorily by such discontinuous functions. Certain mathematical difficulties may be introduced as a result, but this is due to the model rather than the physics of the situation. However, these problems can usually be overcome without too much trouble.

We should again stress a point that was mentioned earlier, because if this point is not well understood a lot of confusion can result. The

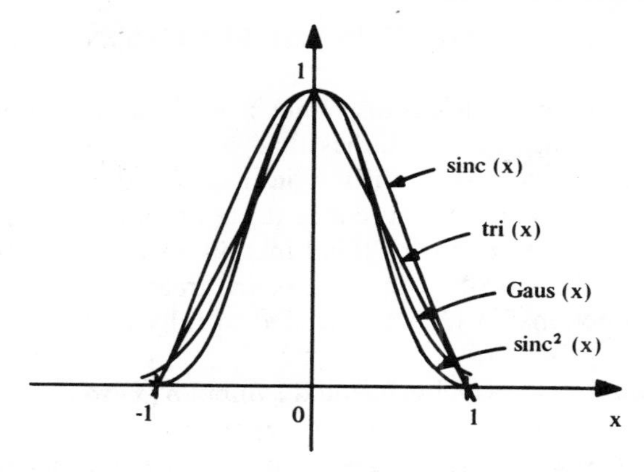

Figure 3-10 Similarities of the triangle, sinc, sinc^2, and Gaussian functions.

arguments of the above functions are always dimensionless, and thus x_0 and b must always have the same units as x. From time to time b will have a value of unity, and unless care is taken the units can get mixed up or lost. For example, suppose we wish to represent the transmittance $t(x)$ of a slit that is 1 cm wide. We should write

$$t(x) = \text{rect}\left(\frac{x}{1\,\text{cm}}\right), \tag{3.11}$$

but if we specify that x is to be measured in centimeters, we might change this to

$$t(x) = \text{rect}(x). \tag{3.12}$$

In this form, the units are implicit. For a slit 2 cm wide, where again we have specified that x is to be measured in centimeters, we would write

$$t(x) = \text{rect}\left(\frac{x}{2}\right). \tag{3.13}$$

If we now choose to measure x in millimeters, then Eq. (3.13) would have to be changed to

$$t(x) = \text{rect}\left(\frac{x}{20}\right). \tag{3.14}$$

The main thing to remember is that the scaling factors will have the correct values only if the units have been properly taken into account.

3-2 THE IMPULSE FUNCTION

The impulse function, which is also known as Dirac's delta function, is of such great importance in our studies that we treat it separately. We shall not dwell on the mathematical difficulties associated with its use, of which there are many, but shall instead concentrate primarily on its applications. For more rigorous treatments of this function, refer to Lighthill (Ref. 3-4) and Bracewell (Ref. 3-5); both discuss in great detail the concepts of "generalized functions," which are needed to fully understand the impulse function.

We shall employ the widely accepted symbol $\delta(x)$ to denote the impulse function, which is frequently used to represent such quantities as point sources, point masses, point charges, or any other quantities that are highly localized in some coordinate system. We know, of course, that real physical quantities cannot truly exhibit point characteristics, but often such an idealization is not only permissible but desirable as well. The reason for this, as we shall see later, is that in many cases a system can be completely characterized by its response to an impulsive input. This is known as the *impulse response* (Green's function) of the system. If the input to a system is quite complicated, it may be very difficult to calculate the output by brute-force methods. On the other hand, if the system's impulse response is known, and if the system is a linear system, we need only to decompose the complicated input into a superposition of a large number of delta functions, each appropriately weighted and positioned. The net response to the complicated input is then determined by adding together the responses to all of the individual delta functions. If this is confusing, don't worry about it just yet; we shall cover this topic again in much greater detail in Chapters 5 and 8.

Now let us discuss the characteristics of the impulse function, which in the usual sense is not really a function at all. It is often described as having an infinite height and zero width such that its area is equal to unity, but such a description is not particularly satisfying. A more formal definition is obtained by first forming a sequence of functions, such as $|1/b|\mathrm{Gaus}(x/b)$, and then defining $\delta(x)$ to be

$$\delta(x)=\lim_{b\to 0}\frac{1}{|b|}\mathrm{Gaus}\left(\frac{x}{b}\right). \tag{3.15}$$

As $|b|$ gets smaller and smaller the members of the sequence become taller and narrower, but their area remains constant as illustrated in Fig. 3-11. This definition, however, is not very satisfying either.

Perhaps the best way to think about the delta function is from a physical point of view; we need only consider it to be so narrow that making it any

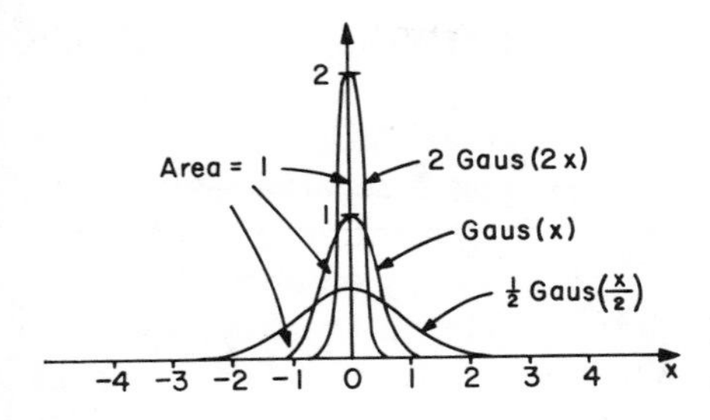

Figure 3-11 A sequence of constant-area Gaussian functions.

narrower would in no way influence the results in which we are interested. To illustrate we consider the following two examples. In the first a rectangular voltage pulse $v_1(t)$ is applied to the input terminals of the *R-C* circuit shown in Fig. 3-12. We choose the width of this pulse to be T, and its height to be proportional to T^{-1} such that its area remains constant as T is varied. The output voltage $v_2(t)$ will vary with time, and its exact form will depend on the relative values of T and the product RC as illustrated in Fig. 3-13. (Don't be concerned if your understanding of *R-C* circuits is limited; this example is included only because it provides a very nice demonstration of some of the important characteristics of the delta function.)

If T is much larger than RC, the capacitor will be almost completely charged to the voltage A/T before the pulse ends, at which time it will begin to discharge back to zero. This is shown in Fig. 3-13(a). If we shorten the pulse so that $T \ll RC$, the capacitor will not have a chance to become fully charged before the pulse ends. Thus the output voltage behaves as illustrated in Fig. 3-13(b), and it can be seen that there is a considerable difference between this output and the preceding one. If we now make T still shorter, as in Fig. 3-13(c), we note very little change in the shape of the output. In fact, as we continue to make T shorter and shorter, the only noticeable change is in the time it takes the output to reach a maximum, and this time is just equal to T. If this interval is too short to be resolved by our measuring device, the input is effectively behaving as a delta function and decreasing its duration further will have no observable effect on the output, which now closely resembles the impulse response of the circuit.

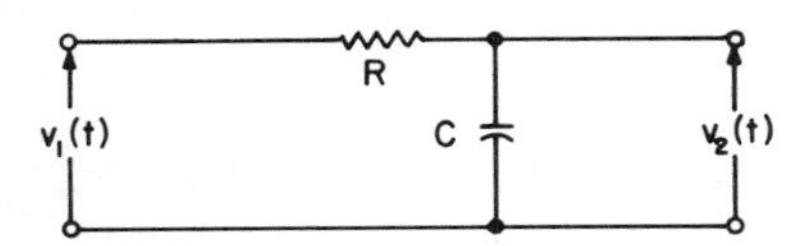

Figure 3-12 An *R-C* circuit for demonstrating the nature of delta functions.

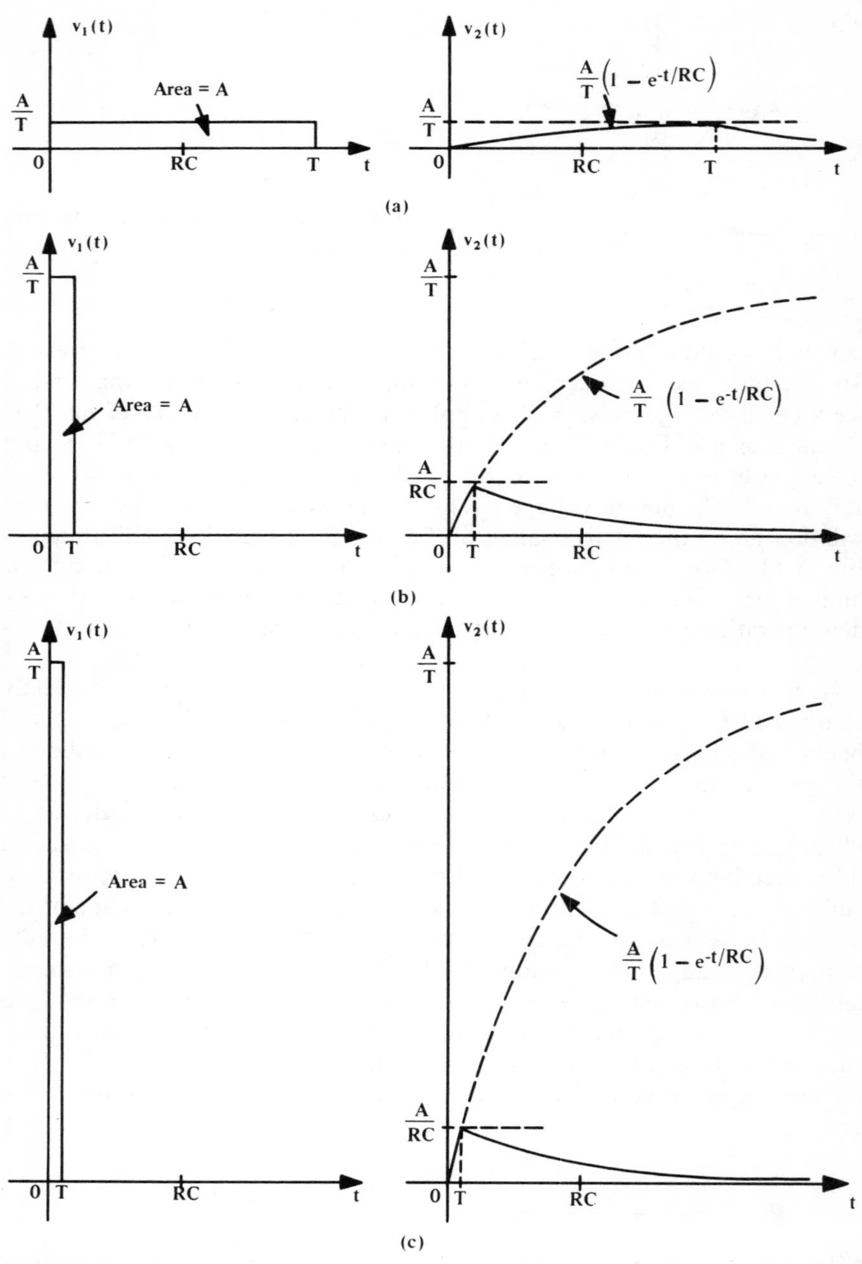

Figure 3-13 Input-output relationships for *R-C* circuit of Fig. 3-12. (*a*) $T \gg RC$. (*b*) $T \ll RC$. (*c*) Input approaching a delta function.

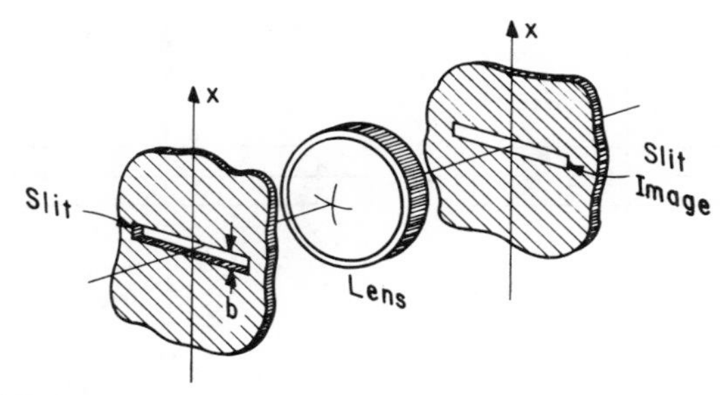

Figure 3-14 Optical imaging system for demonstrating nature of delta functions.

For the second example, consider the imaging system of Fig. 3-14. A slit of width b, located in an opaque screen, is transilluminated by a monochromatic plane wave of light, and an image of this slit is cast onto an observation screen. For simplicity, we shall concern ourselves only with variations of the image irradiance in a direction perpendicular to the slit. The irradiance of the slit illumination is made proportional to b^{-1}, such that the product of this irradiance and the slit width is constant. The image irradiance will depend both on the width of the slit and the resolving power of the lens. If b is large enough, the image will resemble the slit very closely, but as b is decreased, the image will begin to look more and more like the Fraunhofer diffraction pattern of the slit. Finally a point will be reached beyond which the image will remain unchanged as b is decreased. At this point the slit may be considered to be a delta function (line source) as far as the imaging system is concerned. This example is illustrated in Fig. 3-15, and once again don't be too concerned if you don't understand how we arrived at these results. The main point of this example is that we really don't need to worry about the exact width or shape of the delta function; we only need to think of it as a function so narrow that making it any narrower will have no effect on any observation of interest.

With the above discussions in mind, we shall use the following properties as our definition of the delta function: given the real constant x_0 and the arbitrary complex-valued function $f(x)$, which is continuous at the point $x=x_0$, then

$$\delta(x-x_0)=0, \qquad x\neq x_0 \tag{3.16}$$

$$\int_{x_1}^{x_2} f(\alpha)\delta(\alpha-x_0)\,d\alpha=f(x_0), \qquad x_1<x_0<x_2. \tag{3.17}$$

If $f(x)$ is discontinuous at the point $x=x_0$, Eq. (3.17) may still be used but the value of $f(x_0)$ is taken to be the average of the limiting values as x_0 is

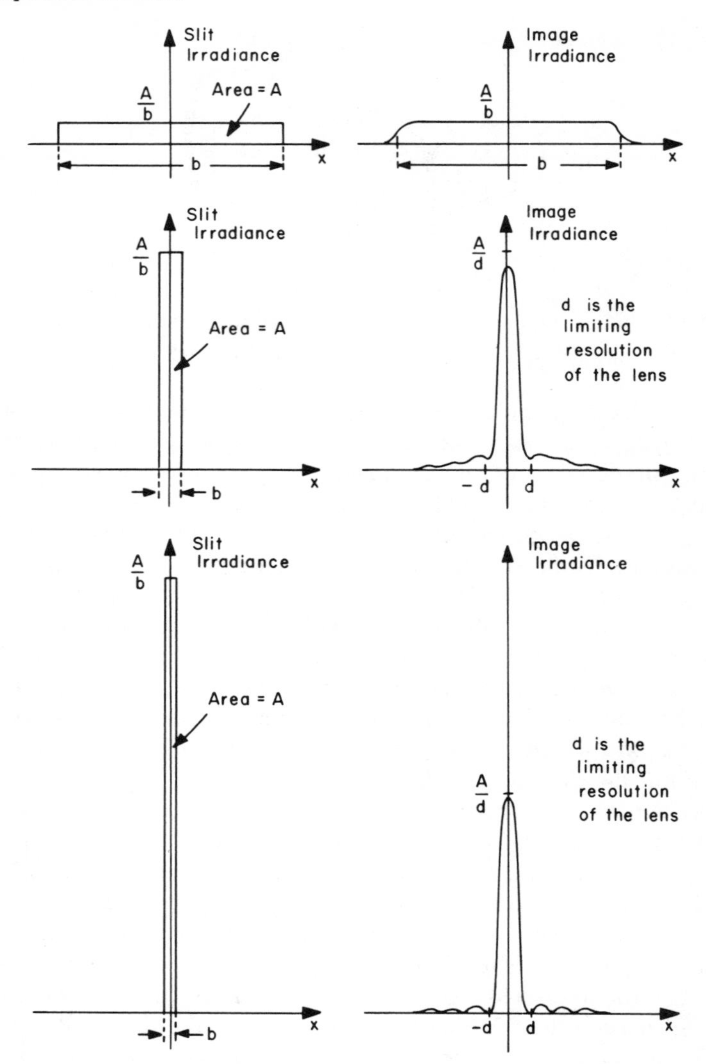

Figure 3-15 Object-image relationships for imaging system of Fig. 3-14.

approached from above and from below. Note that we do not specify the value of $\delta(x-x_0)$ at the point $x=x_0$. Many authors define this value to be infinite, but it is not really necessary because any additional properties that are needed can be derived from Eqs. (3.16) and (3.17) alone. Also note that from a mathematical point of view, we do not care about the exact form of the delta function itself but only about its behavior under integration.

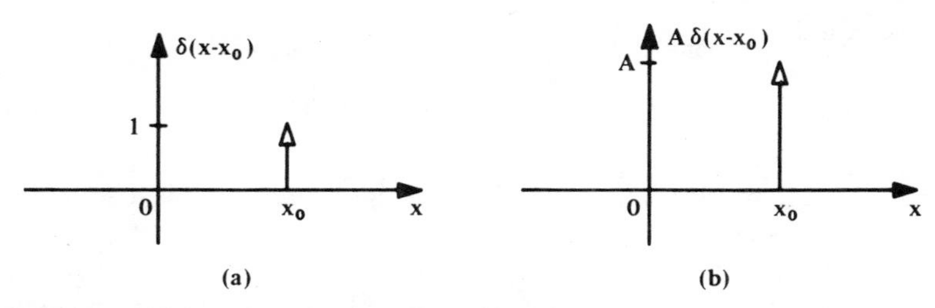

Figure 3-16 Graphical representation of delta function located at $x=x_0$. (*a*) Area equal to unity. (*b*) Area equal to A.

On graphs we will *represent* $\delta(x-x_0)$ as a spike of unit height located at the point $x=x_0$, but observe that the *height of the spike* corresponds to the *area of the delta function*. Such a representation is shown in Fig. 3-16(a). To represent the function $A\,\delta(x-x_0)$ we would use a spike of height A, as illustrated in Fig. 3-16(b), because the area of this function is now A instead of unity.

The impulse function often appears with a scaled argument, or in combination with another function, and it is important to determine the meaning of such expressions. To do so, we first multiply the expression in question by some "sufficiently smooth" test function $\theta(x)$, and then use Eq. (3.17) to evaluate the integral of this product. For example, suppose we wished to determine the meaning of the expression $\delta[(x-x_0)/b]$. At first glance it might appear that it is exactly the same as $\delta(x-x_0)$, because "you obviously can't scale something whose width is zero to begin with." However, let us proceed as outlined above. First we multiply $\delta[(x-x_0)/b]$ by the test function $\theta(x)$, which we require to be continuous at the point $x=x_0$. Then, letting $x_1=-\infty$ and $x_2=+\infty$ to simplify the notation, and making the change of variable $\alpha=b\beta$, we find from Eq. (3.17) that

$$\begin{aligned}\int_{-\infty}^{\infty}\theta(\alpha)\delta\left(\frac{\alpha-x_0}{b}\right)d\alpha &= |b|\int_{-\infty}^{\infty}\theta(b\beta)\delta\left(\beta-\frac{x_0}{b}\right)d\beta\\ &= |b|\theta(b\beta)\big|_{\beta=x_0/b}\\ &= |b|\theta(x_0).\end{aligned} \tag{3.18}$$

But this is exactly the result we would have obtained if we had started with the function $|b|\delta(x-x_0)$ rather than $\delta[(x-x_0)/b]$. Thus, since these two expressions yield the same result when Eq. (3.17) is applied, we say that they are equal, i.e.,

$$\delta\left(\frac{x-x_0}{b}\right)=|b|\delta(x-x_0). \tag{3.19}$$

We note that our intuition proved to be incorrect in this case, and in general it is best not to rely on intuition.

In the above development we equated two quantities because they produced the same results when an integral operation was performed. Obviously, such an approach cannot be used in every situation; for example, the striking of a window by a bird and a brick may produce the same net result, but it is not fair to conclude that a bird and a brick are the same thing. Nevertheless, we may use this kind of procedure to "equate" various expressions containing delta functions as long as we realize it is always their behavior under integration that will be of interest in the end.

Properties of Delta Functions

We now list several of the most useful properties of the delta function, all of which may easily be derived from the defining properties. Given the real constants a, b, x_0, and A and the arbitrary complex-valued function $f(x)$, we obtain the following:

1. *Defining properties*

$$\delta(x-x_0)=0, \qquad x \neq x_0,$$

$$\int_{x_1}^{x_2} f(\alpha)\delta(\alpha-x_0)\,d\alpha=f(x_0), \qquad x_1<x_0<x_2.$$

This second property is often called the "sifting" property, because it "sifts" out a single value of the function $f(x)$.

2. *Scaling properties*

$$\delta\left(\frac{x-x_0}{b}\right)=|b|\delta(x-x_0),$$

$$\delta(ax-x_0)=\frac{1}{|a|}\delta\left(x-\frac{x_0}{a}\right),$$

$$\delta(-x+x_0)=\delta(x-x_0),$$

$$\delta(-x)=\delta(x).$$

Note that $\delta(x)$ behaves as if it were an even function.

3. *Properties in products*

$$f(x)\delta(x-x_0)=f(x_0)\delta(x-x_0),$$

$$x\delta(x-x_0)=x_0\delta(x-x_0),$$

$$\delta(x)\delta(x-x_0)=0, \quad x_0\neq 0$$

$$\delta(x-x_0)\delta(x-x_0) \quad \text{is not defined.}$$

4. *Integral properties*

$$\int_{-\infty}^{\infty} A\delta(\alpha-x_0)d\alpha=A,$$

$$\int_{-\infty}^{\infty} \delta(\alpha-x_0)d\alpha=1,$$

$$\int_{-\infty}^{\infty} \delta(\alpha-x_0)\delta(x-\alpha)d\alpha=\delta(x-x_0).$$

Once again we mention the very important role played by the impulse function in the analysis of linear systems. If the system is a linear system and if its impulse response is known, then in theory the output of this system can be determined for almost any input, no matter how complex. This rather amazing property of linear systems is a result of the following: almost any arbitrary function can be decomposed into a linear combination of delta functions, each of which produces its own impulse response. Thus, by application of the *superposition principle*, which is discussed in Chapter 5, the overall response to the arbitrary input can be found by adding all of the individual impulse responses.

3-3 RELATIVES OF THE IMPULSE FUNCTION

In this section we shall define and discuss several functions that are closely related to the delta function. Some of these functions may seem a bit strange at this point, but they will be quite helpful later on.

The Even and Odd Impulse Pairs

We use the symbol $\delta\delta(x)$ to denote the even-impulse pair, which we define to be

$$\delta\delta(x)=\left[\delta(x+1)+\delta(x-1)\right]. \tag{3.20}$$

It consists of a pair of delta functions located at the points $x = \pm 1$, each having an area of unity. In a similar fashion, the odd-impulse pair $\delta_{\delta}(x)$ is defined to be

$$\delta_{\delta}(x) = [\delta(x+1) - \delta(x-1)], \tag{3.21}$$

and the only difference between it and the even-impulse pair is in the sign of the second delta function. These functions are illustrated in Fig. 3-17. Our definitions are similar to those of Bracewell (Ref. 3-6), except that in his definitions the delta functions are located at the points $x = \pm\frac{1}{2}$ rather than at the points $x = \pm 1$, and the area of each is $\frac{1}{2}$ rather than unity.

By using the various properties of the delta function discussed in the previous section, it may be shown that

$$\delta\delta\left(\frac{x-x_0}{b}\right) = |b|[\delta(x-x_0+b) + \delta(x-x_0-b)],$$

$$\delta_{\delta}\left(\frac{x-x_0}{b}\right) = |b|[\delta(x-x_0+b) - \delta(x-x_0-b)]. \tag{3.22}$$

It may also be shown that

$$f(x)\delta\delta\left(\frac{x-x_0}{b}\right) = |b|[f(x_0-b)\delta(x-x_0+b) + f(x_0+b)\delta(x-x_0-b)],$$

$$f(x)\delta_{\delta}\left(\frac{x-x_0}{b}\right) = |b|[f(x_0-b)\delta(x-x_0+b) - f(x_0+b)\delta(x-x_0-b)], \tag{3.23}$$

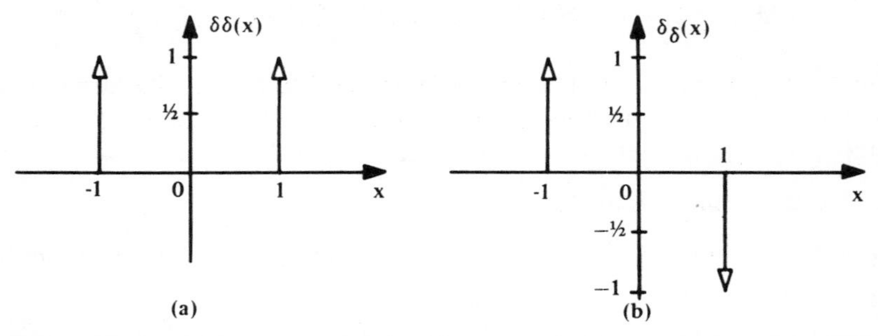

Figure 3-17 Graphical representation of impulse pairs. (*a*) Even. (*b*) Odd.

from which we obtain

$$\int_{-\infty}^{\infty} f(\alpha)\delta\delta\left(\frac{\alpha-x_0}{b}\right)d\alpha=|b|\left[f(x_0-b)+f(x_0+b)\right],$$

$$\int_{-\infty}^{\infty} f(\alpha)\delta_\delta\left(\frac{\alpha-x_0}{b}\right)d\alpha=|b|\left[f(x_0-b)-f(x_0+b)\right]. \tag{3.24}$$

Thus the impulse pairs may be used to sample a function in two places, i.e., to "sift out" its value at two points. In the above, we have made the usual assumptions that x_0 is a real number, that b is a nonzero real number, and that $f(x)$ is continuous at the points $x=x_0\pm b$.

The following properties of the impulse pairs are immediately obvious:

$$\delta\delta(-x)=\delta\delta(x),$$

$$\delta_\delta(-x)=-\delta_\delta(x). \tag{3.25}$$

It is also easy to see that

$$\int_{-\infty}^{\infty}\delta\delta\left(\frac{\alpha}{b}\right)d\alpha=2|b|,$$

$$\int_{-\infty}^{\infty}\delta_\delta\left(\frac{\alpha}{b}\right)d\alpha=0. \tag{3.26}$$

Some of the above properties are shown in Figs. 3-18 and 3-19, in which we assume that both x_0 and b are positive.

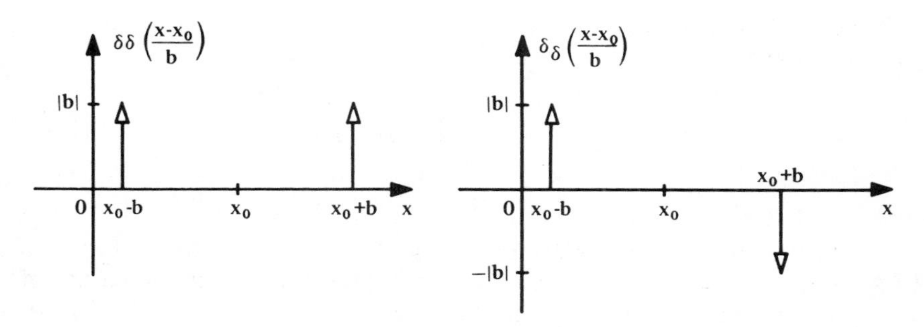

Figure 3-18 Scaled and shifted impulse pairs.

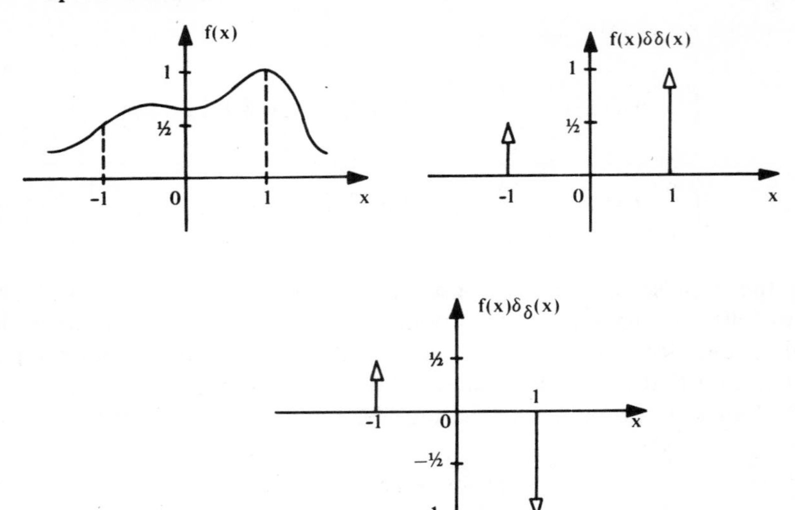

Figure 3-19 Product of a function with the impulse pairs.

The Comb Function

It is often helpful to have a special function that represents an array of delta functions, and following Goodman (Ref. 3-7), we shall call our special function for this purpose the "comb" function. It is defined by

$$\operatorname{comb}(x) = \sum_{n=-\infty}^{\infty} \delta(x-n), \tag{3.27}$$

where n takes on only integer values. Thus the comb function is an array of unit-area delta functions spaced one unit apart. It derives its name from its graphical representation, which looks very much like a comb [see Fig. 3-20(a)]. It is easy to show that the scaled and shifted version of the comb function is given by

$$\operatorname{comb}\left(\frac{x-x_0}{b}\right) = |b| \sum_{n=-\infty}^{\infty} \delta(x-x_0-nb), \tag{3.28}$$

which is illustrated in Fig. 3-20(b).

Our comb function is exactly the same as Bracewell's "shah" function (Ref. 3-8), but as explained earlier in this chapter, the descriptive nature of our notation seems more desirable than the III(x) symbol used by Bracewell.

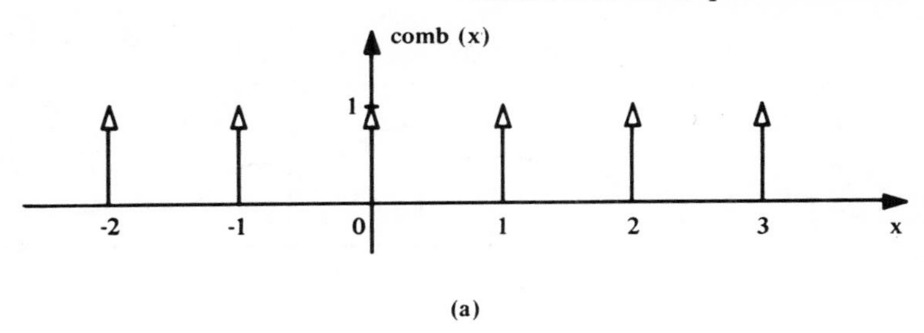

(a)

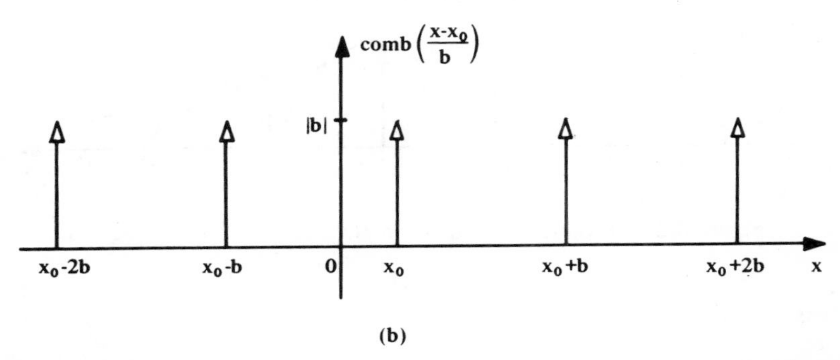

(b)

Figure 3-20 Graphical representation of the comb function. (*a*) Centered at the origin. (*b*) Scaled and shifted.

If we now choose $x_0=0$, Eq. (3.28) becomes

$$\operatorname{comb}\left(\frac{x}{b}\right)=|b|\sum_{n=-\infty}^{\infty}\delta(x-nb), \tag{3.29}$$

and we see that $\operatorname{comb}(x/b)$ is an array of delta functions spaced $|b|$ units apart, each having an area of $|b|$. As the magnitude of the scaling factor is increased, the separation between the delta functions becomes greater and the area of each increases. As $|b|$ becomes smaller, both the spacing and the area of the delta functions decreases. Hence the product of the number of delta functions in an interval and the area of each remains fixed. Rearranging Eq. (3.29), we obtain

$$\frac{1}{|b|}\operatorname{comb}\left(\frac{x}{b}\right)=\sum_{n=-\infty}^{\infty}\delta(x-nb), \tag{3.30}$$

which is just an array of unit-area delta functions spaced by $|b|$ units.

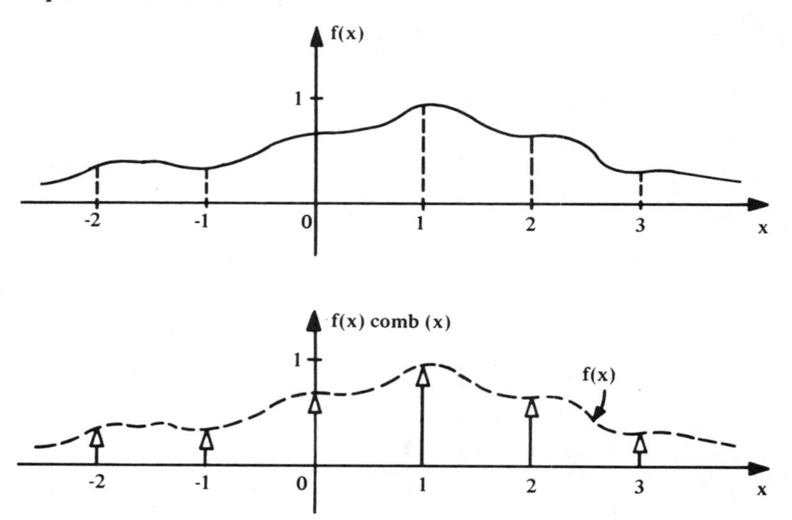

Figure 3-21 Product of a function $f(x)$ and the comb function.

It is not difficult to show that

$$f(x)\left[\frac{1}{|b|}\operatorname{comb}\left(\frac{x-x_0}{b}\right)\right]=\sum_{n=-\infty}^{\infty} f(x_0+nb)\delta(x-x_0-nb), \qquad (3.31)$$

and thus the comb function may be used to sample another function at regular intervals. This property, which is illustrated in Fig. 3-21, will be most useful when we study sampling theory in Chapter 8.

The comb function itself is of infinite extent, and thus cannot be used to represent any real physical quantity. However, when combined with another function that limits its extent, it can be used for such a purpose. Suppose, for example, we are given an opaque screen containing several very narrow, regularly spaced slits. We wish to describe the irradiance distribution $I(x)$ on one side of this screen when it is illuminated from the other side. (Again we consider this to be a one-dimensional problem, concerning ourselves only with the variation of irradiance in a direction perpendicular to the slits.) Let us assume that there are to be nine slits, spaced 2 cm apart, and that these slits are so narrow that they behave as line sources when transilluminated. If the spatial coordinate x is measured in centimeters, we might represent the transmitted irradiance by the function

$$I(x)=A[\delta(x+8)+\delta(x+6)+\delta(x+4)+\delta(x+2)+\delta(x)$$
$$+\delta(x-2)+\delta(x-4)+\delta(x-6)+\delta(x-8)], \qquad (3.32)$$

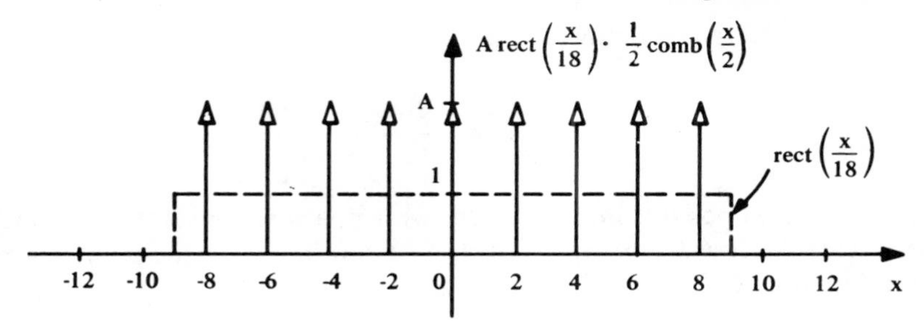

Figure 3-22 A truncated comb function.

where A is a constant determined by the strength of the illumination and the width of the slits. This is quite a cumbersome expression, however, and a more compact form is desirable. It is easy to see that the expression

$$I(x)=A\sum_{n=-4}^{4}\delta(x-2n) \tag{3.33}$$

is just what we are looking for. We may find it advantageous to go one step further and write Eq. (3.33) as

$$\begin{aligned} I(x)&=A\,\mathrm{rect}\left(\frac{x}{18}\right)\sum_{n=-\infty}^{\infty}\delta(x-2n)\\ &=A\,\mathrm{rect}\left(\frac{x}{18}\right)\left[\tfrac{1}{2}\mathrm{comb}\left(\frac{x}{2}\right)\right], \end{aligned} \tag{3.34}$$

because by writing it this way, as a product of a rectangle function and a comb function, certain mathematical operations that we might wish to perform on this function are simplified. The function $I(x)$ is illustrated in Fig. 3-22.

Derivatives of the Impulse Function

We now introduce the derivatives $\delta^{(k)}(x)$ of the impulse function, where

$$\delta^{(k)}(x)=\frac{d^k\delta(x)}{dx^k}. \tag{3.35}$$

As was done for the impulse function itself, the derivatives are described in terms of their integral properties. First, we choose some function $f(x)$

whose kth derivative is defined to be

$$f^{(k)}(x)=\frac{d^k f(x)}{dx^k}, \tag{3.36}$$

where k is a real positive integer. Then, with x_0 a real constant, we require that at least the first k derivatives of $f(x)$ be continuous at the point $x=x_0$, and define the derivatives of the impulse function by the properties

$$\delta^{(k)}(x-x_0)=0, \qquad x\neq x_0 \tag{3.37}$$

$$\int_{x_1}^{x_2} f(\alpha)\delta^{(k)}(\alpha-x_0)\,d\alpha=(-1)^k f^{(k)}(x_0), \qquad x_1<x_0<x_2 \tag{3.38}$$

where

$$f^{(k)}(x_0)=\left.\frac{d^k f(x)}{dx^k}\right|_{x=x_0}. \tag{3.39}$$

It is not easy to visualize the behavior of these functions, nor is it easy to represent them graphically. For example, let us consider the first derivative, which is often called a *doublet*. According to our definitions,

$$\delta^{(1)}(x-x_0)=0, \qquad x\neq x_0 \tag{3.40}$$

$$\int_{x_1}^{x_2} f(\alpha)\delta^{(1)}(\alpha-x_0)\,d\alpha=-f^{(1)}(x_0), \qquad x_1<x_0<x_2 \tag{3.41}$$

and we see that $\delta^{(1)}(x-x_0)$ is zero everywhere except at $x=x_0$, and it has the property of sifting out the negative slope of the function $f(x)$ at the point $x=x_0$. However, it is still difficult to form a mental picture of this function. Perhaps the best way to do so is to look once again at the delta function itself. The delta function is often written as the derivative of the step function, i.e.,

$$\delta(x-x_0)=\frac{d}{dx}\,\mathrm{step}(x-x_0), \tag{3.42}$$

where the area of $\delta(x-x_0)$ is just equal to the magnitude of the discontinuity of the step function (unity in this case). Suppose that we now express the delta function as

$$\delta(x)=\lim_{b\to 0}\frac{1}{b}\,\mathrm{rect}\left(\frac{x}{b}\right), \tag{3.43}$$

so that

$$\delta^{(1)}(x) = \lim_{b\to 0} \frac{d}{dx}\frac{1}{b}\operatorname{rect}\left(\frac{x}{b}\right), \tag{3.44}$$

where b is taken to be positive to simplify the development. Then writing the rectangle function as

$$\operatorname{rect}\left(\frac{x}{b}\right) = \operatorname{step}\left(x+\frac{b}{2}\right) - \operatorname{step}\left(x-\frac{b}{2}\right), \tag{3.45}$$

and using Eq. (3.42), we have

$$\delta^{(1)}(x) = \lim_{b\to 0} \frac{1}{b}\left[\delta\left(x+\frac{b}{2}\right) - \delta\left(x-\frac{b}{2}\right)\right]. \tag{3.46}$$

The expression in the brackets consists of a positive delta function of area b^{-1} located at $x=-b/2$, and a negative delta function of area b^{-1} located at $x=b/2$. The areas of these delta functions are just equal to the discontinuities of the rectangle function, and they tend to infinity as $b\to 0$. Also, the separation of these delta functions tends to zero as $b\to 0$.

Thus the doublet is made up of a positive and a negative delta function, each located at the origin and of infinite area, and it is usually represented graphically by a pair of half-arrows of unit height as shown in Fig. 3-23. It should be pointed out, however, that the height of these spikes no longer corresponds to the area of each delta function. It is interesting to note the similarity between the odd impulse pair and the quantity in the brackets of Eq. (3.46), which suggests that we might also define the doublet by the expression

$$\delta^{(1)}(x) = \lim_{b\to 0} \frac{1}{b}\delta\delta\left(\frac{x}{b}\right). \tag{3.47}$$

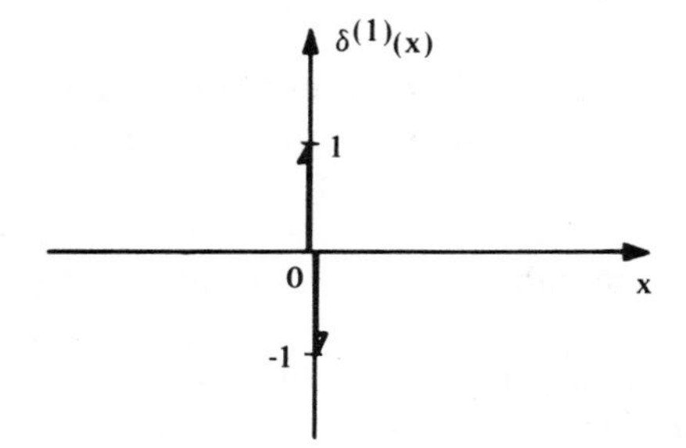

Figure 3-23 The derivative of the delta function.

Although similar approaches may be used to investigate the higher order derivatives of the delta function, we shall not do so because of the difficulties involved.

The following additional properties may be "derived" from Eq. (3.38)

$$\delta^{(k)}\left(\frac{x-x_0}{b}\right)=b^k|b|\delta^{(k)}(x-x_0), \tag{3.48}$$

$$\delta^{(k)}(-x)=(-1)^k\delta^{(k)}(x). \tag{3.49}$$

Thus we see that $\delta^{(k)}(x)$ possesses even symmetry if k is even and odd symmetry if k is odd.

In addition we have

$$\frac{(-1)^k(x-x_0)^k}{k!}\delta^{(k)}(x-x_0)=\delta(x-x_0), \tag{3.50}$$

$$-x\delta^{(1)}(x)=\delta(x), \tag{3.51}$$

but care must be exercised in using these results. For example, although Eq. (3.51) is valid, we cannot rearrange it to read $\delta^{(1)}(x)=-\delta(x)/x$ because the latter expression is not defined in the one-dimensional case.

One more property of interest is that the total area under any of the derivatives is identically zero, i.e.,

$$\int_{-\infty}^{\infty}\delta^{(k)}(\alpha)d\alpha=0. \tag{3.52}$$

As will be seen later, the delta function derivatives can be used to advantage in evaluating the Fourier transforms of the derivatives of a function.

3-4 TWO-DIMENSIONAL FUNCTIONS

In this section we develop the notation for several two-dimensional functions. Such notation will be quite useful in our studies of optical problems.

Rectangular Coordinates

Since we will be dealing primarily with separable functions, which can be written as the product of one-dimensional functions, we draw heavily on the notation already established for one-dimensional functions. We specify that x_0 and y_0 are real constants and that b and d are real, non-zero

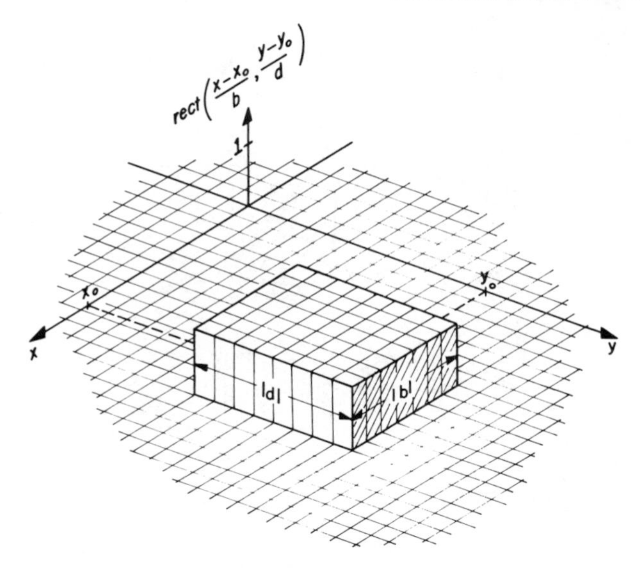

Figure 3-24 The two-dimensional rectangle function.

constants. In the figures, we shall assume all four of these constants to be positive unless otherwise noted.

Rectangle Function We define the two-dimensional rectangle function to be the product of one-dimensional rectangle functions, i.e.,

$$\text{rect}\left(\frac{x-x_0}{b}, \frac{y-y_0}{d}\right) = \text{rect}\left(\frac{x-x_0}{b}\right)\text{rect}\left(\frac{y-y_0}{d}\right). \tag{3.53}$$

This function, which is illustrated in Fig. 3-24, is often used to describe the transmittance function of a rectangular aperture. It has a "volume" of $|bd|$, as may easily be seen from the figure.

The Triangle Function The two-dimensional triangle function is given by

$$\text{tri}\left(\frac{x-x_0}{b}, \frac{y-y_0}{d}\right) = \text{tri}\left(\frac{x-x_0}{b}\right)\text{tri}\left(\frac{y-y_0}{d}\right). \tag{3.54}$$

The graph of this function is shown in Fig. 3-25, where we have chosen $x_0 = y_0 = 0$ to simplify the drawing. At first it might seem that the triangle function should be shaped like a pyramid, but as may be seen in the figure, this is not the case. The profile in a direction perpendicular to either axis is always triangular, but the profile along a diagonal is made up of two

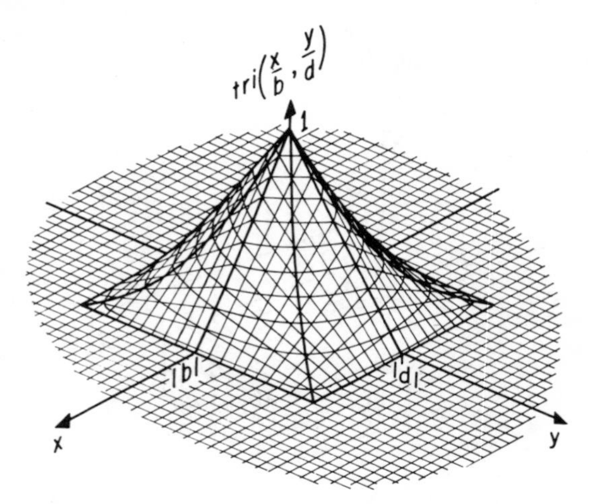

Figure 3-25 The two-dimensional triangle function.

parabolic segments. In addition, the shape of the constant-height contours is not preserved in going from top to bottom. Even though it may not be immediately obvious, the volume of this function is just equal to $|bd|$, a result easily obtained from Eq. (2.31). Thus, we see that things are not quite what they seem to be at first glance. One of the uses of the triangle function is in representing the *optical transfer function* of an incoherent imaging system whose limiting pupil is rectangular.

The Sinc Function We define the two-dimensional sinc function to be

$$\mathrm{sinc}\left(\frac{x-x_0}{b},\frac{y-y_0}{d}\right)=\mathrm{sinc}\left(\frac{x-x_0}{b}\right)\mathrm{sinc}\left(\frac{y-y_0}{d}\right). \tag{3.55}$$

This function describes the *coherent impulse response* of an imaging system with a rectangular pupil function; it is illustrated in Fig. 3-26, where we have let $x_0=y_0=0$ and $b=2d$ to simplify the drawing. The sinc^2 function is, of course, just

$$\mathrm{sinc}^2\left(\frac{x-x_0}{b},\frac{y-y_0}{d}\right)=\mathrm{sinc}^2\left(\frac{x-x_0}{b}\right)\mathrm{sinc}^2\left(\frac{y-y_0}{d}\right), \tag{3.56}$$

and because its graph looks quite similar to that of the sinc function itself, we do not show this graph here. The sinc^2 function describes the *incoherent impulse response* of an imaging system with a rectangular pupil function. Both the sinc and sinc^2 functions have a volume equal to $|bd|$.

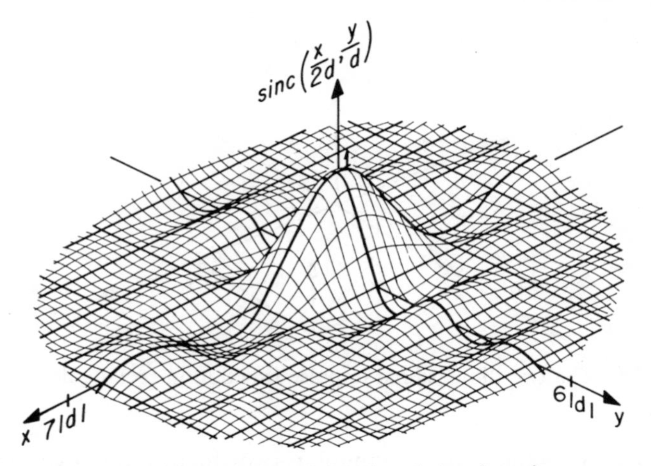

Figure 3-26 The two-dimensional sinc function.

The Gaussian Function This function is also defined as the product of one-dimensional functions, i.e.,

$$\mathrm{Gaus}\left(\frac{x-x_0}{b},\frac{y-y_0}{d}\right)=\mathrm{Gaus}\left(\frac{x-x_0}{b}\right)\mathrm{Gaus}\left(\frac{y-y_0}{d}\right), \qquad (3.57)$$

and it is shown in Fig. 3-27 with $x_0=y_0=0$ and $b=2d$. The volume of the Gaussian function is once again equal to $|bd|$.

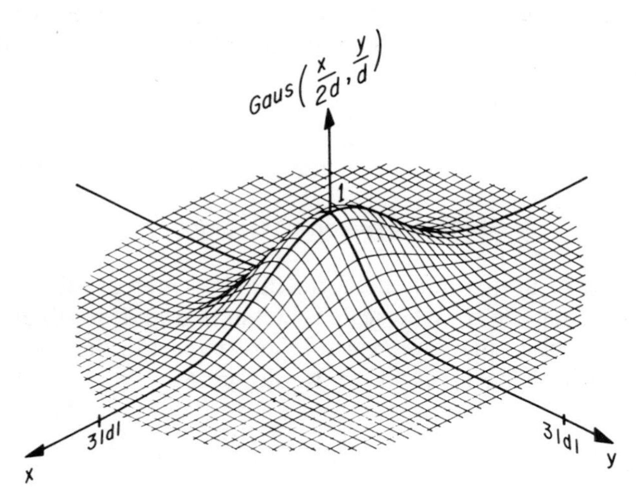

Figure 3-27 The two-dimensional Gaussian function.

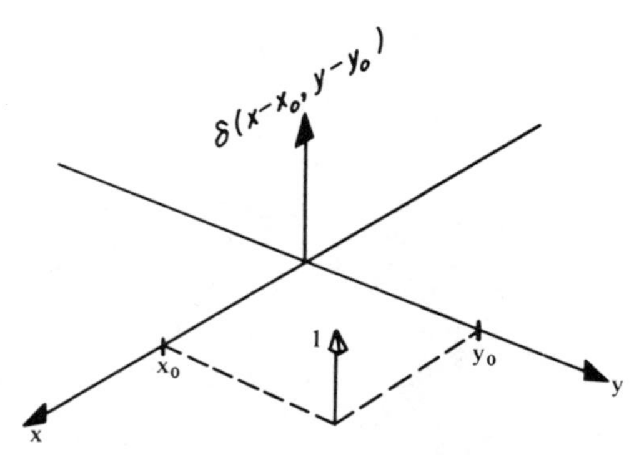

Figure 3-28 The two-dimensional impulse function in rectangular coordinates.

The Impulse Function In rectangular coordinates the two-dimensional delta function is defined by

$$\delta(x-x_0, y-y_0)=\delta(x-x_0)\delta(y-y_0). \tag{3.58}$$

Since we have already discussed the one-dimensional delta function in great detail, we shall not dwell on $\delta(x-x_0, y-y_0)$ here. This function has unit volume, and is represented graphically by a spike of unit height located at the point (x_0, y_0) as shown in Fig. 3-28.

We stress that the *height* of the spike in a graphical representation corresponds to the *volume* of the delta function. Obviously, the scaled delta function $\delta(x/b, y/d)$ has a volume of $|bd|$, and would be depicted by a spike of height $|bd|$. The two-dimensional delta function is very useful in describing such quantities as point sources of light, the transmittance of pinholes, etc.

The Comb Function The two-dimensional comb function, sometimes called the "bed of nails" function, is the doubly-periodic function given by

$$\text{comb}(x, y)=\text{comb}(x)\text{comb}(y), \tag{3.59}$$

and is an array of unit-volume delta functions located at integral values of x and y. This function may be scaled and shifted and, to obtain an array of unit-volume delta functions spaced $|b|$ units apart in the x-direction and $|d|$ units apart in the y-direction, we would write

$$\frac{1}{|bd|}\text{comb}\left(\frac{x}{b}, \frac{y}{d}\right)=\frac{1}{|b|}\text{comb}\left(\frac{x}{b}\right)\frac{1}{|d|}\text{comb}\left(\frac{y}{d}\right). \tag{3.60}$$

The array given by this equation is illustrated in Fig. 3-29.

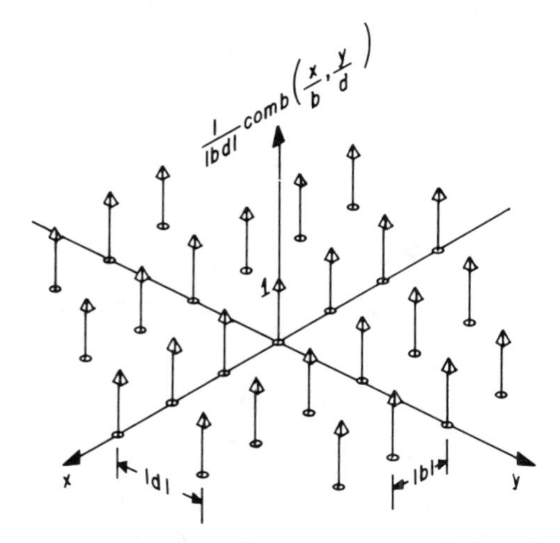

Figure 3-29 The two-dimensional comb function.

Just as the one-dimensional comb function may be used for sampling one-dimensional functions, the two-dimensional comb function may be used to sample two-dimensional functions. In addition, it may be used to represent an array of point sources, the transmittance of an array of pinholes, etc.

Polar Coordinates

In most optical systems, the various lenses and stops are circular and exhibit radial symmetry about the optical axis of the system, at least to a first approximation. As a result, it is often desirable to express certain functions in polar coordinates when dealing with such systems. In this section we develop the notation for several radially symmetric functions that are frequently needed in the analysis of optical problems. By radially symmetric functions, we mean functions that vary only with the radial distance r and have no angular dependence at all. Note that r is a real nonnegative variable.

The Cylinder Function The cylinder function, which can be used to describe the transmittance of a circular aperture, is defined by

$$\mathrm{cyl}\left(\frac{r}{d}\right)=\begin{cases}1, & 0\leqslant r<\dfrac{d}{2}\\ \frac{1}{2}, & r=\dfrac{d}{2}\\ 0, & r>\dfrac{d}{2}\end{cases} \tag{3.61}$$

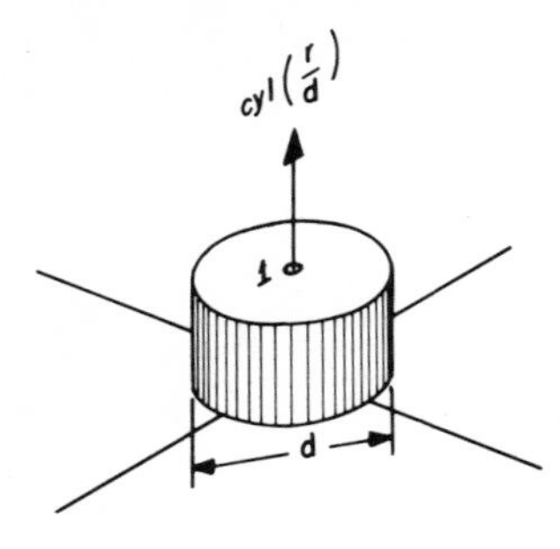

Figure 3-30 The cylinder function.

Here d is the diameter of the aperture and is a real positive constant. Thus the cylinder function exhibits no dependence on θ and is radially symmetric about the origin. Note that this function seems to be one-dimensional, and indeed it is strictly from a mathematical point of view. Hence, care must be exercised when performing mathematical operations on such a function. The cylinder function is illustrated in Fig. 3-30.

The volume of the cylinder function is easily found by integrating over the $r-\theta$ plane, i.e.,

$$\int_0^{2\pi}\int_0^{\infty}\mathrm{cyl}\left(\frac{\alpha}{d}\right)\alpha\,d\alpha\,d\beta=2\pi\int_0^{d/2}\alpha\,d\alpha$$

$$=2\pi\left(\frac{\alpha^2}{2}\right)\Bigg]_0^{d/2}$$

$$=\frac{\pi d^2}{4},\tag{3.62}$$

as expected. Here α and β are the dummy variables of integration corresponding to r and θ, respectively.

The Sombrero Function Another function frequently encountered in optics is what we shall call the *sombrero function*; it is the polar-coordinate counterpart of the two-dimensional sinc function given by Eq. (3.55). It is defined to be

$$\mathrm{somb}\left(\frac{r}{d}\right)=\frac{2J_1\left(\frac{\pi r}{d}\right)}{\left(\frac{\pi r}{d}\right)},\tag{3.63}$$

where $J_1(\cdot)$ is the first-order Bessel function of the first kind. A graph of

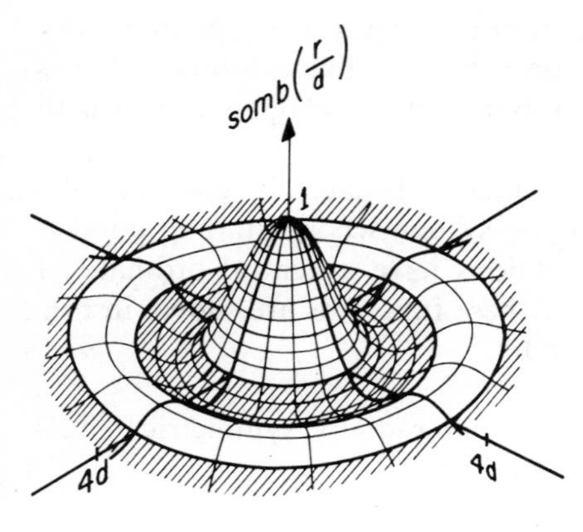

Figure 3-31 The sombrero function.

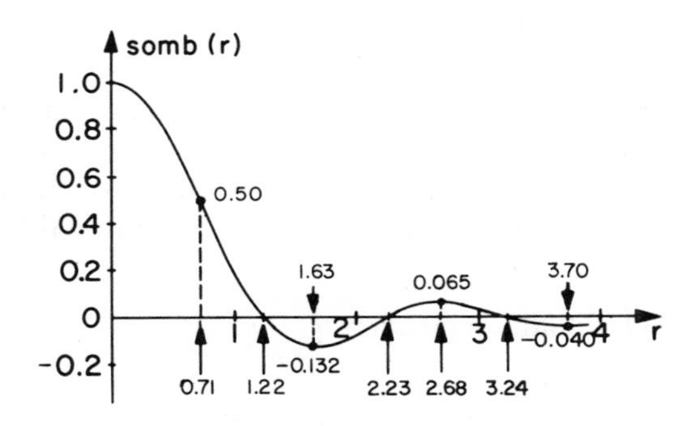

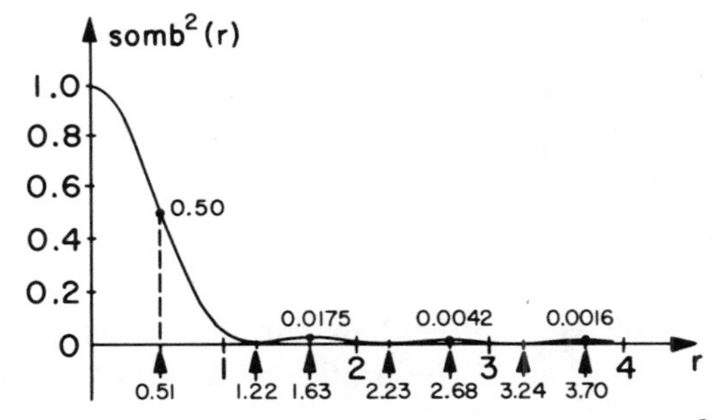

Figure 3-32 Radial profiles of somb(r) and somb^2(r).

somb(r/d) is shown in Fig. 3-31, and from this figure it is apparent why we chose to name it the sombrero function: it looks like a sombrero! It is used to describe the coherent impulse response of an imaging system with a circular limiting pupil.

The somb$^2(r/d)$ is also an important function because it represents the incoherent impulse response of an imaging system whose limiting pupil is circular. Both the somb and somb^2 functions have a central ordinate of unity and a volume of $4d^2/\pi$; profiles of these functions are shown in Fig. 3-32 and they are tabulated in the Appendix.

The Gaussian Function The two-dimensional radially symmetric Gaussian function is written in polar coordinates as

$$\text{Gaus}\left(\frac{r}{d}\right) = e^{-\pi(r/d)^2}, \tag{3.64}$$

and has exactly the same form as a one-dimensional Gaussian function. The difference, of course, lies in how these functions are used to represent physical quantities and in how various mathematical operations are performed on them. While the area under the one-dimensional function is

$$\begin{aligned}\int_{-\infty}^{\infty} \text{Gaus}(\alpha/d)\,d\alpha &= \int_{-\infty}^{\infty} e^{-\pi(\alpha/d)^2}\,d\alpha \\ &= |d|,\end{aligned} \tag{3.65}$$

the volume under Gaus(r/d) is

$$\begin{aligned}\int_0^{2\pi}\int_0^{\infty} \text{Gaus}(\alpha/d)\alpha\,d\alpha\,d\beta &= 2\pi\int_0^{\infty} e^{-\pi(\alpha/d)^2}\alpha\,d\alpha \\ &= d^2.\end{aligned} \tag{3.66}$$

A graph of the radially symmetric Gaussian function is shown in Fig. 2-12 and will not be repeated here. The profile of this function is illustrated by Fig. 3-9.

The Impulse Function We have seen that in rectangular coordinates a unit-volume impulse located at (x_0, y_0) is just the product of one-dimensional delta functions given by

$$\delta(x-x_0, y-y_0) = \delta(x-x_0)\delta(y-y_0). \tag{3.67}$$

The corresponding expression in polar coordinates is

$$\delta(\mathbf{r}-\mathbf{r}_0)=\frac{\delta(r-r_0)}{r_0}\delta(\theta-\theta_0), \tag{3.68}$$

where $\mathbf{r}$ and $\mathbf{r}_0$ are vectors associated with the points (r,θ) and (r_0,θ_0), respectively, $r_0=\sqrt{x_0^2+y_0^2}$ and $\theta_0=\tan^{-1}(y_0/x_0)$. This expression is valid for $r_0>0$ and $0\leqslant\theta_0\leqslant 2\pi$, and it can be derived as follows: denoting increments in the radial and azimuthal directions by Δr and $\Delta\theta$, respectively, we write

$$\begin{aligned}
\delta(\mathbf{r}-\mathbf{r}_0) &= \lim_{\substack{\Delta r\to 0\\ \Delta\theta\to 0}} \frac{1}{r_0\Delta r\,\Delta\theta}\operatorname{Gaus}\left(\frac{r-r_0}{\Delta r},\frac{\theta-\theta_0}{\Delta\theta}\right)\\
&=\frac{1}{r_0}\left[\lim_{\Delta r\to 0}\frac{1}{\Delta r}\operatorname{Gaus}\left(\frac{r-r_0}{\Delta r}\right)\right]\left[\lim_{\Delta\theta\to 0}\frac{1}{\Delta\theta}\operatorname{Gaus}\left(\frac{\theta-\theta_0}{\Delta\theta}\right)\right]\\
&=\frac{\delta(r-r_0)}{r_0}\delta(\theta-\theta_0).
\end{aligned} \tag{3.69}$$

The volume of this delta function is unity, as it should be.

For a delta function located at the origin the representation takes the form

$$\delta(\mathbf{r})=\frac{\delta(r)}{\pi r}, \tag{3.70}$$

which may seem strange at first. However, it can be shown that this function possesses the properties required of a two-dimensional delta function. Figure 3-33 illustrates both of the above delta functions.

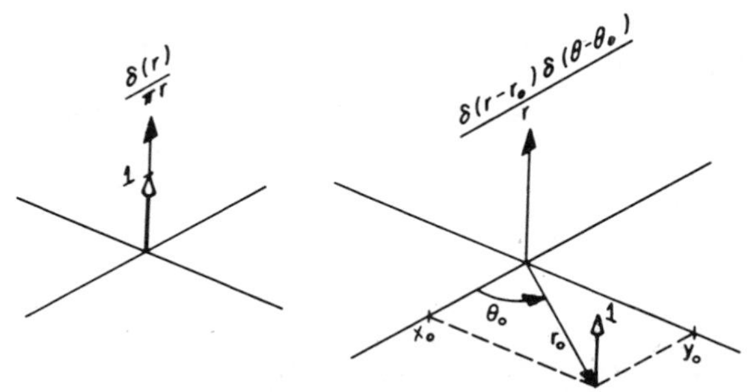

Figure 3-33 The two-dimensional delta function in polar coordinates.

It should be pointed out that the function $\delta(r)/\pi r$ appears to be one-dimensional function, and care must be taken to insure that it is not used improperly. We recall that it is only the integral properties of delta functions that are of interest, and that by themselves they are meaningless.

Shifting of Radially Symmetric Functions Sometimes it is necessary to describe a radially symmetric function whose center has been shifted from the origin to some new point (r_0, θ_0). Then, in general, we must replace r by $[r^2 + r_0^2 - 2rr_0\cos(\theta - \theta_0)]^{\frac{1}{2}}$ in the argument of the shifted function. For example, if the center of the function $f(\mathbf{r}) = g(r)$ were shifted to the point (r_0, θ_0) we would have

$$f(\mathbf{r}-\mathbf{r}_0) = g\left(\sqrt{r^2 + r_0^2 - 2rr_0\cos(\theta - \theta_0)}\,\right), \tag{3.71}$$

which is no longer radially symmetric. Note that we cannot just blindly substitute $(r - r_0)$ for r and $(\theta - \theta_0)$ for θ in the original function. It might be advantageous to switch to rectangular coordinates, because with $h(x, y) = g\left(\sqrt{x^2 + y^2}\,\right)$ we could write Eq. (3.71) more compactly as

$$\begin{aligned} f(\mathbf{r}-\mathbf{r}_0) &= h(x - x_0, y - y_0) \\ &= g\left(\sqrt{(x - x_0)^2 + (y - y_0)^2}\,\right). \end{aligned} \tag{3.72}$$

Figure 3-34 shows a "top view" of unshifted and shifted versions of the function $f(\mathbf{r}) = \text{cyl}(r)$.

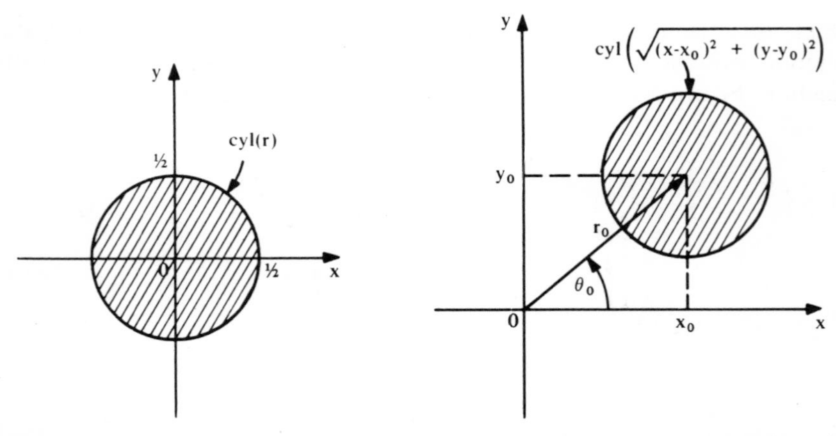

Figure 3-34 The cylinder function centered at the origin and centered at the point (x_0, y_0).

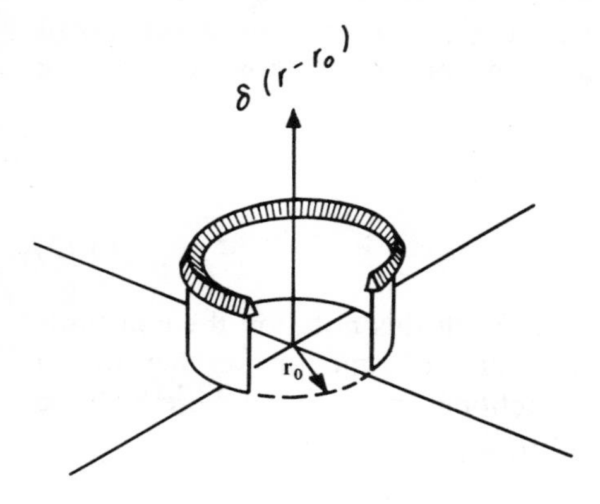

Figure 3-35 The function $\delta(r-r_0)$.

It is also instructive to consider the function $\delta(r-r_0)$, which has the same form as a one-dimensional delta function. Since we are dealing with a two-dimensional function, it might at first appear that $\delta(r-r_0)$ denotes the usual two-dimensional, unit-volume delta function that has been shifted by some amount r_0. However, a little thought will reveal that it is the "ringlike" delta function illustrated in Fig. 3-35, and that its volume is

$$\int_0^{2\pi}\int_0^{\infty}\delta(\alpha-r_0)\alpha\,d\alpha\,d\beta=2\pi\int_0^{\infty}\alpha\delta(\alpha-r_0)d\alpha$$
$$=2\pi r_0. \tag{3.73}$$

It should now be apparent that care must be exercised when dealing with radially symmetric functions and when attempting to shift them to some new location. A great deal of time and frustration can be saved by heeding this warning.

3-5 TWO-DIMENSIONAL FUNCTIONS OF THE FORM $f[w_1(x, y), w_2(x, y)]$

In addition to scaling and shifting, it is often useful to be able to rotate and/or skew a two-dimensional function $f(x, y)$ in a particular coordinate system. All of these operations may be performed by replacing the independent variables x and y with suitably chosen functions $w_1(x, y)$ and $w_2(x, y)$, respectively. The resulting function

$$g(x,y)=f[w_1(x,y), w_2(x,y)] \tag{3.74}$$

is then, in general, a scaled, shifted, rotated, and/or skewed version of $f(x, y)$. For the most part, we shall concentrate on those versions of $f(x, y)$ for which

$$w_1(x,y) = a_1x + b_1y + c_1,$$
$$w_2(x,y) = a_2x + b_2y + c_2, \tag{3.75}$$

where the a_i, b_i, and c_i are real constants. With this notation the constants b_1 and b_2 no longer correspond to the widths of functions as they did in earlier sections; the widths of the functions encountered here will be determined by various combinations of a_1, b_1, a_2, and b_2.

$f(a_1x+b_1y+c_1)$

We begin our discussions by considering the two-dimensional function

$$h(x,y) = f(x) = \mathrm{rect}(x) \tag{3.76}$$

illustrated in Fig. 3-36, a separable function with no dependence on y.

Next we define

$$g(x,y) = h(a_1x + b_1y + c_1, y) = f(a_1x + b_1y + c_1) = \mathrm{rect}(a_1x + b_1y + c_1), \tag{3.77}$$

which we observe to be inseparable in x and y. To understand the behavior of this function, we need only recall how the rectangle function depends on its argument:

$$\mathrm{rect}(a_1x + b_1y + c_1) = \begin{cases} 1, & -\frac{1}{2} < a_1x + b_1y + c_1 < \frac{1}{2} \\ \frac{1}{2}, & a_1x + b_1y + c_1 = \pm\frac{1}{2} \\ 0, & \text{otherwise.} \end{cases} \tag{3.78}$$

Note that the expression

$$a_1x + b_1y + c_1 = \pm\tfrac{1}{2} \tag{3.79}$$

describes two lines of slope $-a_1/b_1$, y-axis intercepts of $(\pm\frac{1}{2} - c_1)/b_1$ and

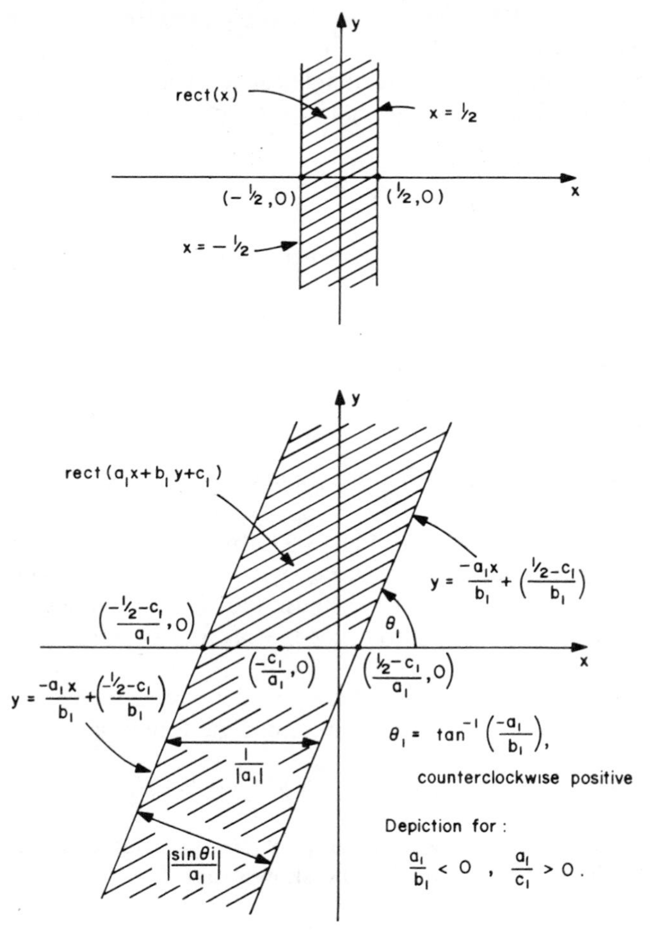

Figure 3-36 The functions $\text{rect}(x)$ and $\text{rect}(a_1 x + b_1 y + c_1)$.

x-axis intercepts of $(\pm\frac{1}{2} - c_1)/a_1$. The function $g(x, y)$ is shown in Fig. 3-36, where the angle θ_1 is given by

$$\theta_1 = \tan^{-1}\left(\frac{-a_1}{b_1}\right). \tag{3.80}$$

We have adopted the usual convention in which the positive direction for θ_1 is counterclockwise. If a_1 and b_1 have opposite signs, the lines given by Eq. (3.79) have a positive slope; if they have the same sign, the slope of these lines is negative. In addition, the x-axis intercepts of these lines are shifted to the right if a_1 and c_1 have opposite signs and to the left if they have the same sign.

Although we chose a simple rectangle function for our present example, other functions behave similarly under this type of transformation.

$f(a_1x+b_1y+c_1,\ a_2x+b_2y+c_2)$

Next we consider the function

$$\begin{aligned} f(x,y) &= \mathrm{rect}(x,y) \\ &= \mathrm{rect}(x)\mathrm{rect}(y) \end{aligned} \tag{3.81}$$

and its scaled, shifted, and skewed version

$$\begin{aligned} g(x,y) &= f(a_1x+b_1y+c_1,\ a_2x+b_2y+c_2) \\ &= \mathrm{rect}(a_1x+b_1y+c_1)\mathrm{rect}(a_2x+b_2y+c_2). \end{aligned} \tag{3.82}$$

These functions are shown in Fig. 3-37. In this figure the angle $\theta_2 = \tan^{-1}(-a_2/b_2)$, and the positive direction for θ_2 is also counterclockwise. Thus, if a_2 and b_2 have opposite signs, the lines bounding the second rectangle function have a positive slope; if a_2 and b_2 have the same sign, their slope is negative.

For a function $f(x, y)$ that is centered on the origin, such as $\mathrm{rect}(x, y)$, it may be convenient to introduce an intermediate function

$$u(x,y) = f(a_1x+b_1y, a_2x+b_2y), \tag{3.83}$$

which is a scaled and skewed version of $f(x, y)$, but one that remains centered on the origin. It can then be shown that

$$\begin{aligned} g(x,y) &= f(a_1x+b_1y+c_1, a_2x+b_2y+c_2) \\ &= u(x-x_0, y-y_0), \end{aligned} \tag{3.84}$$

where

$$x_0 = \frac{b_1c_2 - b_2c_1}{a_1b_2 - a_2b_1}, \qquad y_0 = \frac{a_2c_1 - a_1c_2}{a_1b_2 - a_2b_1}. \tag{3.85}$$

This procedure is depicted in Fig. 3-38, from which it can be seen that $g(x,y)$ has the shape of $u(x,y)$ and is centered at the point (x_0, y_0).

The above technique can also be useful for functions $f(x,y)$ that are not initially centered on the origin; however, $g(x,y)$ will no longer be centered

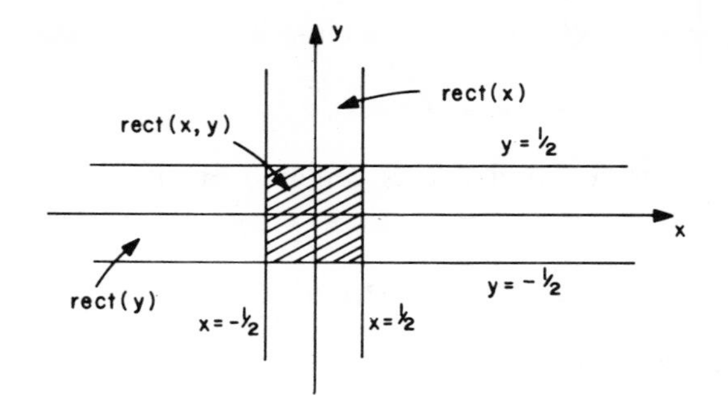

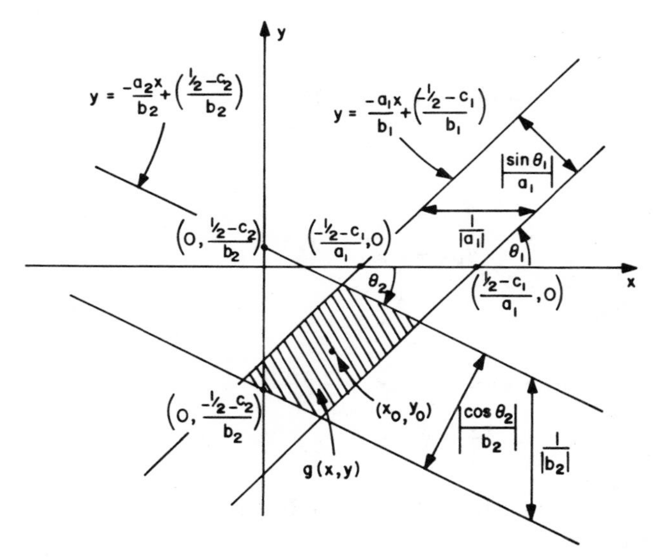

Figure 3-37 The functions $\mathrm{rect}(x,y)$ and $\mathrm{rect}(a_1x+b_1y+c_1, a_2x+b_2y+c_2)$.

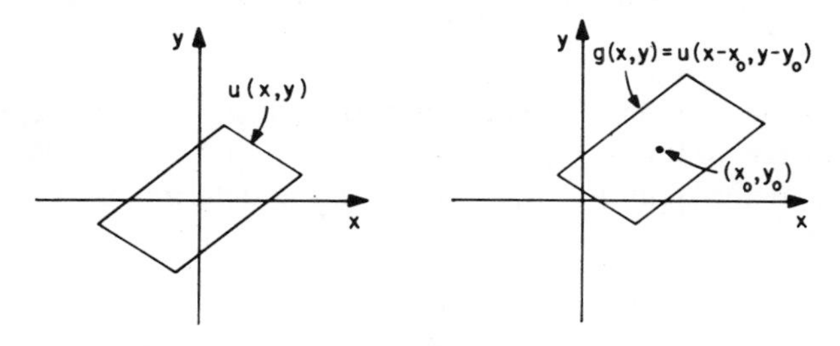

Figure 3-38 Scaled, skewed, and shifted versions of $\mathrm{rect}(x,y)$.

at the point (x_0, y_0). To illustrate, let us choose

$$f(x,y)=t(x-x_f, y-y_f), \tag{3.86}$$

where $t(x, y)$ is some function that is centered on the origin. If we then define

$$v(x,y)=t(a_1x+b_1y, a_2x+b_2y), \tag{3.87}$$

it can be shown that

$$\begin{aligned} u(x,y)&=f(a_1x+b_1y, a_2x+b_2y)\\ &=t(a_1x+b_1y-x_f,\ a_2x+b_2y-y_f)\\ &=v(x-x_u, y-y_u), \end{aligned} \tag{3.88}$$

where

$$x_u=\frac{b_2x_f-b_1y_f}{a_1b_2-a_2b_1},\quad y_u=\frac{a_1y_f-a_2x_f}{a_1b_2-a_2b_1}. \tag{3.89}$$

Finally we obtain

$$\begin{aligned} g(x,y)&=u(x-x_0, y-y_0)\\ &=v(x-x_u-x_0, y-y_u-y_0), \end{aligned} \tag{3.90}$$

where x_0 and y_0 are given by Eq. (3.85). Figure 3-39 shows this result when $t(x,y)=\text{rect}(x,y)$.

It is not difficult to show that the volume of $g(x, y)$ is given by

$$\iint_{-\infty}^{\infty} g(\alpha,\beta)\,d\alpha\,d\beta=\frac{1}{|a_1b_2-a_2b_1|}\iint_{-\infty}^{\infty} f(\alpha,\beta)\,d\alpha\,d\beta; \tag{3.91}$$

it is equal to the volume of $f(x, y)$ divided by $|a_1b_2-a_2b_1|$. Note that it is also equal to the volumes of $u(x, y)$ [Eq. (3.83)] and $v(x, y)$ [Eq. (3.87)].

Observe that the quantity $(a_1b_2-a_2b_1)$ is the determinant of the pair of equations

$$\begin{aligned} a_1x+b_1y&=-c_1,\\ a_2x+b_2y&=-c_2, \end{aligned} \tag{3.92}$$

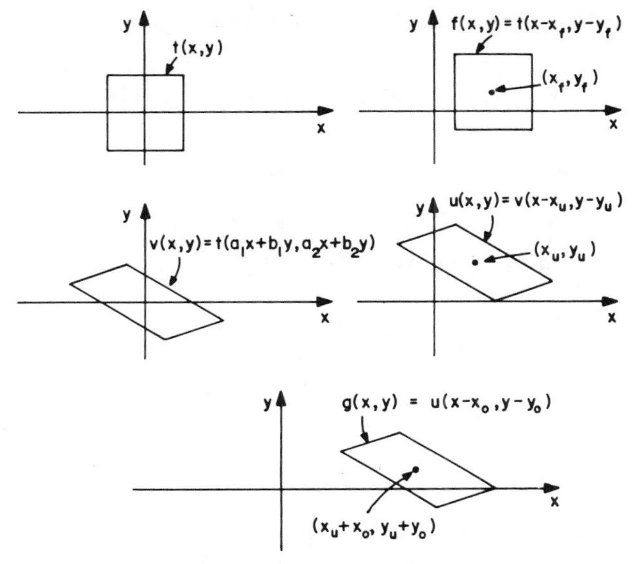

Figure 3-39 Determination of scaled, skewed, and shifted version of a function $f(x,y)$.

and because we will encounter it frequently we denote it by

$$a_1b_2 - a_2b_1 = D. \tag{3.93}$$

We now consider the cylinder function of diameter d,

$$f(x, y) = \text{cyl}\left(\frac{\sqrt{x^2+y^2}}{d}\right), \tag{3.94}$$

and its modified version

$$\begin{aligned} g(x, y) &= f(a_1x + b_1y + c_1,\ a_2x + b_2y + c_2) \\ &= \text{cyl}\left(\frac{1}{d}\sqrt{(a_1x + b_1y + c_1)^2 + (a_2x + b_2y + c_2)^2}\right). \end{aligned} \tag{3.95}$$

Since the equation describing the circle that bounds $f(x, y)$ is

$$x^2 + y^2 - \left(\frac{d}{2}\right)^2 = 0, \tag{3.96}$$

the corresponding equation for the curve bounding $g(x, y)$ is

$$(a_1x+b_1y+c_1)^2+(a_2x+b_2y+c_2)^2-\left(\frac{d}{2}\right)^2=0. \tag{3.97}$$

If we perform the squaring operations and collect terms, Eq. (3.97) becomes

$$(a_1^2+a_2^2)x^2+2(a_1b_1+a_2b_2)xy+(b_1^2+b_2^2)y^2$$
$$+2(a_1c_1+a_2c_2)x+2(b_1c_1+b_2c_2)y+c_1^2+c_2^2-\frac{d^2}{4}=0, \tag{3.98}$$

which is the equation for a general conic. By defining

$$A=a_1^2+a_2^2, \quad C=b_1^2+b_2^2, \quad J=2(b_1c_1+b_2c_2),$$
$$B=2(a_1b_1+a_2b_2), \quad I=2(a_1c_1+a_2c_2), \quad K=c_1^2+c_2^2-\frac{d^2}{4}, \tag{3.99}$$

we may then simplify Eq. (3.98):

$$Ax^2+Bxy+Cy^2+Ix+Jy+K=0. \tag{3.100}$$

It can be shown (see Ref. 3-9) that if $4AC>B^2$, Eq. (3.100) is the equation of an ellipse whose center is located at

$$x_0=\frac{BJ-2CI}{4AC-B^2}=\frac{b_1c_2-b_2c_1}{D},$$
$$y_0=\frac{BI-2AJ}{4AC-B^2}=\frac{a_2c_1-a_1c_2}{D}, \tag{3.101}$$

where D is given by Eq. (3.93). If we erect a new coordinate system x_e, y_e centered at the point (x_0, y_0) and rotated clockwise through an angle θ_e, the axes of the ellipse will coincide with the axes of this new coordinate system. The angle θ_e is defined to be

$$\theta_e=\tfrac{1}{2}\tan^{-1}\left(\frac{B}{A-C}\right)$$
$$=\tfrac{1}{2}\tan^{-1}\left[\frac{2(a_1b_1+a_2b_2)}{(a_1^2+a_2^2)-(b_1^2+b_2^2)}\right], \tag{3.102}$$

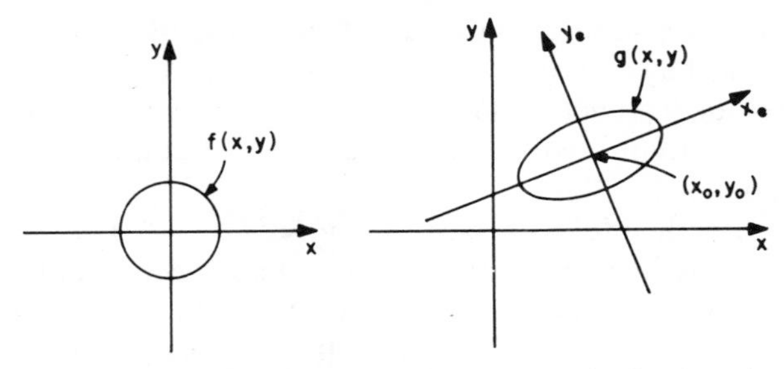

Figure 3-40 Scaled, skewed, and shifted cylinder function.

and we require it to have a value between 0° and 90°: if $B/(A-C)$ is negative, $90° < 2\theta_e \leqslant 180°$ and $45° < \theta_e \leqslant 90°$; if $B/(A-C)$ is positive, $0° < 2\theta_e \leqslant 90°$ and $0° < \theta_e \leqslant 45°$.

The lengths of the semiaxes corresponding to the x_e-axis and y_e-axis, respectively, are found to be

$$r_1 = \frac{d}{2}\left[A\cos^2\theta_e + B\sin\theta_e\cos\theta_e + C\sin^2\theta_e\right]^{-\frac{1}{2}},$$

$$r_2 = \frac{d}{2}\left[A\sin^2\theta_e - B\sin\theta_e\cos\theta_e + C\cos^2\theta_e\right]^{-\frac{1}{2}}, \tag{3.103}$$

where d is the diameter of the original cylinder function. Figure 3-40 illustrates a typical example of a scaled, shifted, and skewed cylinder function.

The technique of defining the intermediate function

$$u(x,y) = f(a_1x + b_1y, a_2x + b_2y)$$

$$= \mathrm{cyl}\left(\frac{1}{d}\sqrt{(a_1x + b_1y)^2 + (a_2x + b_2y)^2}\right) \tag{3.104}$$

is also useful here; then $g(x, y) = u(x - x_0, y - y_0)$ as before. In addition, the volume of $g(x, y)$ is still equal to that of $f(x, y)$ divided by $|D| = |a_1b_2 - a_2b_1|$ as specified by Eq. (3.91).

$\delta(a_1x + b_1y + c_1)$

Because of its unique characteristics, we shall consider the delta function $\delta(a_1x + b_1y + c_1)$ separately. First we set $a_1 = 1$, $b_1 = 0$, and $c_1 = -x_0$ so that we may study the function

$$g(x,y) = \delta(x - x_0). \tag{3.105}$$

It behaves as a one-dimensional delta function in the x-direction, and is independent of the y-variable. In other words, for every value of y, $g(x, y)$ is a unit-area delta function located at the point $x = x_0$; it has a volume that is highly concentrated in the x-direction and uniformly distributed in the y-direction.

It may also be helpful to regard $g(x, y) = \delta(x - x_0)$ as a *line mass* (Ref. 3-10) lying on the line $x = x_0$ and having a *mass density* along that line of $dm/dy = 1$. If we then integrate this density in the y-direction, we obtain the *total mass* associated with $g(x, y)$ [this is also equal to the double integral of $g(x, y)$]. Still another point of view is as follows: $g(x, y)$ has a *strength* (Ref. 3-11) that is localized at x_0 in the x-direction, it has a *strength per unit length* of unity in the y-direction, and its *overall strength* is given by an appropriate integration. It is immaterial which of the above views we adopt; we simply choose the one that we are most comfortable with at the time. If we let dm/dy represent the y-distribution of volume, mass, or strength, then the total volume, mass, or strength is given by

$$m = \int_{-\infty}^{\infty} \left(\frac{dm}{d\beta} \right) d\beta$$

$$= \iint_{-\infty}^{\infty} g(\alpha, \beta) \, d\alpha \, d\beta. \tag{3.106}$$

For the present example, m has a value of infinity because $g(x, y)$ is infinitely long in the y-direction.

We now study the more general delta function

$$g(x, y) = \delta(a_1 x + b_1 y + c_1). \tag{3.107}$$

We note that this function is zero wherever its argument is not zero, and it may therefore be regarded as a line mass lying on the line described by

$$a_1 x + b_1 y + c_1 = 0. \tag{3.108}$$

We may solve Eq. (3.108) for either x or y and write

$$\delta(a_1 x + b_1 y + c_1) = \frac{1}{|a_1|} \delta\left(x + \frac{b_1 y}{a_1} + \frac{c_1}{a_1} \right)$$

$$= \frac{1}{|b_1|} \delta\left(y + \frac{a_1 x}{b_1} + \frac{c_1}{b_1} \right) \tag{3.109}$$

These line masses have associated densities in the x- and y-directions,

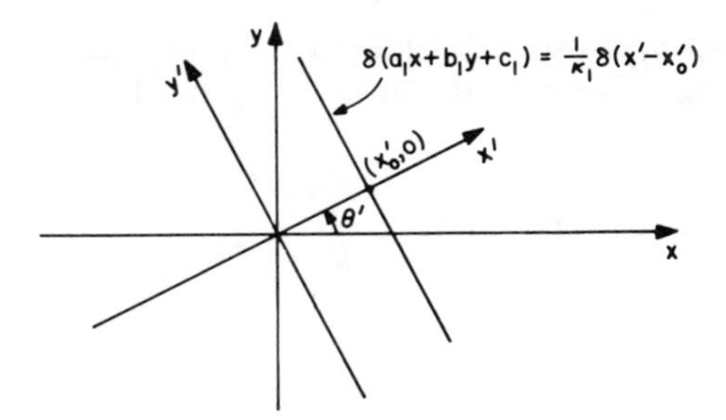

Figure 3-41 The function $\delta(a_1x+b_1y+c_1)$.

respectively, of

$$\frac{dm}{dx}=\frac{1}{|a_1|}, \qquad \frac{dm}{dy}=\frac{1}{|b_1|}. \tag{3.110}$$

It is sometimes convenient to define a new coordinate system x',y' oriented such that the line mass is parallel with the y'-axis and perpendicular to the x'-axis as shown in Fig. 3-41. If we make the substitutions

$$x'=x\cos\theta'+y\sin\theta', \qquad k_1=\sqrt{a_1^2+b_1^2},$$

$$y'=-x\sin\theta'+y\cos\theta', \qquad \cos\theta'=\frac{a_1}{k_1},$$

$$\theta'=\tan^{-1}\left(\frac{b_1}{a_1}\right), \qquad \sin\theta'=\frac{b_1}{k_1}, \tag{3.111}$$

and solve for x and y, we find that

$$x=\frac{a_1x'-b_1y'}{k_1}, \qquad y=\frac{b_1x'+a_1y'}{k_1}. \tag{3.112}$$

Consequently, we may express $g(x,y)$ in the new coordinate system as

$$\begin{aligned}
g'(x',y')&=g\left(\frac{a_1x'-b_1y'}{k_1},\frac{b_1x'+a_1y'}{k_1}\right)\\
&=\delta\left(\frac{a_1^2x'-a_1b_1y'}{k_1}+\frac{b_1^2x'+a_1b_1y'}{k_1}+c_1\right)\\
&=\delta(k_1x'+c_1)\\
&=\frac{1}{k_1}\delta(x'-x_0'),
\end{aligned} \tag{3.113}$$

where we have set $c_1/k_1 = -x'_0$. In the x', y' coordinate system, $g'(x', y')$ represents a line mass lying on the line $x' = x'_0$ and having a density in the y'-direction of

$$\frac{dm}{dy'} = \frac{1}{k_1}. \tag{3.114}$$

$f(x, y)\delta(a_1x + b_1y + c_1)$

This product may be considered to be a line mass lying along the line $a_1x + b_1y + c_1 = 0$, as before, but now the density depends on the function $f(x, y)$. Using previous results we obtain

$$\begin{aligned} g(x,y) &= f(x,y)\delta(a_1x + b_1y + c_1) \\ &= \frac{1}{|a_1|} f\left(\frac{-b_1y}{a_1} - \frac{c_1}{a_1}, y\right)\delta\left(x + \frac{b_1y}{a_1} + \frac{c_1}{a_1}\right) \\ &= \frac{1}{|b_1|} f\left(x, \frac{-a_1x}{b_1} - \frac{c_1}{b_1}\right)\delta\left(y + \frac{a_1x}{b_1} + \frac{c_1}{b_1}\right), \end{aligned} \tag{3.115}$$

where the coefficients of the latter two delta functions are the densities in the x- and y-directions, in that order. In the rotated coordinate system x', y' we obtain

$$\begin{aligned} g'(x',y') &= g\left(\frac{a_1x' - b_1y'}{k_1}, \quad \frac{b_1x' + a_1y'}{k_1}\right) \\ &= f\left(\frac{a_1x' - b_1y'}{k_1}, \quad \frac{b_1x' + a_1y'}{k_1}\right)\frac{1}{k_1}\delta(x' - x'_0) \\ &= \frac{1}{k_1} f\left(\frac{a_1x'_0 - b_1y'}{k_1}, \quad \frac{b_1x'_0 + a_1y'}{k_1}\right)\delta(x' - x'_0), \end{aligned} \tag{3.116}$$

where $k_1 = \sqrt{a_1^2 + b_1^2}$ and $c_1/k_1 = -x'_0$ as before. The density along this line mass, located on the line $x' = x'_0$, is

$$\frac{dm}{dy'} = \frac{1}{k_1} f\left(\frac{a_1x'_0 - b_1y'}{k_1}, \quad \frac{b_1x'_0 + a_1y'}{k_1}\right), \tag{3.117}$$

and the total mass (or volume) is found by integrating Eq. (3.117) over y'.

Example

Consider the function

$$g(x,y)=\text{rect}(y)\delta(x-2y-1). \tag{3.118}$$

We may regard it as being either a line mass of the form given, with density in the y-direction of

$$\frac{dm}{dy}=\text{rect}(y), \tag{3.119}$$

or a line mass of the form

$$g(x,y)=\frac{1}{2}\text{rect}\left(\frac{x-1}{2}\right)\delta\left(y-\frac{x}{2}+\frac{1}{2}\right), \tag{3.120}$$

with a density in the x-direction of

$$\frac{dm}{dx}=\frac{1}{2}\text{rect}\left(\frac{x-1}{2}\right). \tag{3.121}$$

In the appropriate x', y' coordinate system, we find

$$g'(x',y')=\frac{1}{\sqrt{5}}\text{rect}\left(\frac{y'-2/\sqrt{5}}{\sqrt{5}}\right)\delta\left(x'-\frac{1}{\sqrt{5}}\right). \tag{3.122}$$

The density of this function in the y'-direction is

$$\frac{dm}{dy'}=\frac{1}{\sqrt{5}}\text{rect}\left(\frac{y'-2/\sqrt{5}}{\sqrt{5}}\right), \tag{3.123}$$

and the total mass, found by integrating Eqs. (3.119), (3.121), or (3.123), is $m=1$. ■

Example

Let us examine the function

$$g(x,y)=\text{sinc}(x,y)\delta(x-y).$$

Any of the following expressions are valid:

$$g(x,y)=\operatorname{sinc}^2(y)\delta(x-y), \qquad \frac{dm}{dy}=\operatorname{sinc}^2(y),$$

$$g(x,y)=\operatorname{sinc}^2(x)\delta(y-x), \qquad \frac{dm}{dx}=\operatorname{sinc}^2(x),$$

$$g'(x',y')=\frac{1}{\sqrt{2}}\operatorname{sinc}^2\left(\frac{y'}{\sqrt{2}}\right)\delta(x'), \quad \frac{dm}{dy'}=\frac{1}{\sqrt{2}}\operatorname{sinc}^2\left(\frac{y'}{\sqrt{2}}\right). \tag{3.124}$$

By proper integration of any of these expressions, the total mass (volume) is found to be $m=1$. ■

$\delta(a_1x+b_1y+c_1, a_2x+b_2y+c_2)$

We may use Eq. (3.115) to show that

$$\begin{aligned}
g(x,y)&=\delta(a_1x+b_1y+c_1,\ a_2x+b_2y+c_2)\\
&=\frac{1}{|a_1|}\delta\left(x+\frac{b_1y}{a_1}+\frac{c_1}{a_1}\right)\delta\left(\frac{-a_2b_1y-a_2c_1}{a_1}+b_2y+c_2\right)\\
&=\frac{1}{|a_1|}\delta\left(x+\frac{b_1y}{a_1}+\frac{c_1}{a_1}\right)\frac{|a_1|}{|a_1b_2-a_2b_1|}\delta\left(y-\frac{a_2c_1-a_1c_2}{a_1b_2-a_2b_1}\right)\\
&=\frac{1}{|a_1b_2-a_2b_1|}\delta\left(x-\frac{b_1c_2-b_2c_1}{a_1b_2-a_2b_1}\right)\delta\left(y-\frac{a_2c_1-a_1c_2}{a_1b_2-a_2b_1}\right)\\
&=\frac{1}{|D|}\delta(x-x_0,\ y-y_0),
\end{aligned} \tag{3.125}$$

where x_0, y_0, and D are as defined earlier. Thus, $g(x, y)$ is simply a two-dimensional delta function, of volume $|D|^{-1}$, located at the point (x_0, y_0). Or, it may be regarded as a point mass (of mass $|D|^{-1}$) located at (x_0, y_0).

$f(x, y)\delta(a_1x+b_1y+c_1,\ a_2x+b_2y+c_2)$

With x_0, y_0, and D as defined previously, it may be shown that

$$\begin{aligned}
g(x,y)&=f(x,y)\delta(a_1x+b_1y+c_1,\ a_2x+b_2y+c_2)\\
&=\frac{1}{|D|}f(x_0,y_0)\delta(x-x_0,\ y-y_0),
\end{aligned} \tag{3.126}$$

which is simply a delta function of volume (mass) $|D|^{-1}f(x_0, y_0)$.

comb($a_1x+b_1y+c_1$, $a_2x+b_2y+c_2$)

By replacing c_1 and c_2 in Eq. (3.125) with c_1-n and c_2-m, respectively, it can be shown that

$$\begin{aligned} g(x,y) &= \operatorname{comb}(a_1x+b_1y+c_1, a_2x+b_2y+c_2) \\ &= \sum_{n=-\infty}^{\infty}\sum_{m=-\infty}^{\infty} \delta(a_1x+b_1y+c_1-n)\delta(a_2x+b_2y+c_2-m) \\ &= \frac{1}{|D|}\sum_{n=-\infty}^{\infty}\sum_{m=-\infty}^{\infty} \delta\left(x-x_0-\frac{b_2n}{D}+\frac{b_1m}{D}\right) \\ &\quad \times \delta\left(y-y_0+\frac{a_2n}{D}-\frac{a_1m}{D}\right), \end{aligned} \tag{3.127}$$

which may be regarded as a skewed array of delta functions, or point masses, each having a volume (mass) of $|D|^{-1}$. Again, x_0, y_0, and D are as defined in Eqs. (3.85) and (3.93).

A function $f(x, y)$ is called *skew periodic* (Ref. 3-11) if it can be put in the form

$$f(x,y)=f(x+\alpha_1n+\alpha_2m, y+\beta_1n+\beta_2m), \tag{3.128}$$

where α_1, α_2, β_1, and β_2 are real constants and n and m are integers. Clearly, then, the $g(x,y)$ of Eq. (3.127) is skew periodic with $\alpha_1=-b_2/D$, $\alpha_2=b_1/D$, $\beta_1=a_2/D$, and $\beta_2=-a_1/D$. Such a function is depicted in Fig. 3-42, where we have chosen $x_0=y_0=0$ for simplicity. Note that the lines corresponding to constant values of n and m form a skewed *lattice*, and that the point masses of $g(x,y)$ are located at the intersections of these lines. Related to this lattice is another gridwork, called the *reciprocal lattice*, that will be of great interest to us later. If we are given the function

$$\begin{aligned} g(x,y) &= \operatorname{comb}(a_1x+b_1y, a_2x+b_2y) \\ &= \frac{1}{|D|}\sum_{n=-\infty}^{\infty}\sum_{m=-\infty}^{\infty} \delta\left(x-\frac{b_2n}{D}+\frac{b_1m}{D}\right) \\ &\quad \times \delta\left(y+\frac{a_2n}{D}-\frac{a_1m}{D}\right) \end{aligned} \tag{3.129}$$

and its associated lattice, the related reciprocal lattice will be associated

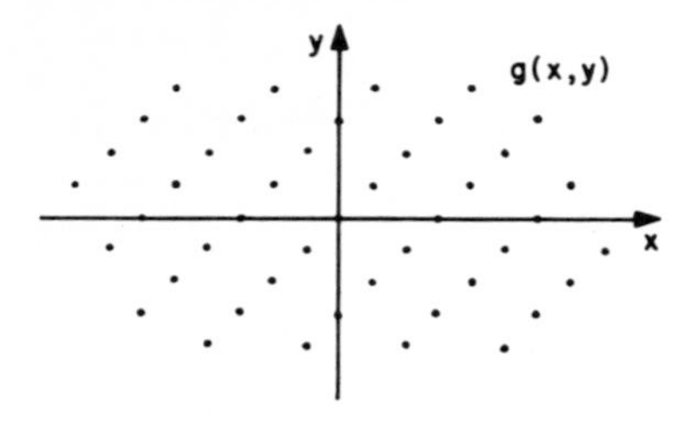

Figure 3-42 A skew periodic comb function.

with the function

$$G(\xi,\eta)=\frac{1}{|D|}\operatorname{comb}\left(\frac{b_2\xi}{D}-\frac{a_2\eta}{D},\ \frac{-b_1\xi}{D}+\frac{a_1\eta}{D}\right)$$

$$=\sum_{n=-\infty}^{\infty}\ \sum_{m=-\infty}^{\infty}\delta(\xi-a_1n-a_2m)\delta(\eta-b_1n-b_2m), \qquad (3.130)$$

where ξ and η are frequency variables that correspond to the x- and y-variables, respectively. The relationship between these two lattices is illustrated in Fig. 3-43; note that the lines of the reciprocal lattice for which n is constant are perpendicular to those of the original lattice for which m is constant, and vice versa. Also note that each of these lattices is the reciprocal lattice of the other. The perpendicular spacings of these various sets of parallel lines are as follows:

	Perpendicular spacing of lines for which	
	n is constant	m is constant
Original lattice	$\dfrac{1}{k_1}$	$\dfrac{1}{k_2}$
Reciprocal lattice	$\dfrac{\lvert D\rvert}{k_2}$	$\dfrac{\lvert D\rvert}{k_1}$

(3.131)

where $k_1=\sqrt{a_1^2+b_1^2}$ and $k_2=\sqrt{a_2^2+b_2^2}$.

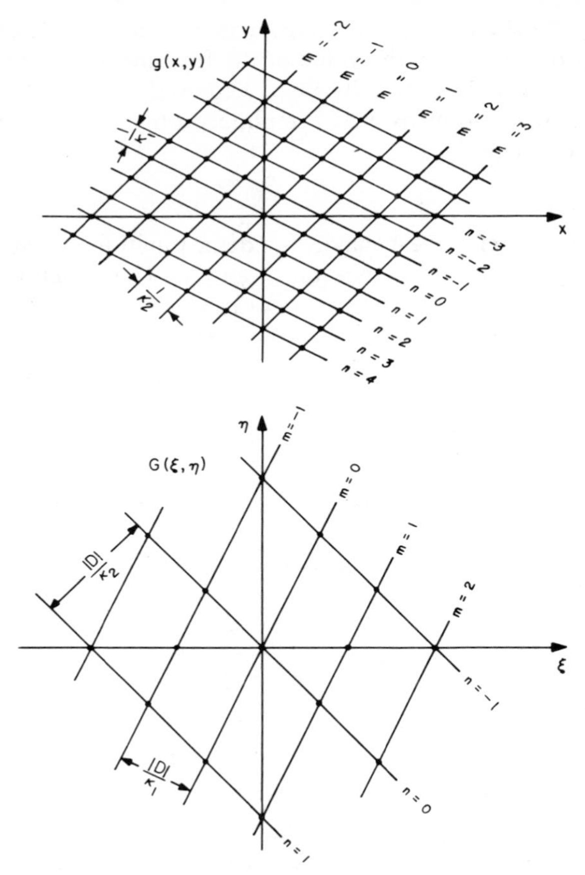

Figure 3-43 Relationship between the original lattice and the reciprocal lattice of a skew periodic comb function.

$f(x, y)\mathrm{comb}(a_1x+b_1y+c_1,\ a_2x+b_2y+c_2)$

Using the results of previous sections, we may show that

$$
\begin{aligned}
g(x,y) &= f(x,y)\mathrm{comb}(a_1x+b_1y+c_1,\ a_2x+b_2y+c_2)\\
&= \frac{1}{|D|}\sum_{n=-\infty}^{\infty}\sum_{m=-\infty}^{\infty} f\left(x_0+\frac{b_2n}{D}-\frac{b_1m}{D},\ y_0-\frac{a_2n}{D}+\frac{a_1m}{D}\right)\\
&\quad\times\delta\left(x-x_0-\frac{b_2n}{D}+\frac{b_1m}{D}\right)\delta\left(y-y_0+\frac{a_2n}{D}-\frac{a_1m}{D}\right),
\end{aligned}
\tag{3.132}
$$

which is simply a skewed array of point masses. The mass at each point of the array is proportional to the value of the function $f(x, y)$ at that point; again, x_0, y_0 and D are as given by Eqs. (3.85) and (3.93). The above result will be useful in two-dimensional sampling problems, which we will encounter in later chapters.

$\delta[w(x, y)]$

We now discuss briefly the behavior of delta functions whose arguments have a more general dependence on x and y. The function $\delta[w(x, y)]$ is a line mass lying on the curve $s = w(x, y) = 0$, and its density along that curve is determined in the following manner (see Ref. 3-10): let $d\ell = \sqrt{(dx)^2 + (dy)^2}$ be an incremental distance along the curve $w(x, y) = 0$, and let

$$w_x = \frac{\partial w(x,y)}{\partial x}, \qquad w_y = \frac{\partial w(x,y)}{\partial y}. \tag{3.133}$$

Then

$$\frac{dm}{d\ell} = \frac{1}{\sqrt{w_x^2 + w_y^2}}, \tag{3.134}$$

so that

$$\delta[w(x,y)] = \frac{1}{\sqrt{w_x^2 + w_y^2}}\delta(s). \tag{3.135}$$

In addition, if we solve $w(x, y) = 0$ for x and denote the ith solution by x_i, we may regard $\delta[w(x, y)]$ as the line mass

$$\delta[w(x,y)] = \sum_i \frac{1}{|w_x|}\delta(x - x_i). \tag{3.136}$$

Similarly, if y_i denotes the ith y solution for $w(x, y) = 0$, we have

$$\delta[w(x,y)] = \sum_i \frac{1}{|w_y|}\delta(y - y_i). \tag{3.137}$$

Example

Let $w(x, y)=\sqrt{x^2+y^2}-r_0$, such that $\delta[w(x, y)]$ corresponds to the "ring-like" delta function of Fig. 3-35. We have

$$w_x=\frac{x}{\sqrt{x^2+y^2}}, \qquad x_1, x_2=\pm\sqrt{r_0^2-y^2},$$
$$w_y=\frac{y}{\sqrt{x^2+y^2}}, \qquad y_1, y_2=\pm\sqrt{r_0^2-x^2}. \tag{3.138}$$

Consequently, we may write

$$\begin{aligned}\delta[w(x,y)]&=\left|\frac{\sqrt{x^2+y^2}}{x}\right|\left[\delta\left(x-\sqrt{r_0^2-y^2}\right)+\delta\left(x+\sqrt{r_0^2-y^2}\right)\right]\\ &=\frac{r_0}{\sqrt{r_0^2-y^2}}\delta\left(x-\sqrt{r_0^2-y^2}\right)+\frac{r_0}{\sqrt{r_0^2-y^2}}\delta\left(x+\sqrt{r_0^2-y^2}\right),\end{aligned} \tag{3.139}$$

which is valid for all $|y|<r_0$. Also, for $|x|<r_0$ we obtain

$$\delta[w(x,y)]=\frac{r_0}{\sqrt{r_0^2-x^2}}\delta\left(y-\sqrt{r_0^2-x^2}\right)+\frac{r_0}{\sqrt{r_0^2-x^2}}\delta\left(y+\sqrt{r_0^2-x^2}\right). \tag{3.140}$$

Finally, we note that

$$\begin{aligned}w_x^2+w_y^2&=\frac{x^2}{x^2+y^2}+\frac{y^2}{x^2+y^2}\\ &=1,\end{aligned} \tag{3.141}$$

so that with $r=\sqrt{x^2+y^2}$ and $s=r-r_0$, Eq. (3.135) yields

$$\delta[w(x,y)]=\delta(r-r_0), \tag{3.142}$$

which agrees with our initial assumption.

REFERENCES

3-1 R. Bracewell, *The Fourier Transform and Its Applications*, McGraw-Hill, New York, 1956, p. 62.

3-2 *Ibid.*, p. 54.

3-3 *Ibid.*, p. 59.

3-4 M. J. Lighthill, *Introduction to Fourier Analysis and Generalised Functions*, Cambridge University Press, London, 1958.

3-5 Bracewell, *op. cit.*, p. 69.

3-6 *Ibid.*, p. 79.

3-7 J. W. Goodman, *Introduction to Fourier Optics*, McGraw-Hill, New York, 1968, p. 15.

3-8 Bracewell, *op. cit.*, p. 77.

3-9 K. O. May, *Elementary Analysis*, Wiley, New York, 1952, p. 478.

3-10 A. Papoulis, *Systems and Transforms with Applications in Optics*, McGraw-Hill, New York, 1968, p. 40.

3-11 D. C. Champeney, *Fourier Transforms and Their Physical Applications*, Academic, London, 1973, p. 50.

3-12 Papoulis, *op. cit.*, p. 116.

PROBLEMS

3-1. Given that $f(x) = \mathrm{rect}(x+2) + \mathrm{rect}(x-2)$. Sketch the following functions:

a. $f(x)$.
b. $g(x) = f(x-1)$.
c. $h(x) = f(x)\mathrm{sgn}(x)$.
d. $p(x) = h(x-1)$.

3-2 Sketch the following functions:

a. $f(x) = \mathrm{rect}\left(\frac{x}{4}\right) - \mathrm{rect}\left(\frac{x}{2}\right)$.
b. $g(x) = 2\,\mathrm{tri}\left(\frac{x}{2}\right) - \mathrm{tri}(x)$.
c. $h(x) = 2\,\mathrm{tri}\left(\frac{x}{2}\right) - 2\,\mathrm{tri}(x)$.
d. $p(x) = \left[\mathrm{rect}\left(\frac{x}{4}\right) - \mathrm{tri}\left(\frac{x}{2}\right)\right]\mathrm{sgn}(x)$.

3-3. Given the positive real constants b and x_0, and the function $f(x) = \mathrm{tri}(x)\mathrm{step}(x)$, sketch the following:

a. $f(x)$.
b. $f\left(\frac{x}{b}\right)$.
c. $f(x+x_0)$.
d. $f(x-x_0)$.
e. $f(-x) = f\left(\frac{x}{-1}\right)$.
f. $f\left(\frac{x}{-b}\right)$.

g. $f(-x+x_0)=f\left(\frac{x-x_0}{-1}\right)$.

h. $f(-x-x_0)=f\left(\frac{x+x_0}{-1}\right)$.

i. $f\left(\frac{x+x_0}{b}\right)$.

j. $f\left(\frac{x-x_0}{b}\right)$.

k. $f\left(\frac{-x+x_0}{b}\right)=f\left(\frac{x-x_0}{-b}\right)$.

l. $f\left(\frac{-x-x_0}{b}\right)=f\left(\frac{x+x_0}{-b}\right)$.

3-4. With ξ a real parameter and b and x_0 real constants, show that:

a. $\int_{-\infty}^{\infty} \delta(\alpha)e^{j2\pi\xi\alpha}d\alpha=1$.

b. $\int_{-\infty}^{\infty} \delta\left(\frac{\alpha-x_0}{b}\right)e^{j2\pi\xi\alpha}d\alpha=|b|e^{j2\pi\xi x_0}$.

3-5. With $f(x)$ an arbitrary function and a, b, and x_0 real constants, show that the following expressions are correct:

a. $f(x)\delta\left(\frac{x-x_0}{b}\right)=|b|f(x_0)\delta(x-x_0)$.

b. $f(x)\delta(ax-x_0)=\frac{1}{|a|}f\left(\frac{x_0}{a}\right)\delta\left(x-\frac{x_0}{a}\right)$.

3-6. With $f(x)$ an arbitrary function and b and x_0 real constants, show that the following expressions are correct:

a. $f(x)\delta\delta\left(\frac{x-x_0}{b}\right)=|b|[f(x_0-b)\delta(x-x_0+b)$

$+f(x_0+b)\delta(x-x_0-b)]$.

b. $f(x)\text{comb}\left(\frac{x-x_0}{b}\right)=|b|\sum_{n=-\infty}^{\infty} f(x_0+nb)\delta(x-x_0-nb)$.

3-7. Graph axial profiles [i.e., $f(x,0)$ and $f(0,y)$, etc.] for the following functions:

a. $f(x,y)=\text{sinc}\left(\frac{x}{2},y\right)$.

b. $g(x,y)=\text{rect}\left(\frac{x}{2},\frac{y}{4}\right)-\text{rect}\left(x,\frac{y}{2}\right)$.

c. $h(x,y)=\text{somb}\left(\sqrt{x^2+\left(\frac{y}{2}\right)^2}\right)$.

d. $p(x,y)=\text{cyl}\left(\frac{\sqrt{x^2+y^2}}{2}\right)-\text{cyl}\left(\sqrt{x^2+y^2}\right)$.

3-8. Given the function $f(x,y)=\text{rect}(x,y)$, sketch a "top view" of the following functions:

a. $u(x,y)=f(5x-0.5y,\ 0.25y)$.

b. $g(x,y)=f(5x-0.5y+2,\ 0.25y)+f(5x-0.5y-2,\ 0.25y)$.

c. Show that $g(x,y)$ can be represented by the expression $g(x,y)=u(x+x_0,y)+u(x-x_0,y)$, and find the value of x_0.

3-9. Given the delta function $\delta(3x+4y-5)$, find a representation for this delta function in the primed coordinate system described by Eq. (3.111) and depicted in Fig. 3-41.

3-10. Given the function $g(x, y)=\text{comb}(x+2y, -x+y)$:

a. Represent $g(x, y)$ as a doubly infinite sum as in Eq. (3.129).
b. Find the associated function $G(\xi, \eta)$ given by Eq. (3.130).
c. Sketch "top views" of the lattice associated with $g(x, y)$ and the related reciprocal lattice associated with $G(\xi, \eta)$.

CHAPTER 4

HARMONIC ANALYSIS

Until now we have represented the various physical quantities of interest by their *time-domain* or *space-domain* descriptions; in this chapter we shall see how these quantities may be described in the *temporal-frequency* and *spatial-frequency domains*. We assume that the reader is reasonably familiar with the *Fourier-series expansion* and has at least some background concerning the general theory of orthogonal expansions. Because our primary objective here is to develop a basic understanding of the concepts associated with frequency-domain descriptions, our treatment will not display a high degree of mathematical rigor. Those wishing a more rigorous treatment should consult any of a number of books on advanced mathematics; e.g., see (Ref. 4-1).

4-1 ORTHOGONAL EXPANSIONS

It is frequently desirable to decompose a given function into some combination of more elementary functions, called *basis functions*, the motivation being that by doing so certain mathematical operations may be greatly simplified. In addition, such a decomposition may allow more insight to be gained concerning the nature of a problem than might otherwise be gained.

Let us assume that we would like to determine the effects of a particular physical system on some arbitrary input signal, which is represented by the time function $f(t)$. In other words, we want to determine the output of the system when the input is $f(t)$. A system can usually be described by a mathematical operator, and its effect on the input signal characterized by the effect this operator has on the function $f(t)$; however, even when we know the form of the operator explicitly, it will generally be quite difficult to perform the required mathematical operation unless $f(t)$ is one of a few functions possessing certain special properties. But suppose there exists a set of complex-valued basis functions $\{\psi_n(t)\}$, $n=1,2,3,\ldots$, each of which possesses these special properties, and that the arbitrary function $f(t)$ can be expressed as a linear superposition of them. We should be able to find the net result of operating directly on $f(t)$ by first operating on each of the component functions separately and then adding together the results of the individual operations. Of course, both the operator and the function $f(t)$ must satisfy certain conditions for this approach to be valid, but most of the systems and signals with which we are concerned meet these requirements.

By choosing a suitable set of basis functions $\{\psi_n(t)\}$, we can represent almost any arbitrary function $f(t)$ on some interval (t_1, t_2) by an expansion of the form

$$f(t) = c_1\psi_1(t) + c_2\psi_2(t) + c_3\psi_3(t) + \ldots$$
$$= \sum_{n=1}^{\infty} c_n\psi_n(t), \tag{4.1}$$

where the complex-valued coefficients c_n appropriately weight each term of the series. Depending on the situation, this series may have either a finite or an infinite number of terms. For such an expansion to be useful, we must not only be able to find the appropriate set of functions $\{\psi_n(t)\}$, but we must also be able to calculate the coefficients c_n. In addition, we would like to be able to determine the value of each coefficient without being required to know the value of any other coefficient. All of these desirable properties can be realized if the basis functions $\psi_n(t)$ form a *complete, orthogonal set* on the interval (t_1, t_2).

The orthogonality of functions is defined as follows: if μ_m is a real constant, and if

$$\int_{t_1}^{t_2} \psi_n(\alpha)\psi_m^*(\alpha)\, d\alpha = \mu_m \delta_{nm} \tag{4.2}$$

for all m and n, then the functions $\psi_n(t)$ are said to be *orthogonal* on the interval (t_1, t_2). The symbol δ_{nm} is called the Kronecker delta, which has the properties

$$\delta_{nm} = \begin{cases} 0, & n \neq m \\ 1, & n = m. \end{cases} \tag{4.3}$$

Note that the integral of Eq. (4.2) is equal to the area under the product of $\psi_n(t)$ and $\psi_m^*(t)$. If $n \neq m$, this product has as much negative area as positive area, and thus the total area is zero. However, for $n=m$, the product $|\psi_m(t)|^2$ is never negative and the total area is equal to μ_m.

If Eq. (4.2) is satisfied, and if $\mu_m=1$ for every m, then we have

$$\int_{t_1}^{t_2}\psi_n(\alpha)\psi_m^*(\alpha)\,d\alpha=\delta_{nm}, \tag{4.4}$$

and the functions $\psi_n(t)$ are said to be *orthogonal* and *normal*, or *orthonormal*, on the interval (t_1,t_2). We note that any orthogonal set of functions can be normalized by properly weighting each member of the set.

We shall discuss the completeness of a set of orthogonal functions presently, but first let us see how the coefficients are computed. To find the ith coefficient c_i, we first multiply both sides of Eq. (4.1) by the function $\psi_i^*(t)$ and then integrate between the limits t_1 and t_2. Thus

$$\int_{t_1}^{t_2} f(\alpha)\psi_i^*(\alpha)\,d\alpha=\int_{t_1}^{t_2}c_1\psi_1(\alpha)\psi_i^*(\alpha)\,d\alpha+\ldots$$

$$+\int_{t_1}^{t_2}c_i\psi_i(\alpha)\psi_i^*(\alpha)\,d\alpha+\ldots \tag{4.5}$$

but because the functions $\psi_n(t)$ are orthogonal, we see from Eq. (4.2) that all of the terms on the right side of this expression vanish except the ith term, leaving simply

$$\int_{t_1}^{t_2} f(\alpha)\psi_i^*(\alpha)\,d\alpha=c_i\mu_i. \tag{4.6}$$

Rearranging this equation, we obtain

$$c_i=\frac{1}{\mu_i}\int_{t_1}^{t_2} f(\alpha)\psi_i^*(\alpha)\,d\alpha, \tag{4.7}$$

and it is now apparent that each coefficient is independent of all other coefficients. In addition, the coefficients are unique for a given set of basis functions and a given $f(t)$.

Now let us consider the completeness condition. Consider any arbitrary function $g(t)$ that is square integrable and not identically zero on the interval (t_1,t_2), i.e.,

$$\int_{t_1}^{t_2} g^2(\alpha)\,d\alpha<\infty$$

$$g(t)\not\equiv 0,\quad t_1\leqslant t\leqslant t_2. \tag{4.8}$$

Then the set of functions $\{\psi_n(t)\}$, orthogonal on the same interval, is said to be *complete* if there exists no $g(t)$ such that

$$\int_{t_1}^{t_2} g(\alpha)\psi_n^*(\alpha)\,d\alpha = 0 \tag{4.9}$$

for all members $\psi_n(t)$ of the set. In other words, *any* function $g(t)$ satisfying Eq. (4.8) will have a nontrivial expansion and can thus be represented on the interval (t_1, t_2) by a superposition of the functions $\psi_n(t)$.

In many cases the representation of a function by an infinite series proves to be quite cumbersome, and to help get around this problem the function is approximated by a truncated expansion, which contains only a finite number of terms. Let $\hat{f}(t)$ be the truncated version of $f(t)$, such that

$$\hat{f}(t) = \sum_{n=1}^{N} \hat{c}_n \psi_n(t), \tag{4.10}$$

where $N < \infty$. A question now arises concerning the coefficients $\hat{c}_n$ of this series: should they be recalculated, or can the previously determined coefficients for the infinite series be used here as well? It is a rather amazing result that the coefficients calculated for the infinite series will also yield the "best" approximation for the function $\hat{f}(t)$, where by "best" we mean that the *integral squared error* will be minimized, i.e.,

$$\int_{t_1}^{t_2} \left[f(\alpha) - \hat{f}(\alpha) \right]^2 d\alpha = \text{minimum}. \tag{4.11}$$

As more and more terms are included in the expansion, it seems intuitively obvious that $\hat{f}(t)$ will more closely approximate $f(t)$ and the integral squared error will become smaller and smaller. But it is certainly less obvious that, even when we include only a few terms in the series, the coefficients given by Eq. (4.7) are still the best coefficients according to the criterion specified by Eq. (4.11). Thus we have

$$\hat{c}_n = c_n, \quad n = 1, 2, \ldots, N. \tag{4.12}$$

Now let us briefly review the important properties of the expansions just discussed. Given the functions $\psi_n(t)$, which form a complete orthogonal set on some interval (t_1, t_2), almost any arbitrary function $f(t)$ can be expanded on this interval according to

$$f(t) = \sum_{n=1}^{\infty} c_n \psi_n(t). \tag{4.13}$$

This series is said to *converge in the mean* to the function $f(t)$ at every point in the interval for which the expansion is valid. Even though the orthogonality condition allows us to determine the coefficients c_n easily and independently of one another, the completeness condition ensures us that $f(t)$ may be any function satisfying the conditions of Eq. (4.8). In addition, if the series is truncated, the infinite-series coefficients will still yield the best approximation to the function $f(t)$ in the sense of minimizing the integral-squared error. Again, for a more complete and rigorous development, the reader is referred to a book dealing with advanced mathematics.

Until now our discussion of orthogonal expansions has been in rather general terms. This was done purposely so that we could concentrate on the nature of all such expansions rather than on the characteristics of a single expansion utilizing a particular set of basis functions. At this point, however, we choose some specific expansions to develop a more physical picture of the abstract notions encountered.

Although there are several different sets of functions that satisfy the desired orthogonality and completeness conditions, one of the most important for our study of linear systems is the set of complex exponentials

$$\psi_n(t) = e^{j2\pi n\nu_0 t}, n = 0, \pm 1, \pm 2, \ldots . \quad (4.14)$$

These functions, which form the basis for the Fourier-series expansion, are orthogonal over any interval equal to an integral number of periods of the first-order terms, or fundamental components. That is,

$$t_1 - t_2 = k\left(\frac{1}{\nu_0}\right), k = 1,2,3, \ldots . \quad (4.15)$$

The primary reason for their importance is that they are *eigenfunctions of linear shift-invariant operators*, which are the operators used to describe *linear shift-invariant systems*. Although this topic will be discussed in considerable detail in Chapter 5, we mention its significance here: if the input to such a system is an eigenfunction of the system, then the output will be equal to the product of the input and a complex constant of proportionality. This means that the output will have the same form, or shape, as the input, but its magnitude may be changed and it may be shifted along the abscissa. In other words, the eigenfunctions of a linear shift-invariant system are passed by the system unchanged in form, but possibly changed in magnitude and position.

In addition to the complex exponentials, the functions

$$\psi_n(t) = \sin(2\pi n\nu_0 t), \quad n = 0, 1, 2, \ldots$$

$$\psi_n(t) = \cos(2\pi n\nu_0 t), \quad n = 0, 1, 2, \ldots \quad (4.16)$$

form complete, orthogonal sets over the intervals given by Eq. (4.15), and they too are eigenfunctions of certain kinds of systems. Of course, these functions are related to the complex exponentials by Euler's formula [Eq. (2.42)].

Still other functions forming complete, orthogonal sets are the Walsh functions, $\text{sal}(n,t)$ and $\text{cal}(n,t)$, a few of which are shown in Fig. 4-1 (Ref. 4-2). As can be seen, they are rectangular-wave functions having only the values $+1$ and -1. We shall not discuss them further there, but these functions form the basis for the Hadamard transform and are finding applications in the fields of communications and information processing (Refs. 4-2, 4-3, and 4-4). In particular, the Walsh functions are useful in applications requiring digital processing of two-dimensional signals, e.g., the processing of photographic images.

At this point, for the sake of thoroughness, we include a set of functions that are orthogonal but not complete. For example, consider the functions

$$x_n(t) = \text{rect}\left(t - \frac{2n+1}{2}\right), \quad n = 0, 1, 2, 3, 4, \tag{4.17}$$

which are shown in Fig. 4-2. These functions are quite obviously orthogonal on the interval $(0, 5)$, but they are not complete; it is not hard to visualize a number of functions that cannot be adequately represented by linear combinations of them.

So far we have discussed only expansions in terms of denumerable sets of functions $\{\psi_n(t)\}$. It may also be possible to decompose an arbitrary

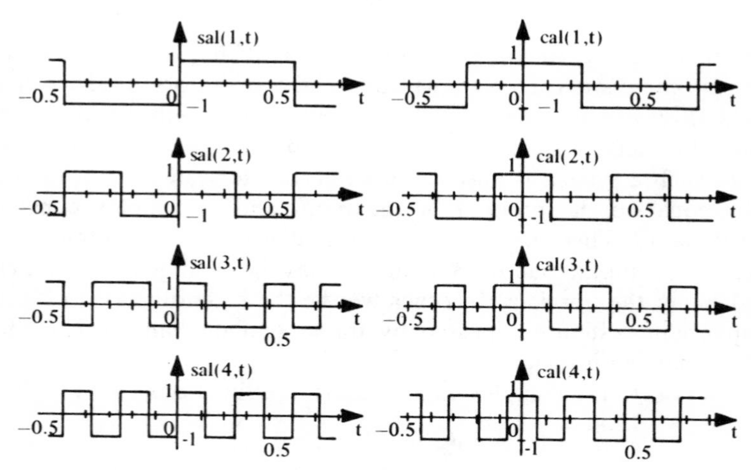

Figure 4-1 The Walsh functions $\text{sal}(n,t)$ and $\text{cal}(n,t)$.

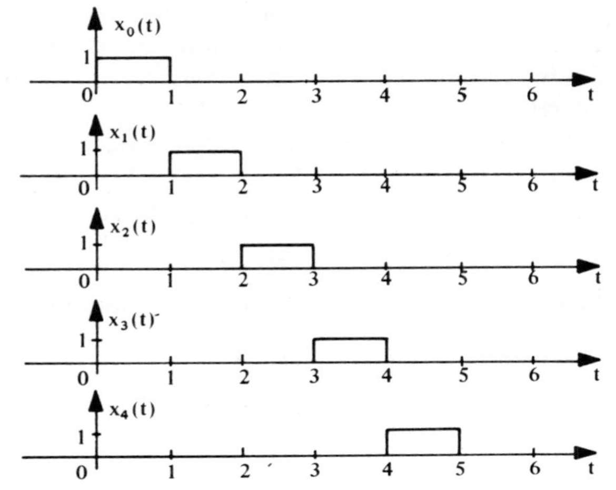

Figure 4-2 An incomplete set of orthogonal functions.

function $f(t)$ into a linear combination of the functions $\psi(t;\nu)$; here the parameter ν, which replaces the index symbol n, may take on any real value (recall that n was allowed to take on only discrete values). Thus the functions $\psi(t;\nu)$ form a continuum rather than a denumerable set, and an expansion in terms of them is written as an integral rather than a sum, i.e.,

$$f(t)=\int_{-\infty}^{\infty} W(\nu)\psi(t;\nu)\,d\nu, \tag{4.18}$$

where $W(\nu)$ is a weighting function that closely corresponds to the coefficients c_n of Eq. (4.1).

The orthogonality condition for the functions $\psi(t;\nu)$ is given by the infinite integral

$$\int_{-\infty}^{\infty} \psi(\alpha;\nu)\psi^*(\alpha;\nu')\,d\alpha=\mu(\nu)\,\delta(\nu-\nu'), \tag{4.19}$$

where the prime notation merely indicates that ν and ν' are not identically the same [similar to the n and m of Eq. (4.2)]. Note that the interval over which orthogonality holds is now infinite. If $\mu(\nu)=1$ for all ν, then

$$\int_{-\infty}^{\infty} \psi(\alpha;\nu)\psi^*(\alpha;\nu')\,d\alpha=\delta(\nu-\nu'), \tag{4.20}$$

and these functions are orthonormal on $(-\infty, \infty)$. As previously stated, any orthogonal system of functions can be normalized, so without loss of generality we now choose our basis functions to be orthonormal and proceed to calculate the weighting function $W(\nu)$. We first multiply both sides of Eq. (4.18) by $\psi^*(t;\nu')$, and then integrate from $-\infty$ to ∞. Thus

$$\int_{-\infty}^{\infty} f(\alpha)\psi^*(\alpha;\nu')\,d\alpha = \int_{-\infty}^{\infty}\left[\int_{-\infty}^{\infty} W(\nu)\psi(\alpha;\nu)\,d\nu\right]\psi^*(\alpha;\nu')\,d\alpha, \quad (4.21)$$

and by interchanging the order of integration and using Eq. (4.20) we find that

$$\int_{-\infty}^{\infty} f(\alpha)\psi^*(\alpha;\nu')\,d\alpha = \int_{-\infty}^{\infty} W(\nu)\left[\int_{-\infty}^{\infty} \psi(\alpha;\nu)\psi^*(\alpha;\nu')\,d\alpha\right]d\nu$$

$$= \int_{-\infty}^{\infty} W(\nu)\,\delta(\nu-\nu')\,d\nu. \quad (4.22)$$

But from the defining property of the delta function, the right side of this equation reduces to

$$\int_{-\infty}^{\infty} W(\nu)\,\delta(\nu-\nu')\,d\nu = W(\nu'), \quad (4.23)$$

so that finally we obtain

$$W(\nu') = \int_{-\infty}^{\infty} f(\alpha)\psi^*(\alpha;\nu')\,d\alpha. \quad (4.24)$$

The prime notation no longer has any significance in this expression, so we omit it for convenience. Hence the weighting function $W(\nu)$ is given by

$$W(\nu) = \int_{-\infty}^{\infty} f(\alpha)\psi^*(\alpha;\nu)\,d\alpha. \quad (4.25)$$

Note the similarity between this equation and the equation used for computing the infinite-series coefficients. Once again the the orthogonality of the functions permits the weighting function to be calculated with relative ease, whereas their completeness ensures that almost any arbitrary function will have a nontrivial expansion.

The complex exponentials

$$\psi(t;\nu) = e^{j2\pi\nu t} \quad (4.26)$$

form a complete, orthogonal system of functions, and they are without question the most important basis functions for our studies here; not only are they eigenfunctions of linear shift-invariant operators, but they are also the functions upon which the Fourier transform is based. After the following section, they will be the primary basis functions for any expansions we make.

In addition to the complex exponentials of Eq. (4.26), the functions

$$\psi(t;\nu)=\sin(2\pi\nu t)$$

$$\psi(t;\nu)=\cos(2\pi\nu t) \tag{4.27}$$

form complete, orthogonal systems, and, because they are closely related to the complex exponentials, we use them a great deal in our work. It is important to understand that the parameter ν, called the temporal frequency of the function, may take on any real value in the above expressions. Therefore, until the value of ν is specified, each of these expressions represents a family of time functions rather than a single, specific function. Once the value of ν is specified, say $\nu=\nu_0$, then these expressions represent single time functions; for example, $\sin(2\pi\nu_0 t)$ is a single sinusoidal function of time, and it has a period of $T=\nu_0^{-1}$.

4-2 THE FOURIER SERIES

We choose to discuss the Fourier-series before tackling the Fourier integral primarily because it is the more easily understood of the two; hence, the transition from the series to the integral is made with less difficulty than the other way around. We assume that the reader has some familiarity with the Fourier series but none with the Fourier integral.

The Fourier series is an expansion of the type given by Eq. (4.1) and is used primarily to represent periodic functions. As we shall see, however, it can also be used to represent nonperiodic functions. The basis functions for this series are either the complex exponentials of Eq. (4.14) or the sine and cosine functions of Eq. (4.16). We shall use the complex exponential form of the series because it leads more directly to the Fourier integral and, as a result, the index symbol n of Eq. (4.1) will range from $-\infty$ to ∞ rather than from 1 to ∞.

Let us now consider the expansions of periodic functions. The function $f(t)$, which is periodic with period $T=\nu_0^{-1}$, can be represented everywhere by a Fourier series if it satisfies a set of conditions known as the Dirichlet conditions. These conditions are stated in many different ways, but they effectively reduce to the following: in any finite interval $t_1 \leqslant t \leqslant t_2$, $f(t)$ must (1) be single valued, (2) have a finite number of maxima and minima, (3) have at most a finite number of finite, and no infinite, discontinuities, and (4) be absolutely integrable, i.e.,

$$\int_{t_1}^{t_2} |f(\alpha)|\, d\alpha < \infty. \tag{4.28}$$

If these conditions are satisfied, then the Fourier series

$$f(t) = \sum_{n=-\infty}^{\infty} c_n e^{j2\pi n \nu_0 t} \tag{4.29}$$

converges uniformly to $f(t)$ at every point of continuity, and at every point of discontinuity it converges to the average of the limiting values of $f(t)$ as the discontinuity is approached from above and from below. In general, any one of these conditions may be relaxed to some degree, the penalty usually being that one or more of the remaining requirements must be tightened. We point out that any function that accurately represents a real physical quantity will satisfy the Dirichlet conditions.

From Eq. (4.29) it is apparent that the Fourier series decomposes a function into a linear combination of complex exponentials, with the complex-valued coefficients c_n assigning the proper weight to each term. As we will see later, the terms of the series for $n = \pm 1$ may be combined to form a sinusoidal function of frequency ν_0, called the *fundamental component* of the original function. (Recall that ν_0 is the fundamental frequency of this function.) Similarly, the terms for $n = \pm 2$ may be combined to form a sinusoid of frequency $2\nu_0$, which is known as the *second-harmonic component* of the function, and in general, the terms for $n = \pm m$ comprise its mth *harmonic component*, with frequency $m\nu_0$. It is a common mistake to state that the function $f(t)$ consists of a large number of "frequencies." What should be said is that $f(t)$ is composed of a large number of *sinusoidal components* having harmonically related frequencies. Each component is merely characterized by a particular frequency, and should not itself be referred to as "a frequency."

With $f(t)$ expanded as in Eq. (4.29), the coefficients are found to be

$$c_n = \frac{1}{T} \int_t^{t+T} f(\alpha) e^{-j2\pi n \nu_0 \alpha}\, d\alpha, \tag{4.30}$$

where the integration is taken over one period $T = \nu_0^{-1}$. For sufficiently large n, these coefficients always decrease in magnitude at least as fast as n^{-1}, but if $f(t)$ has one or more discontinuites, they can decrease no faster than this. If the function itself is everywhere continuous but its derivative is discontinuous at one or more points, the coefficients decrease as n^{-2}. Practically speaking, this means that the smoother the function, the faster its Fourier series converges (Ref. 4-1).

Now let us briefly consider truncated Fourier series. If a function $f(t)$ is approximated by only a few terms of its Fourier series, we reasonably expect the approximation to differ somewhat from $f(t)$. We also expect the approximation to become better as more and more terms are included, and this is generally true except in the vicinity of a discontinuity. To illustrate, consider the rectangular-wave function of Fig. 4-3(a). Near each discontinuity, the truncated series representation of this function will exhibit an overshoot no matter how many terms are included. Not only that, the magnitude of the overshoot eventually stabilizes at approximately 9% of the magnitude of the discontinuity as the number of terms included becomes larger and larger (Ref. 4-5). The overshoot occurs nearer and

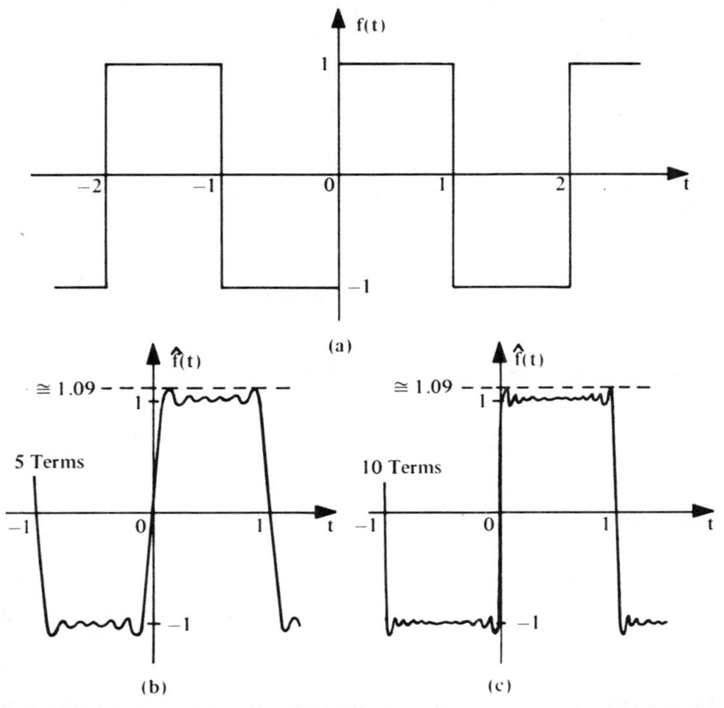

Figure 4-3 Representation of a rectangular-wave function by a truncated Fourier series. (*a*) Rectangle-wave function. (*b*) Series with 5 terms. (*c*) Series with 10 terms.

nearer to the discontinuity, but its magnitude stays the same as shown in Fig. 4-3(b) and (c). This is called *Gibbs phenomenon.*

As mentioned earlier, the Fourier series may also be used to represent a nonperiodic function; however, this representation is valid only in a finite interval and not everywhere. Consider the nonperiodic function

$$g(t)=\mathrm{tri}\left(\frac{t-t_0}{b}\right). \tag{4.31}$$

To represent this function by a Fourier series, we first form the periodic function $g'(t)$, with period $T \geqslant 2b$, such that

$$g(t)=g'(t)\mathrm{rect}\left(\frac{t-t_0}{T}\right). \tag{4.32}$$

(Here the primed notation does not indicate a derivative.) We then find the Fourier series for $g'(t)$, and use it to represent $g(t)$ *only in the interval* $t_0-b \leqslant t \leqslant t_0+b$; outside this interval the series represents $g'(t)$ and not $g(t)$. The obvious choice for $g'(t)$ in this case is

$$g'(t)=\sum_{n=-\infty}^{\infty}\mathrm{tri}\left(\frac{t-t_0-nT}{b}\right), \tag{4.33}$$

and both $g(t)$ and $g'(t)$ are illustrated in Fig. 4-4. The Fourier series for $g'(t)$ is

$$g'(t)=\sum_{n=-\infty}^{\infty} c_n' e^{j2\pi n\nu_0 t}, \tag{4.34}$$

where again $\nu_0=T^{-1}$, with the coefficients given by

$$c_n'=\frac{1}{T}\int_t^{t+T} g'(\alpha)e^{-j2\pi n\nu_0\alpha}d\alpha. \tag{4.35}$$

We finally obtain

$$\begin{aligned} g(t)&=\sum_{n=-\infty}^{\infty} c_n' e^{j2\pi n\nu_0 t},\ t_0-b \leqslant t \leqslant t_0+b \\ &=0,\ \text{elsewhere}. \end{aligned} \tag{4.36}$$

Thus the coefficients computed for the periodic function $g'(t)$ can be used in the representation of the nonperiodic function $g(t)$.

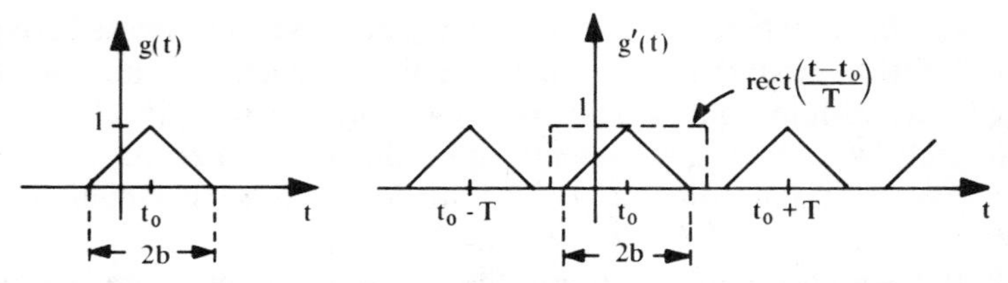

Figure 4-4 Fourier series representation of a nonperiodic function.

4-3 THE FOURIER INTEGRAL

To represent a nonperiodic function $g(t)$ everywhere, an integral expansion of the form given by Eq. (4.18) is required. The Fourier integral is just what we need to do the job, decomposing the function into a linear combination of the complex exponentials of Eq. (4.26). If $g(t)$ satisfies the Dirichlet conditions, with the strengthened requirement that it must be absolutely integrable on the *infinite* interval $(-\infty, \infty)$, we may represent it by the integral

$$g(t)=\int_{-\infty}^{\infty} G(\nu)e^{j2\pi\nu t}d\nu. \tag{4.37}$$

Here $G(\nu)$ corresponds to the weighting function $W(\nu)$ of Eq. (4.18) and is given by

$$G(\nu)=\int_{-\infty}^{\infty} g(\alpha)e^{-j2\pi\nu\alpha}d\alpha. \tag{4.38}$$

Note the similarity between these two integrals, which are called *Fourier integrals*.* (An apology is offered for the inconsistent use of dummy integration variables, but it is felt that in some cases less confusion will arise if a dummy variable is not used.)

The weighting function $G(\nu)$ is known as the *Fourier transform* of $g(t)$; note that $g(t)$ is the time-domain representation of some quantity, whereas $G(\nu)$ describes it in the *temporal-frequency domain*. The function $G(\nu)$ is also often called the *complex temporal-frequency spectrum*, or simply *frequency spectrum*, of $g(t)$. It plays the same role as the coefficients of the Fourier series, assigning the proper weights to the various components, but in general it is a piecewise-continuous function of the frequency variable ν.

*Other definitions, slightly different from these, are often used. The various choices are discussed in Chapter 7.

In addition, the components no longer have harmonically related frequencies, and the component lying between the frequency ν and $\nu + d\nu$ is assigned an infinitesimal weight $G(\nu)d\nu$. Nevertheless, if all of these appropriately weighted components are added together in the proper fashion, the resulting function is just the original $g(t)$, which is often referred to as the *inverse Fourier transform* of $G(\nu)$.

With the strengthened condition that a function must be absolutely integrable on the interval $(-\infty, \infty)$, it is apparent that periodic functions cannot possess Fourier transforms as they are defined here. However, since it is desirable to associate some sort of frequency spectra with periodic as well as nonperiodic functions, we modify the existence conditions to allow the Fourier transforms of periodic functions to be included. We shall postpone a detailed discussion of this maneuver until Chapter 7 and proceed to investigate the behavior of such transforms, accepting for the time being that they are legitimate.

We first represent the periodic function $f(t)$ by its Fourier series, i.e.,

$$f(t) = \sum_{n=-\infty}^{\infty} c_n e^{j2\pi n\nu_0 t}, \tag{4.39}$$

where again ν_0 is the fundamental frequency of $f(t)$. The Fourier transform of $f(t)$, which we shall designate by $F(\nu)$, can then be written as

$$\begin{aligned} F(\nu) &= \int_{-\infty}^{\infty} f(\alpha) e^{-j2\pi\nu\alpha} d\alpha \\ &= \int_{-\infty}^{\infty} \left[\sum_{n=-\infty}^{\infty} c_n e^{j2\pi n\nu_0 \alpha} \right] e^{-j2\pi\nu\alpha} d\alpha, \end{aligned} \tag{4.40}$$

and, by interchanging the order of summation and integration, we obtain

$$F(\nu) = \sum_{n=-\infty}^{\infty} c_n \left[\int_{-\infty}^{\infty} e^{-j2\pi(\nu - n\nu_0)\alpha} d\alpha \right]. \tag{4.41}$$

But from the orthogonality of the complex exponentials it may be shown that

$$\int_{-\infty}^{\infty} e^{-j2\pi(\nu - n\nu_0)\alpha} d\alpha = \delta(\nu - n\nu_0), \tag{4.42}$$

with the final result that

$$F(\nu) = \sum_{n=-\infty}^{\infty} c_n \delta(\nu - n\nu_0). \tag{4.43}$$

Thus the Fourier transform of a periodic function consists of an array of weighted delta functions, separated by frequency intervals equal to the fundamental frequency ν_0. In other words, *periodic functions have discrete spectra* (sometimes referred to as line spectra).

In contrast, *the spectra of nonperiodic functions are not discrete*; they are at least piecewise-continuous functions of the frequency variable ν. We shall see examples of each type of spectra in the next section.

The frequency spectrum of an arbitrary function $f(t)$ is in general complex valued. As a result, it is often advantageous to put it in the form

$$F(\nu)=A(\nu)e^{-j\Phi(\nu)}, \tag{4.44}$$

where $A(\nu)$ is known as the *amplitude spectrum* of $f(t)$ and $\Phi(\nu)$ is called its *phase spectrum*. Both $A(\nu)$ and $\Phi(\nu)$ are real-valued functions, and we shall allow each of them to take on negative as well as positive values. Some authors define $A(\nu)$ to be the modulus of $F(\nu)$, but as will become apparent later, it is frequently easier to describe $F(\nu)$ by allowing $A(\nu)$ to be bipolar.

4-4 SPECTRA OF SOME SIMPLE FUNCTIONS

To help develop a better physical feeling for frequency-domain representations, we now calculate and graph the spectra of a few simple functions.

Sinusoidal Functions

We shall start with the function

$$f(t)=A\cos(2\pi\nu_0 t), \tag{4.45}$$

a graph of which is shown in Fig. 4-5. We first use Eq. (4.30) to calculate the coefficients c_n, and then Eq. (4.43) to find the spectrum $F(\nu)$. Thus, with $T=\nu_0^{-1}$ a positive constant as usual,

$$\begin{aligned} c_n &= \frac{1}{T}\int_t^{t+T} A\cos(2\pi\nu_0\alpha)e^{-j2\pi n\nu_0\alpha}d\alpha \\ &= \frac{1}{T}\int_t^{t+T} \frac{A}{2}\left[e^{j2\pi\nu_0\alpha}+e^{-j2\pi\nu_0\alpha}\right]e^{-j2\pi n\nu_0\alpha}d\alpha, \end{aligned} \tag{4.46}$$

and once again utilizing the orthogonality of the complex exponentials, we obtain

$$c_n=\begin{cases} A/2, & n=\pm 1 \\ 0, & n\neq \pm 1 \end{cases}. \tag{4.47}$$

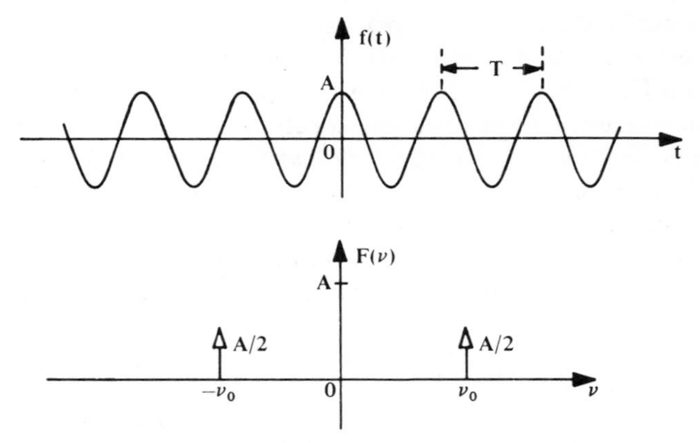

Figure 4-5 Cosine function and its spectrum.

Finally,

$$F(\nu) = \sum_{n=-\infty}^{\infty} c_n \delta(\nu - n\nu_0)$$

$$= \frac{A}{2}\left[\delta(\nu - \nu_0) + \delta(\nu + \nu_0)\right]$$

$$= \frac{A}{2\nu_0}\delta\delta\left(\frac{\nu}{\nu_0}\right), \tag{4.48}$$

and we see that the spectrum of $A\cos(2\pi\nu_0 t)$ consists simply of two delta functions, located at $\nu = \pm\nu_0$, each having an area of $A/2$ as illustrated in Fig. 4-5. Hence this function is composed of a single positive-frequency exponential component and a single negative-frequency exponential component, which is in agreement with the notion we developed in Chapter 2 (see Fig. 2-25). It is easy to show that the function $\sin(2\pi\nu_0 t)$ has a similar spectrum, except that the odd impulse pair is involved rather than the even impulse pair.

Let us now demonstrate that the spectrum of Eq. (4.48) is indeed the correct spectrum by showing that the original function is recovered when Eq. (4.37) is applied. We have

$$\int_{-\infty}^{\infty} F(\nu) e^{j2\pi\nu t} d\nu = \int_{-\infty}^{\infty} \frac{A}{2\nu_0}\delta\delta\left(\frac{\nu}{\nu_0}\right) e^{j2\pi\nu t} d\nu$$

$$= \frac{A}{2}\int_{-\infty}^{\infty}\left[\delta(\nu - \nu_0) + \delta(\nu + \nu_0)\right] e^{j2\pi\nu t} d\nu, \tag{4.49}$$

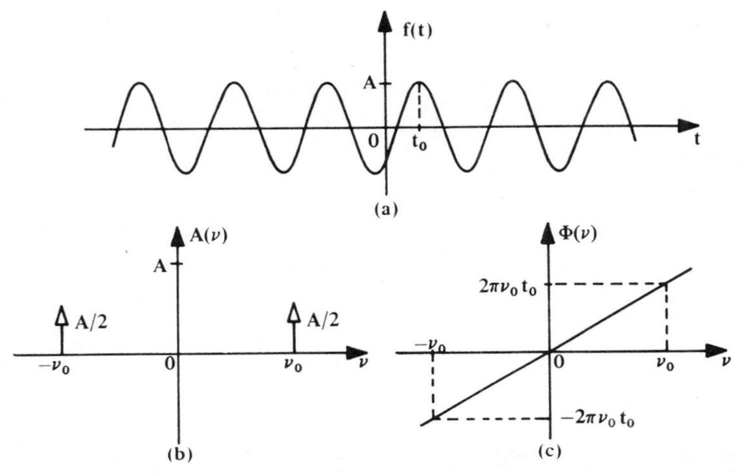

Figure 4-6 Fourier transform of a shifted cosine function. (*a*) Shifted cosine function. (*b*) Amplitude spectrum. (*c*) Phase spectrum.

but from the defining property of the delta function we obtain

$$\int_{-\infty}^{\infty} F(\nu)e^{j2\pi\nu t}d\nu = \frac{A}{2}\left[e^{j2\pi\nu_0 t} + e^{-j2\pi\nu_0 t}\right]$$

$$= A\cos(2\pi\nu_0 t). \tag{4.50}$$

Voilà!

You will note that the phase spectrum of $A\cos(2\pi\nu_0 t)$ is identically zero, but this occurs only in a few special cases. To investigate the relationship between a function and its phase spectrum, let us consider the following shifted version of our original function:

$$f(t) = A\cos(2\pi\nu_0 t - \theta_0)$$

$$= A\cos 2\pi\nu_0(t - t_0), \tag{4.51}$$

where $\theta_0 = 2\pi\nu_0 t_0$. This function is shown in Fig. 4-6. It is easy to determine that the coefficients c_n are given by

$$c_n = \begin{cases} (A/2)e^{\mp j\theta_0}, & n = \pm 1 \\ 0, & n \neq \pm 1, \end{cases} \tag{4.52}$$

so that

$$F(\nu)=\frac{A}{2}\left[\delta(\nu-\nu_0)e^{-j\theta_0}+\delta(\nu+\nu_0)e^{j\theta_0}\right]$$

$$=\frac{A}{2}\left[\delta(\nu-\nu_0)e^{-j2\pi\nu_0 t_0}+\delta(\nu+\nu_0)e^{j2\pi\nu_0 t_0}\right]$$

$$=\frac{A}{2}\left[\delta(\nu-\nu_0)+\delta(\nu+\nu_0)\right]e^{-j2\pi\nu t_0}$$

$$=\frac{A}{2\nu_0}\delta\delta\left(\frac{\nu}{\nu_0}\right)e^{-j2\pi\nu t_0}, \qquad (4.53)$$

and we see that the amplitude spectrum is the same as it was for the unshifted cosine function, while the phase spectrum is now given by

$$\Phi(\nu)=2\pi\nu t_0=\frac{\theta_0\nu}{\nu_0}. \qquad (4.54)$$

The amplitude and phase spectra are shown in Fig. 4-6(b) and (c). Of course, the delta functions of the amplitude spectrum sift out only those values of $\Phi(\nu)$ for which $\nu=\pm\nu_0$. Thus the spectrum of the displaced cosine function also consists of two exponential components, but the phase of these components is shifted by $\pm\theta_0$, which is just enough to shift the function along the time axis by an amount t_0. It is now apparent that the phase spectrum is somehow related to the amount of shift exhibited by a function, but that is not the entire story; we shall come back to this point in Chapter 7.

The higher the frequency of a sinusoidal function, the farther out its spectrum extends along the frequency axis, a relationship that is demonstrated in Fig. 4-7. Conversely, as the frequency becomes smaller the spectrum becomes narrower, and for the zero-frequency case the spectrum is simply a single delta function at the origin.

Now let us look at the sum of several sinusoidal functions, i.e., let us define $f(t)$ to be

$$f(t)=A_0+A_1\cos(2\pi\nu_1 t)+A_2\cos(2\pi\nu_2 t)$$

$$+A_3\cos(2\pi\nu_3 t)+A_4\cos(2\pi\nu_4 t), \qquad (4.55)$$

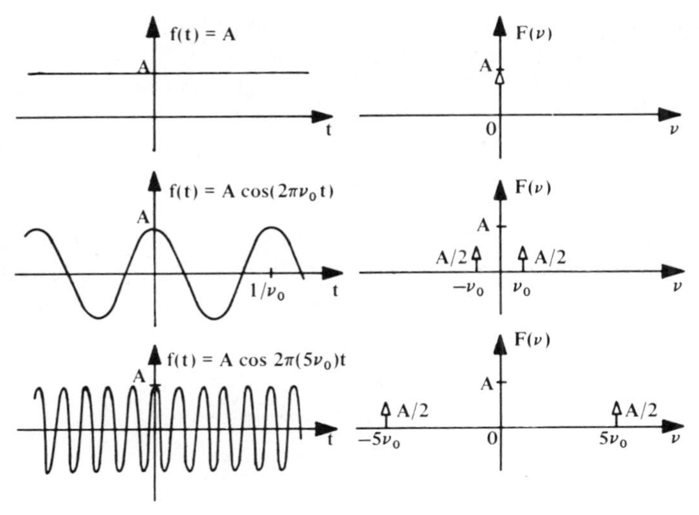

Figure 4-7 Relationship between the frequency of a cosine function and its spectrum.

where A_i is the magnitude and ν_i the frequency of the ith term. Here we assume that the terms are arranged in the order of increasing frequency to simplify the development. The spectrum of this function is found to be

$$F(\nu) = A_0\delta(\nu) + \frac{A_1}{2\nu_1}\delta\delta\left(\frac{\nu}{\nu_1}\right) + \frac{A_2}{2\nu_2}\delta\delta\left(\frac{\nu}{\nu_2}\right)$$

$$+ \frac{A_3}{2\nu_3}\delta\delta\left(\frac{\nu}{\nu_3}\right) + \frac{A_4}{2\nu_4}\delta\delta\left(\frac{\nu}{\nu_4}\right) \tag{4.56}$$

and its graph is displayed in Fig. 4-8. At this point a very interesting and

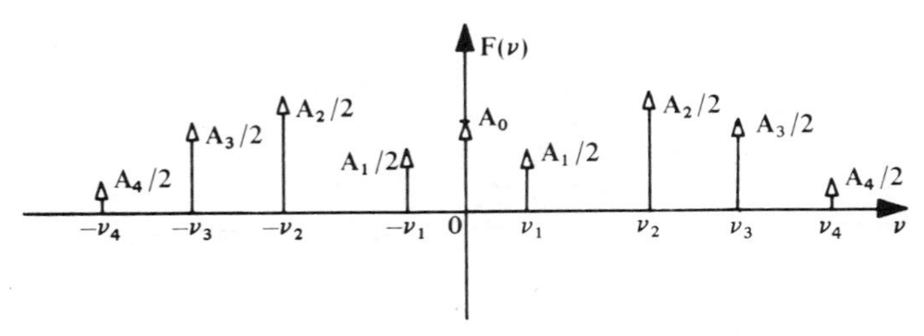

Figure 4-8 Spectrum of the sum of several cosine functions.

important observation can be made: the overall width of the spectrum $F(\nu)$ is directly proportional to the frequency of the highest-frequency component of $f(t)$. Physically this means that functions containing only slowly varying components have narrow spectra, whereas functions with rapidly varying components have spectra with a broad overall width.

Rectangle Wave

The extension to nonsinusoidal periodic functions is straightforward, and to illustrate we choose the rectangle-wave function of Fig. 4-9. With a bit of manipulation, the spectrum of this function can be put into the form

$$\begin{aligned} F(\nu) &= \frac{A}{2}\,\mathrm{sinc}\left(\frac{\nu}{2\nu_0}\right)\sum_{n=-\infty}^{\infty}\delta(\nu - n\nu_0) \\ &= \frac{A}{2}\sum_{n=-\infty}^{\infty}\mathrm{sinc}\left(\frac{n}{2}\right)\delta(\nu - n\nu_0), \end{aligned} \tag{4.57}$$

where again $\nu_0 = T^{-1}$ is the fundamental frequency of $f(t)$. It is apparent that the phase spectrum is zero in this case, and the graph of the amplitude

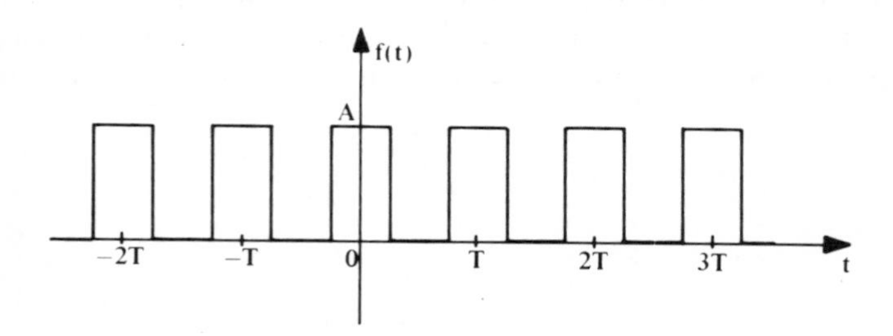

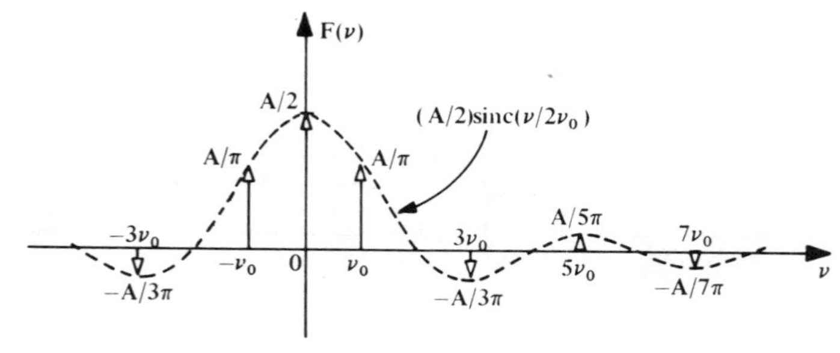

Figure 4-9 A rectangle-wave function and its spectrum.

spectrum is shown in Fig. 4-9. The expansion of $f(t)$ may be written as

$$
\begin{aligned}
f(t) &= \frac{A}{2} \sum_{n=-\infty}^{\infty} \operatorname{sinc}\left(\frac{n}{2}\right) e^{j2\pi n\nu_0 t} \\
&= \frac{A}{2} + \frac{A}{\pi}\left[e^{j2\pi\nu_0 t} + e^{-j2\pi\nu_0 t}\right] \\
&\quad - \frac{A}{3\pi}\left[e^{j2\pi(3\nu_0)t} + e^{-j2\pi(3\nu_0)t}\right] \\
&\quad + \frac{A}{5\pi}\left[e^{j2\pi(5\nu_0)t} + e^{-j2\pi(5\nu_0)t}\right] \\
&\quad - \frac{A}{7\pi}\left[e^{j2\pi(7\nu_0)t} + e^{-j2\pi(7\nu_0)t}\right] + \ldots . \qquad (4.58)
\end{aligned}
$$

Combining the exponentials in the brackets,

$$
\begin{aligned}
f(t) = \frac{A}{2} + \frac{2A}{\pi}\Big[& \cos 2\pi\nu_0 t - \tfrac{1}{3}\cos 2\pi(3\nu_0)t \\
& + \tfrac{1}{5}\cos 2\pi(5\nu_0)t - \tfrac{1}{7}\cos 2\pi(7\nu_0)t + \ldots \Big]. \qquad (4.59)
\end{aligned}
$$

and you will note that this particular example contains only odd-harmonic components. By graphing the various components and adding them together, one at a time, it becomes apparent that with each additional term the sum more closely resembles $f(t)$. This is illustrated in Fig. 4-10. Another graphical method that can be useful in visualizing the Fourier decomposition of $f(t)$ is the phasor-addition method presented in Chapter 2. To simplify the implementation of this method, we first define the complex-valued function $h(t)$ to be

$$
\begin{aligned}
h(t) = \frac{A}{2} + \frac{2A}{\pi}\Big[& e^{j2\pi\nu_0 t} - \tfrac{1}{3}e^{j2\pi(3\nu_0)t} \\
& + \tfrac{1}{5}e^{j2\pi(5\nu_0)t} - \tfrac{1}{7}e^{j2\pi(7\nu_0)t} + \ldots \Big]. \qquad (4.60)
\end{aligned}
$$

Thus

$$
f(t) = \operatorname{Re}\{h(t)\}, \qquad (4.61)
$$

and the value of $f(t)$ at any time is determined by adding together the phasor components of $h(t)$ and projecting this sum on the axis of reals, as depicted in Fig. 4-11.

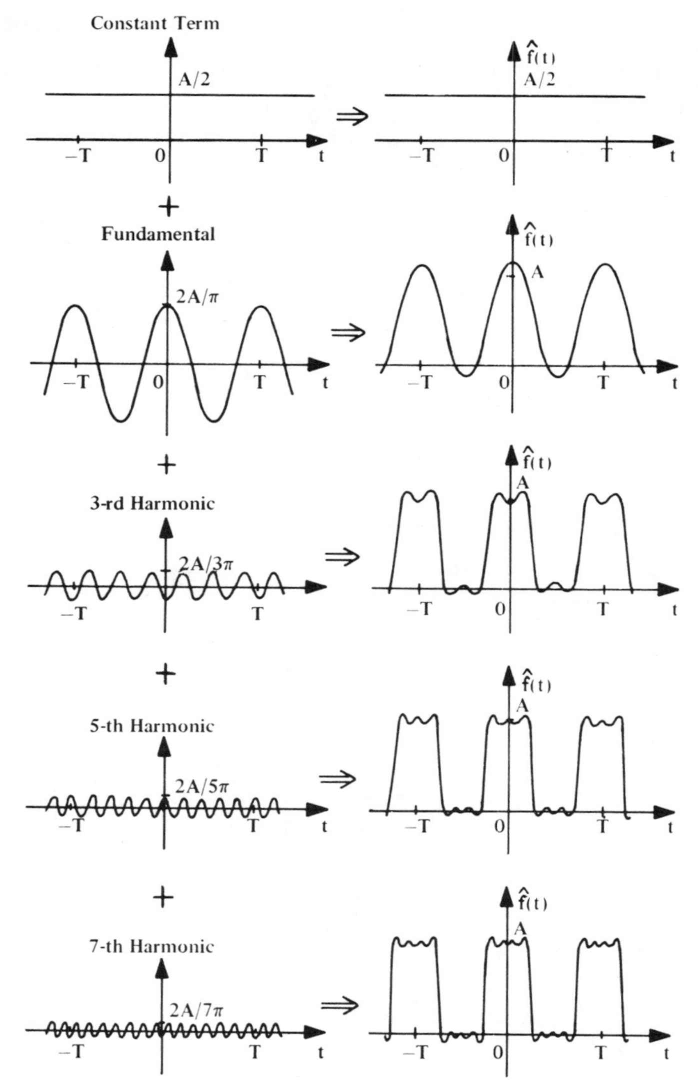

Figure 4-10 Fourier decomposition of a rectangle-wave function.

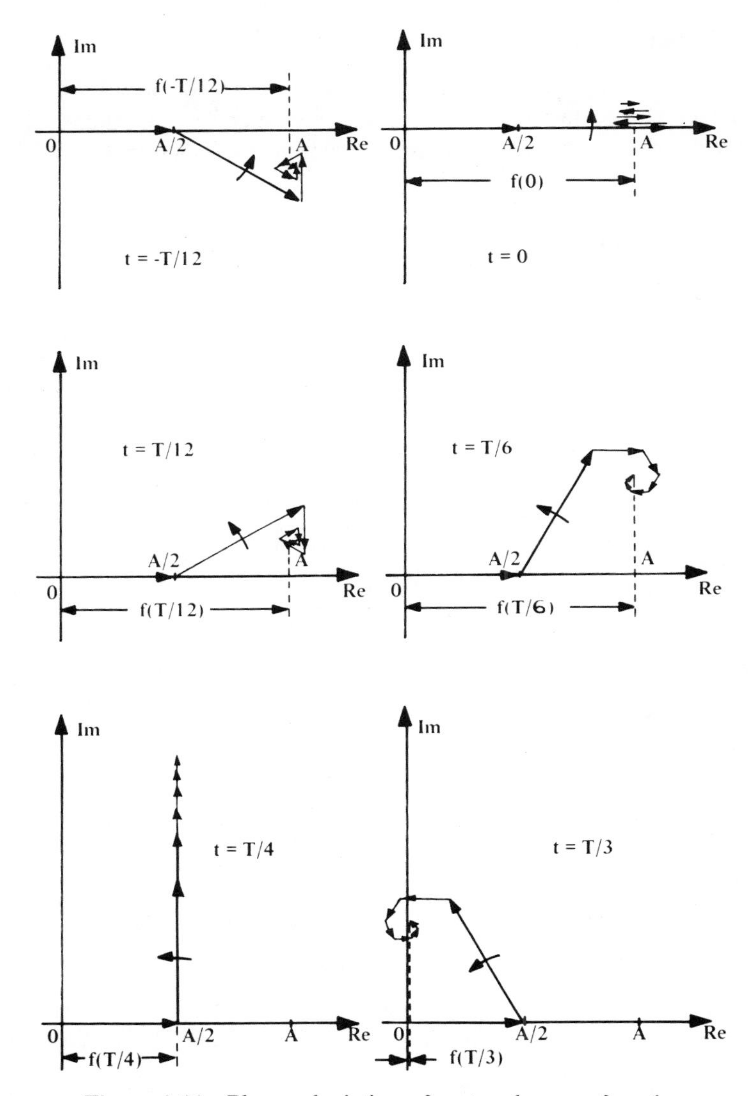

Figure 4-11 Phasor depiction of rectangle-wave function.

As we did for the sine wave, let us see what effect a shift has on the spectrum of the rectangle wave. Suppose we specify the time shift to be one-quarter of a period in the positive direction as shown in Fig. 4-12. The spectrum is now found to be

$$F(\nu) = \frac{A}{2}\operatorname{sinc}\left(\frac{\nu}{2\nu_0}\right)e^{-j(\pi\nu/2\nu_0)}\sum_{n=-\infty}^{\infty}\delta(\nu - n\nu_0)$$

$$= \frac{A}{2}\sum_{n=-\infty}^{\infty}\operatorname{sinc}\left(\frac{n}{2}\right)e^{-j(n\pi/2)}\delta(\nu - n\nu_0), \tag{4.62}$$

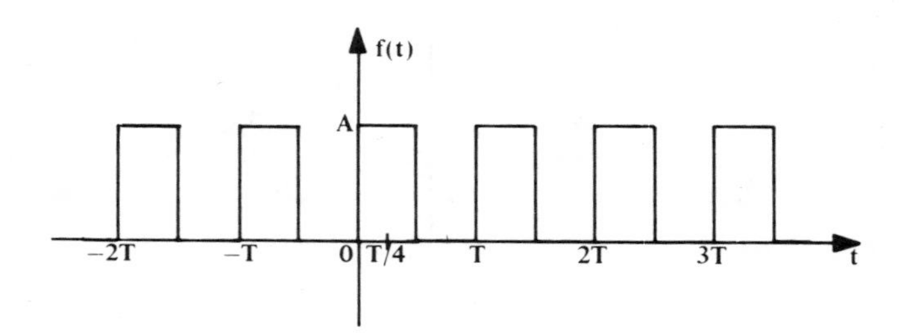

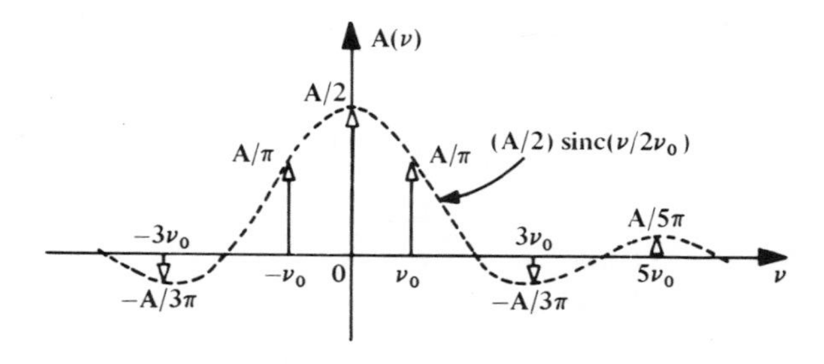

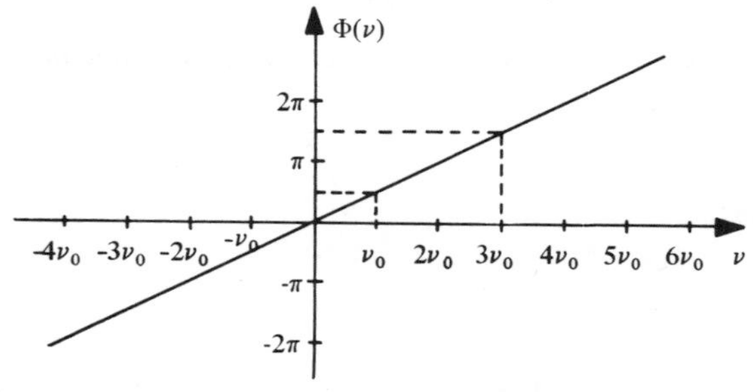

Figure 4-12 Amplitude and phase spectra of a shifted rectangle-wave function.

The amplitude and phase spectra, respectively, are given by

$$A(\nu)=\frac{A}{2}\operatorname{sinc}\left(\frac{\nu}{2\nu_0}\right)\sum_{n=-\infty}^{\infty}\delta(\nu-n\nu_0), \tag{4.63}$$

$$\Phi(\nu)=\frac{\pi\nu}{2\nu_0}, \tag{4.64}$$

and we see that, although the amplitude spectrum is the same as for the unshifted rectangle wave, the phase spectrum is no longer zero. *The phase of each harmonic component is shifted by an amount proportional to its frequency, and this causes all components to be shifted in time by the proper amount* ($T/4$ in this example). Again the delta functions of the amplitude spectrum sift out only those values of $\Phi(\nu)$ for which $\nu=\pm n\nu_0$, and thus the mth harmonic component is shifted in phase by $m\pi/2$ radians. Note that a phase shift of $\pi/2$ radians is required to displace the first-harmonic component one-quarter of a period in time, but a phase shift of $3\pi/2$ radians is required to displace the third harmonic by this amount, etc.

The series expansion for this shifted function may be written as

$$f(t)=\frac{A}{2}+\frac{2A}{\pi}\Big[\cos 2\pi\nu_0(t-T/4)-\tfrac{1}{3}\cos 2\pi(3\nu_0)(t-T/4)$$
$$+\tfrac{1}{5}\cos 2\pi(5\nu_0)(t-T/4)-\ldots\Big], \tag{4.65}$$

or we may again specify $f(t)$ by

$$f(t)=\operatorname{Re}\{h(t)\}, \tag{4.66}$$

where now

$$h(t)=\frac{A}{2}+\frac{2A}{\pi}\left\{\exp\left[j\left(2\pi\nu_0 t-\frac{\pi}{2}\right)\right]-\tfrac{1}{3}\exp\left(j\left[2\pi(3\nu_0)t-\frac{3\pi}{2}\right]\right)\right.$$
$$\left.+\tfrac{1}{5}\exp\left(j\left[2\pi(5\nu_0)t-\frac{5\pi}{2}\right]\right)-\ldots\right\}. \tag{4.67}$$

Figures 4-13 and 4-14 show the same graphical treatment for the shifted function as was previously shown for the unshifted function.

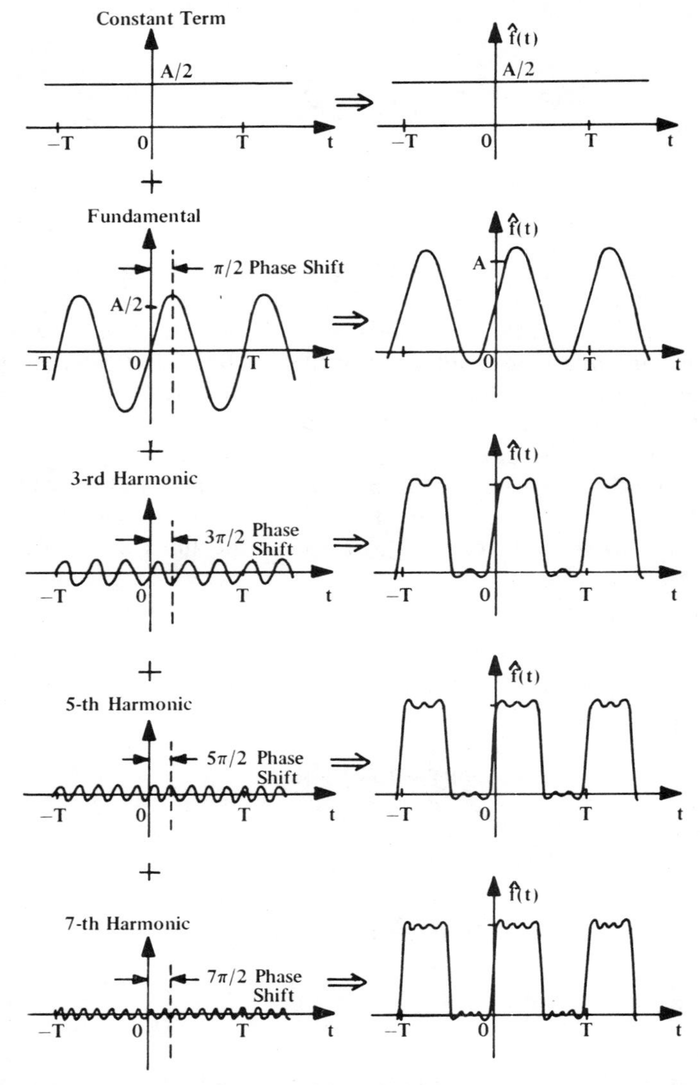

Figure 4-13 Fourier decomposition of shifted rectangle-wave functior

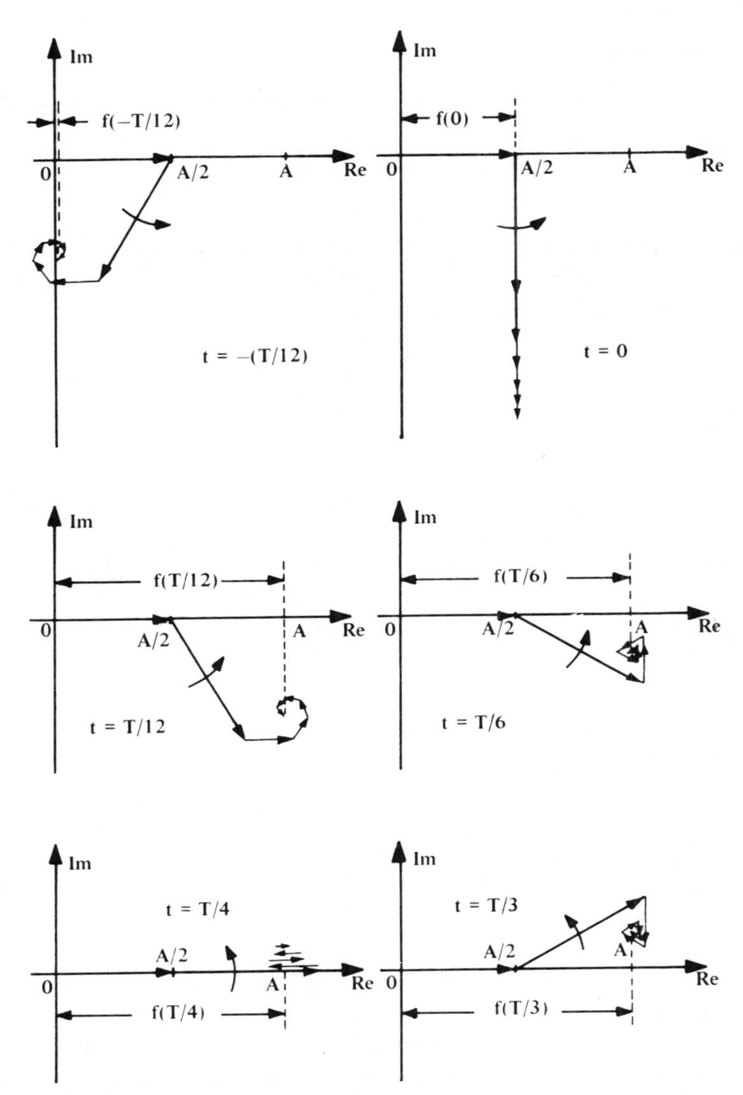

Figure 4-14 Phasor depiction of shifted rectangle-wave function.

Rectangular Pulse

For our first nonperiodic example, we choose the single rectangular pulse

$$g(t) = A\,\mathrm{rect}\left(\frac{t}{T/2}\right). \tag{4.68}$$

This is not only an extremely simple function, but it also provides a good comparison with the periodic rectangle wave example presented above. The spectrum of $g(t)$ is calculated by using Eq. (4.38), i.e.,

$$\begin{aligned}
G(\nu) &= \int_{-\infty}^{\infty} g(\alpha) e^{-j2\pi\nu\alpha} d\alpha \\
&= A \int_{-T/4}^{T/4} e^{-j2\pi\nu\alpha} d\alpha \\
&= \frac{A}{-j2\pi\nu} e^{-j2\pi\nu\alpha} \Big]_{-T/4}^{T/4} \\
&= \frac{A}{-j2\pi\nu} \left[e^{-j\pi\nu T/2} - e^{+j\pi\nu T/2} \right] \\
&= \frac{AT}{2} \frac{\sin\left(\frac{\pi\nu T}{2}\right)}{\left(\frac{\pi\nu T}{2}\right)} \\
&= \frac{AT}{2} \mathrm{sinc}\left(\frac{\nu T}{2}\right).
\end{aligned} \tag{4.69}$$

This function, along with $g(t)$, is graphed in Fig. 4-15, and we see that it is a continuous spectrum, having nonzero components in every finite frequency interval. Note that as the width of $g(t)$ is increased, the width of its spectrum decreases, and conversely, as T is decreased, $G(\nu)$ becomes wider. This agrees with the contention that slowly varying functions have narrow spectra, whereas the spectra of more rapidly varying functions extend to higher frequencies. It is interesting to compare this spectrum with that for the unshifted rectangle wave, which is a discrete spectrum consisting only of delta functions. These delta functions have areas governed by the same (except for a multiplicative constant) sinc function that specifies the spectrum of the pulse everywhere. In addition, the phase spectrum is zero in each case.

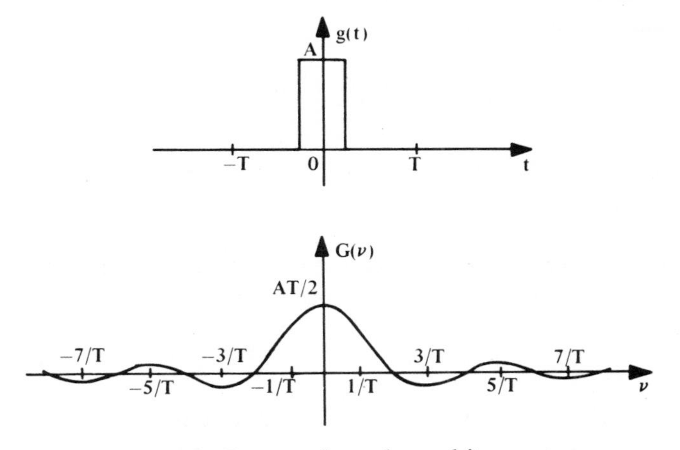

Figure 4-15 Rectangular pulse and its spectrum.

Now consider the function

$$g(t) = A\,\mathrm{rect}\left(\frac{t-t_0}{T/2}\right), \tag{4.70}$$

which is just a shifted version of the previous $g(t)$. The spectrum of this

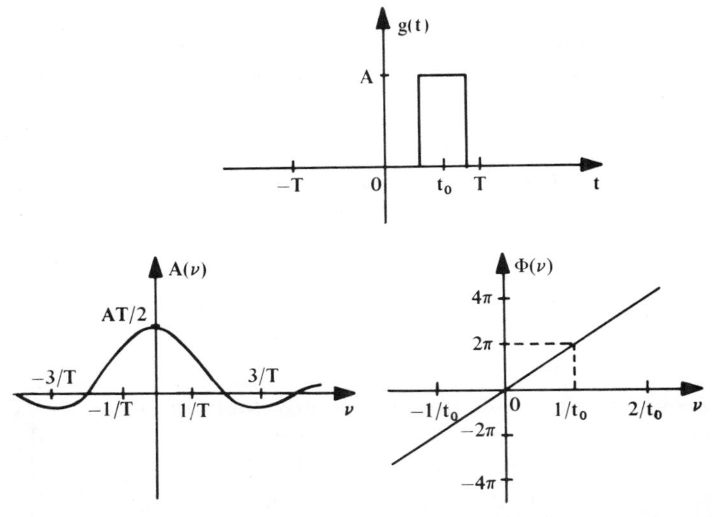

Figure 4-16 Amplitude and phase spectra of a shifted rectangular pulse.

new function is found to be

$$G(\nu)=\frac{AT}{2}\operatorname{sinc}\left(\frac{\nu T}{2}\right)e^{-j2\pi\nu t_0}, \tag{4.71}$$

from which it is clear that the amplitude spectrum is the same as for the unshifted pulse, whereas the phase spectrum is not. The phase of each exponential component is shifted by an amount proportional to its frequency, the constant of proportionality being $2\pi t_0$, which in turn causes the function to be displaced from the origin by an amount t_0. The function $g(t)$ and its spectrum are shown in Fig. 4-16.

4-5 SPECTRA OF TWO-DIMENSIONAL FUNCTIONS

We have seen how the spectrum of a one-dimensional function tells us something about the behavior of that function: a discrete spectrum is associated with a periodic function, whereas a nondiscrete spectrum is characteristic of a nonperiodic function; a narrow overall spectrum implies a relatively smooth function, but a broad overall spectrum is produced by a function having rapid oscillations; etc. We may not be able to determine the exact form of the function from a casual observation of its spectrum, but we can obtain a great deal of information about its behavior.

We now extend our notions about spectra to two-dimensional functions, and, because we will be dealing primarily with optical systems in later chapters, we choose to deal with functions of the spatial coordinates x and y. If a two-dimensional function $f(x,y)$ satisfies the Dirichlet conditions, it can be decomposed into a linear combination of complex exponentials according to

$$f(x,y)=\iint_{-\infty}^{\infty} F(\xi,\eta)\,e^{j2\pi(\xi x+\eta y)}\,d\xi\,d\eta. \tag{4.72}$$

Here $F(\xi,\eta)$ is given by

$$F(\xi,\eta)=\iint_{-\infty}^{\infty} f(\alpha,\beta)\,e^{-j2\pi(\alpha\xi+\beta\eta)}\,d\alpha\,d\beta, \tag{4.73}$$

and is known as the *two-dimensional Fourier transform* of $f(x,y)$, ξ and η being the spatial-frequency variables corresponding to the x- and y-directions, respectively. This function is also called the *complex spatial-frequency spectrum* of $f(x,y)$, or more simply, its spectrum, and it is the *spatial frequency domain* representation of $f(x,y)$.

The concept of a spatial frequency may be somewhat difficult to grasp at first (it was for the author), particularly for those who have previously associated the terms frequency and frequency spectra only with such time-varying phenomena as radio waves, sound waves, etc. However, once the initial confusion is overcome, you will wonder why such a simple thing could cause any trouble at all. The temporal frequency of a time-varying sine wave describes the number of oscillations made by the function per unit time; for a function that varies sinusoidally with some spatial coordinate, the spatial frequency associated with the function in that same direction indicates the number of repetitions the function makes per unit distance. It's really that simple!

To illustrate, we shall determine the spectra of some spatially varying functions. Consider a large plowed field, with furrows running north and south as shown in Fig. 4-17(a). We erect a coordinate system with the y-axis parallel to the furrows and the x-axis perpendicular to them, and for simplicity we assume that the field is level and the profile of the furrows is sinusoidal. If we now designate the height of the surface above some arbitrary reference surface by $h(x,y)$, we may write

$$h(x,y)=A+B\cos(2\pi\xi_0 x), \tag{4.74}$$

where A is the average height above the reference surface, $2B$ is the depth

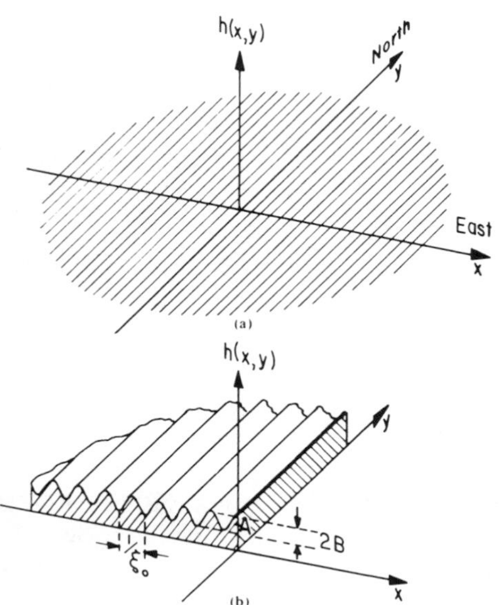

Figure 4-17 Example of spatially varying function. (*a*) Plowed field. (*b*) Idealized surface-height variations of plowed field.

of the furrows and ξ_0 is the number of furrows per unit distance. Note that $h(x,y)$ can be put in the form

$$h(x,y)=f(x)\,g(y), \tag{4.75}$$

where

$$\begin{aligned} f(x)&=A+B\cos(2\pi\xi_0 x)\\ g(y)&=1, \end{aligned} \tag{4.76}$$

and we see that it is separable in x and y. Profiles of $h(x,y)$ along each axis are shown in Fig. 4-17(b).

The spatial-frequency spectrum of $h(x,y)$ is given by

$$\begin{aligned} H(\xi,\eta)&=\iint\limits_{-\infty}^{\infty} h(\alpha,\beta)\,e^{-j2\pi(\xi\alpha+\eta\beta)}d\alpha\,d\beta\\ &=\int_{-\infty}^{\infty} f(\alpha)\,e^{-j2\pi\xi\alpha}d\alpha\int_{-\infty}^{\infty} g(\beta)\,e^{-j2\pi\eta\beta}d\beta\\ &=F(\xi)G(\eta), \end{aligned} \tag{4.77}$$

where $F(\xi)$ and $G(\eta)$ are the one-dimensional transforms of $f(x)$ and $g(y)$, respectively, and it is apparent that the transform of a separable function is itself separable. Thus, with

$$\begin{aligned} F(\xi)&=A\delta(\xi)+\frac{B}{2\xi_0}\delta\delta\left(\frac{\xi}{\xi_0}\right)\\ G(\eta)&=\delta(\eta), \end{aligned} \tag{4.78}$$

we obtain

$$\begin{aligned} H(\xi,\eta)&=\left[A\delta(\xi)+\frac{B}{2\xi_0}\delta\delta\left(\frac{\xi}{\xi_0}\right)\right]\delta(\eta)\\ &=A\delta(\xi,\eta)+\frac{B}{2}\left[\delta(\xi-\xi_0,\eta)+\delta(\xi+\xi_0,\eta)\right], \end{aligned} \tag{4.79}$$

and we see that the two-dimensional spectrum of $h(x,y)$ consists of a delta function at the origin, which indicates the average height above the reference surface, and two others located at $\xi=\pm\xi_0$ and $\eta=0$, as shown in Fig. 4-18. The latter two tell us that the height of the field varies

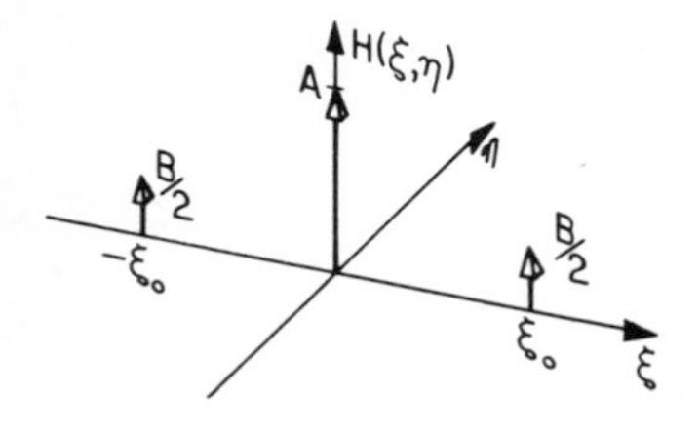

Figure 4-18 Spatial-frequency spectrum of plowed field surface-height function.

sinusoidally in the x-direction, with a frequency of ξ_0, and that it does not have any variations in the y-direction. Of course, this result is an idealization because of our assumptions concerning the profile of the furrows and the fact that we neglected the finite size of the field. Nevertheless, this example serves nicely to demonstrate the basic concepts we wish to consider at this time.

For our last example, let us again consider the plowed field, but with the furrows now running in a northeasterly-southwesterly direction as

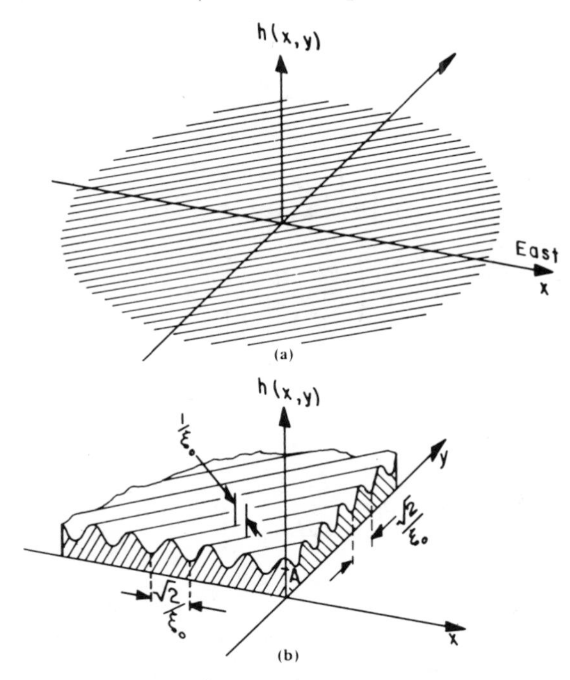

Figure 4-19 Plowed field with furrow direction rotated 45°. (*a*) Plowed field. (*b*) Idealized surface-height variations of plowed field.

illustrated in Fig. 4-19(a). Therefore,

$$h(x,y)=A+B\cos\left[2\pi\left(\frac{\xi_0}{\sqrt{2}}\right)(x-y)\right], \tag{4.80}$$

where ξ_0 has the same value as before. Note that $h(x,y)$ is no longer separable, and that the profiles along both axes are now sinusoidal as depicted in Fig. 4-19(b). To calculate the spectrum of this function, we use Euler's formula for the cosine function, which allows us to write $h(x,y)$ as the sum of separable functions. Thus

$$\begin{aligned}
H(\xi,\eta)=\iint_{-\infty}^{\infty}\Bigg\{&A+\frac{B}{2}\exp\left[-j2\pi\left(\frac{\xi_0}{\sqrt{2}}\right)\alpha\right]\exp\left[j2\pi\left(\frac{\xi_0}{\sqrt{2}}\right)\beta\right]\\
&+\frac{B}{2}\exp\left[j2\pi\left(\frac{\xi_0}{\sqrt{2}}\right)\alpha\right]\exp\left[-j2\pi\left(\frac{\xi_0}{\sqrt{2}}\right)\beta\right]\Bigg\}\\
&\times\exp\left[-j2\pi(\xi\alpha+\eta\beta)\right]d\alpha\, d\beta\\
=A\delta(\xi)\delta(\eta)+\frac{B}{2}&\left[\delta\left(\xi+\frac{\xi_0}{\sqrt{2}}\right)\delta\left(\eta-\frac{\xi_0}{\sqrt{2}}\right)\right.\\
&\left.+\delta\left(\xi-\frac{\xi_0}{\sqrt{2}}\right)\delta\left(\eta+\frac{\xi_0}{\sqrt{2}}\right)\right]\\
=A\delta(\xi,\eta)+\frac{B}{2}&\left[\delta\left(\xi+\frac{\xi_0}{\sqrt{2}},\eta-\frac{\xi_0}{\sqrt{2}}\right)\right.\\
&\left.+\delta\left(\xi-\frac{\xi_0}{\sqrt{2}},\eta+\frac{\xi_0}{\sqrt{2}}\right)\right],
\end{aligned} \tag{4.81}$$

and we see that this spectrum is the same as the previous one except that it has been rotated 45° in a clockwise direction. This is shown in Fig. 4-20. Once again the delta function at the origin is related to the average height

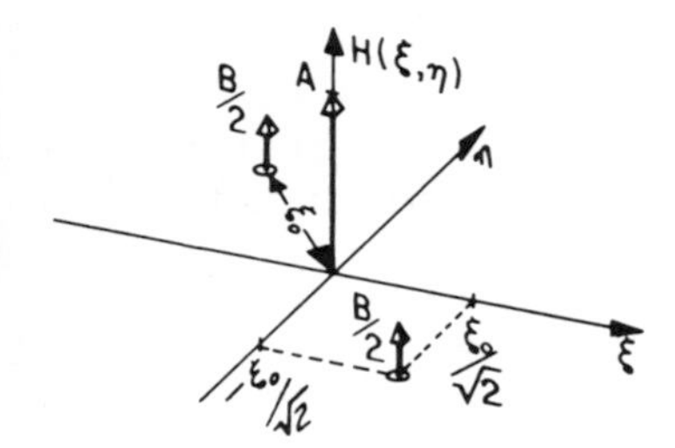

Figure 4-20 Spatial-frequency spectrum of rotated surface height function.

of the field above the reference surface and the other two delta functions indicate a sinusoidal variation of the surface height. However, these latter delta functions are now situated along a line in the frequency plane that corresponds to the direction of northwest-southeast in the field, and, because they are still displaced from the origin by an amount ξ_0, the sinusoidal oscillations are in the northwesterly-southeasterly direction and have a frequency of ξ_0. It should be clear that there are no variations in the direction parallel to the furrows, just as in the first example, but the coordinates $\xi = \mp\xi_0/\sqrt{2}$ and $\eta = \pm\xi_0/\sqrt{2}$ of the delta functions reveal that there are sinusoidal variations, at a reduced frequency of $\xi_0/\sqrt{2}$, in both the x-direction and the y-direction. This agrees with our physical understanding of the problem, because if we were to drive a tractor across the field in an easterly direction, striking the furrows at an angle of 45°, we know that the undulations would occur less frequently than if we drove in a direction perpendicular to the furrows.

More complicated functions have more complicated spectra, but we shall postpone discussions of such functions until a later chapter; the examples presented here should do for the time being.

REFERENCES

4-1 C. R. Wylie, Jr., *Advanced Engineering Mathematics*, McGraw-Hill, New York, 1951.

4-2 H. H. Harmuth, "Applications of Walsh Functions in Communications," *IEEE Spectrum*, **6**(11): 82 (1969).

4-3 W. K. Pratt, J. Kane, and H. C. Andrews, "Hadamard Transform Image Coding," *Proc. IEEE*, **57**(1): 58 (1969).

4-4 H. C. Andrews, *Computer Techniques in Image Processing*, Academic, New York, 1970.

4-5 R. Bracewell, *The Fourier Transform and Its Applications*, McGraw-Hill, New York, 1965, p. 211.

PROBLEMS

Sketch the function specified for each part of the following problems. Then calculate and sketch the amplitude and phase spectra of each.

4-1. Do this problem for the function $f(t)=\sum_{n=-\infty}^{\infty}\text{rect}(t-2n)$.

a. $f(t)$.
b. $g(t)=f(t-0.25)$.
c. $h(t)=1-f(t)=f(t-1)$.
d. $p(t)=2f(t)-1$.

4-2. Do this problem for the function $f(t)=\sum_{n=-\infty}^{\infty}\text{rect}(t-5n)$.

a. $f(t)$.
b. $g(t)=f(t-2)$.
c. $h(t)=1-f(t-2.5)$.

4-3. Do this problem for the function $f(t)=\sum_{n=-\infty}^{\infty}\text{tri}(t-2n)$.

a. $f(t)$.
b. $g(t)=f(t-0.5)$.
c. $h(t)=1-f(t)=f(t-1)$.

4-4. Do this problem for the function $f(t)=\sum_{n=-\infty}^{\infty}\text{tri}(t-3n)$.

a. $f(t)$.
b. $g(t)=f(t-1)$.
c. $h(t)=1-f(t-1.5)$.

4-5. Do this problem for the rectangle function $f(t)=\text{rect}(0.5t)$.

a. $f(t)$.
b. $g(t)=f(t-1)$.
c. $h(t)=f(t+2)+f(t-2)$.

4-6. Do this problem for the triangle function $p(t)=\text{tri}(t)$.

a. $p(t)$.
b. $r(t)=p(t+1)$.
c. $s(t)=p(t)+p(t-4)$.

4-7. For this problem, use $f(t)$ and $p(t)$ from Problems 4-5 and 4-6.

a. $u(t)=0.5f(t)+p(t)$.
b. $v(t)=f(t)-p(t)$.
c. $w(t)=v(t)\,\text{sgn}(t)$.

CHAPTER 5

MATHEMATICAL OPERATORS AND PHYSICAL SYSTEMS

For our purposes, a physical system may be thought of as any device that exhibits some sort of response when a stimulus is applied. The stimulus is often called the *input* to the system, whereas the response is known as its *output*. In general, a system may have multiple input and output terminals, and the number of each need not be the same, but we shall deal primarily with systems having a single input terminal and a single output terminal. In addition, we shall not concern ourselves particularly with the internal workings of systems, but only with their terminal properties, i.e., their input-output relationships.

A record player is an example of a combination mechanical-electrical-acoustical system, the input being the mechanical vibrations of the stylus in the grooves of the record and the output the audio signal produced by the loudspeakers. An aneroid barometer is an example of a simple mechanical system; here the input is the pressure of the surrounding air and the output is the position of the pointer on the dial. A camera is an example of an optical system, for which the irradiance distribution of some object is the input and the density of the resulting photographic negative is the output. We could mention many other kinds of systems, but the three examples above should suffice to illustrate what we mean by the word "system."

5-1 SYSTEM REPRESENTATION BY MATHEMATICAL OPERATORS

In the analysis of a physical system, it is necessary to find a model that appropriately describes the behavior of the system in a mathematical sense. Such a model will always be an idealization, for fairly obvious reasons, but it can still be a very useful analysis tool as long as its limitations are understood and taken into account. The modeling of systems is often discussed in terms of *mathematical operators*, and this is the approach we shall take (see Ref. 5-1).

Consider the set of functions $\{f_1(x), f_2(x), \ldots, f_n(x)\}$. According to some rule, which we leave unspecified for the time being, let us assign to every element of this set the corresponding element of a second set $\{g_1(x), g_2(x), \ldots, g_n(x)\}$. In other words, for $i=1,2,\ldots,n$, let us assign the element $g_1(x)$ to the element $f_1(x)$. If the operator $\mathcal{S}\{\ \}$ is used to denote the rule by which this assignment is made, the process can be represented by the expression

$$\mathcal{S}\{f_i(x)\}=g_i(x), \qquad i=1,2,\ldots,n, \tag{5.1}$$

and we say that the first set of functions is mapped, or transformed, into the second set by the operator $\mathcal{S}\{\ \}$. Such an operator is frequently used to describe the behavior of a system by considering its input and output signals to be elements of two sets of functions as discussed above. Thus, the effect of a system is to map a set of input signals into a set of output signals. The rule governing this transformation might be determined by any of a number of things, including a differential equation, an integral equation, a graph, a table of numbers, etc. For example, the differential equation

$$\frac{d^2}{dx^2} g_i(x)+Kg_i(x)=f_i(x) \tag{5.2}$$

might be used as the model for a specific system, the forcing function $f_i(x)$ representing the input to the system and the solution $g_i(x)$ representing its output. Pictorially, the operator representation of a system is shown in Fig. 5-1.

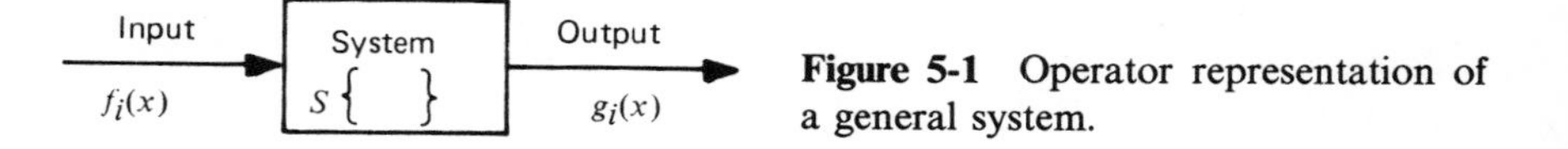

Figure 5-1 Operator representation of a general system.

In general, the elements of the input set need not be included in the output set and vice versa. However, for most of our work the input and output sets will be comprised exclusively of functions possessing Fourier transforms, either one-dimensional or two-dimensional as the case may be, and as a result every element found in one set will also be found in the other. In addition, we shall deal chiefly with systems for which each input is mapped into a unique output, although these systems will not necessarily yield different outputs for different inputs. As such, they are frequently called many-to-one systems.

5-2 SOME IMPORTANT TYPES OF SYSTEMS

It is virtually impossible to make a complete analysis of a general system, and it is only when the system exhibits certain desirable characteristics that much headway can be made. The most important of these properties will now be discussed.

Linear Systems

Consider a system characterized by the operator $\mathcal{S}\{\ \}$, such that for two arbitrary input signals $f_1(x)$ and $f_2(x)$ we have

$$\mathcal{S}\{f_1(x)\} = g_1(x),$$

$$\mathcal{S}\{f_2(x)\} = g_2(x). \tag{5.3}$$

Then for two arbitrary complex constants a_1 and a_2, the system is said to be *linear* if, for an input equal to the sum $[a_1 f_1(x) + a_2 f_2(x)]$, the output is given by

$$\begin{aligned}\mathcal{S}\{a_1 f_1(x) + a_2 f_2(x)\} &= \mathcal{S}\{a_1 f_1(x)\} + \mathcal{S}\{a_2 f_2(x)\} \\ &= a_1 \mathcal{S}\{f_1(x)\} + a_2 \mathcal{S}\{f_2(x)\} \\ &= a_1 g_1(x) + a_2 g_2(x).\end{aligned} \tag{5.4}$$

Thus, the *principle of superposition* applies for linear systems in that the overall response to a linear combination of stimuli is simply the same linear combination of the individual responses. This result, which is illustrated in Fig. 5-2, is of great importance in the analysis of linear systems because the response to a complicated input can be found by first decomposing this input into a linear combination of elementary functions,

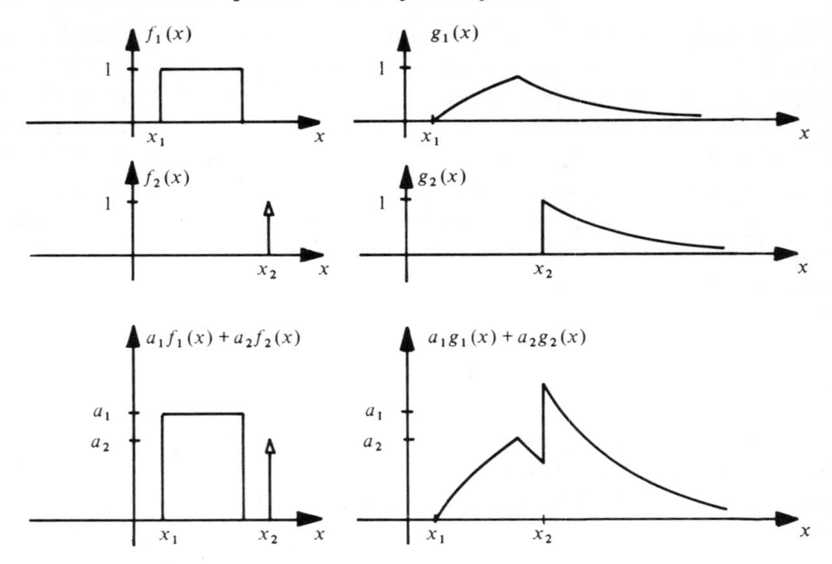

Figure 5-2 The superposition principle applied to a linear system.

finding the response to each elementary function, and then taking the same linear combination of these elementary responses. Physically, linearity implies that the behavior of the system is independent of the magnitude of the input. This, of course, is an idealization, but it is often a good one over a limited range of the input.

The principle of superposition does not apply for nonlinear systems, and the analysis of such systems is in general much more difficult. Suppose, for example, that a system is to be modeled by a differential equation; it will be a linear differential equation if the system is linear, and a nonlinear differential equation if the system is nonlinear. The relative difficulty of these two cases will be apparent to those familiar with the solutions of differential equations.

Note that our definition of linearity does not require the output to have the same shape (or form) as the input, i.e., we do not require that

$$g(x)=af(x), \tag{5.5}$$

where a is an arbitrary constant. (We have now dropped the subscripts for compactness.) Such a system is linear, and in fact is a special type of what are called linear, memoryless systems, but this behavior is not characteristic of all linear systems. Thus the definition of Eq. (5.4) includes systems other than the simple amplifiers and attenuators that Eq. (5.5) describes.

Shift-Invariant Systems

A system is said to be *shift invariant* (fixed, stationary, time invariant, space invariant, isoplanatic) if the only effect caused by a shift in the position of the input is an equal shift in the position of the output. In other words, if a system is shift invariant, and if

$$\mathcal{S}\{f(x)\}=g(x), \tag{5.6}$$

then

$$\mathcal{S}\{f(x-x_0)\}=g(x-x_0), \tag{5.7}$$

where x_0 is a real constant. Thus the magnitude and shape of the output are unchanged as the input is shifted along the x-axis; only the location of the output is changed. This is illustrated in Fig. 5-3.

Shift invariance implies that the *behavior of the system* is not a function of the independent variable. This, too, is an idealization when applied to physical systems, but again it may be a very good one over some finite range. To illustrate, most electronic systems are not truly shift invariant (time invariant in this case) because the aging or failure of various components causes their performance to change with time. However, over periods of a few hours or a few days the performance of such systems will usually be stable enough for them to be considered shift invariant.

Systems that are both *linear and shift invariant* are particularly easy to analyze, and these are the kinds of systems we shall be concerned with in most of our work. We shall call them *LSI systems* for short, and denote them generically by the operator $\mathcal{L}\{\ \}$. Thus, if $\mathcal{L}\{f_i(x)\}=g_i(x)$, then by combining Eqs. (5.4) and (5.7) we find that

$$\mathcal{L}\{a_1f_1(x-x_1)+a_2f_2(x-x_2)\}=a_1g(x-x_1)+a_2g(x-x_2), \tag{5.8}$$

where x_1 and x_2 are real constants.

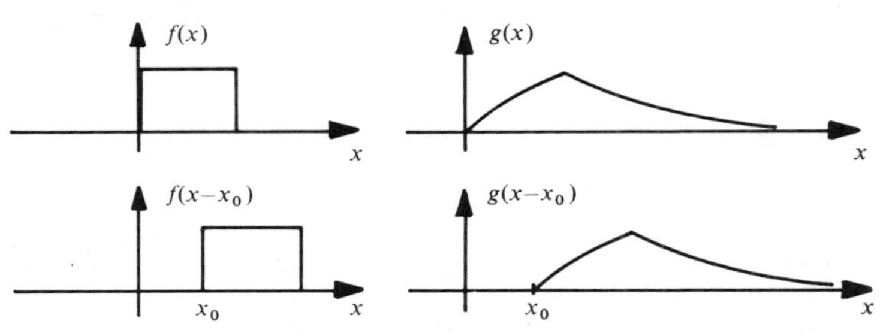

Figure 5-3 Input-output relationships for a shift-invariant system.

Causal Systems

Many authors of electrical engineering texts state that a system is *causal* if the value of the output at any point does not depend on future values of the input. Many also state that only causal systems can exist in the real world. However, it is implicit in these statements that they apply only to systems with time-varying inputs and outputs. Therefore, since we are interested not only in systems with time-varying inputs and outputs, but also in those for which the input and output signals are functions of spatial coordinates, we must question the general application of these notions about causality.

Consider a system for which the input and output are time functions, i.e., the simple R-C circuit of Fig. 3-12. We choose the input voltage to be a delta function located at the time origin. With the assumption that there have been no previous inputs and that no energy is stored in the capacitor, the output voltage will behave as shown in Fig. 5-4. At all times prior to $t=0$, it is exactly zero, and in no way is it influenced by the delta function input that has yet to occur. At $t=0$, when the input occurs, the output voltage jumps abruptly to a value of A/RC; it then begins to decay toward zero as time goes on. Thus, at any time t, the value of the output is completely determined by past values of the input and does not depend on the future behavior of the input. This makes sense physically, because "we all know" that a system cannot respond to a stimulus before the stimulus is applied.

Now, in contrast, consider the optical system of Fig. 3-14. With the assumption that the slit is located at the point $x=0$, and that it is sufficiently narrow to be considered a line source, i.e., a delta function in one dimension, the output behaves in a fashion similar to that shown in Fig. 5-5. From this figure it is apparent that even though the input is nonzero only at the origin, it influences the output for both positive and negative values of x. At first it may seem strange that such a system can exhibit a response "before" the stimulus occurs; however, the mystery vanishes when we realize that the use of the term "before" is improper here. The input and output are no longer functions of time, and it is now

Figure 5-4 Input-output relationships for a causal system.

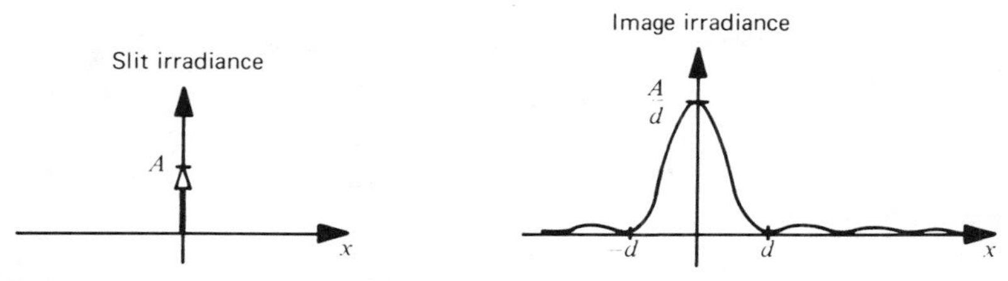

Figure 5-5 Input-output relationships for a system with noncausal behavior.

inappropriate to talk about their past or future values with respect to some observation point, or to use the terms "before" and "after" with reference to various points along the axis of the independent variable. In other words, we must now discard the somewhat natural association of the values of the independent variable with a temporal sequence of events. Thus, according to a mathematical definition of causality rather than a physical one, the optical system under discussion is a *noncausal system*, and the notion that only causal systems can exist in the real world is seen to be misleading.

Closely associated with the concept of causality is the consideration of *initial conditions*, i.e., the conditions of energy storage within a system when the input is applied. Although such a consideration is quite important in general, we shall deal exclusively with systems for which energy storage is of no concern. Therefore, we omit further discussion of initial conditions and how they are handled; the details are available elsewhere for those who are interested (e.g., Ref. 5-2).

Systems with Memory

A system is said to have *memory* if, instead of being a function of the input, the output is a *functional* of the input. Such a system is also referred to as a dynamic system. Conversely, a *memoryless* (zero-memory, instantaneous) system is one for which the output is a *function* of the input. To illustrate the difference, we consider an example of each kind of system. For a system with memory, the output is related to the input by the operator expression

$$g(x)=\mathcal{S}\{f(x)\}, \tag{5.9}$$

and its value at any point $x=x_1$ is dependent on the values of $f(x)$ over a wide range of x, possibly everywhere. The operation denoted by $\mathcal{S}\{\ \}$ must first be performed, and then the result evaluated at the point $x=x_1$,

i.e.,

$$g(x_1)=\mathcal{S}\{f(x)\}|_{x=x_1}. \tag{5.10}$$

Suppose, for example, that $\mathcal{S}\{\ \}=d/dx$ and $f(x)=\exp\{-\pi x^2\}$. The output is given by

$$g(x)=-2\pi x\,e^{-\pi x^2} \tag{5.11}$$

and at the point $x=x_1$,

$$g(x_1)=-2\pi x_1 e^{-\pi x_1^2}. \tag{5.12}$$

We obviously would obtain an incorrect result by first substituting x_1 in $f(x)$ and then taking the derivative of $f(x_1)$.

On the other hand, the output of a memoryless system may be written as

$$g(x)=u[f(x);x], \tag{5.13}$$

where $u[\ ]$ denotes a function. The output of this system at the point $x=x_1$ depends only on the value of the input at that point, and possibly on x_1 itself, but it does not "remember" the past behavior of the input nor "foresee" its future. Thus

$$g(x_1)=u[f(x_1);x_1]. \tag{5.14}$$

To demonstrate, let us consider the function

$$u[f(x);x]=a(x)f(x)$$

where $a(x)$ is some arbitrary function. Thus the output becomes

$$g(x)=a(x)f(x), \tag{5.15}$$

and at the point $x=x_1$ it has the value

$$g(x_1)=a(x_1)f(x_1). \tag{5.16}$$

From the above discussion, we see that a memoryless system simply maps the value of the input at a single point into the value of the output at that point, whereas a system with memory maps the value of the input at many points into the value of the output at a single point. (Basically, this is the difference between a function and a functional.)

There are other considerations related to the classification of systems, such as stability, controllability, etc., but they are of little importance in our work here. The classifications discussed above are by far the most important for us, and unless otherwise stated, we will be concerned only with linear, shift-invariant, noncausal systems with memory. In addition, they will be many-to-one systems with a single input terminal and a single output terminal.

5-3 THE IMPULSE RESPONSE

When the input to a system is a single delta function, the output is called the *impulse response* of the system (also known as a Green's function). For a general system, the impulse response depends on the point at which the input delta function is applied; i.e., for an input impulse located at $x=x_0$, the impulse response is denoted by $h(x;x_0)$. In other words, if the system is characterized by

$$\mathcal{S}\{f(x)\}=g(x), \tag{5.17}$$

we denote the output by $g(x)=h(x;x_0)$ when $f(x)=\delta(x-x_0)$. Thus

$$\mathcal{S}\{\delta(x-x_0)\}=h(x;x_0). \tag{5.18}$$

Now let us consider an LSI system, characterized by the operator $\mathcal{L}\{\ \}$. We see from Eq. (5.18) that for an impulsive input applied at the origin, the output is given by

$$\mathcal{L}\{\delta(x)\}=h(x;0), \tag{5.19}$$

whereas for one applied at $x=x_0$ it is described by

$$\mathcal{L}\{\delta(x-x_0)\}=h(x;x_0). \tag{5.20}$$

But because the system is shift invariant, the second of these responses must be identical to the first except that it is shifted by an amount x_0 along the x-axis as required by Eq. (5.7). Hence,

$$\mathcal{L}\{\delta(x-x_0)\}=h(x-x_0;0), \tag{5.21}$$

and by combining this result with Eq. (5.20) we obtain

$$h(x;x_0)=h(x-x_0;0). \tag{5.22}$$

It is apparent from this expression that the impulse response of a shift-invariant system depends only on the separation of the observation point x and the point x_0 at which the input is applied, and not on the value of either by itself. Thus we change our notation for the impulse response of such a system to the more concise form

$$\mathcal{L}\{\delta(x-x_0)\}=h(x-x_0). \tag{5.23}$$

As a result, when the input is applied at the origin, the impulse response reduces simply to

$$\mathcal{L}\{\delta(x)\}=h(x). \tag{5.24}$$

As we shall see in Chapter 6, LSI systems are completely characterized by their impulse responses. This is a curious but important property of these systems, and one that we shall use a great deal in our studies. Figures 5-4 and 5-5 show some typical impulse responses. It is interesting to note that the response of an LSI system to a delta function input is not likely to look much like a delta function itself.

5-4 COMPLEX EXPONENTIALS: EIGENFUNCTIONS OF LINEAR SHIFT-INVARIANT SYSTEMS

As mentioned in Chapter 4, when the input to a system is an eigenfunction of the system, the output is simply the product of the input and a complex constant of proportionality. To express this mathematically, we let $\psi(x;\xi_0)$ be an eigenfunction of the LSI system $\mathcal{L}\{\ \}$, where ξ_0 is an arbitrary real constant. Then if $\psi(x;\xi_0)$ is the input to the system, the output is given by

$$\mathcal{L}\{\psi(x;\xi_0)\}=H(\xi_0)\psi(x;\xi_0). \tag{5.25}$$

The complex-valued constant of proportionality $H(\xi_0)$ is called the *eigenvalue* associated with the *eigenfunction* $\psi(x;\xi_0)$, and it depends on the value of the constant ξ_0. If we now write

$$H(\xi_0)=A(\xi_0)e^{-j\Phi(\xi_0)}, \tag{5.26}$$

where $A(\xi_0)$ describes the attenuation (or gain) of $H(\xi_0)$, and $\Phi(\xi_0)$ its phase, Eq. (5.25) becomes

$$\mathcal{L}\{\psi(x;\xi_0)\}=A(\xi_0)e^{-j\Phi(\xi_0)}\psi(x;\xi_0), \tag{5.27}$$

which leads to an extremely important conclusion: in passing through the system, an eigenfunction may be attenuated (or amplified) and its phase may be shifted, but it remains unchanged in form.

The complex exponential $\exp\{j2\pi\xi_0 x\}$ is an eigenfunction of any LSI system, as we shall now show. With $\exp\{j2\pi\xi_0 x\}$ the input to such a system, the output may be written as

$$\mathcal{L}\{e^{j2\pi\xi_0 x}\} = g(x;\xi_0). \tag{5.28}$$

Now suppose that a shifted version of this exponential is applied to the system, i.e., let the input be $\exp\{j2\pi\xi_0(x-x_0)\}$, where x_0 is also a real constant. Then by linearity,

$$\begin{aligned}\mathcal{L}\{e^{j2\pi\xi_0(x-x_0)}\} &= \mathcal{L}\{e^{-j2\pi\xi_0 x_0}e^{j2\pi\xi_0 x}\}\\ &= e^{-j2\pi\xi_0 x_0}\mathcal{L}\{e^{j2\pi\xi_0 x}\}\\ &= e^{-j2\pi\xi_0 x_0}g(x;\xi_0).\end{aligned} \tag{5.29}$$

But by shift invariance,

$$\mathcal{L}\{e^{j2\pi\xi_0(x-x_0)}\} = g(x-x_0;\xi_0), \tag{5.30}$$

and by combining Eqs. (5.29) and (5.30) we obtain

$$g(x-x_0;\xi_0) = e^{-j2\pi\xi_0 x_0}g(x;\xi_0). \tag{5.31}$$

From this relationship it can be seen that $g(x;\xi_0)$ must be of the form

$$g(x;\xi_0) = H(\xi_0)e^{j2\pi\xi_0 x}, \tag{5.32}$$

where $H(\xi_0)$ represents some complex-valued constant. Thus

$$\mathcal{L}\{e^{j2\pi\xi_0 x}\} = H(\xi_0)e^{j2\pi\xi_0 x}, \tag{5.33}$$

and we see that the complex exponential $\exp\{j2\pi\xi_0 x\}$ is indeed an eigenfunction of an LSI system, with $H(\xi_0)$ the associated eigenvalue. It is interesting to note that the value of $H(\xi_0)$ can be found by setting $x=0$ in Eq. (5.32), i.e.,

$$H(\xi_0) = g(0;\xi_0). \tag{5.34}$$

This curious result reveals that $H(\xi_0)$ has a value equal to that of the output at the origin, assuming, of course, that the input is $\exp\{j2\pi\xi_0 x\}$.

In the above development, the eigenvalue $H(\xi_0)$ obviously is a function of ξ_0, the frequency of the input eigenfunction. But since the value of ξ_0 is completely arbitrary, we might just as well leave it unspecified and describe the eigenvalue as a function of the general frequency parameter ξ. Denoted in this fashion, $H(\xi)$ is often called the *transfer function* (frequency response) of the system and, as we shall see later, it is just the Fourier transform of the impulse response $h(x)$. It describes the attenuation and phase shift experienced by an exponential eigenfunction in passing through the system, and it is a function of the frequency of this eigenfunction.

We have seen what happens when the input to *any* LSI system is a complex exponential of the form $\exp\{j2\pi\xi_0 x\}$. Now let us consider a special LSI system: one having a *real-valued impulse response*. Such a system transforms a real-valued input into a real-valued output, and is therefore the kind of system most commonly encountered in practice. The *transfer function* $H(\xi)$ of this kind of system is *hermitian*, i.e., the real part of $H(\xi)$ is even and the imaginary part is odd, a property which can be expressed by

$$H(\xi) = H^*(-\xi). \tag{5.35}$$

We already know what the output of this system will be when the input is $\exp\{j2\pi\xi_0 x\}$; now let us find the output when the input is a cosine (or sine) function. For an input of $\cos(2\pi\xi_0 x)$, the output is

$$\begin{aligned}\mathcal{L}\{\cos(2\pi\xi_0 x)\} &= \mathcal{L}\{\tfrac{1}{2}[e^{j2\pi\xi_0 x} + e^{-j2\pi\xi_0 x}]\}\\ &= \tfrac{1}{2}\mathcal{L}\{e^{j2\pi\xi_0 x}\} + \tfrac{1}{2}\mathcal{L}\{e^{j2\pi(-\xi_0)x}\}\\ &= \tfrac{1}{2}H(\xi_0)e^{j2\pi\xi_0 x} + \tfrac{1}{2}H(-\xi_0)e^{j2\pi(-\xi_0)x},\end{aligned} \tag{5.36}$$

but since $H(-\xi_0) = H^*(\xi_0)$, we may write

$$\mathcal{L}\{\cos(2\pi\xi_0 x)\} = \tfrac{1}{2}[H(\xi_0)e^{j2\pi\xi_0 x}] + \tfrac{1}{2}[H(\xi_0)e^{j2\pi\xi_0 x}]^*, \tag{5.37}$$

and from Eq. (2.52) we obtain

$$\mathcal{L}\{\cos(2\pi\xi_0 x)\} = \text{Re}\{H(\xi_0)e^{j2\pi\xi_0 x}\}. \tag{5.38}$$

Then with $H(\xi_0) = A(\xi_0)\exp\{-j\Phi(\xi_0)\}$,

$$\mathcal{L}\{\cos(2\pi\xi_0 x)\} = A(\xi_0)\cos[2\pi\xi_0 x - \Phi(\xi_0)], \tag{5.39}$$

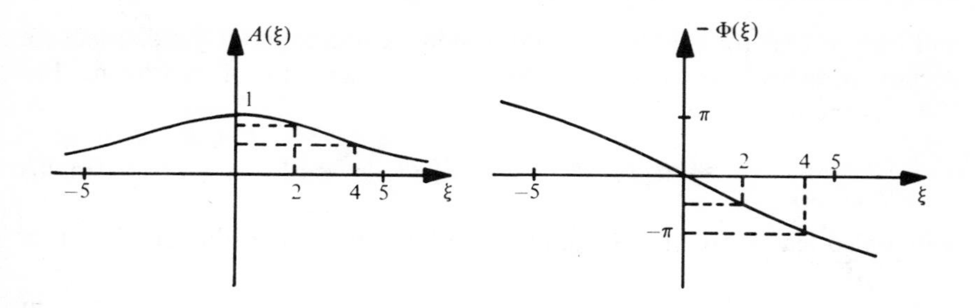

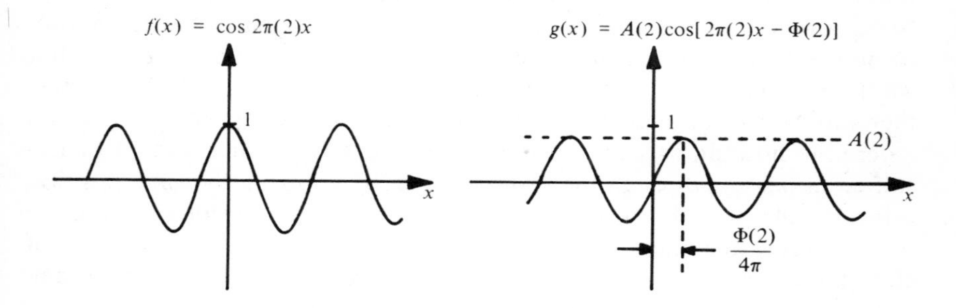

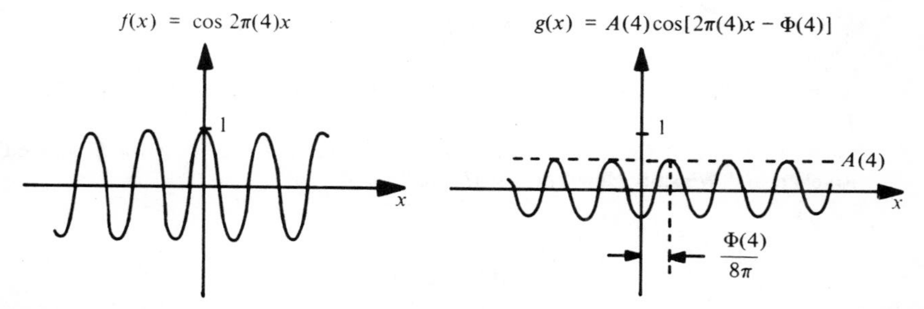

Figure 5-6 Effects of the amplitude and phase transfer functions on cosinusoidal input signals.

and we see that for this special LSI system, a cosine input yields a cosine output, possibly attenuated and shifted. Another way of expressing Eq. (5.38) is the following:

$$\mathcal{L}\{\text{Re}\{e^{j2\pi\xi_0 x}\}\} = \text{Re}\{\mathcal{L}\{e^{j2\pi\xi_0 x}\}\}, \tag{5.40}$$

which also leads directly to Eq. (5.39). These results are illustrated in Fig. 5-6.

By now you should be starting to understand the reasons for our discussions of such topics as Fourier decomposition, superposition, eigenfunctions, etc. To strengthen this understanding, consider the problem of finding the output of an LSI system for an arbitrary input signal. It should be apparent that we can decompose this input signal into its Fourier components, which are complex exponentials of the form $\exp\{j2\pi\xi x\}$. But we know that these exponentials are eigenfunctions of LSI systems; therefore, if we also know the transfer function of the system, we can determine how much each Fourier component is attenuated and phase-shifted in passing through the system. Then, by applying the superposition principle, the overall response can be found by adding together all of these individually attenuated and shifted Fourier components. The details, of course, are somewhat more involved than indicated here, but the approach just outlined is basically the one we will use in our studies of optical systems.

REFERENCES

5-1 A. Papoulis, *The Fourier Integral and Its Applications*, McGraw-Hill, New York, 1962, p. 82.

5-2 G. R. Cooper and C. D. McGillem, *Methods of Signal and System Analysis*, Holt, Rinehart and Winston, New York, 1967, pp. 20, 194.

PROBLEMS

Given a system $\mathcal{S}\{\ \}$ with input $f(x)$ and output $g(x)$ as shown in Fig. 5-1:

5-1. Assume the system to be characterized by the operator

$$\mathcal{S}\{f(x)\} = \left[a\left(\frac{d^2}{dx^2}\right) + b\left(\frac{d}{dx}\right) + c\right] f(x),$$

where a, b, and c are arbitrary constants.

a. Is the system linear? Shift invariant?
b. Calculate and sketch the output for $a=(2\pi)^{-1}$, $b=1$, $c=4\pi$, and $f(x)=\text{Gaus}(x)$.
c. Repeat part (b) for $f(x)=2\ \text{Gaus}(x-2)$.

5-2. Assume the system to be characterized by the operator

$$\mathcal{S}\{f(x)\}=0.5\int_{-\infty}^{x} f(\alpha)d\alpha.$$

a. Is the system linear? Shift invariant?
b. Calculate and sketch the output for $f(x)=\text{rect}\left(\frac{x-2}{4}\right)$.
c. Repeat part (b) for $f(x)=2\ \text{rect}\left(\frac{x+2}{4}\right)$.

5-3. Assume the system to be characterized by the operator

$$\mathcal{S}\{f(x)\}=a[f(x)]^2+bf(x),$$

where a and b are arbitrary constants.

a. Is the system linear? Shift invariant?
b. Calculate and sketch the output for $f(x)=\text{Gaus}(x)$.
c. Repeat part (b) for $f(x)=2\ \text{Gaus}(x)$.
d. Repeat part (b) for $f(x)=\text{Gaus}(x-2)$.

5-4. Assume the system to be a spectrum analyzer characterized by the operator

$$\mathcal{S}\{f(x)\}=\int_{-\infty}^{\infty} f(\alpha)e^{-j2\pi\xi\alpha}d\alpha\bigg|_{\xi=\frac{x}{a}}$$

$$=F\left(\frac{x}{a}\right),$$

where $F(\xi)$ is the Fourier integral of $f(x)$ and the constant a has the units of x^2. In other words, the input is mapped into a scaled version of its Fourier spectrum.

a. Is the system linear? Shift invariant?
b. Calculate and sketch the output for $f(x)=\text{rect}(x)$.
c. Repeat part (b) for $f(x)=2\ \text{rect}(x)$.
d. Repeat part (b) for $f(x)=\text{rect}(x-2)$.

CHAPTER 6

CONVOLUTION

The concept of convolution embraces many fields in the physical sciences and, as Bracewell points out (Ref. 6-1), it is known by many names, e.g., composition product, superposition integral, running mean, Duhamel integral, etc. As for our work here, convolution plays such a central role in the analysis of LSI systems that we devote an entire chapter to it and to other similar operations. As will be shown, the output of an LSI system is given by the convolution of the input with the system's impulse response; in theory, then, if we know the impulse response of the system, we can calculate the output of the system for any input simply by performing a convolution. This is an extremely powerful result, and one we shall use repeatedly.

6-1 THE CONVOLUTION OPERATION

At first we deal only with real-valued functions to minimize confusion about the nature of convolution; in Section 6-3 we extend this operation to include complex-valued functions.

The convolution of two real-valued functions $f(x)$ and $h(x)$, for which we use the shorthand notation $f(x)*h(x)$, is defined by the integral

$$f(x)*h(x)=\int_{-\infty}^{\infty} f(\alpha)h(x-\alpha)d\alpha. \qquad (6.1)$$

This integral is clearly a function of the independent variable x, and we therefore represent it by a third function $g(x)$, i.e.,

$$f(x)*h(x)=g(x). \qquad (6.2)$$

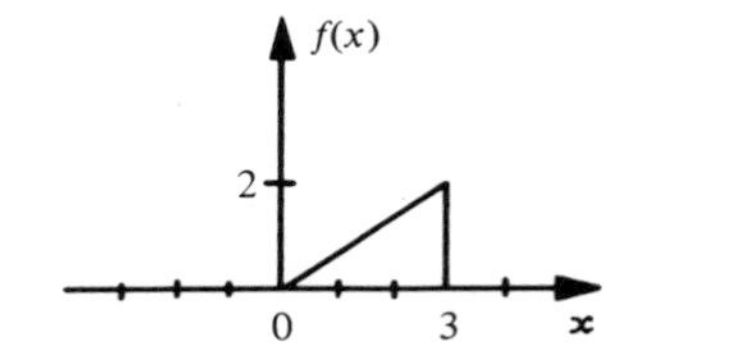

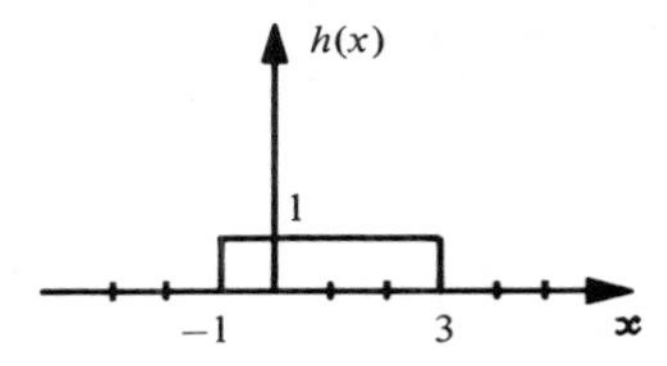

Figure 6-1 Functions used to illustrate convolution by graphical methods.

The convolution operation may be viewed simply as one of finding the *area of the product* of $f(\alpha)$ and $h(x-\alpha)$ as x is allowed to vary. Hence, even though some of the details can be a bit tricky, it should pose no conceptual problems (but it always does).

To investigate this operation, let us consider the convolution $f(x)*h(x) = g(x)$ of the two functions shown in Fig. 6-1. We could obtain the desired result by direct integration, but such an approach offers little insight into the nature of the problem and is quite involved even for these relatively simple functions. Therefore, we shall start with the graphical method outlined below and illustrated in Fig. 6-2.

Graphical Procedure for Convolution

1. First we graph the function $f(\alpha)$, using the dummy integration variable α for the horizontal coordinate.
2. Next, having chosen a convenient value for x, say $x=0$, we graph the function $h(x-\alpha)=h(-\alpha)$ below that of $f(\alpha)$. Note that $h(-\alpha)$ is simply a mirror image of the function $h(\alpha)$, i.e., it is the reflection of $h(\alpha)$ about the origin.
3. The product $f(\alpha)h(x-\alpha)=f(\alpha)h(-\alpha)$ is then found and graphed.
4. Next, the area of this product is calculated, the value of which is equal to the value of the convolution for the particular x chosen. Note that it is the *area of the product* of the two functions that is of concern; only for special cases, such as the present one, is this also equal to their common areas, or areas of overlap. Thus, for $x=0$, we have

$$\int_{-\infty}^{\infty} f(\alpha)h(-\alpha)d\alpha = g(0). \tag{6.3a}$$

5. Returning to Step (2), a new value is chosen for x, say $x=1$, and the function $h(x-\alpha)=h(1-\alpha)$ is graphed. To obtain $h(1-\alpha)$, we merely shift the function $h(-\alpha)$ one unit to the right. The product $f(\alpha)h(x-\alpha)= f(\alpha)h(1-\alpha)$ is then found and graphed, and the area of this product

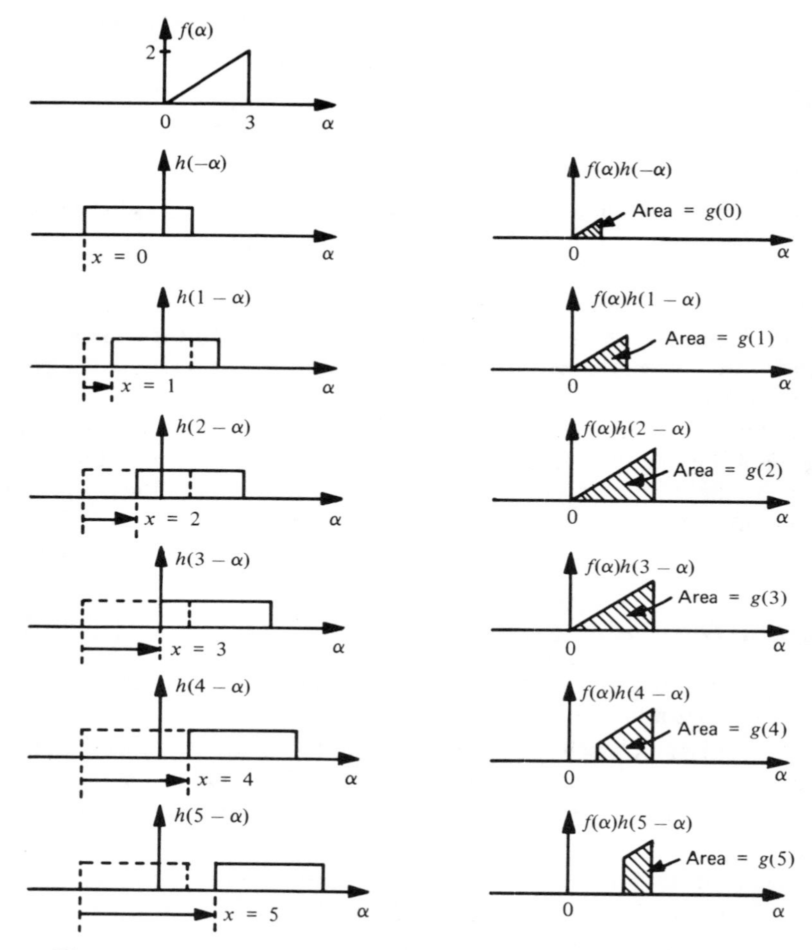

Figure 6-2 Graphical method for convolving functions of Fig. 6-1.

calculated to obtain $g(1)$, i.e.,

$$\int_{-\infty}^{\infty} f(\alpha)h(1-\alpha)d\alpha = g(1). \tag{6.3b}$$

6. The above process is then repeated for still another value of x, and another, etc., until the required areas have been calculated for all x.
7. Finally, the computed areas are used to graph the entire function $g(x)$ as shown in Fig. 6-3.

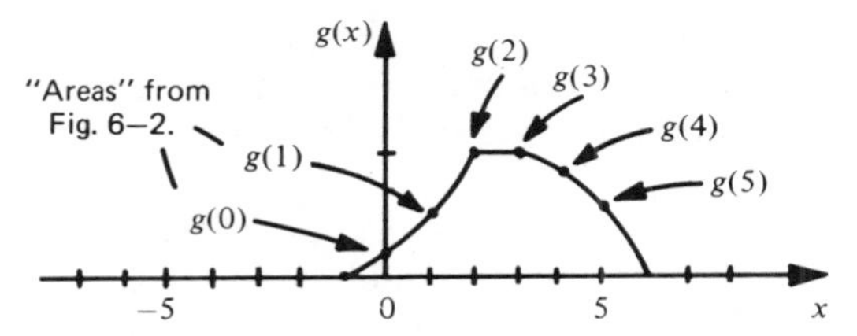

Figure 6-3 Resulting convolution of functions shown in Fig. 6-1.

Note: There are two foolproof ways for finding the function $h(x-\alpha)$. For the first method, we start out by graphing $h(\alpha)$. Next we graph $h(-\alpha)$, and finally we shift $h(-\alpha)$ by an amount x to find $h(x-\alpha)$. The shift is to the right for positive x and to the left for negative x. With the second method, we again start out by graphing $h(\alpha)$. Then we shift this function by an amount x to obtain $h(\alpha-x)$, the shift being to the right for positive x and to the left for negative x. Finally, we reflect $h(\alpha-x)$ about the point $\alpha=x$ to obtain $h(x-\alpha)$.

As an aid in performing graphical convolutions, Bracewell (Ref. 6-1) suggests the construction of a movable piece of paper on which the second function is graphed backward as illustrated in Fig. 6-4. As this piece of

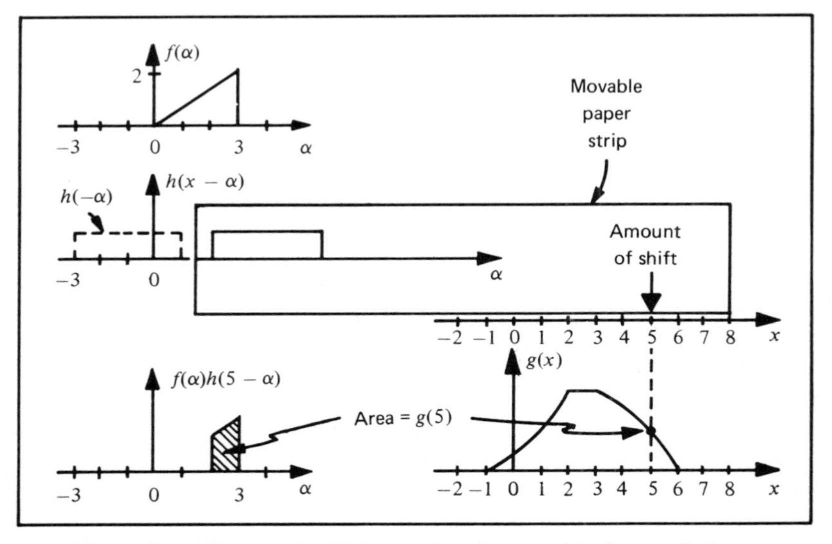

Figure 6-4 Suggested aid for performing graphical convolution.

paper is slid along the α-axis, corresponding to a change in the value of x, the entire process may be visualized with ease. With practice this procedure can often be accomplished mentally, thus eliminating the need for the movable paper.

In the above example, the graphical approach was the easiest to use because all of the required areas could be determined by inspection. Nevertheless, it will be instructive to carry out this convolution by direct integration as well.

Convolution by Direct Integration

The main difficulties here lie in setting up the problem correctly, keeping track of the limits of integration, etc. Therefore, in order to simplify things as much as possible, we shall divide the range of the independent variable x into five distinct intervals as indicated below.

1. $x \leqslant -1$. The product $f(\alpha)h(x-\alpha)$ is identically zero everywhere in this interval, with the result that $g(x)=0$.
2. $-1 < x \leqslant 2$. In this interval the convolution integral may be evaluated as follows:

$$g(x)=\int_{-\infty}^{\infty} f(\alpha)h(x-\alpha)d\alpha$$

$$=\int_0^3 \frac{2\alpha}{3} h(x-\alpha)d\alpha$$

$$=\frac{2}{3}\int_0^{x+1} \alpha\, d\alpha$$

$$=\frac{(x+1)^2}{3}. \tag{6.4a}$$

Thus, for $-1 < x \leqslant 2$, the function $g(x)$ is simply a segment of a parabola.
3. $2 < x \leqslant 3$. Here the upper limit of integration becomes a constant, i.e.,

$$g(x)=\frac{2}{3}\int_0^3 \alpha\, d\alpha$$

$$=3. \tag{6.4b}$$

Thus, $g(x)$ is constant in this region.

4. $3<x\leqslant 6.$ Now the integral takes the form

$$g(x)=\frac{2}{3}\int_{x-3}^{3}\alpha\,d\alpha$$

$$=3-\frac{(x-3)^2}{3}, \tag{6.4c}$$

which is again seen to be a parabola.

5. $x>6.$ As in the interval $x\leqslant -1$, the product $f(\alpha)h(x-\alpha)$ is identically zero here, and hence $g(x)=0$.

Finally, we specify the entire function:

$$g(x)=\begin{cases} 0, & x\leqslant -1 \\ \frac{1}{3}(x+1)^2, & -1<x\leqslant 2 \\ 3, & 2<x\leqslant 3 \\ 3-\frac{1}{3}(x-3)^2, & 3<x\leqslant 6 \\ 0, & x>6. \end{cases} \tag{6.4d}$$

The graph of this function is shown in Fig. 6-3.

Convolution of Several Functions

We now consider the convolution of more than two functions. Suppose, for example, that the function $f(x)$ is itself a convolution, i.e.,

$$f(x)=v(x)*w(x)$$

$$=\int_{-\infty}^{\infty}v(\beta)w(x-\beta)d\beta. \tag{6.5}$$

Then the convolution $g(x)=f(x)*h(x)$ would become

$$g(x)=[v(x)*w(x)]*h(x)$$

$$=\left[\int_{-\infty}^{\infty}v(\beta)w(x-\beta)d\beta\right]*h(x)$$

$$=\int_{-\infty}^{\infty}\left[\int_{-\infty}^{\infty}v(\beta)w(\alpha-\beta)d\beta\right]h(x-\alpha)d\alpha. \tag{6.6}$$

Thus, the notion of convolution can be extended to include any number of functions. At this point it is natural to raise the following question: "Is the convolution operation associative?" In other words, is it correct to write

$$[v(x)*w(x)]*h(x)=v(x)*[w(x)*h(x)]? \tag{6.7}$$

We shall defer discussion of this question until Sec. 6-3.

6-2 EXISTENCE CONDITIONS

The convolution $g(x)=f(x)*h(x)$ does not exist for all possible functions $f(x)$ and $h(x)$, but it will exist if these functions satisfy certain conditions, as outlined below. We do not attempt to state these existence conditions in a completely rigorous fashion; rather, we try to impart an intuitive feeling for them. It should be pointed out that we normally are concerned with functions for which the existence conditions are satisfied, for example, functions representing real physical quantities. Occasionally, however, it may be advantageous to convolve functions that do not truly represent such quantities; for instance, the step function cannot represent a real physical quantity because of its semi-infinite extent, but there are situations for which it adequately describes the behavior of certain phenomena. We therefore wish to include it as a candidate for convolution.

Given below is a set of conditions that are *sufficient* for the existence of the convolution $g(x)=f(x)*h(x)$. Assuming that $f(x)$ and $h(x)$ are single-valued and reasonably well behaved, this convolution will exist provided that (a) *both* $f(x)$ and $h(x)$ are absolutely integrable on the semi-infinite interval $(-\infty,0)$; *or* (b) *both* $f(x)$ and $h(x)$ are absolutely integrable on the semi-infinite interval $(0,\infty)$; *or* (c) *either* $f(x)$ *or* $h(x)$ (or both) are absolutely integrable on the infinite interval $(-\infty,\infty)$. To illustrate, we list several examples for which these conditions are satisfied and several for which they are not:

Convolution Exists	Convolution Does Not Exist
$\text{step}(x)*\text{step}(x)$	$\text{step}(x)*\text{step}(-x)$
$1*\text{rect}(x)$	$1*\text{step}(x)$
$\text{tri}(x)*\text{sgn}(x)$	$\text{ramp}(x)*\text{sgn}(x)$
$\text{rect}(x)*\cos(2\pi\xi_0 x)$	$\cos(2\pi\xi_0 x)*\cos(2\pi\xi_0 x)$

It should be pointed out that although the above conditions are sufficient, they are not necessary. That is, there are functions that do not satisfy any of these conditions but for which the convolution integral exists. One such example is the convolution of $\cos(2\pi\xi_0 x)$ with step(x). In addition, the given set of conditions excludes the convolution of any two periodic functions; this is misleading because convolutions of such functions can exist, although when they do they are identically zero. We shall have more to say about periodic functions momentarily. The main point here is that although the given conditions can be very helpful, they may exclude some perfectly legitimate functions and must therefore be applied with care.

The above existence conditions can be extended to include the convolution of more than two functions, as we now demonstrate. Consider the convolution of the n functions $f_1(x), f_2(x), \ldots, f_n(x)$; i.e.,

$$g(x) = \left(\{[f_1(x)*f_2(x)]*\ldots\}*f_{n-1}(x)\right)*f_n(x). \tag{6.8}$$

This convolution will exist provided that (a) *all* of the functions $f_i(x)$ are absolutely integrable on the semi-infinite interval $(-\infty, 0)$; *or* (b) *all* of the functions $f_i(x)$ are absolutely integrable on the semi-infinite interval $(0, \infty)$; *or* (c) *at least all but one* of the functions $f_i(x)$ are absolutely integrable on the infinite interval $(-\infty, \infty)$. Again we point out that this set of conditions must be applied with caution, because it may exclude certain allowable operations.

Let us return to the problem of convolving two periodic functions. In general, we can draw the following conclusions about such an operation. The convolution of two periodic functions (a) will not exist if *both* functions have nonzero average values; (b) will not exist if their periods are commensurate; (c) will exist and be identically zero if either of the functions has an average value of zero *and* if their periods are incommensurate. To illustrate, let us assume that $p(x)$ is any real-valued periodic function whose average value is *zero*, that $q(x)$ is any real-valued periodic function whose average value is *nonzero*, and that the ratio of the periods of $p(x)$ and $q(x)$ forms a rational number. Then, with a and b real constants whose ratio forms an irrational number,

$$p(x)*p(x) \textit{ does not exist,}$$

$$q(x)*q(x) \textit{ does not exist,}$$

$$p(ax)*p(bx) = 0,$$

$$q(ax)*q(bx) \textit{ does not exist,}$$

$$p(ax)*q(bx) = 0.$$

Although these results are informative, they are not extremely useful because even when the convolution does exist it is identically zero. Therefore, it is sometimes necessary to consider a different approach for the convolution of periodic functions. If, for example, we were to convolve only a finite segment of one function with the other function, a meaningful result might be obtained. Let us proceed as follows: with $p(x)$ and $q(x)$ as given above, we truncate $p(x)$ by multiplying it by $\text{rect}(x/d)$, where d is a real constant. Then the convolution

$$\left[p(x)\text{rect}\left(\frac{x}{d}\right)\right]*q(x)=\int_{-\infty}^{\infty}p(\alpha)\text{rect}\left(\frac{\alpha}{d}\right)q(x-\alpha)d\alpha$$

$$=\int_{-d/2}^{d/2}p(\alpha)q(x-\alpha)d\alpha \tag{6.9}$$

in general exists, is periodic, and has the same period as $q(x)$. If we truncate $q(x)$ rather than $p(x)$, the convolution will have a different form and a period equal to that of $p(x)$. This approach can be quite useful, and it more nearly corresponds to the type of convolution that is likely to be encountered in solving real-world problems.

Still another way of defining the convolution operation for troublesome functions is

$$p(x)*q(x)=\lim_{b\to\infty}\frac{1}{b}\int_{-b/2}^{b/2}p(\alpha)q(x-\alpha)d\alpha. \tag{6.10}$$

It may be that this limit exists, whereas the convolution according to the original definition does not.

By now it should be clear that even though convolution is conceptually simple, there are certain cases for which it is an extremely complex operation. One cannot simply feed two functions into the "convolution machine," turn the crank, and expect that whatever comes out will be meaningful. The various aspects of convolution discussed in this section must be taken into consideration in order to properly understand it and to obtain useful results from it.

6-3 PROPERTIES OF CONVOLUTION

In this section we discuss some of the most important properties of the convolution operation. In doing so, we assume that all of the functions to be convolved are well behaved and that all of the convolutions exist.

Commutative Property

Given the convolution

$$f(x)*h(x)=\int_{-\infty}^{\infty} f(\alpha)h(x-\alpha)d\alpha,$$

we make the change of variable $x-\alpha=\beta$, with the result that

$$\begin{aligned} f(x)*h(x) &= -\int_{\infty}^{-\infty} f(x-\beta)h(\beta)d\beta \\ &= \int_{-\infty}^{\infty} h(\beta)f(x-\beta)d\beta \\ &= h(x)*f(x). \end{aligned} \tag{6.11}$$

We see, therefore, that it does not matter which function is folded and shifted; the result is the same either way.

Distributive Property

It is easy to show that, with a and b arbitrary constants

$$[av(x)+bw(x)]*h(x)=a[v(x)*h(x)]+b[w(x)*h(x)]. \tag{6.12}$$

This property, also called the linearity property of convolution, is important in the analysis of linear, shift-invariant systems.

Shift Invariance

Given the convolution

$$\begin{aligned} f(x)*h(x) &= \int_{-\infty}^{\infty} f(\alpha)h(x-\alpha)d\alpha \\ &= g(x), \end{aligned}$$

then

$$\begin{aligned} f(x-x_0)*h(x) &= \int_{-\infty}^{\infty} f(\alpha-x_0)h(x-\alpha)d\alpha \\ &= \int_{-\infty}^{\infty} f(\beta)h(x-x_0-\beta)d\beta \\ &= g(x-x_0), \end{aligned} \tag{6.13}$$

where we have let $\alpha - x_0 = \beta$. Similarly,

$$f(x)*h(x-x_0)=g(x-x_0). \tag{6.14}$$

Thus, if either function is shifted by an amount x_0, the resulting convolution is simply shifted by the same amount, but remains unchanged in magnitude and form.

Associative Property

Consider the convolution of Eq. (6.6), i.e.,

$$[v(x)*w(x)]*h(x)=\int_{-\infty}^{\infty}\left[\int_{-\infty}^{\infty}v(\beta)w(\alpha-\beta)d\beta\right]h(x-\alpha)d\alpha.$$

By interchanging the order of integration we obtain

$$[v(x)*w(x)]*h(x)=\int_{-\infty}^{\infty}v(\beta)\left[\int_{-\infty}^{\infty}w(\alpha-\beta)h(x-\alpha)d\alpha\right]d\beta, \tag{6.15}$$

but we see that the inner integral is simply the convolution $w(x-\beta)*h(x)$. Therefore, if we let $w(x)*h(x)=u(x)$, we may use Eq. (6.13) to write

$$w(x-\beta)*h(x)=u(x-\beta), \tag{6.16}$$

and thus Eq. (6.15) becomes

$$\begin{aligned}[v(x)*w(x)]*h(x)&=\int_{-\infty}^{\infty}v(\beta)u(x-\beta)d\beta\\&=v(x)*u(x)\\&=v(x)*[w(x)*h(x)].\end{aligned} \tag{6.17}$$

It is apparent, then, that the order in which the individual convolutions are performed is immaterial, and combining Eqs. (6.11) and (6.17) we may write

$$\begin{aligned}[v(x)*w(x)]*h(x)&=v(x)*w(x)*h(x)\\&=v(x)*h(x)*w(x)\\&=h(x)*v(x)*w(x), \text{ etc.}\end{aligned} \tag{6.18}$$

Convolution Properties of Delta Functions

One of the properties used to define the delta function, Eq. (3.17), may be written in the form of a convolution. To do so, the limits of integration are taken to be $\pm\infty$, and, because it is entirely arbitrary, the real constant x_0 is replaced by the real variable x. Thus, for any real-valued function $f(x)$,

$$\int_{-\infty}^{\infty} f(\alpha)\delta(\alpha - x)d\alpha = f(x). \tag{6.19}$$

However, since $\delta(x)$ is an even function, we may write Eq. (6.19) as

$$\int_{-\infty}^{\infty} f(\alpha)\delta(x - \alpha)d\alpha = f(x), \tag{6.20}$$

or

$$f(x)*\delta(x) = f(x). \tag{6.21}$$

This is a very important result; convolution of a delta function with *any* other function merely reproduces the other function.

To investigate convolution properties of delta-function derivatives, we start with Eqs. (3.38) and (3.48) and follow a procedure similar to that used above for the delta function itself. We obtain the result

$$f(x)*\delta^{(k)}(x) = f^{(k)}(x). \tag{6.22}$$

Thus, convolution with the derivative of a delta function is equivalent to a differentiation operation.

Convolution of Derivatives of Functions

Suppose we are given the convolution $f(x)*h(x) = g(x)$ and we wish to find the convolution of $f^{(m)}(x)$ with $h^{(n)}(x)$, i.e., the convolution of the mth derivative of $f(x)$ with the nth derivative of $h(x)$. From Eq. (6.22) we may write

$$\begin{aligned} f^{(m)}(x) &= f(x)*\delta^{(m)}(x), \\ h^{(n)}(x) &= h(x)*\delta^{(n)}(x), \end{aligned} \tag{6.23}$$

and using the commutative property we obtain

$$\begin{aligned} f^{(m)}(x)*h^{(n)}(x) &= \left[f(x)*\delta^{(m)}(x) \right]*\left[h(x)*\delta^{(n)}(x) \right] \\ &= f(x)*h(x)*\delta^{(m)}(x)*\delta^{(n)}(x) \\ &= g(x)*\delta^{(m+n)}(x) \\ &= g^{(m+n)}(x). \end{aligned} \tag{6.24}$$

Thus the convolution of the mth and nth derivatives of two functions, respectively, is given by the $(m+n)$th derivative of their convolution.

Smoothing

Given the convolution $g(x)=f(x)*h(x)$, it is generally true that $g(x)$ will be smoother than either $f(x)$ or $h(x)$. The fine structure of each tends to be washed out, the sharp peaks and valleys tend to be rounded, etc., but the amount of smoothing depends on the exact nature of the two functions being convolved. To explore this property, we consider the following situations.

Let $f(x)$ represent any arbitrary function and let $h(x)$ be a delta function. We already know that $g(x)=f(x)*\delta(x)=f(x)$; thus no smoothing occurs and $g(x)$ is an exact reproduction of $f(x)$. Suppose now, that $h(x)$ is no longer a delta function, but that it is still fairly narrow with respect to the fine structure of $f(x)$. In this case, $f(x)$ will be reproduced with reasonably high fidelity regardless of the exact shape of $h(x)$, although some smoothing will likely take place. As $h(x)$ becomes still wider, the convolution can be thought of as a still-smoother version of $f(x)$; eventually, if $h(x)$ becomes too wide, the fine detail of $f(x)$ will be completely washed out. These various cases are illustrated in Fig. 6-5.

Another general statement can be made about the convolution operation: the width of a convolution is approximately equal to the sum of the widths of the functions being convolved, as can be seen in Fig. 6-5. This statement becomes exact if both of the functions being convolved have *compact support*, which means that both are identically zero outside some finite interval (e.g., the rectangle and triangle functions have compact support). If the functions $f_1(x)$ and $f_2(x)$ have compact support, and if their widths are b_1 and b_2, respectively, then their convolution $f_3(x)=f_1(x)*f_2(x)$ will have compact support and its width will be $b_3=b_1+b_2$. This is illustrated in Fig. 6-6.

There are special cases where neither smoothing nor widening occur. One such case is the convolution of the sinc function with itself, for which we find that

$$\text{sinc}(x)*\text{sinc}(x)=\text{sinc}(x). \tag{6.25}$$

This very interesting result can be extended as follows. Let $f(x)$ be any *band-limited* function, i.e., a function whose Fourier transform $F(\xi)$ has compact support. Furthermore, let us define the total width of $F(\xi)$ to be W, such that $F(\xi)\equiv 0$ for $|\xi|\geqslant W/2$. Then if $W\leqslant|b|^{-1}$ we find that

$$f(x)*\frac{1}{|b|}\text{sinc}\left(\frac{x}{b}\right)=f(x), \tag{6.26}$$

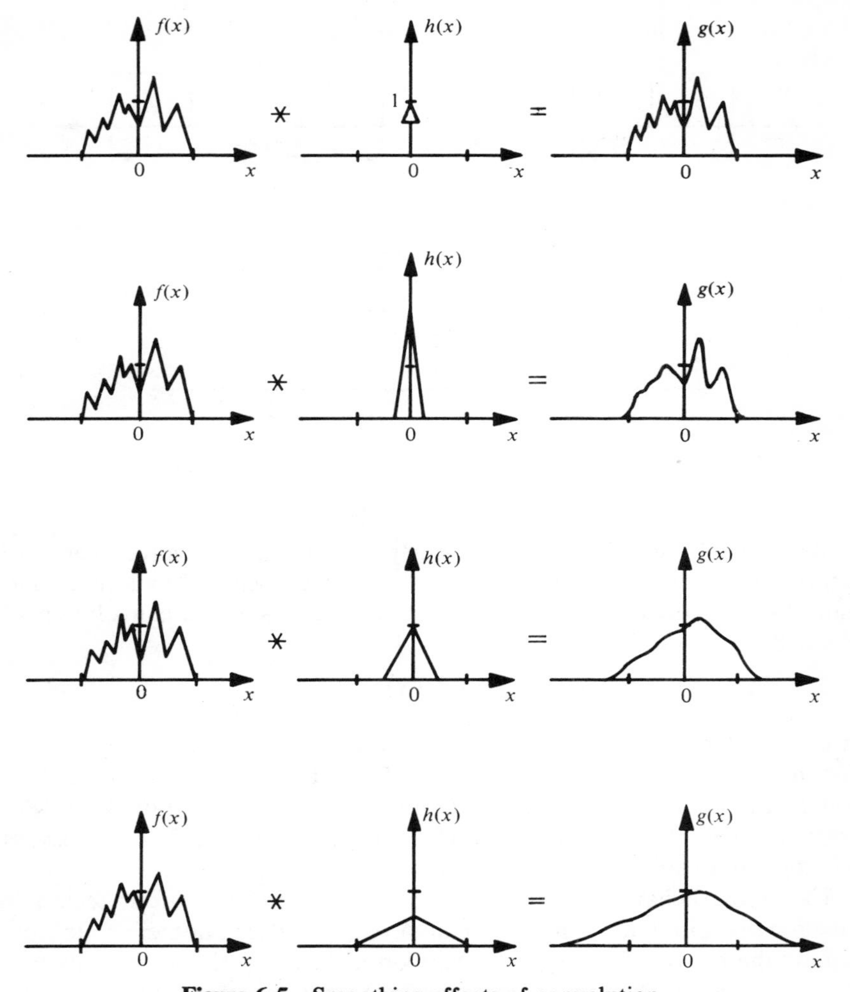

Figure 6-5 Smoothing effects of convolution.

as will be shown in a later chapter. Thus, when the above conditions are met, the convolution of a band-limited function with a sinc function reproduces the band-limited function exactly; no smoothing and no spreading take place.

Repeated Convolution

The convolution of a large number of functions generally yields a function that is much smoother than any of the individual functions involved in the

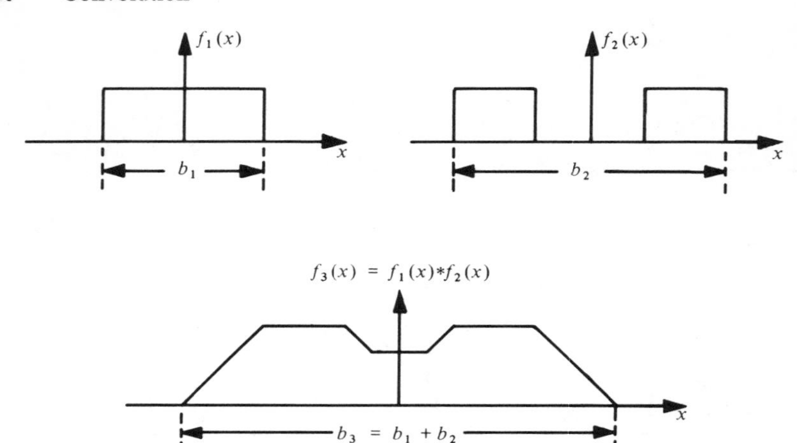

Figure 6-6 Convolution of two functions with compact support.

convolution. In addition, as the number of functions convolved becomes larger and larger, the resulting convolution often begins to look more and more like a Gaussian function. For example, consider the convolution of Eq. (6.8):

$$g(x) = f_1(x)*f_2(x)*\ldots*f_n(x).$$

It can be shown that, if the functions $f_1(x), f_2(x), \ldots, f_n(x)$ satisfy certain conditions, then as $n \rightarrow \infty$ $g(x)$ will tend toward a Gaussian function. This is a nonrigorous statement of the *Central Limit Theorem*. For those who are interested, Bracewell (Ref. 6-1) and Papoulis (Ref. 6-2) both discuss this topic in detail.

The stated behavior is illustrated in Fig. 6-7, in which four rectangle functions of different widths are convolved one after another. Note how rapidly the convolution begins to take on a Gaussian shape, even though the individual functions are themselves far from Gaussian. In this example, the individual functions all possess even symmetry, but the same sort of behavior can also be observed with functions that have no symmetry whatsoever. To illustrate, the repeated self convolution of the function $f(x) = \exp(-x)\text{step}(x)$ becomes quite Gaussian in nature after only four or five repetitions, as shown in Fig. 6-8. Bracewell (Ref. 6-1) gives an expression for this particular repeated convolution: if the convolution operation is performed n times, the resulting function has the form $(n!)^{-1}x^n\exp(-x)\text{step}(x)$.

There is another interesting observation that can be made concerning repeated convolution: if all of the functions to be convolved are identically

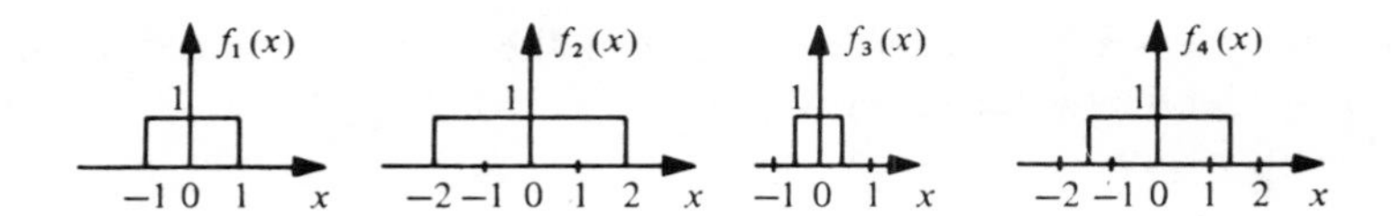

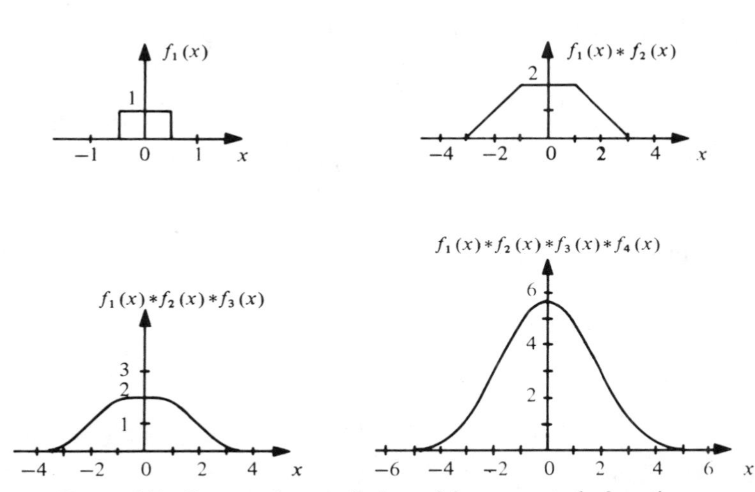

Figure 6-7 Repeated convolution of four rectangle functions.

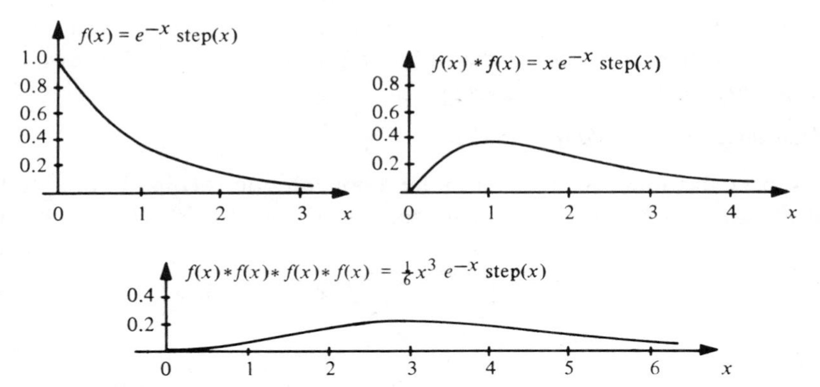

Figure 6-8 Repeated convolution of the function $\exp\{-x\}$step(x).

zero everywhere to the left (or right) of the origin, their convolution will also be identically zero everywhere to the left (or right) of the origin. This behavior is illustrated in Fig. 6-8.

Scaling Property

Given the convolution $f(x)*h(x)=g(x)$, we might yield to our intuition and write $f(x/b)*h(x/b)=g(x/b)$. However, this result is incorrect. To illustrate, let us define $v(x)=f(x/b)$ and $w(x)=h(x/b)$. Then

$$\begin{aligned} v(x)*w(x) &= \int_{-\infty}^{\infty} v(\alpha)w(x-\alpha)\,d\alpha \\ &= \int_{-\infty}^{\infty} f\left(\frac{\alpha}{b}\right)h\left(\frac{x-\alpha}{b}\right)d\alpha, \end{aligned} \tag{6.27}$$

and with the change of variable $\alpha=b\beta$

$$\begin{aligned} v(x)*w(x) &= |b|\int_{-\infty}^{\infty} f(\beta)h\left(\frac{x}{b}-\beta\right)d\beta \\ &= |b|\,g\left(\frac{x}{b}\right). \end{aligned} \tag{6.28}$$

Finally, we obtain

$$f\left(\frac{x}{b}\right)*h\left(\frac{x}{b}\right)=|b|\,g\left(\frac{x}{b}\right). \tag{6.29}$$

Thus, we see that indiscriminate substitution in the shorthand representation for the convolution operation can lead to incorrect conclusions. It is important, therefore, to use the defining integral whenever functions with scaled arguments are convolved.

Area under a Convolution

It is easy to show that the area under a convolution is equal to the product of the areas of the functions being convolved, i.e., if $g(x)=f(x)*h(x)$, then

$$\begin{aligned} \int_{-\infty}^{\infty} g(\beta)\,d\beta &= \int_{-\infty}^{\infty}\left[\int_{-\infty}^{\infty} f(\alpha)h(\beta-\alpha)\,d\alpha\right]d\beta \\ &= \int_{-\infty}^{\infty} f(\alpha)\left[\int_{-\infty}^{\infty} h(\beta-\alpha)\,d\beta\right]d\alpha \\ &= \left[\int_{-\infty}^{\infty} f(\alpha)\,d\alpha\right]\left[\int_{-\infty}^{\infty} h(\beta)\,d\beta\right]. \end{aligned} \tag{6.30}$$

Convolution of Complex-Valued Functions

So far we have only considered convolutions of real-valued functions; now we wish to extend this operation to include complex-valued functions as well. Suppose that $f_1(x)$ and $f_2(x)$ are two complex-valued functions described by

$$f_i(x)=a_i(x)e^{j\phi_i(x)}, \quad i=1,2, \tag{6.31}$$

or by

$$f_i(x)=v_i(x)+jw_i(x), \quad i=1,2, \tag{6.32}$$

where $a_i(x)$, $\phi_i(x)$, $v_i(x)$, and $w_i(x)$ are all real-valued functions. Then, one form of their convolution is given by

$$\begin{aligned} f_3(x)&=f_1(x)*f_2(x)\\ &=\int_{-\infty}^{\infty} a_1(\alpha)e^{j\phi_1(\alpha)}a_2(x-\alpha)e^{j\phi_2(x-\alpha)}d\alpha\\ &=\int_{-\infty}^{\infty} a_1(\alpha)a_2(x-\alpha)e^{j[\phi_1(\alpha)+\phi_2(x-\alpha)]}d\alpha. \end{aligned} \tag{6.33}$$

Without knowing more about $a_i(x)$ and $\phi_i(x)$, this is about as far as we can go except to say that $f_3(x)$ will, in general, be complex valued. That this is so may be seen by convolving the functions in the form given by Eq. (6.32), i.e.,

$$\begin{aligned} f_3(x)&=[v_1(x)+jw_1(x)]*[v_2(x)+jw_2(x)]\\ &=v_1(x)*v_2(x)+jv_1(x)*w_2(x)+jw_1(x)*v_2(x)-w_1(x)*w_2(x)\\ &=[v_1(x)*v_2(x)-w_1(x)*w_2(x)]+j[v_1(x)*w_2(x)+w_1(x)*v_2(x)]\\ &=v_3(x)+jw_3(x). \end{aligned} \tag{6.34}$$

The conditions for the existence of such a convolution must now be applied to the above expression term by term.

6-4 CONVOLUTION AND LINEAR SHIFT-INVARIANT SYSTEMS

Let us consider an arbitrary LSI system characterized by the operator $\mathcal{L}\{\ \}$ and having an impulse response $h(x)$. For an input $f(x)$, the output

$g(x)$ is given by

$$g(x) = \mathfrak{L}\{f(x)\}, \tag{6.35}$$

as discussed in Chapter 5. Suppose we were now to represent $f(x)$ as a linear combination of appropriately weighted and located delta functions, as in Eq. (6.20). In other words, suppose we were to describe $f(x)$ by

$$f(x) = \int_{-\infty}^{\infty} f(\alpha)\delta(x-\alpha)d\alpha, \tag{6.36}$$

where the delta function located at the point $x = \alpha$ is assigned an infinitesimal area of $f(\alpha)d\alpha$. Each component delta function is therefore weighted, according to its location, by the value of the input function at that location. Then, substituting this expression into Eq. (6.35), we have

$$g(x) = \mathfrak{L}\left\{\int_{-\infty}^{\infty} f(\alpha)d\alpha\,\delta(x-\alpha)\right\}, \tag{6.37}$$

but by the linearity of $\mathfrak{L}\{\ \}$, we may interchange the order of the integration and $\mathfrak{L}\{\ \}$ operations to obtain

$$\begin{aligned} g(x) &= \int_{-\infty}^{\infty} \mathfrak{L}\{f(\alpha)d\alpha\,\delta(x-\alpha)\} \\ &= \int_{-\infty}^{\infty} f(\alpha)d\alpha\,\mathfrak{L}\{\delta(x-\alpha)\}. \end{aligned} \tag{6.38}$$

The last step was taken by noting that the quantity $f(\alpha)d\alpha$ describes only the area of each component delta function, and is therefore a constant as far as the operator $\mathfrak{L}\{\ \}$ is concerned. From Eq. (5.23) we recognize that $\mathfrak{L}\{\delta(x-\alpha)\}$ is simply the response of the system to an impulse located at the point α, so that Eq. (6.38) becomes

$$\begin{aligned} g(x) &= \int_{-\infty}^{\infty} f(\alpha)d\alpha\,h(x-\alpha) \\ &= \int_{-\infty}^{\infty} f(\alpha)h(x-\alpha)d\alpha. \end{aligned} \tag{6.39}$$

But this is just the convolution of $f(x)$ and $h(x)$, and we therefore arrive at the following very important conclusion: *the output of an LSI system is given by the convolution of the input with the impulse response of the system*, i.e.,

$$g(x) = f(x) * h(x). \tag{6.40}$$

Now let us review the above development from a physical point of view. Effectively, we started out by decomposing the input function $f(x)$ into a linear combination of elementary functions—delta functions in this case—each appropriately weighted according to its location. Then, knowing the response of the system to a single delta function, we applied each of the input component delta functions to the system separately and determined their separate responses. Finally, invoking the linearity of the system, we calculated the overall response to the entire input by simply adding together all of the individual responses. This, in words, describes the mathematical operation of Eq. (6.40).

Because the behavior of an LSI system is described by a convolution, the output of such a system is generally a smoothed version of the input, the degree of smoothing depending strongly on the *width* of the impulse response. The slowly varying components of the input are reproduced with little degradation, but any structure significantly finer than the width of the impulse response is severely attenuated. In other words, the system can follow the slowly changing portions of the input, but it is unable to resolve the rapidly oscillating portions. Not only is the resolving capability of a system highly dependent on the width of the impulse response, it also depends strongly on the *form* of the impulse response. We shall postpone a detailed discussion of these considerations until a later chapter, but we point out here that it is because of them that audio amplifiers display an inability to amplify the higher-frequency portions of an input signal, that images formed by an optical system contain less detail than the original scene, etc. In any case, an LSI system is *completely* characterized by its impulse response; if the impulse response is known, a unique output can be determined (in theory, at least) for every input.

Example

To aid in visualizing the transformation of an input signal into an output signal, we have chosen an example for which the system has the idealized impulse response shown in Fig. 6-9, and for which the input is a time-varying signal consisting of the four delta functions

$$f(t)=\delta(t+2)+\delta(t-1)+\delta(t-3)+\delta(t-6), \tag{6.41}$$

which are also shown in Fig. 6-9. The output is given by

$$\begin{aligned} g(t)&=f(t)*h(t)\\ &=\left[\delta(t+2)+\delta(t-1)+\delta(t-3)+\delta(t-6)\right]*h(t)\\ &=h(t+2)+h(t-1)+h(t-3)+h(t-6), \end{aligned} \tag{6.42}$$

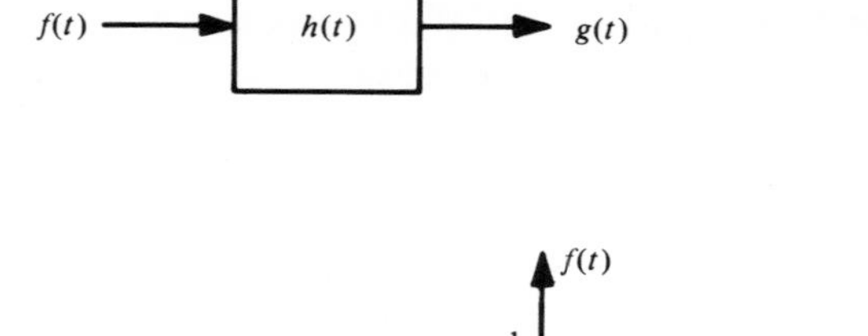

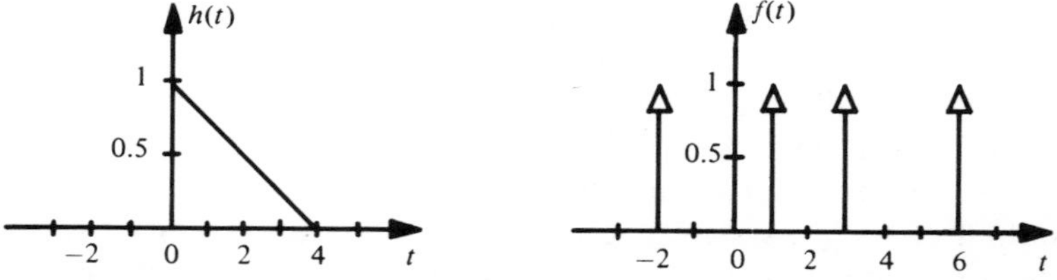

Figure 6-9 An example involving a linear shift invariant system.

and it is apparent that each term of this expression represents the response to an individual delta function component of the input, whereas the sum of the terms represents the response to the entire input. Both the individual responses and the overall response are illustrated in Fig. 6-10. Note that the overall response is both smoother and wider than the input signal $f(t)$. ■

In the preceding discussion, we took the point of view that each part of the input produces an appropriately weighted and shifted version of $h(t)$, the sum of which yields the output $g(t)$. We now wish to consider a second, equally valid point of view. When a system has an impulse response similar to that shown in Fig. 6-9, i.e., one rising abruptly to a maximum and then decaying slowly back to zero, it should be clear that

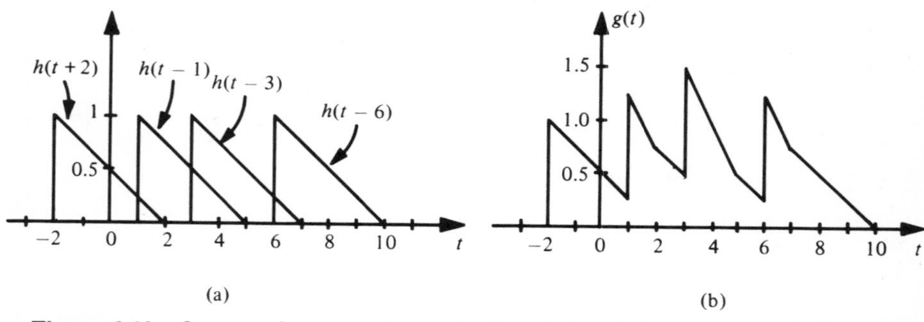

Figure 6-10 Output of system shown in Fig. 6-9. (*a*) Responses to individual input delta functions. (*b*) Overall response to combined input delta functions.

the most recently occurring portions of the input will have the greatest influence on the value of the output at any particular observation time. On the other hand, it may not be clear how a convolution operation can produce this sort of behavior.

To see this, let us write the output as

$$g(t)=\int_{-\infty}^{\infty} f(\tau)h(t-\tau)d\tau, \tag{6.43}$$

which shows that the output has a value equal to the area under the product $f(\tau)h(t-\tau)$. Thus the function $h(t-\tau)$, a reflected and shifted version of the impulse response, is seen to play the role of a time-dependent weighting factor, the effect of which is to weight each part of the input according to the interval between its time of occurrence and the time at which the observation is made. In other words, at any observation time t, the effective contribution made by that portion of the input occurring at time τ is determined not only by the value of the input at τ but also by that of the impulse response evaluated at the time difference $t-\tau$. Because $h(t-\tau)$ has been reflected, those portions of the input occurring in the more distant past are devalued by a greater amount than those occurring more recently.

We now illustrate this point of view, choosing an observation time of $t=4$. The weighting factor $h(4-\tau)$ is shown in Fig. 6-11(a), and the product $f(\tau)h(4-\tau)$ is shown in Fig. 6-11(b). We see that the effective contribution due to the delta function located at $t=-2$ is zero because it occurred more than four units in the past. The delta function occurring at $t=1$, three units in the past, contributes a value of only 0.25 to the integral, whereas that occurring at $t=3$, just one unit in the past, makes a contribution of 0.75. Finally, the impulse at $t=6$ has not yet occurred, and thus its contribution is zero. Adding together the various contributions, we find that $g(4)=1$, which agrees with the previous result.

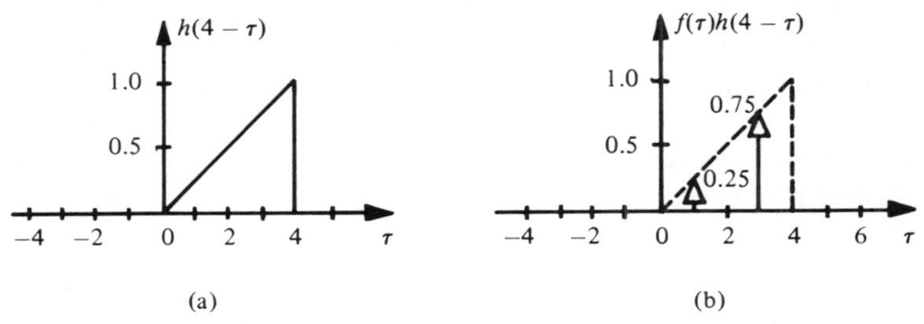

Figure 6-11 Alternative method for determining output of system shown in Fig. 6-9. (*a*) Weighting factor $h(4-\tau)$. (*b*) Product $f(\tau)h(4-\tau)$.

6-5 CROSS CORRELATION AND AUTOCORRELATION

Given the two complex-valued functions $f(x)$ and $g(x)$, we define the *cross correlation of* $f(x)$ *with* $g(x)$ to be

$$f(x)\star g(x)=\int_{-\infty}^{\infty} f(\alpha)\, g(\alpha-x)d\alpha. \tag{6.44}$$

Note that although this operation is similar to convolution, there is one very important difference: the function $g(x)$ is not folded as in convolution. If we make the change of variable $\beta=\alpha-x$, Eq. (6.44) becomes

$$f(x)\star g(x)=\int_{-\infty}^{\infty} f(\beta+x)\, g(\beta)d\beta, \tag{6.45}$$

which leads us to the conclusion that, in general,

$$f(x)\star g(x)\neq g(x)\star f(x). \tag{6.46}$$

Thus, unlike convolution, the cross-correlation operation does not commute. Nevertheless, it can be expressed in terms of a convolution by noting that

$$\begin{aligned}\int_{-\infty}^{\infty} f(\alpha)\, g(\alpha-x)d\alpha &= \int_{-\infty}^{\infty} f(\alpha)\, g\left(\frac{x-\alpha}{-1}\right)d\alpha\\ &= f(x)*g\left(\frac{x}{-1}\right)\\ &= f(x)*g(-x).\end{aligned} \tag{6.47}$$

Consequently, we obtain

$$f(x)\star g(x)=f(x)*g(-x). \tag{6.48}$$

The conditions for the existence of this operation are the same as for convolution, but they must now be applied to $f(x)$ and $g(-x)$ rather than to $f(x)$ and $g(x)$ directly. If $f(x)=g(x)$, we refer to this operation as the *autocorrelation operation*.

Extending the above development, we now define the *complex cross correlation of* $f(x)$ *with* $g(x)$ to be

$$\begin{aligned}\gamma_{fg}(x)&=f(x)\star g^*(x)\\ &=\int_{-\infty}^{\infty} f(\alpha)\, g^*(\alpha-x)d\alpha.\end{aligned} \tag{6.49}$$

A change of variable leads to a second form for $\gamma_{fg}(x)$:

$$\gamma_{fg}(x)=\int_{-\infty}^{\infty} f(\beta+x)\,g^*(\beta)\,d\beta. \tag{6.50}$$

Note that by interchanging the arguments of f and g in the above equations, a folded version of the complex cross correlation function is obtained, i.e.,

$$\int_{-\infty}^{\infty} f(\alpha-x)\,g^*(\alpha)\,d\alpha=\int_{-\infty}^{\infty} f(\beta)\,g^*(\beta+x)\,d\beta$$
$$=\gamma_{fg}(-x). \tag{6.51}$$

It may also be shown that

$$\gamma_{fg}(x)=f(x)*g^*(-x). \tag{6.52}$$

Now let us write the expression for the complex cross correlation of $g(x)$ with $f(x)$. We have

$$\begin{aligned}\gamma_{gf}(x)&=g(x)\star f^*(x)\\ &=\int_{-\infty}^{\infty} g(\beta+x)\,f^*(\beta)\,d\beta\\ &=\left[\int_{-\infty}^{\infty} f(\beta)\,g^*(\beta+x)\,d\beta\right]^*,\end{aligned} \tag{6.53}$$

and from Eq. (6.51) we find that

$$\gamma_{gf}(x)=\gamma_{fg}^*(-x). \tag{6.54}$$

In a similar fashion, it may be shown that

$$\gamma_{fg}(x)=\gamma_{gf}^*(-x). \tag{6.55}$$

Because the cross-correlation operation is not commutative, it is important to pay particular attention to the order in which the functions are written and to note which of the functions is conjugated. An error here can produce a folded or conjugated version of the desired result.

Suppose we now let $g(x)=f(x)$, so that Eq. (6.49) becomes

$$\begin{aligned}\gamma_{ff}(x)&=f(x)\star f^*(x)\\ &=\int_{-\infty}^{\infty} f(\alpha)\,f^*(\alpha-x)\,d\alpha.\end{aligned} \tag{6.56}$$

This is known as the *complex autocorrelation of* $f(x)$. The double subscript notation is redundant here, and we shall therefore denote the complex autocorrelation function of $f(x)$ simply by $\gamma_f(x)$. If we write $f(x)$ in the form $f(x)=a(x)\exp\{j\phi(x)\}$, the complex autocorrelation becomes

$$\begin{aligned}\gamma_f(x)&=\left[a(x)e^{j\phi(x)}\right]\star\left[a(x)e^{-j\phi(x)}\right]\\&=\int_{-\infty}^{\infty}a(\alpha)a(\alpha-x)\exp\{j[\phi(\alpha)-\phi(\alpha-x)]\}d\alpha,\end{aligned}\tag{6.57}$$

where $a(x)$ and $\phi(x)$ are real valued as usual. On the other hand, if we put $f(x)$ in the form $f(x)=v(x)+jw(x)$, then

$$\begin{aligned}\gamma_f(x)&=\left[v(x)+jw(x)\right]\star\left[v(x)-jw(x)\right]\\&=v(x)\star v(x)+w(x)\star w(x)+j\left[w(x)\star v(x)-v(x)\star w(x)\right].\end{aligned}\tag{6.58}$$

But since $v(x)$ and $w(x)$ are both real valued, we have $v(x)\star v(x)=\gamma_v(x), w(x)\star w(x)=\gamma_w(x)$, etc., so that

$$\gamma_f(x)=\gamma_v(x)+\gamma_w(x)+j\left[\gamma_{wv}(x)-\gamma_{vw}(x)\right].\tag{6.59}$$

Note that if $f(x)$ is real valued, i.e., $w(x)\equiv 0$, then $\gamma_f(x)$ is real valued. If $v(x)\equiv 0$, such that $f(x)$ is imaginary valued, $\gamma_f(x)$ is again real valued. Finally, if $f(x)$ is complex valued, $\gamma_f(x)$ is complex valued.

From Eq. (6.52), we find that

$$\gamma_f(x)=f(x)*f^*(-x),\tag{6.60}$$

and from Eq. (6.54) we obtain

$$\gamma_f(x)=\gamma_f^*(-x).\tag{6.61}$$

Thus, the *complex autocorrelation function is hermitian*. This function has still another important property: *its modulus is maximum at the origin*, i.e.,

$$|\gamma_f(x)|\leqslant\gamma_f(0).\tag{6.62}$$

To show this, we use Schwarz's Inequality. Given two complex-valued functions $g(\alpha)$ and $h(\alpha)$, Schwarz's Inequality may be expressed as (Ref. 6-3)

$$\left|\int_a^b g(\alpha)h(\alpha)d\alpha\right|\leqslant\left[\int_a^b|g(\alpha)|^2d\alpha\right]^{1/2}\left[\int_a^b|h(\alpha)|^2d\alpha\right]^{1/2}.\tag{6.63}$$

By choosing the limits of integration to be infinite, and by setting $g(\alpha)=f(\alpha)$ and $h(\alpha)=f^*(\alpha-x)$, we obtain

$$\left|\int_{-\infty}^{\infty} f(\alpha)f^*(\alpha-x)\,d\alpha\right| \leqslant \left[\int_{-\infty}^{\infty}|f(\alpha)|^2 d\alpha\right]^{1/2}\left[\int_{-\infty}^{\infty}|f^*(\alpha-x)|^2 d\alpha\right]^{1/2} \tag{6.64}$$

We assume, of course, that each of these integrals exists. It is clear that a shift in the origin cannot change the value of the last integral in this expression, and therefore

$$\left[\int_{-\infty}^{\infty}|f(\alpha)|^2 d\alpha\right]^{1/2} = \left[\int_{-\infty}^{\infty}|f^*(\alpha-x)|^2 d\alpha\right]^{1/2}. \tag{6.65}$$

As a consequence, Eq. (6.64) becomes

$$\left|\int_{-\infty}^{\infty} f(\alpha)f^*(\alpha-x)\,d\alpha\right| \leqslant \int_{-\infty}^{\infty}|f(\alpha)|^2 d\alpha,$$

from which it follows that $|\gamma_f(x)| \leqslant \gamma_f(0)$.

We summarize the important properties of the complex autocorrelation function in Table 6-1. It is interesting to note that $\gamma_f(x)$ can be either real or complex valued, but it cannot be imaginary valued. That this is so may easily be seen by referring to Eq. (6.54), which we rewrite as

$$\gamma_f(x) = \mathrm{Re}\{\gamma_f(x)\} + j\mathrm{Im}\{\gamma_f(x)\}. \tag{6.66}$$

In order for $\gamma_f(x)$ to be imaginary valued, we must have $\mathrm{Re}\{\gamma_f(x)\}\equiv 0$. We recall that $\mathrm{Im}\{\gamma_f(x)\}$ is an odd function, which means it is zero at the origin, i.e., $\mathrm{Im}\{\gamma_f(0)\}=0$. But we also recall that $|\gamma_f(x)| \leqslant \gamma_f(0)$, and this condition is satisfied only for the trivial case for which $\gamma_f(x)\equiv 0$. It follows that $\gamma_f(x)$ cannot be purely imaginary.

Table 6-1

Properties of the Autocorrelation Function

$f(x)$	$\gamma_f(x)$
complex valued	hermitian, $\lvert\gamma_f(x)\rvert \leqslant \gamma_f(0)$
real valued	real and even, $\gamma_f(x) \leqslant \gamma_f(0)$
imaginary valued	real and even, $\gamma_f(x) \leqslant \gamma_f(0)$

We point out that the above special properties of the autocorrelation function do not, in general, hold for either convolution or for cross correlation. They do not even hold for self-convolution, except in special cases. To illustrate, consider an arbitrary real-valued function $f(x)$. The autocorrelation function $\gamma_f(x)$ will be an even function and will have a maximum at the origin. On the other hand, the self-convolution $f(x)*f(x)$ will generally be neither an even function, nor will it have a maximum at the origin. To convince yourself that these statements are correct, try one or two simple examples.

We now briefly discuss the conditions for the existence of an autocorrelation function. Since the autocorrelation of $f(x)$ can be expressed as $\gamma_f(x)=f(x)*f^*(-x)$, the existence conditions are the same as for convolution, but they must now be applied to $f(x)$ and $f^*(-x)$. We can therefore draw the following conclusions:

1. The autocorrelation of a function will exist if this function is absolutely integrable on the infinite interval $(-\infty,\infty)$.
2. The autocorrelation of a periodic function does not exist.

When dealing with functions that have no autocorrelation in the strict sense, it is often helpful to use a truncated version, as in Eq. (6.9), or to redefine the autocorrelation as a limit, as in Eq. (6.10).

REFERENCES

6-1 R. Bracewell, *The Fourier Transform and Its Applications*, McGraw-Hill, New York, 1965.

6-2 A. Papoulis, *The Fourier Integral and Its Applications*, McGraw-Hill, New York, 1962, pp. 227–239.

6-3 J. Thomas, *An Introduction to Statistical Communication Theory*, Wiley, New York, 1969.

PROBLEMS

6-1. Perform the following convolutions and sketch the result in each case:

a. $f(x)=\text{step}(x)*\text{step}(x)$.
b. $g(x)=\text{step}(x)*\text{ramp}(x)$.
c. $h(x)=\text{step}(x)*\text{rect}(x)$.
d. $m(x)=\text{step}(x)*\text{tri}(x)$.
e. $n(x)=\text{rect}(x)*\text{rect}(x)$.

f. $p(x)=\text{rect}(x)*\text{rect}(x-2)$.

g. $r(x)=\text{rect}(x)*\text{rect}\left(\frac{x}{4}\right)$.

h. $s(x)=\text{step}(x)*[\text{rect}(x+2)-\text{rect}(x-2)]$.

i. $t(x)=\text{rect}(x)*\text{rect}(x)*\text{rect}(x)$.

j. $u(x)=\text{tri}(x)*\text{tri}(x)$.

k. $v(x)=\text{step}(x)*[e^{-x}\text{step}(x)]$.

l. $w(x)=\text{rect}(x-0.5)*[e^{-x}\text{step}(x)]$.

6-2. With $f(x)$ and $g(x)$ arbitrary functions and b, d, x_1, and x_2 real constants, show the following:

a. $\delta(x-x_1)*f(x)=f(x-x_1)$.

b. $\delta(x)*f(x-x_2)=f(x-x_2)$.

c. $\delta\left(\frac{x-x_1}{b}\right)*f\left(\frac{x-x_2}{d}\right)=|b|f\left(\frac{x-x_1-x_2}{d}\right)$.

d. In general, $f\left(\frac{x}{b}\right)*g\left(\frac{x}{d}\right)\neq f\left(\frac{x}{d}\right)*g\left(\frac{x}{b}\right)$ if $b\neq d$. (A counter example will suffice.)

e. $\frac{1}{|b|}\delta\delta\left(\frac{x}{b}\right)*f(x)=f(x+b)+f(x-b)$.

f. $\frac{1}{|b|}\delta\delta\left(\frac{x}{b}\right)*g(x)=g(x+b)-g(x-b)$.

g. $\frac{1}{|b|}\text{comb}\left(\frac{x}{b}\right)*f(x)=\sum_{n=-\infty}^{\infty} f(x-nb)$.

6-3. Perform the following convolutions and sketch the results in each case:

a. $f(x)=\frac{1}{2}\delta\delta\left(\frac{x}{2}\right)*\text{rect}(x)$.

b. $g(x)=\frac{1}{2}\delta\delta\left(\frac{-x}{2}\right)*\text{rect}(x)$.

c. $h(x)=\frac{1}{4}\text{comb}\left(\frac{x}{4}\right)*\text{rect}\left(\frac{x}{2}\right)$.

d. $p(x)=\delta\delta(x)*\delta\delta(x)$.

e. $r(x)=\delta\delta(x)*\delta_{\delta}(x)$.

f. $s(x)=\frac{1}{2}\delta\delta\left(\frac{x}{2}\right)*\delta\delta(x)$.

g. $u(x)=\delta^{(1)}(x)*\text{tri}(x)$.

h. $v(x)=\delta^{(1)}(x)*\text{Gaus}(x)$.

6-4. Given an LSI system with impulse response $h(x)=\exp\{-x\}\text{step}(x)$, find the output $g_i(x)$ for each of the following input signals $f_i(x)$. Sketch both $f_i(x)$ and $g_i(x)$.

a. $f_1(x)=\text{step}(x-5)$.

b. $f_2(x)=\text{rect}(10x-0.5)$.

c. $f_3(x) = \text{rect}(0.1x - 0.5)$.
d. $f_4(x) = 10\delta\delta(10x)$.
e. $f_5(x) = h(x)$.

6-5. Sketch each of the following functions, then find and sketch the self-convolution and autocorrelation of each; i.e., sketch $f(x)$, $f(x)*f(x)$, $f(x)\star f(x)$, $g(x)$, $g(x)*g(x)$, $g(x)\star g(x)$, etc.

a. $f(x) = \text{rect}(x+2) + \text{rect}(x-2)$.
b. $g(x) = f(x-2.5)$.
c. $h(x) = f(x)\text{sgn}(x)$.
d. $p(x) = h(x-2.5)$.
e. $r(x) = 0.5\,\text{rect}(0.5x+1) + 2\,\text{rect}(x-1)$.
f. $s(x) = \text{tri}(x-1) + 2\delta(x-4)$.
g. $u(x) = \delta(x) + \delta(x-4) + \delta(x-7) + \delta(x-9)$.
h. What important conclusions can you draw about the convolution and autocorrelation operations relative to the symmetry (or lack thereof) of the functions involved?

6-6. Evaluate the following convolution integrals:

a. $\text{Gaus}\left(\frac{x}{3}\right)*\text{Gaus}\left(\frac{x}{4}\right) = \int_{-\infty}^{\infty} \exp\left[-\pi\left(\frac{\alpha}{3}\right)^2\right]\exp\left[-\pi\left(\frac{x-\alpha}{4}\right)^2\right]d\alpha.$

b. $\text{sinc}(3x)*\text{sinc}(2x) = \int_{-\infty}^{\infty} \frac{\sin\pi 3\alpha}{\pi 3\alpha}\left[\frac{\sin\pi 2(x-\alpha)}{\pi 2(x-\alpha)}\right]d\alpha.$

c. $\text{sinc}(x)*\text{sinc}^2(0.5x) = \int_{-\infty}^{\infty} \frac{\sin\pi\alpha}{\pi\alpha}\left[\frac{\sin\pi 0.5(x-\alpha)}{\pi 0.5(x-\alpha)}\right]^2 d\alpha.$

d. $\frac{2}{1+(2\pi x)^2}*\frac{2}{1+(2\pi x)^2} = \int_{-\infty}^{\infty}\left[\frac{2}{1+4\pi^2\alpha^2}\right]\left[\frac{2}{1+4\pi^2(x-\alpha)^2}\right]d\alpha.$

CHAPTER 7

THE FOURIER TRANSFORM

The Fourier transform was introduced briefly in Chapter 4, and an attempt was made to indicate its importance in the study of LSI systems. We noted, for example, that almost any reasonably well-behaved function $f(x)$ can be decomposed into a linear combination of complex exponentials of the form $\exp\{j2\pi\xi x\}$ by performing a Fourier transformation. Later, in Chapter 5, we learned that these complex exponentials are eigenfunctions of LSI systems, and that the response of such a system to one of its eigenfunctions is just that same eigenfunction, only possibly changed in height and shifted in phase. We then pointed out that this change of height and shift of phase is governed by the transfer function of the system, which is simply the Fourier transform of the impulse response. Thus we have seen that the study of LSI systems is closely related to the study of Fourier transforms.

In the present chapter, we study the Fourier transform in much greater detail than in the previous chapters. Included are discussions of various interpretations of the Fourier transform, derivations of many of its important properties, and calculations of several elementary Fourier transform pairs. In addition, we discuss some relatively advanced aspects of the transform operations and list a number of transform pairs that are not normally found in textbooks such as this one.

7-1 INTRODUCTION TO THE FOURIER TRANSFORM

Given an arbitrary, complex-valued function $f(x)$, we form the integral

$$F(\xi)=\int_{-\infty}^{\infty} f(\alpha)e^{-j2\pi\xi\alpha}d\alpha, \qquad (7.1)$$

where ξ is a real variable. If this integral exists for all values of ξ, then the function $F(\xi)$ is called the *Fourier transform* of $f(x)$ The conditions required for the existence of $F(\xi)$ were discussed in Chapter 4, but we shall briefly review them here. The function $f(x)$ will have a Fourier transform if it (a) is single valued, (b) has a finite number of maxima and minima in any finite interval, (c) has at most a finite number of finite discontinuities and no infinite discontinuities in any finite interval, and (d) is absolutely integrable on the interval $(-\infty, \infty)$, i.e.,

$$\int_{-\infty}^{\infty} |f(\alpha)|\,d\alpha < \infty.$$

As Bracewell points out (Ref. 7-1), the above conditions are automatically satisfied if the function $f(x)$ accurately describes a real physical quantity. It is clear, however, that many functions used in day-to-day engineering work do not satisfy them. Included in this category are functions having impulsive behavior, functions that are constant over their entire domain, and all periodic functions. Such functions, of course, cannot accurately represent real physical quantities, but they are frequently found to be such good approximations, and their simple form so useful, that to discard them would introduce unnecessary difficulties. Furthermore, even though these functions may not have Fourier transforms in the strict sense, many of them do possess what are called generalized Fourier transforms. To begin with we shall deal only with functions that satisfy the existence conditions, and we postpone the discussion of generalized transforms until later.

As we learned in Chapter 4, Eq. (7.1) can be inverted to obtain

$$f(x) = \int_{-\infty}^{\infty} F(\beta) e^{j2\pi\beta x}\,d\beta, \tag{7.2}$$

assuming, of course, that $F(\beta)$ exists. We call $f(x)$ the *inverse Fourier transform* of $F(\xi)$, and these two functions are known as a *Fourier transform pair*. In Eq. (7.2) $F(\xi)$ serves as a complex-valued weighting factor, prescribing the correct magnitude and phase shift for all of the exponential components, or Fourier components, of which $f(x)$ is comprised. Conversely, in Eq. (7.1) the roles of $F(\xi)$ and $f(x)$ were interchanged; in that expression $f(x)$ prescribed the magnitude and phase shift for all of the (inverse) Fourier components of $F(\xi)$.

Both the integral defining the Fourier transform and the one defining its inverse are frequently referred to simply as *Fourier integrals*, and no distinction is made between them. Although this can initially cause some confusion, it soon becomes clear that there is no fundamental difference between these two integrals; in fact, the *only* difference is in the sign of the exponent of their respective kernels. Consequently, if the transform and inverse transform operation are performed on the same function, the resulting expressions can differ at most by the sign of the argument. To illustrate, let us compare the Fourier transform of $f(x) = \text{rect}(x)$ with the inverse transform of $G(\xi) = \text{rect}(\xi)$. We already know that

$$F(\xi) = \int_{-\infty}^{\infty} \text{rect}(\alpha) e^{-j2\pi\xi\alpha} d\alpha$$

$$= \text{sinc}(\xi), \tag{7.3}$$

so let us concentrate on evaluating

$$g(x) = \int_{-\infty}^{\infty} \text{rect}(\beta) e^{j2\pi\beta x} d\beta. \tag{7.4}$$

First we make the change of variable $x = -\gamma$, and by doing so we obtain

$$g(-\gamma) = \int_{-\infty}^{\infty} \text{rect}(\beta) e^{-j2\pi\beta\gamma} d\beta. \tag{7.5}$$

But this is just the integral of Eq. (7.3), from which we find

$$g(-\gamma) = \text{sinc}(\gamma). \tag{7.6}$$

Finally, with $-\gamma = x$,

$$g(x) = \text{sinc}(-x), \tag{7.7}$$

and we see that the only difference in the form of Eqs. (7.3) and (7.7) is in the sign of the argument. (Actually, in this case, there is no difference because the sinc function is an even function.)

The disparity in the two Fourier integrals can also be explained by considering the complex exponentials $\exp\{\pm j2\pi\xi x\}$ to represent phasors in the complex plane, as discussed in Chapter 2. Thus, the basis functions we have chosen to synthesize $f(x)$ are phasors that rotate in a counterclockwise direction with increasing argument. On the other hand, the basis functions selected for the synthesis of $F(\xi)$ are phasors that rotate clockwise with increasing argument. We could just as well have picked things the other way around, associating the clockwise-rotating phasors with the inverse transform and those rotating counterclockwise with the transform.

The Fourier transform and its inverse are frequently defined by

$$F(\omega)=K_1\int_{-\infty}^{\infty} f(\alpha)e^{-j\omega\alpha}d\alpha \tag{7.8}$$

$$f(x)=K_2\int_{-\infty}^{\infty} F(\beta)e^{j\beta x}d\beta, \tag{7.9}$$

where K_1 and K_2 are real constants such that $K_1K_2=(2\pi)^{-1}$. The rationale behind the use of these definitions is implied to be that, because the kernels do not explicitly contain a factor of 2π, they are less cumbersome than ours. However, this advantage is offset by the requirement to include the constants K_1 and K_2; in addition, unless K_1 and K_2 are chosen to be equal, Eqs. (7.8) and (7.9) do not possess the symmetry that our definitions have. Another consideration has to do with the units of the frequency variable employed. The ξ of our definitions has units of cycles per unit of x, whereas the ω of Eq. (7.8) has dimensions of radians per unit of x. In our studies we generally find it advantageous to specify the frequency variable in cycles per unit of x (e.g., Hz, cycles per millimeter, etc.), and therefore our definitions seem preferable on all counts. However, we once again point out that there is nothing sacred about them.

To avoid the necessity of writing the entire integral each time we wish to perform a transform or inverse-transform operation, we now specify some shorthand notation for these operations. Assuming that $f(x)$ and $F(\xi)$ are a Fourier transform pair, as defined by Eqs. (7.1) and (7.2), we use any of the following to denote their relationship:

$$\begin{aligned} &F(\xi)=\mathfrak{F}\{f(x)\}, &&f(x)=\mathfrak{F}^{-1}\{F(\xi)\},\\ &F(\xi)\overset{\mathfrak{F}}{\leftarrow} f(x), &&f(x)\overset{\mathfrak{F}}{\rightarrow}F(\xi),\\ &F(\xi)\overset{\mathfrak{F}^{-1}}{\rightarrow} f(x), &&f(x)\overset{\mathfrak{F}^{-1}}{\leftarrow}F(\xi). \end{aligned} \tag{7.10}$$

In the last four expressions, the arrow always points in the direction of the operation indicated above it. Thus the expressions of the second row are read in either of two ways: $F(\xi)$ is the Fourier transform of $f(x)$, or the transform of $f(x)$ is $F(\xi)$. Similarly, those in the last row can be taken to mean either that the inverse transform of $F(\xi)$ is $f(x)$, or that $f(x)$ is the inverse transform of $F(\xi)$. It is important to remember that the arrow symbol merely indicates the transform-pair relationship between $f(x)$ and $F(\xi)$, and that it must not be treated as an equality sign; to do so will lead to nothing but trouble. There are other short-hand methods for denoting transform operations, such as overbars, double-headed arrows, etc., but the ones listed above are normally sufficient. We point out that there is no single expression that can be used effectively in every situation, and as you will discover, each form has its time and place in the world of analysis.

We have adopted the lower-case letter/upper-case letter method for denoting a function and its Fourier transform (i.e., the transform of $f(x)$ is denoted by $F(\xi)$, that of $g(x)$ is denoted by $G(\xi)$, etc.), and we shall continue to use this method whenever it is feasible to do so. However, there may be situations when its use is neither practical nor possible. Suppose, for example, we wished to specify the Fourier transform of the function $g(x)$, but found we had already used $G(\cdot)$ to represent some other quantity; or suppose that we wanted to describe the Fourier transform of a function already denoted by an upper-case letter, e.g., the function $E(x)$. One way of getting around this problem is to use the same letter and same case for both the function and the transform, but to place a tilde over the one representing the transform. Thus, for the above situations we would write

$$g(x) \stackrel{\mathscr{F}}{\rightarrow} \tilde{g}(\xi), \qquad g(x) \stackrel{\mathscr{F}^{-1}}{\leftarrow} \tilde{g}(\xi),$$

$$E(x) \stackrel{\mathscr{F}}{\rightarrow} \tilde{E}(\xi), \qquad E(x) \stackrel{\mathscr{F}^{-1}}{\leftarrow} \tilde{E}(\xi). \tag{7.11}$$

There are still other methods used for denoting functions and their transforms, but the two presented in this section should be adequate.

Generalized Fourier Transforms

As mentioned earlier, functions that are constant over their entire domain, functions containing impulses, and periodic functions, to name a few, do not possess Fourier transforms in the strict sense because they violate at least one of the existence conditions. Nevertheless, it is often possible to associate *generalized Fourier transforms* with such functions by employing a limiting procedure (Ref. 7-2). The function under consideration is first defined as the limit of a sequence of functions, each member of which has an ordinary Fourier transform. Then the transform of each member of this sequence is found and a corresponding sequence of transforms is obtained. If the limit of the latter sequence exists, it is called the *generalized Fourier transform* of the original function.

To illustrate, let us consider the impulse function $\delta(x)$. It can be argued that this function does not possess a Fourier transform because it has an infinite discontinuity at the origin. Therefore, let us express $\delta(x)$ as the limit of a sequence of rectangle functions, each of which has a well-defined

Fourier transform, i.e.,

$$\delta(x) = \lim_{b \to 0} \frac{1}{|b|} \operatorname{rect}\left(\frac{x}{b}\right). \tag{7.12}$$

From Eq. (4.69) we can obtain

$$\frac{1}{|b|} \operatorname{rect}\left(\frac{x}{b}\right) \xrightarrow{\mathcal{F}} \operatorname{sinc}(b\xi), \tag{7.13}$$

and it is easy to see that

$$\lim_{b \to 0} \operatorname{sinc}(b\xi) = 1. \tag{7.14}$$

Finally, we write

$$\begin{aligned} \mathcal{F}\{\delta(x)\} &= \mathcal{F}\left\{ \lim_{b \to 0} \frac{1}{|b|} \operatorname{rect}\left(\frac{x}{b}\right)\right\} \\ &= \lim_{b \to 0} \mathcal{F}\left\{ \frac{1}{|b|} \operatorname{rect}\left(\frac{x}{b}\right)\right\} \\ &= \lim_{b \to 0} \operatorname{sinc}(b\xi) \\ &= 1. \end{aligned} \tag{7.15}$$

Thus, the generalized transform of the unit-area delta function is a constant of unit height.

Generalized transforms can be found for many other functions that do not have ordinary Fourier transforms. In fact, a large number of functions encountered in engineering problems possess Fourier transforms only in the generalized sense (e.g., the delta function, the sine and cosine functions, the step function). Generalized transforms are manipulated in the same fashion as ordinary transforms, and in the future we shall make no distinction between the two; it is to be understood that the term "Fourier transform" means "generalized Fourier transform" if we are dealing with functions that violate the existence conditions.

We could have omitted the previous discussion by agreeing to accept, without question, the validity of the equations defining the impulse function [Eqs. (3.16) and (3.17)], because any generalized transforms obtained by the limiting process described above can also be obtained by direct application of these equations. However, for the sake of completeness, we felt that an introduction to the concepts of generalized transforms was desirable. For more details, see Ref. 7-3.

Inversion Formula

We now prove that the inverse Fourier transform of a Fourier transform produces the original function, i.e., that

$$\mathscr{F}^{-1}\{\mathscr{F}\{f(x)\}\}=f(x). \tag{7.16}$$

To begin with,

$$\mathscr{F}\{f(x)\}=\int_{-\infty}^{\infty} f(\alpha)e^{-j2\pi\xi\alpha}d\alpha. \tag{7.17}$$

Hence,

$$\begin{aligned}\mathscr{F}^{-1}\{\mathscr{F}\{f(x)\}\}&=\int_{-\infty}^{\infty}\left[\int_{-\infty}^{\infty} f(\alpha)e^{-j2\pi\beta\alpha}d\alpha\right]e^{j2\pi\beta x}d\beta\\&=\int_{-\infty}^{\infty} f(\alpha)\left[\int_{-\infty}^{\infty} e^{-j2\pi\beta(\alpha-x)}d\beta\right]d\alpha.\end{aligned} \tag{7.18}$$

But by the orthogonality of the complex exponentials (see Chapter 4), we know that

$$\int_{-\infty}^{\infty} e^{-j2\pi\beta(\alpha-x)}d\beta=\delta(\alpha-x). \tag{7.19}$$

Finally, from the defining properties of the delta function, we obtain

$$\begin{aligned}\mathscr{F}^{-1}\{\mathscr{F}\{f(x)\}\}&=\int_{-\infty}^{\infty} f(\alpha)\delta(\alpha-x)d\alpha\\&=f(x),\end{aligned} \tag{7.20}$$

and the proof is complete. In a similar manner, it may be shown that $\mathscr{F}\{\mathscr{F}^{-1}\{F(\xi)\}\}=F(\xi)$.

We now see that there is a one-to-one correspondence between a function and its Fourier transform. Every transformable function possesses a unique Fourier transform, the inverse of which is exactly the original function. As a result, any such function can be represented equally well in its own domain and in the domain of its Fourier transform; the choice is governed by the details of the particular situation.

7-2 INTERPRETATIONS OF THE FOURIER TRANSFORM

In this section we attempt to develop a better understanding of the mechanics of the Fourier transformation, and we start by considering the Fourier transform of an arbitrary complex-valued function $f(x)$. Aside from the various shorthand forms, the transform of $f(x)$ is most concisely expressed as

$$F(\xi)=\int_{-\infty}^{\infty} f(\alpha)e^{-j2\pi\xi\alpha}d\alpha. \tag{7.21}$$

However, from a physical point of view, this form is perhaps not the easiest to interpret. We therefore use Euler's formula to write

$$F(\xi)=\int_{-\infty}^{\infty} f(\alpha)\cos(2\pi\xi\alpha)d\alpha - j\int_{-\infty}^{\infty} f(\alpha)\sin(2\pi\xi\alpha)d\alpha, \tag{7.22}$$

and if we now express $f(x)$ in terms of its real and imaginary components, i.e.,

$$f(x)=v(x)+jw(x), \tag{7.23}$$

Eq. (7.22) becomes

$$F(\xi)=\int_{-\infty}^{\infty} v(\alpha)\cos(2\pi\xi\alpha)d\alpha + \int_{-\infty}^{\infty} w(\alpha)\sin(2\pi\xi\alpha)d\alpha$$
$$+j\left[\int_{-\infty}^{\infty} w(\alpha)\cos(2\pi\xi\alpha)d\alpha - \int_{-\infty}^{\infty} v(\alpha)\sin(2\pi\xi\alpha)d\alpha\right]. \tag{7.24}$$

Thus, the Fourier transform operation can be performed by evaluating four integrals and properly adding together the results. Note that although $F(\xi)$ is in general complex valued, the integrands of each term are real valued. Consequently, there should be no conceptual difficulties encountered in calculating $F(\xi)$.

If we now require that $f(x)$ be real valued, i.e., $w(x)=0$, then $f(x)=v(x)$ and Eq. (7.24) reduces to

$$F(\xi)=\int_{-\infty}^{\infty} f(\alpha)\cos(2\pi\xi\alpha)d\alpha - j\int_{-\infty}^{\infty} f(\alpha)\sin(2\pi\xi\alpha)d\alpha. \tag{7.25}$$

Thus, only two integrals need be evaluated. If $f(x)$ is both real valued and even, the integrand of the second term of Eq. (7.25) is odd, which causes it to vanish, and we are left with

$$F(\xi)=\int_{-\infty}^{\infty} f(\alpha)\cos(2\pi\xi\alpha)d\alpha. \tag{7.26}$$

Hence the entire process reduces to one of evaluating a single integral, the integrand of which is real valued. Of course, the original expression for $F(\xi)$ [Eq. (7.1)] also has only one integral to evaluate, but that evaluation is not as easy to visualize physically because the integrand is complex valued. For other real, imaginary, even, and odd combinations, results similar to the above are obtained, but we shall wait until Sec. 7-3 to discuss them further.

We now require $f(x)$ to be both real and even. As a result, we can investigate the mechanics of the Fourier transform by evaluating the single integral of Eq. (7.26) and we need not concern ourselves with the more general formula given by Eq. (7.24). Nevertheless, an arbitrary $f(x)$ would require at most four calculations of the type we are about to make. For our development, let us choose $f(x)=\text{rect}(x)$, which is one of the simplest functions available. As we already know, its Fourier transform is given by

$$\begin{aligned} F(\xi) &= \int_{-\infty}^{\infty} \text{rect}(\alpha)\cos(2\pi\xi\alpha)d\alpha \\ &= \text{sinc}(\xi), \end{aligned} \tag{7.27}$$

and we may interpret this expression as follows: at any particular frequency ξ, the value of $F(\xi)$ is just equal to the area of the product $\text{rect}(x)\cos(2\pi\xi x)$. For example, at $\xi=\xi_1$, we have

$$F(\xi_1)=\int_{-\infty}^{\infty} \text{rect}(\alpha)\cos(2\pi\xi_1\alpha)d\alpha, \tag{7.28}$$

whereas at $\xi=\xi_2$

$$F(\xi_2)=\int_{-\infty}^{\infty} \text{rect}(\alpha)\cos(2\pi\xi_2\alpha)d\alpha, \tag{7.29}$$

at ξ_3

$$F(\xi_3)=\int_{-\infty}^{\infty} \text{rect}(\alpha)\cos(2\pi\xi_3\alpha)d\alpha, \tag{7.30}$$

etc.

We may view the integrand $\text{rect}(x)\cos(2\pi\xi x)$ as the two-dimensional function represented graphically by the surface of Fig. 7-1. The rectangle

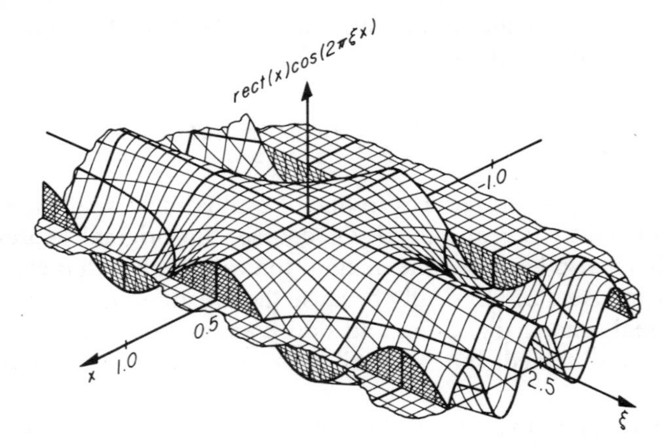

Figure 7-1 Two-dimensional representation of the integrand of the Fourier integral of rect(x).

function acts as an envelope and is modulated by a cosine function, the frequency of which increases linearly with $|\xi|$. If we were to slice this surface along lines perpendicular to the ξ-axis, the profiles generated would be cosine functions whose frequency depends on the location of the slice and whose extent in the x-direction is governed by rect(x). Thus, for any particular frequency ξ, corresponding to a given slice, the value of $F(\xi)$ is the area of the associated profile. This point of view is illustrated in Fig. 7-2, which shows several profiles of rect$(x)\cos(2\pi\xi x)$ and their relationship to the resulting transform $F(\xi) = \text{sinc}(\xi)$.

With $\xi = 0$, the cosine function is unity for all x, and thus $F(0)$ is just equal to the area of rect(x). Then, as $|\xi|$ increases from zero, the cosine function begins to oscillate, the frequency of these oscillations growing linearly with $|\xi|$, and more and more positive and negative lobes of $\cos(2\pi\xi x)$ are included within the rectangle function envelope as $|\xi|$ gets larger and larger. Whenever the areas of the positive and negative lobes are equal, which occurs for integral values of ξ, we have $F(\xi) = 0$. Between these zeros, $F(\xi)$ alternates from positive to negative values as additional positive and negative lobes are introduced under rect(x).

The fluctuations of $F(\xi)$ are related to the abruptness with which the lobes of the cosine function influence the value of the integral as a function of ξ. Since the rectangle function has significant discontinuities at $x = \pm 0.5$, this effect is quite pronounced. In addition, $F(\xi)$ dies away rather slowly because of the discontinuities in $f(x)$. Such behavior can be visualized readily with the aid of Figs. 7-1 and 7-2. The transform of a smoother function would exhibit less-pronounced fluctuations but would

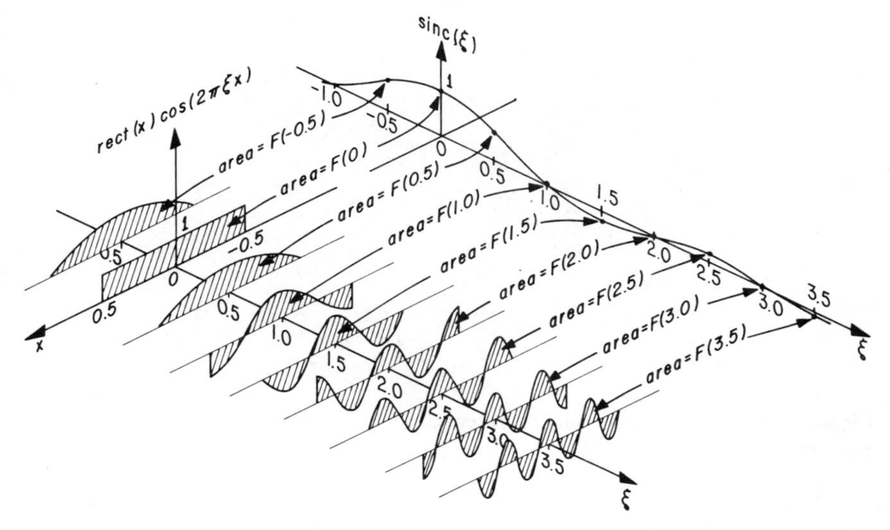

Figure 7-2 The Fourier integral of rect(x) is equal to the area of the profiles of its integrand.

die out more rapidly with increasing $|\xi|$. We shall have more to say about this shortly.

First, however, let us consider the inverse transform of the function $F(\xi)=\text{sinc}(\xi)$. Since sinc(ξ) is both real and even, we can invoke the previously used arguments to show that

$$f(x)=\int_{-\infty}^{\infty}\text{sinc}(\beta)\cos(2\pi\beta x)d\beta. \tag{7.31}$$

Thus, at any point x, $f(x)$ has a value equal to the area of sinc(ξ) $\cos(2\pi\xi x)$. As before, we consider this product to be a two-dimensional function, and we represent it by the surface of Fig. 7-3. If we now slice this surface along lines perpendicular to the x-axis, the profile generated at any point x has an area equal to the value of $f(x)$ at that point. This is illustrated in Fig. 7-4, which also shows the inverse transform $f(x)=$ rect(x).

The previously discussed relationship between the smoothness of a function and the behavior of its transform also applies here, but now the roles of $f(x)$ and $F(\xi)$ are reversed. In the present case it is the smoothness of $F(\xi)$ that determines the oscillatory behavior of $f(x)$, as well as its asymptotic behavior. It is interesting to note that, with the exception of the discontinuities at $x=\pm 0.5$, rect(x) has no fluctuations at all because sinc(ξ) is such a smooth function.

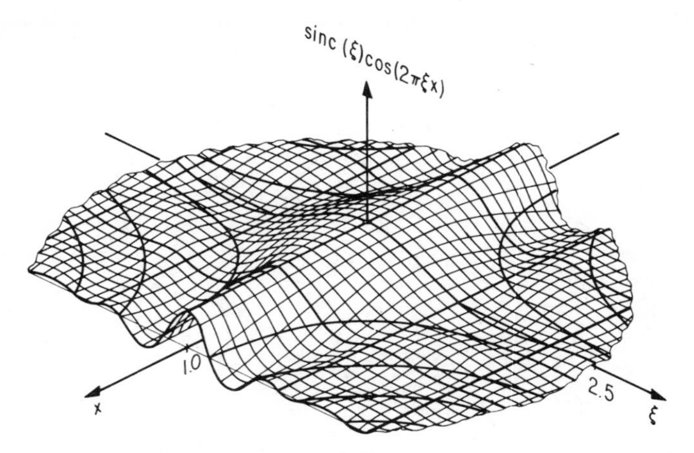

Figure 7-3 Two-dimensional representation of the integrand of the inverse Fourier transform of sinc(ξ).

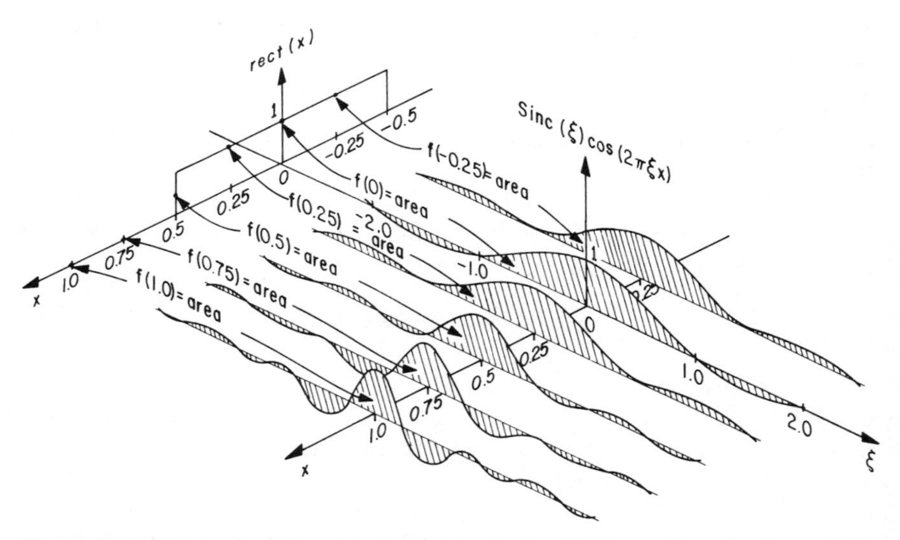

Figure 7-4 The inverse Fourier transform of sinc(ξ) is equal to the area of the profiles of its integrand.

It should be clear that the surfaces of Figs. 7-1 and 7-3 are very closely related. The profile of the first surface along the line $\xi=0$ is just $f(x)$, and the profile of the second surface along the line $x=0$ is just $F(\xi)$. Thus, these two profiles form a Fourier transform pair. Many other interesting observations can also be made concerning these figures.

Now let us return to the discussion of smoothness and its effect on Fourier transformation. A little thought will show that if a function $f(x)$ exhibits rapid oscillations, sharp peaks, discontinuities, etc., such behavior must be due to those Fourier components whose periods are at least as small as the scale of the fine structure of $f(x)$. In other words, the finer the structure of $f(x)$ is, the higher the frequency its Fourier components will have and the broader its spectrum $F(\xi)$ will be. Some examples illustrating this are shown in Fig. 7-5.

The smoother a function is, the more rapidly its transform will approach zero with increasing frequency. This behavior is put on more rigorous footing by the following statements (Ref. 7-4). For sufficiently large $|\xi|$, the Fourier transform of a function having no discontinuities decreases in magnitude at least as fast as $|\xi|^{-1}$. However, if the function has one or more discontinuities, its transform can decrease no faster than this. More generally, if a function and its first k derivatives are continuous, then the

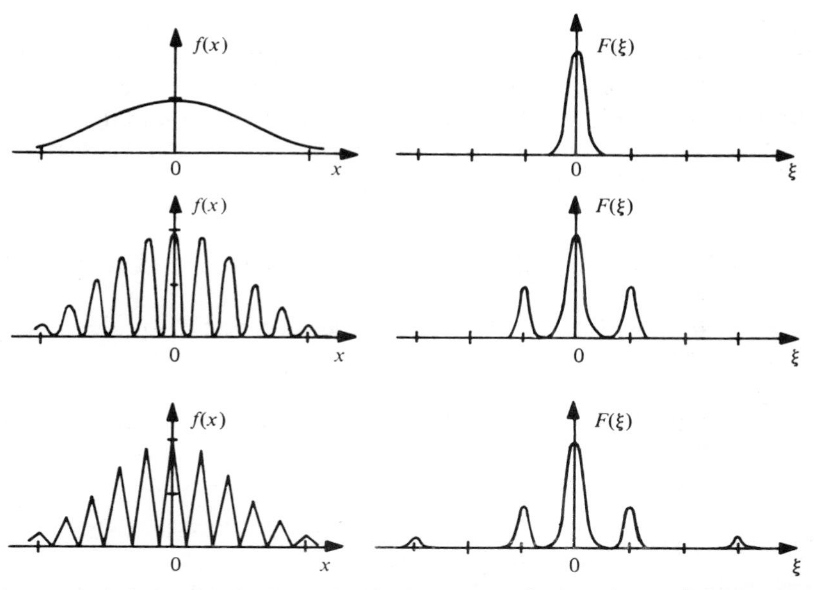

Figure 7-5 Relationship between the fine structure of a function and the nature of its spectrum.

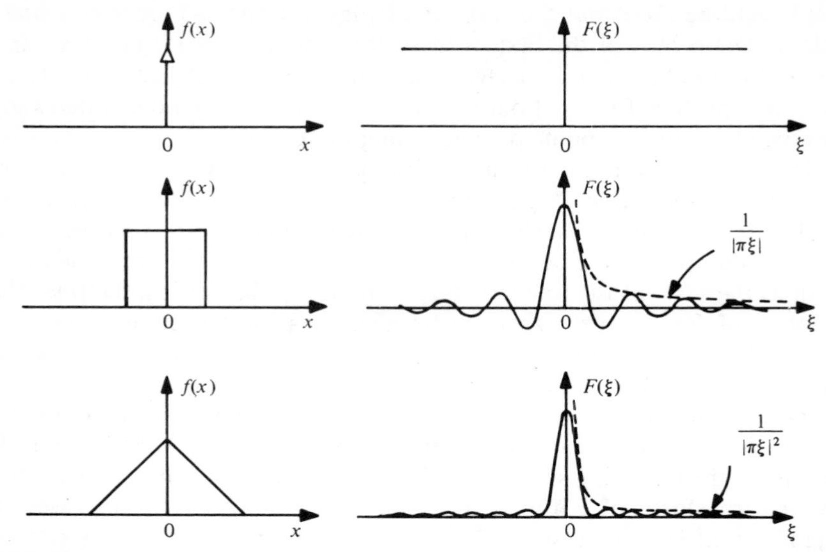

Figure 7-6 Effect of the smoothness of a function on the behavior of its Fourier transform for large values of $|\xi|$.

magnitude of its transform will decrease at least as fast as $|\xi|^{-(k+2)}$ for sufficiently large $|\xi|$. In terms of impulsive behavior, we have the following (Ref. 7-1). If $f(x)$ contains one or more impulses, $|F(\xi)|$ behaves as $|\xi|^0$ (i.e., a constant) for large $|\xi|$. If $f(x)$ is not impulsive, and if its mth derivative is the first to become impulsive, then, in general, $|F(\xi)|$ behaves as $|\xi|^{-m}$ at infinity. Several examples are shown in Fig. 7-6.

7-3 PROPERTIES OF THE FOURIER TRANSFORM

In this section we derive several important properties of the Fourier transform, and we also present several others without derivation. Knowing the Fourier transforms of only a few functions, transforms of many other functions can be found quite easily by applying these properties. Table 7-1 lists the symmetry properties of the Fourier transform; Table 7-2 contains a fairly extensive summary of other properties.

Symmetry Properties

Many problems involving Fourier transforms can be simplified by taking advantage of their symmetry properties. This was demonstrated earlier in the chapter when we found that the Fourier transform of an even,

Table 7-1

Symmetry Properties of Fourier Transforms

$f(x)$	$F(\xi)$
Complex, no symmetry	Complex, no symmetry
Hermitian	Real, no symmetry
Antihermitian	Imaginary, no symmetry
Complex, even	Complex, even
Complex, odd	Complex, odd
Real, no symmetry	Hermitian
Real, even	Real, even
Real, odd	Imaginary, odd
Imaginary, no symmetry	Antihermitian
Imaginary, even	Imaginary, even
Imaginary, odd	Real, odd

real-valued function could be found by evaluating a single integral with a real-valued integrand. We also discovered that the resulting transform was itself real and even.

The summary of symmetry properties listed in Table 7-1 can be obtained by visual inspection of Eqs. (7.21), (7.22), or (7.24). If a function is neither even nor odd, we shall say that it has no symmetry. If its real part is even and its imaginary part odd, it will be called hermitian, whereas for the opposite case it will be called antihermitian. It is interesting to note that the even and odd parts of any transform pair are themselves transform pairs.

Linearity Property

If $f(x)\xrightarrow{\mathcal{F}}F(\xi)$ and $h(x)\xrightarrow{\mathcal{F}}H(\xi)$, and if A_1 and A_2 are arbitrary constants, then the Fourier transform of the sum $[A_1f(x)+A_2h(x)]$ is

$$\begin{aligned}\mathcal{F}\{A_1f(x)+A_2h(x)\}&=\int_{-\infty}^{\infty}[A_1f(\alpha)+A_2h(\alpha)]e^{-j2\pi\xi\alpha}d\alpha\\&=A_1\int_{-\infty}^{\infty}f(\alpha)e^{-j2\pi\xi\alpha}d\alpha+A_2\int_{-\infty}^{\infty}h(\alpha)e^{-j2\pi\xi\alpha}d\alpha\\&=A_1F(\xi)+A_2H(\xi).\end{aligned}\tag{7.32}$$

Thus, *the transform of a sum of two functions is the sum of their individual transforms*. This result is very important in the study of LSI systems, because it allows us to compute the spectrum of a sum of signals by simply adding together their individual spectra.

Central Ordinate

If Eq. (7.21) is evaluated at $\xi=0$, we have

$$F(0)=\int_{-\infty}^{\infty} f(\alpha)d\alpha. \tag{7.33}$$

Thus, *the area of a function is equal to the central ordinate of its Fourier transform* (see Fig. 7-2). It is also easy to show that

$$f(0)=\int_{-\infty}^{\infty} F(\beta)d\beta. \tag{7.34}$$

These results can be quite useful at times, because they allow the height of a transform to be found without requiring the entire function to be determined.

Scaling Property

If $f(x)\xrightarrow{\mathcal{F}} F(\xi)$ and b is a real, nonzero constant, then

$$\begin{aligned}\mathcal{F}\left\{f\left(\frac{x}{b}\right)\right\} &= \int_{-\infty}^{\infty} f\left(\frac{\alpha}{b}\right)e^{-j2\pi\xi\alpha}d\alpha \\ &= |b|\int_{-\infty}^{\infty} f(\beta)e^{-j2\pi(b\xi)\beta}d\beta \\ &= |b|F(b\xi). \end{aligned} \tag{7.35}$$

By first letting $b>0$, and then $b<0$, it may be seen why the absolute

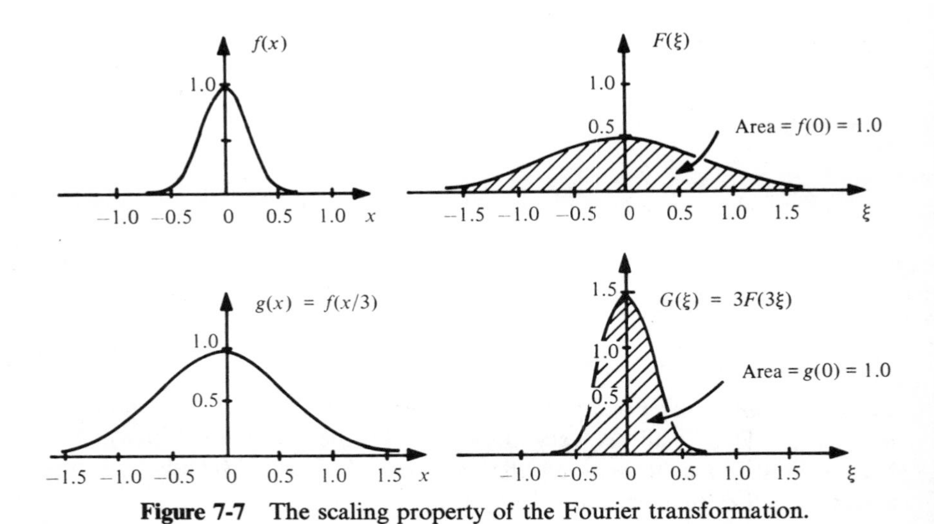

Figure 7-7 The scaling property of the Fourier transformation.

magnitude symbol is required. From this expression we see that if the width of a function is increased (while its height is kept constant), its Fourier transform becomes narrower and taller; if its width is decreased, its transform becomes wider and shorter. This relationship is illustrated in Fig. 7-7. Note that as long as $f(0)$ remains constant, the area of the transform must remain constant.

With $b=-1$, we obtain the interesting result

$$f(-x) \xrightarrow{\mathcal{F}} F(-\xi). \tag{7.36}$$

Shifting Property

If $f(x) \xrightarrow{\mathcal{F}} F(\xi)$ and x_0 is a real constant, possibly zero, then

$$\begin{aligned}\mathcal{F}\{f(x-x_0)\} &= \int_{-\infty}^{\infty} f(\alpha - x_0) e^{-j2\pi\xi\alpha} d\alpha \\ &= \int_{-\infty}^{\infty} f(\beta) e^{-j2\pi\xi(\beta+x_0)} d\beta \\ &= e^{-j2\pi x_0 \xi} \int_{-\infty}^{\infty} f(\beta) e^{-j2\pi\xi\beta} d\beta \\ &= e^{-j2\pi x_0 \xi} F(\xi). \end{aligned} \tag{7.37}$$

Thus, *the Fourier transform of a shifted function is simply the transform of the unshifted function multiplied by an exponential factor having linear phase*. If we let $F(\xi) = A(\xi)\exp\{-j\Phi(\xi)\}$, then

$$f(x-x_0) \xrightarrow{\mathcal{F}} A(\xi)\exp\{-j[\Phi(\xi)+2\pi x_0 \xi]\}, \tag{7.38}$$

and it is clear that the overall phase spectrum of a shifted function will contain a linear-phase term. This term causes each Fourier component to be shifted in phase by an amount proportional to the product of its frequency ξ and the shift distance x_0. Then the sum of all the phase-shifted Fourier components yields the shifted function as discussed in Chapter 4.

Transform of a Conjugate

If $f(x) \xrightarrow{\mathcal{F}} F(\xi)$, the Fourier transform of $f^*(x)$ is given by

$$\begin{aligned}\mathcal{F}\{f^*(x)\} &= \int_{-\infty}^{\infty} f^*(\alpha) e^{-j2\pi\xi\alpha} d\alpha \\ &= \left[\int_{-\infty}^{\infty} f(\alpha) e^{-j2\pi(-\xi)\alpha} d\alpha\right]^* \\ &= F^*(-\xi). \end{aligned} \tag{7.39}$$

Transform of a Transform

Suppose we know that $f(x) \xrightarrow{\mathcal{F}} F(\xi)$, and that we would like to find the Fourier transform of the function $F(x)$. (Remember, the variables x and ξ merely identify the space of concern; $F(\cdot)$ defines the function.) Let us choose $F(x) = g(x)$, and let $g(x) \xrightarrow{\mathcal{F}} G(\xi)$. Therefore,

$$\begin{aligned} \mathcal{F}\{F(x)\} &= G(\xi) \\ &= \int_{-\infty}^{\infty} g(\alpha) e^{-j2\pi\xi\alpha} d\alpha \\ &= \int_{-\infty}^{\infty} F(\alpha) e^{-j2\pi\xi\alpha} d\alpha \\ &= \int_{-\infty}^{\infty} F(\alpha) e^{j2\pi(-\xi)\alpha} d\alpha \\ &= f(-\xi). \end{aligned} \tag{7.40}$$

The last step follows from Eq. (7.2). This result is quite useful, because it effectively provides a method for doubling the size of a table of Fourier transforms without the necessity of actually calculating any new transforms.

Transform of a Convolution

We now derive one of the most important results for the study of LSI systems. Suppose that $g(x) = f(x) * h(x)$ and $g(x) \xrightarrow{\mathcal{F}} G(\xi)$. Then, with $f(x) \xrightarrow{\mathcal{F}} F(\xi)$ and $h(x) \xrightarrow{\mathcal{F}} H(\xi)$,

$$\begin{aligned} G(\xi) &= \mathcal{F}\{f(x) * h(x)\} \\ &= \mathcal{F}\left\{\int_{-\infty}^{\infty} f(\beta) h(x-\beta) d\beta\right\} \\ &= \int_{-\infty}^{\infty} \left[\int_{-\infty}^{\infty} f(\beta) h(\alpha-\beta) d\beta\right] e^{-j2\pi\xi\alpha} d\alpha \\ &= \int_{-\infty}^{\infty} f(\beta) \left[\int_{-\infty}^{\infty} h(\alpha-\beta) e^{-j2\pi\xi\alpha} d\alpha\right] d\beta, \end{aligned} \tag{7.41}$$

where we have interchanged the order of integration. From Eq. (7.37) the inner integral is seen to be equal to $H(\xi)\exp\{-j2\pi\xi\beta\}$, and we obtain

$$\begin{aligned} G(\xi) &= \int_{-\infty}^{\infty} f(\beta)H(\xi)e^{-j2\pi\xi\beta}d\beta \\ &= H(\xi)\int_{-\infty}^{\infty} f(\beta)e^{-j2\pi\xi\beta}d\beta \\ &= F(\xi)H(\xi). \end{aligned} \tag{7.42}$$

In shorthand form this becomes

$$f(x)*h(x) \xrightarrow{\mathcal{F}} F(\xi)H(\xi), \tag{7.43}$$

and we see that *the Fourier transform of a convolution is simply given by the product of the individual transforms*. This is an *extremely powerful* result, and one that we shall use extensively.

Transform of a Product

By using a development similar to that above, it is easy to show that *the Fourier transform of a product is given by the convolution of the individual transforms*, i.e.,

$$f(x)h(x) \xrightarrow{\mathcal{F}} F(\xi)*H(\xi), \tag{7.44}$$

where $f(x) \xrightarrow{\mathcal{F}} F(\xi)$ and $h(x) \xrightarrow{\mathcal{F}} H(\xi)$.

Transform of a Derivative

Given that $f(x) \xrightarrow{\mathcal{F}} F(\xi)$, we write $f(x)$ in the form

$$f(x) = \int_{-\infty}^{\infty} F(\beta)e^{j2\pi\beta x}d\beta. \tag{7.45}$$

Differentiating both sides of this equation k times with respect to x, we obtain

$$\begin{aligned} f^{(k)}(x) &= \frac{d^k}{dx^k}\int_{-\infty}^{\infty} F(\beta)e^{j2\pi\beta x}d\beta \\ &= \int_{-\infty}^{\infty} (j2\pi\beta)^k F(\beta)e^{j2\pi\beta x}d\beta, \end{aligned} \tag{7.46}$$

and comparing this equation with Eq. (7.2), it is clear that $f^{(k)}(x)$ and $(j2\pi\beta)^k F(\beta)$ are a Fourier transform pair.

We now derive this result by another method, one that demonstrates the utility of the delta function derivatives. From Eqs. (6.22) and (7.44) we may write

$$\begin{aligned}\mathcal{F}\{f^{(k)}(x)\} &= \mathcal{F}\{f(x)*\delta^{(k)}(x)\}\\ &= F(\xi)\mathcal{F}\{\delta^{(k)}(x)\},\end{aligned} \tag{7.47}$$

but from Eq. (3.38) we find that

$$\begin{aligned}\mathcal{F}\{\delta^{(k)}(x)\} &= \int_{-\infty}^{\infty}\delta^{(k)}(\alpha)e^{-j2\pi\xi\alpha}d\alpha\\ &= (-1)^k\frac{d^k}{dx^k}\left[e^{j2\pi\xi x}\right]\bigg|_{x=0}\\ &= (-1)^k(-j2\pi\xi)^k\\ &= (j2\pi\xi)^k.\end{aligned} \tag{7.48}$$

Combining Eqs. (7.47) and (7.48) we obtain the result,

$$f^{(k)}(x) \overset{\mathcal{F}}{\rightarrow} (j2\pi\xi)^k F(\xi). \tag{7.49}$$

In a similar fashion, it may be shown that

$$(-j2\pi x)^k f(x) \overset{\mathcal{F}}{\rightarrow} F^{(k)}(\xi). \tag{7.50}$$

Transform of an Integral

Here we wish to find an expression for the Fourier transform of the integral

$$g(x) = \int_{-\infty}^{x} f(\alpha)d\alpha. \tag{7.51}$$

In doing so we make use of a Fourier transform relationship that is not derived until Sec. 7-4; we ask your indulgence in this particular case, because the problem at hand is greatly simplified by using the fact that the transform of the step function is given by

$$\text{step}(x) \overset{\mathcal{F}}{\rightarrow} \frac{1}{j2\pi\xi} + \tfrac{1}{2}\delta(\xi). \tag{7.52}$$

Following Papoulis (Ref. 7-5), we may put g(x) in the form of a convolution, i.e.,

$$g(x)=f(x)*\text{step}(x). \tag{7.53}$$

Therefore, from Eqs. (7.43) and (7.52),

$$\begin{aligned} G(\xi) &= F(\xi)\left[\frac{1}{j2\pi\xi}+\tfrac{1}{2}\delta(\xi)\right] \\ &= \frac{1}{j2\pi\xi}F(\xi)+\frac{F(0)}{2}\delta(\xi), \end{aligned}$$

or

$$\int_{-\infty}^{x} f(\alpha)d\alpha \xrightarrow{\mathfrak{F}} \frac{1}{j2\pi\xi}F(\xi)+\frac{F(0)}{2}\delta(\xi). \tag{7.54}$$

Table of Properties

For easy reference, most of the important properties of the Fourier transform are listed in Table 7-2. In addition, several frequently used special forms are included.

Table 7-2

Properties of Fourier Transforms

A_1 and A_2 arbitrary constants; b and d real nonzero constants; x_0 and ξ_0 real constants; k a positive integer

$g(x)=\int_{-\infty}^{\infty} G(\beta)e^{j2\pi\beta x}d\beta$	$G(\xi)=\int_{-\infty}^{\infty} g(\alpha)e^{-j2\pi\alpha\xi}d\alpha$
$f(\pm x)$	$F(\pm\xi)$
$f^*(\pm x)$	$F^*(\mp\xi)$
$F(\pm x)$	$f(\mp\xi)$
$F^*(\pm x)$	$f^*(\pm\xi)$
$f\left(\frac{x}{b}\right)$	$\lvert b\rvert F(b\xi)$
$\lvert d\rvert f(dx)$	$F\left(\frac{\xi}{d}\right)$
$f(x\pm x_0)$	$e^{\pm j2\pi x_0\xi}F(\xi)$
$e^{\pm j2\pi\xi_0 x}f(x)$	$F(\xi\mp\xi_0)$

Table 7-2 (Continued)

$g(x)=\int_{-\infty}^{\infty} G(\beta)e^{j2\pi\beta x}d\beta$	$G(\xi)=\int_{-\infty}^{\infty} g(\alpha)e^{-j2\pi\alpha\xi}d\alpha$
$f\left(\frac{x \pm x_0}{b}\right)$	$\lvert b\rvert e^{\pm j2\pi x_0 \xi}F(b\xi)$
$\lvert d\rvert e^{\pm j2\pi\xi_0 x}f(dx)$	$F\left(\frac{\xi \mp \xi_0}{d}\right)$
$f^{(k)}(x)$	$(j2\pi\xi)^k F(\xi)$
$(-j2\pi x)^k f(x)$	$F^{(k)}(\xi)$
$\int_{-\infty}^{x} f(\alpha)d\alpha$	$\frac{1}{j2\pi\xi}F(\xi)+\frac{F(0)}{2}\delta(\xi)$
$\frac{1}{-j2\pi x}f(x)+\frac{f(0)}{2}\delta(x)$	$\int_{-\infty}^{\xi} F(\beta)d\beta$
$h(x)$	$H(\xi)$
$A_1 f(x)+A_2 h(x)$	$A_1 F(\xi)+A_2 H(\xi)$
$f(x)*h(x)$	$F(\xi)H(\xi)$
$f(x)h(x)$	$F(\xi)*H(\xi)$
$f(x)\star h(x)$	$F(\xi)H(-\xi)$
$f(x)h(-x)$	$F(\xi)\star H(\xi)$
$\gamma_{fh}(x)=f(x)\star h^*(x)$	$F(\xi)H^*(\xi)$
$f(x)h^*(x)$	$\gamma_{FH}(\xi)=F(\xi)\star H^*(\xi)$
$\gamma_f(x)=f(x)\star f^*(x)$	$\lvert F(\xi)\rvert^2$
$\lvert f(x)\rvert^2$	$\gamma_F(\xi)=F(\xi)\star F^*(\xi)$
$\sum_{n=-\infty}^{\infty} f(x-nd)$	$\frac{1}{\lvert d\rvert}\sum_{n=-\infty}^{\infty} F\left(\frac{n}{d}\right)\delta\left(\xi-\frac{n}{d}\right)$
$\frac{1}{\lvert b\rvert}\sum_{n=-\infty}^{\infty} f\left(\frac{n}{b}\right)\delta\left(x-\frac{n}{b}\right)$	$\sum_{n=-\infty}^{\infty} F(\xi-nb)$

7-4 ELEMENTARY FOURIER TRANSFORM PAIRS

We now derive a number of elementary Fourier transform pairs, making liberal use of the properties derived in the preceeding section. These transform pairs, as well as a number of others, are listed in Table 7-3.

Table 7-3

Elementary Fourier Transform Pairs

x_0 and ξ_0 real constants; k a nonnegative integer
a and c real constants; x and ξ real variables

$g(x)=\int_{-\infty}^{\infty} G(\beta)e^{j2\pi x\beta}d\beta$	$G(\xi)=\int_{-\infty}^{\infty} g(\alpha)e^{-j2\pi\xi\alpha}d\alpha$
1	$\delta(\xi)$
$\delta(x)$	1
$\delta(x\pm x_0)$	$e^{\pm j2\pi x_0\xi}$
$e^{\pm j2\pi\xi_0 x}$	$\delta(\xi\mp\xi_0)$
$\cos(2\pi\xi_0 x)$	$\frac{1}{2\lvert\xi_0\rvert}\delta\delta\left(\frac{\xi}{\xi_0}\right)$
$\frac{1}{2\lvert x_0\rvert}\delta\delta\left(\frac{x}{x_0}\right)$	$\cos(2\pi x_0\xi)$
$\sin(2\pi\xi_0 x)$	$\frac{j}{2\lvert\xi_0\rvert}\delta\delta\left(\frac{\xi}{\xi_0}\right)$
$\frac{j}{2\lvert x_0\rvert}\delta\delta\left(\frac{x}{x_0}\right)$	$-\sin(2\pi x_0\xi)$
$\mathrm{rect}(x)$	$\mathrm{sinc}(\xi)$
$\mathrm{sinc}(x)$	$\mathrm{rect}(\xi)$
$\mathrm{tri}(x)$	$\mathrm{sinc}^2(\xi)$
$\mathrm{sinc}^2(x)$	$\mathrm{tri}(\xi)$
$\mathrm{sgn}(x)$	$\frac{1}{j\pi\xi}$
$\frac{1}{j\pi x}$	$-\mathrm{sgn}(\xi)$
$\mathrm{step}(x)$	$\frac{1}{2}\delta(\xi)+\frac{1}{j2\pi\xi}$

Table 7-3 (Continued)

$g(x)=\int_{-\infty}^{\infty} G(\beta)e^{j2\pi x\beta}d\beta$	$G(\xi)=\int_{-\infty}^{\infty} g(\alpha)e^{-j2\pi\xi\alpha}d\alpha$
$\frac{1}{2}\delta(x)-\frac{1}{j2\pi x}$	$\text{step}(\xi)$
$\text{ramp}(x)$	$\frac{1}{4\pi^2}\left[j\pi\delta^{(1)}(\xi)-\frac{1}{\xi^2}\right]$
$\frac{1}{4\pi^2}\left[\frac{1}{x^2}+j\pi\delta^{(1)}(x)\right]$	$-\text{ramp}(\xi)$
$e^{-\lvert x\rvert}$	$\frac{2}{1+(2\pi\xi)^2}$
$\frac{2}{1+(2\pi x)^2}$	$e^{-\lvert\xi\rvert}$
$e^{-x}\text{step}(x)$	$\frac{1}{1+j2\pi\xi}$
$\frac{1}{1-j2\pi x}$	$e^{-\xi}\text{step}(\xi)$
x^k	$\left(\frac{-1}{j2\pi}\right)^k\delta^{(k)}(\xi)$
$\left(\frac{1}{j2\pi}\right)^k\delta^{(k)}(x)$	ξ^k
$\text{comb}(x)$	$\text{comb}(\xi)$
$\text{Gaus}(x)$	$\text{Gaus}(\xi)$
$\text{sech}(\pi x)$	$\text{sech}(\pi\xi)$
$\frac{1}{\lvert x\rvert^{1/2}}$	$\frac{1}{\lvert\xi\rvert^{1/2}}$
$\cos\pi(x^2-\frac{1}{8})$	$\cos\pi(\xi^2-\frac{1}{8})$
$\sin\pi(x^2-\frac{1}{8})$	$-\sin\pi(\xi^2-\frac{1}{8})$
$\exp[\pm j\pi(x^2-\frac{1}{8})]$	$\exp[\mp j\pi(\xi^2-\frac{1}{8})]$
$\cos(\pi x^2)$	$\cos\pi(\xi^2-\frac{1}{4})$
$\sin(\pi x^2)$	$-\sin\pi(\xi^2-\frac{1}{4})$
$\exp(\pm j\pi x^2)$	$\exp\left(\pm j\frac{\pi}{4}\right)\exp(\mp j\pi\xi^2)$
$\exp\left[-\pi\left(\frac{x^2}{a+jc}\right)\right]$, $a\geqslant 0$, $a^2+c^2<\infty$	$(a+jc)^{1/2}\exp[-\pi(a+jc)\xi^2]$

Transform of $\delta(x)$

The Fourier transform of $\delta(x)$ is

$$\mathcal{F}\{\delta(x)\} = \int_{-\infty}^{\infty} \delta(\alpha) e^{-j2\pi\xi\alpha} d\alpha$$

$$= e^{-j2\pi\xi\alpha}\Big|_{\alpha=0}$$

$$= 1. \tag{7.55}$$

Then, from Eq. (7.32) we find that

$$A\delta(x) \xrightarrow{\mathcal{F}} A, \tag{7.56}$$

and from Eq. (7.37) we obtain

$$\delta(x - x_0) \xrightarrow{\mathcal{F}} e^{-j2\pi x_0 \xi}. \tag{7.57}$$

Transform of a Constant

Combining Eqs. (7.40) and (7.56), the Fourier transform of the constant A is found to be

$$A \xrightarrow{\mathcal{F}} A\delta(\xi). \tag{7.58}$$

Transform of $e^{j2\pi\xi_0 x}$

From Eq. (7.57) we have

$$\delta(x - x_0) \xrightarrow{\mathcal{F}} e^{-j2\pi x_0 \xi},$$

and, making use of the property $\mathcal{F}\{F^*(x)\} = f^*(\xi)$, we find

$$e^{j2\pi\xi_0 x} \xrightarrow{\mathcal{F}} \delta(\xi - \xi_0). \tag{7.59}$$

Transform of $\cos(2\pi\xi_0 x)$

If $\cos(2\pi\xi_0 x)$ is expanded according to Euler's formula, Eqs. (7.32), (7.40), and (7.59) may be used to obtain

$$\mathcal{F}\{\cos(2\pi\xi_0 x)\} = \tfrac{1}{2}\mathcal{F}\{e^{j2\pi\xi_0 x} + e^{-j2\pi\xi_0 x}\}$$

$$= \tfrac{1}{2}[\delta(\xi - \xi_0) + \delta(\xi + \xi_0)]$$

$$= \frac{1}{2|\xi_0|}\delta\delta\left(\frac{\xi}{\xi_0}\right). \tag{7.60}$$

***Transform of* sin($2\pi\xi_0 x$)**

Again Euler's formula is used to expand $\sin(2\pi\xi_0 x)$, with the result

$$\mathcal{F}\{\sin(2\pi\xi_0 x)\} = \frac{1}{2j}\mathcal{F}\{e^{j2\pi\xi_0 x} - e^{-j2\pi\xi_0 x}\}$$

$$= \frac{1}{2j}[\delta(\xi-\xi_0) - \delta(\xi+\xi_0)]$$

$$= \frac{j}{2|\xi_0|}\delta\delta\left(\frac{\xi}{\xi_0}\right). \tag{7.61}$$

***Transform of* sinc(x)**

We have already determined that the transform of $\mathrm{rect}(x)$ is given by $\mathrm{sinc}(\xi)$. Thus, from Eq. (7.40) we obtain

$$\mathcal{F}\{\mathrm{sinc}(x)\} = \mathrm{rect}(-\xi)$$

$$= \mathrm{rect}(\xi). \tag{7.62}$$

The last step follows because $\mathrm{rect}(\xi)$ is even.

***Transform of* tri(x)**

It is easy to show that $\mathrm{tri}(x) = \mathrm{rect}(x) * \mathrm{rect}(x)$, and, therefore, with the aid of Eq. (7.43),

$$\mathcal{F}\{\mathrm{tri}(x)\} = \mathcal{F}\{\mathrm{rect}(x) * \mathrm{rect}(x)\}$$

$$= \mathcal{F}\{\mathrm{rect}(x)\}\mathcal{F}\{\mathrm{rect}(x)\}$$

$$= \mathrm{sinc}^2(\xi). \tag{7.63}$$

***Transform of* $(j\pi x)^{-1}$**

The transform of $(j\pi x)^{-1}$ is written

$$\mathcal{F}\left\{\frac{1}{j\pi x}\right\} = \int_{-\infty}^{\infty}\left(\frac{1}{j\pi\alpha}\right)e^{-j2\pi\xi\alpha}d\alpha$$

$$= \int_{-\infty}^{\infty}\frac{\cos(2\pi\xi\alpha)}{j\pi\alpha}d\alpha - j\int_{-\infty}^{\infty}\frac{\sin(2\pi\xi\alpha)}{j\pi\alpha}d\alpha, \tag{7.64}$$

but the first of these integrals has an odd integrand and vanishes. Thus, with the appropriate substitutions, we may write

$$\mathcal{F}\left\{\frac{1}{j\pi x}\right\} = -2\xi\int_{-\infty}^{\infty} \frac{\sin(2\pi\xi\alpha)}{2\pi\xi\alpha}\,d\alpha$$

$$= -2\xi\int_{-\infty}^{\infty} \operatorname{sinc}(2\xi\alpha)d\alpha$$

$$= \frac{-\xi}{|\xi|}$$

$$= \operatorname{sgn}(-\xi). \tag{7.65}$$

The integral was evaluated by noting that the area of sinc($2\xi x$) is equal to $|2\xi|^{-1}$.

***Transform of* sgn(x)**

Once again making use of Eq. (7.40), the result given by Eq. (7.65) leads to

$$\operatorname{sgn}(x) \xrightarrow{\mathcal{F}} \frac{1}{j\pi\xi}. \tag{7.66}$$

***Transform of* step(x)**

Noting that step(x)$=\frac{1}{2}+\frac{1}{2}$sgn(x), we use Eqs. (7.56) and (7.66) to obtain

$$\mathcal{F}\{\operatorname{step}(x)\} = \mathcal{F}\{\tfrac{1}{2}+\tfrac{1}{2}\operatorname{sgn}(x)\}$$

$$= \tfrac{1}{2}\delta(\xi) + \frac{1}{j2\pi\xi}. \tag{7.67}$$

***Transform of* comb(x)**

To begin with we recognize that comb(x) is a periodic function with period $X=1$. We assume, therefore, that we are permitted to represent it by an expansion of the form given by Eq. (4.29).* Thus we write

$$\operatorname{comb}(x) = \sum_{n=-\infty}^{\infty} c_n e^{j2\pi nx}. \tag{7.68}$$

*In doing so, we are ignoring the Dirichlet condition that prohibits impulsive behavior in the function to be expanded. Nevertheless, this difficulty can be circumvented by using a limiting technique to describe each delta function component of the comb function.

The coefficients c_n are found by applying Eq. (4.30), i.e.,

$$\begin{aligned}
c_n &= \int_{-1/2}^{1/2} f(\alpha) e^{-j2\pi n\alpha} d\alpha \\
&= \int_{-1/2}^{1/2} \mathrm{comb}(\alpha) e^{-j2\pi n\alpha} d\alpha \\
&= \int_{-\infty}^{\infty} \delta(\alpha) e^{-j2\pi n\alpha} d\alpha \\
&= 1,
\end{aligned} \tag{7.69}$$

and thus Eq. (7.68) yields

$$\mathrm{comb}(x) = \sum_{n=-\infty}^{\infty} e^{j2\pi nx}. \tag{7.70}$$

Therefore,

$$\mathcal{F}\{\mathrm{comb}(x)\} = \mathcal{F}\left\{ \sum_{n=-\infty}^{\infty} e^{j2\pi nx} \right\}, \tag{7.71}$$

but by combining the linearity property of the Fourier transform with the result given by Eq. (7.59), we find that

$$\begin{aligned}
\mathcal{F}\{\mathrm{comb}(x)\} &= \sum_{n=-\infty}^{\infty} \mathcal{F}\{ e^{j2\pi nx} \} \\
&= \sum_{n=-\infty}^{\infty} \delta(\xi - n) \\
&= \mathrm{comb}(\xi).
\end{aligned} \tag{7.72}$$

This is a remarkable result; the Fourier transform of a comb function is itself a comb function! However, this is just one of many Fourier transform pairs whose members are identical, and we shall encounter several more such pairs in the sections that follow.

Other Transform Pairs of Interest

We now give, without derivation, several other important transform pairs. All of these pairs were taken from Campbell and Foster (Ref. 7-6).

The following are all identical Fourier transform pairs:

$$\text{Gaus}(x) \xrightarrow{\mathcal{F}} \text{Gaus}(\xi), \tag{7.73}$$

$$\text{sech}(\pi x) \xrightarrow{\mathcal{F}} \text{sech}(\pi \xi), \tag{7.74}$$

$$\frac{1}{|x|^{1/2}} \xrightarrow{\mathcal{F}} \frac{1}{|\xi|^{1/2}}, \tag{7.75}$$

$$\cos\pi\left(x^2 - \tfrac{1}{8}\right) \xrightarrow{\mathcal{F}} \cos\pi\left(\xi^2 - \tfrac{1}{8}\right), \tag{7.76}$$

In Eq. (7.74), $\text{sech}(x) = 2[\exp(x) + \exp(-x)]^{-1}$.

Although not identical, two other pairs that exhibit a great degree of similarity are

$$\sin\pi\left(x^2 - \tfrac{1}{8}\right) \xrightarrow{\mathcal{F}} -\sin\pi\left(\xi^2 - \tfrac{1}{8}\right), \tag{7.77}$$

$$\exp\left[\pm j\pi\left(x^2 - \tfrac{1}{8}\right)\right] \xrightarrow{\mathcal{F}} \exp\left[\mp j\pi\left(\xi^2 - \tfrac{1}{8}\right)\right]. \tag{7.78}$$

From Eq. (7.78) we easily obtain

$$\exp(\pm j\pi x^2) \xrightarrow{\mathcal{F}} \exp\left(\pm j\frac{\pi}{4}\right)\exp(\mp j\pi\xi^2), \tag{7.79}$$

or, in our notation,

$$\text{Gaus}\left(\sqrt{\mp j}\;x\right) \xrightarrow{\mathcal{F}} \exp\left(\pm j\frac{\pi}{4}\right)\text{Gaus}\left(\sqrt{\pm j}\;\xi\right). \tag{7.80}$$

Equation (7.79) can then be used to obtain

$$\cos(\pi x^2) \xrightarrow{\mathcal{F}} \cos\pi\left(\xi^2 - \tfrac{1}{4}\right), \tag{7.81}$$

$$\sin(\pi x^2) \xrightarrow{\mathcal{F}} -\sin\pi\left(\xi^2 - \tfrac{1}{4}\right). \tag{7.82}$$

We now choose two real constants a and c, such that $a \geqslant 0$ and $a^2 + c^2 < \infty$, and form the complex number $a + jc$. We then find (Ref. 7-6)

$$\exp\left[-\pi\left(\frac{x^2}{a+jc}\right)\right] \xrightarrow{\mathcal{F}} (a+jc)^{1/2}\exp\left[-\pi(a+jc)\xi^2\right], \tag{7.83}$$

which can be expressed as

$$\text{Gaus}\left(\frac{x}{\sqrt{a+jc}}\right) \xrightarrow{\mathcal{F}} \sqrt{a+jc}\ \text{Gaus}\left(\sqrt{a+jc}\ \xi\right), \tag{7.84}$$

and with $a=0$ and $c=1$, these last two equations reduce to Eqs. (7.79) and (7.80).

Table of Fourier Transform Pairs

For convenience, we list all of the Fourier transform pairs discussed in the present section in Table 7-3. In addition, we list a few useful pairs that were not discussed. All of the pairs listed can be found in one form or another in standard references on the Fourier transform, e.g., Refs. 7-1, 7-6, and 7-7.

7-5 THE FOURIER TRANSFORM AND LINEAR SHIFT-INVARIANT SYSTEMS

In Chapter 6 we discovered that the output of an LSI system can be found by convolving the input with the system's impulse response; i.e., if an input $f(x)$ is applied to a system whose impulse response is $h(x)$, the output is given by

$$g(x)=f(x)*h(x). \tag{7.85}$$

We also spent a good deal of time discussing this result from a physical point of view.

We now investigate a different, and oftentimes simpler, method for determining the output of such a system, and we do so by exploring the relationship between the spectra of the input and the output. First, we calculate the spectrum of the input and the *transfer function* (system function, frequency response) of the system as follows:

$$F(\xi)=\mathcal{F}\{f(x)\},$$

$$H(\xi)=\mathcal{F}\{h(x)\}. \tag{7.86}$$

Next, by making use of Eqs. (7.85) and (7.43), we find that the output spectrum can be expressed by

$$\begin{aligned}G(\xi) &= \mathcal{F}\{g(x)\}\\ &= \mathcal{F}\{f(x)*h(x)\}\\ &= F(\xi)H(\xi).\end{aligned} \tag{7.87}$$

Thus, we see that *the output spectrum of an LSI system is given simply by the product of the input spectrum and the transfer function*. Of course, once the output spectrum is known, the output itself can be determined by the inverse Fourier transformation

$$\begin{aligned}g(x) &= \mathcal{F}^{-1}\{G(\xi)\}\\ &= \mathcal{F}^{-1}\{F(\xi)H(\xi)\}.\end{aligned} \tag{7.88}$$

At first glance it may seem that this procedure, which involves a direct transformation, a multiplication operation, and an inverse transformation, is much more involved and cumbersome than the evaluation of the convolution of Eq. (7.85). However, this is certainly not the case a great deal of the time. Armed with a basic knowledge of the properties of Fourier transforms and a good table of Fourier transform pairs, one is usually able to solve these kinds of problems much more easily by using the approach just outlined than by using the convolution approach.

Example

An example illustrating this method will be helpful at this point. Let us consider a system whose impulse response is $h(x)=|1/b|\text{sinc}^2(x/b)$, and to which the input $f(x)=A\cos(2\pi\xi_0 x)$ is applied. The input spectrum is found to be

$$\begin{aligned}F(\xi) &= \mathcal{F}\{A\cos(2\pi\xi_0 x)\}\\ &= \frac{A}{2|\xi_0|}\delta\delta\left(\frac{\xi}{\xi_0}\right),\end{aligned} \tag{7.89}$$

and the transfer function is

$$\begin{aligned}H(\xi) &= \mathcal{F}\left\{\frac{1}{|b|}\text{sinc}^2\left(\frac{x}{b}\right)\right\}\\ &= \text{tri}(b\xi).\end{aligned} \tag{7.90}$$

Then, from Eq. (7.87) we obtain

$$G(\xi)=\frac{A}{2|\xi_0|}\delta\delta\left(\frac{\xi}{\xi_0}\right)\text{tri}(b\xi)$$

$$=\frac{A}{2}\left[\delta(\xi+\xi_0)+\delta(\xi-\xi_0)\right]\text{tri}(b\xi)$$

$$=\frac{A}{2}\left[\delta(\xi+\xi_0)\text{tri}(-b\xi_0)+\delta(\xi-\xi_0)\text{tri}(b\xi_0)\right]$$

$$=\frac{A}{2}\text{tri}(b\xi_0)\left[\delta(\xi+\xi_0)+\delta(\xi-\xi_0)\right]$$

$$=\text{tri}(b\xi_0)\frac{A}{2|\xi_0|}\delta\delta\frac{\xi}{\xi_0}. \tag{7.91}$$

Finally, the output itself is found to be

$$g(x)=\mathcal{F}^{-1}\left\{\text{tri}(b\xi_0)\frac{A}{2|\xi_0|}\delta\delta\left(\frac{\xi}{\xi_0}\right)\right\}$$

$$=\text{tri}(b\xi_0)\mathcal{F}^{-1}\left\{\frac{A}{2|\xi_0|}\delta\delta\left(\frac{\xi}{\xi_0}\right)\right\}$$

$$=\text{tri}(b\xi_0)A\cos(2\pi\xi_0 x)$$

$$=\text{tri}(b\xi_0)f(x). \tag{7.92}$$

Thus, in this example, the output has exactly the same form as the input (see Sec. 5-4), and its magnitude is determined by the relative values of b and ξ_0: if $|b\xi_0|<1$, then $g(x)=[1-|\xi_0 b|]f(x)$, whereas if $|b\xi_0|>1$, $g(x)=0$. ■

If we attempt to solve this problem by using the convolution approach, we must evaluate the following integral:

$$g(x)=\frac{A}{b}\int_{-\infty}^{\infty}\cos(2\pi\xi_0\alpha)\left[\frac{\sin\pi\left(\frac{x-\alpha}{b}\right)}{\pi\left(\frac{x-\alpha}{b}\right)}\right]^2 d\alpha. \tag{7.93}$$

This integration can be performed, but the frequency-domain method proves to be much easier.

Now let us look at this approach from a more physical point of view. We know from previous discussions that the Fourier components of any suitable function are also eigenfunctions of LSI systems. Thus, an arbitrary input signal can be decomposed into a linear combination of such eigenfunctions. We also know that the only effect an LSI system can have on one of its eigenfunctions is to possibly alter the magnitude and phase of this eigenfunction, the degree of which is governed by the associated eigenvalue. But the eigenvalue depends on the frequency of the eigenfunction, and therefore the various Fourier components of the input may be altered by different amounts as they pass through the system. (Recall that the eigenvalue for a particular frequency is just equal to the value of the transfer function at that frequency.)

Because the system is linear, and because the eigenfunctions comprising the input are passed through the system unchanged in form, the output can also be expressed as *a linear combination of these same eigenfunctions*. However, because the extent to which these eigenfunctions are attenuated and shifted in phase is frequency-dependent, the output will *not generally be the same linear combination of eigenfunctions* as the input. That this is so may easily be seen by expressing the input and output functions as inverse transforms of their spectra, i.e.,

$$f(x)=\int_{-\infty}^{\infty} F(\beta)e^{j2\pi\beta x}d\beta,$$

$$g(x)=\int_{-\infty}^{\infty} F(\beta)H(\beta)e^{j2\pi\beta x}d\beta. \tag{7.94}$$

Clearly, $g(x)$ can be the same linear combination of eigenfunctions as $f(x)$ only if the transfer function is unity everywhere. In this event, all Fourier components are passed unaltered by the system, and the input and output are identical. For such a system, which cannot exist in the real world, the impulse response is a delta function.

The transfer functions of physical systems actually encountered in practice all tend to approach zero for sufficiently large values of the frequency variable. It can therefore be tacitly assumed that the output of any such system will not contain Fourier components whose frequencies lie above some upper limit, which is often referred to as the *cutoff frequency* of the system. In other words, all physical systems tend to completely *filter out*, or suppress, those Fourier components whose frequencies lie above this limiting value. Depending on the type of system under consideration, those Fourier components of the input whose frequencies lie below the cutoff frequency may be selectively attenuated and shifted in phase; e.g., there may be frequency "bands" in which none of the components are passed by the system, or in which all of the components are passed unaltered, etc., with the parameters of the system governing the exact nature of the filtering that takes place. The topic of filtering will be discussed in much greater detail in Chapter 8.

7-6 RELATED TOPICS

In this section we introduce two additional topics that are related to the Fourier transform in one way or another; the moment theorem and Rayleigh's theorem.

The Moment Theorem

Given a function $f(x)$ and its Fourier transform $F(\xi)$, we employ Eq. (7.50) to write

$$F^{(k)}(\xi)=\int_{-\infty}^{\infty}(-j2\pi\alpha)^k f(\alpha)e^{-j2\pi\xi\alpha}d\alpha. \tag{7.95}$$

By setting $\xi=0$ and dividing both sides of this equation by $(-j2\pi)^k$, we obtain

$$\int_{-\infty}^{\infty}\alpha^k f(\alpha)d\alpha=\frac{F^{(k)}(0)}{(-j2\pi)^k}, \tag{7.96}$$

where

$$F^{(k)}(0)=\left.\frac{d^kF(\xi)}{d\xi^k}\right|_{\xi=0}.$$

The quantity on the left side of Eq. (7.96) is called the *kth moment of $f(x)$*, and we shall denote it by m_k, i.e.,

$$\begin{aligned} m_k&=\int_{-\infty}^{\infty}\alpha^k f(\alpha)d\alpha \\ &=\frac{F^{(k)}(0)}{(-j2\pi)^k}. \end{aligned} \tag{7.97}$$

Thus, *the kth moment of $f(x)$ is proportional to the kth derivative of its Fourier transform, evaluated at the origin*. As explained below, the moments of a function are frequently needed to simplify the solutions of various problems.

The Moment m_0

This moment is given by

$$m_0 = F(0), \tag{7.98}$$

which we already know is simply equal to the area of the function $f(x)$. Clearly, m_0 will always be zero for odd functions, and may be either zero or nonzero for functions that are not odd.

The First Moment m_1

The first moment of the function $f(x)$ has a value proportional to the slope of its Fourier transform at the origin:

$$m_1 = \frac{F^{(1)}(0)}{-j2\pi}. \tag{7.99}$$

From our knowledge of the symmetry properties of Fourier transforms, it is not difficult to conclude that the first moment of any even function must be zero, and that the first moment of an odd function must be nonzero. Going one step farther, any function that does not possess even symmetry will have a nonzero first moment.

The *centroid* (mean abscissa, center of gravity) of a function, which we shall denote by $\bar{x}$, is equal to the first moment divided by the area. Thus, the centroid of $f(x)$ is given by

$$\bar{x} = \frac{m_1}{m_0} = \frac{F^{(1)}(0)}{-j2\pi F(0)}. \tag{7.100}$$

This quantity, when finite, specifies the abscissa about which $f(x)$ is chiefly concentrated, and is important in such fields as mechanics, statistics, etc. Note that when the area of the function is zero, the centroid is not defined.

The Second Moment m_2

From Eq. (7.97), we see that the second moment of $f(x)$ can be written as

$$m_2 = \frac{F^{(2)}(0)}{-4\pi^2}. \tag{7.101}$$

The greater the central curvature of $F(\xi)$, the larger the value of the second moment. When the curvature becomes infinite, such that $F(\xi)$ has a cusp at the origin, m_2 becomes infinite; e.g., the function $f(x) = \mathrm{sinc}^2(x)$ has an infinite second moment because $F(\xi) = \mathrm{tri}(\xi)$ has a cusp at the origin. For a function to have a finite second moment, it must die out more rapidly than x^{-2}.

The mean-square abscissa of $f(x)$ is found by

$$\overline{x^2} = \frac{m_2}{m_0} = \frac{F^{(2)}(0)}{-4\pi^2 F(0)}, \tag{7.102}$$

and is a measure of the dispersion of $f(x)$ about the origin. Its square root $(\overline{x^2})^{1/2}$ is frequently referred to as the radius of gyration of $f(x)$. The *variance* σ^2 is the mean-square abscissa referred to the centroid, and it specifies the dispersion of $f(x)$ about the centroid rather than the origin. It can be shown that the variance is given by

$$\begin{aligned}\sigma^2 &= \overline{x^2} - (\overline{x})^2 \\ &= \frac{F^{(2)}(0)}{-4\pi^2 F(0)} + \frac{1}{4\pi^2}\left[\frac{F^{(1)}(0)}{F(0)}\right]^2.\end{aligned} \tag{7.103}$$

The quantity σ is often called the *standard deviation* of $f(x)$, or the root-mean-square (rms) deviation about the centroid, and it is a measure of the width of $f(x)$. Thus, we see that for a suitable function $f(x)$ the centroid $\overline{x}$ tells us where $f(x)$ is concentrated and σ tells us how much it is spread out about the centroid.

Example

We shall use the methods outlined above to determine the centroid and standard deviation of the function $f(x) = \mathrm{Gaus}(x/b)$. First, we find that

$$\begin{aligned}F(\xi) &= b\,\mathrm{Gaus}(b\xi), \\ F^{(1)}(\xi) &= -2\pi b^3 \xi\,\mathrm{Gaus}(b\xi), \\ F^{(2)}(\xi) &= \left[-2\pi b^3 + 4\pi^2 b^5 \xi^2\right]\mathrm{Gaus}(b\xi),\end{aligned} \tag{7.104}$$

so that

$$\begin{aligned}m_0 &= F(0) = b, \\ m_1 &= \frac{F^{(1)}(0)}{-j2\pi} = 0, \\ m_2 &= \frac{F^{(2)}(0)}{-4\pi^2} = \frac{b^3}{2\pi}.\end{aligned} \tag{7.105}$$

Finally,

$$\bar{x} = \frac{m_1}{m_0} = 0,$$

$$\overline{x^2} = \frac{m_2}{m_0} = \frac{b^2}{2\pi},$$

$$\sigma^2 = \overline{x^2} - (\bar{x})^2 = \frac{b^2}{2\pi}. \tag{7.106}$$

Thus, the centroid of Gaus(x/b) is located at the origin, as expected, and the standard deviation is $\sigma = b/\sqrt{2\pi} \cong 0.4b$. Note that because the centroid lies at the origin, the variance and mean-square abscissa are equal. ■

Example

We repeat the above procedure for the function $f(x) = \delta(x - x_0)$. We obtain

$$F(\xi) = e^{-j2\pi x_0 \xi},$$

$$F^{(1)}(\xi) = -j2\pi x_0 e^{-j2\pi x_0 \xi},$$

$$F^{(2)}(\xi) = -4\pi^2 x_0^2 e^{-j2\pi x_0 \xi}, \tag{7.107}$$

with the result that

$$m_0 = F(0) = 1,$$

$$m_1 = \frac{F^{(1)}(0)}{-j2\pi} = x_0,$$

$$m_2 = \frac{F^{(2)}(0)}{-4\pi^2} = x_0^2. \tag{7.108}$$

Hence,

$$\bar{x} = \frac{m_1}{m_0} = x_0,$$

$$\overline{x^2} = \frac{m_2}{m_0} = x_0^2,$$

$$\sigma^2 = \overline{x^2} - (\bar{x})^2 = 0, \tag{7.109}$$

and in this case the centroid is located at x_0, which again agrees with our understanding of the term, whereas the standard deviation is zero, as it should be for an infinitely narrow delta function. ■

Rayleigh's Theorem

We now derive a theorem that is quite useful from time to time. Given two arbitrary functions $f(x)$ and $g(x)$, we form the product $f(x)g^*(x)$. Expressing each of the factors in terms of its Fourier transform, the area of this product can be written as

$$\int_{-\infty}^{\infty} f(\alpha)g^*(\alpha)d\alpha = \int_{-\infty}^{\infty}\left[\int_{-\infty}^{\infty} F(\beta)e^{j2\pi\alpha\beta}d\beta\right]\left[\int_{-\infty}^{\infty} G(\beta')e^{j2\pi\alpha\beta'}d\beta'\right]^* d\alpha$$

$$= \iint_{-\infty}^{\infty} F(\beta)G^*(\beta')\left[\int_{-\infty}^{\infty} e^{j2\pi(\beta-\beta')\alpha}d\alpha\right] d\beta\, d\beta'$$

$$= \iint_{-\infty}^{\infty} F(\beta)G^*(\beta')\delta(\beta-\beta')d\beta\, d\beta'$$

$$= \int_{-\infty}^{\infty} F(\beta)G^*(\beta)d\beta. \tag{7.110}$$

This result is often referred to as the power theorem, and it equates the area of the product $f(x)g^*(x)$ with that of $F(\xi)G^*(\xi)$.

If we now choose $g(x)=f(x)$, we obtain *Rayleigh's theorem* (energy theorem, Parseval's theorem, Plancherel's theorem), i.e.,

$$\int_{-\infty}^{\infty} |f(\alpha)|^2 d\alpha = \int_{-\infty}^{\infty} |F(\beta)|^2 d\beta. \tag{7.111}$$

Thus the squared modulus of a function and the squared modulus of its transform have equal areas. From a physical point of view, each integral of Eq. (7.111) may represent the energy associated with some process, and the theorem simply states that the computation may be made in either the domain of the function or the domain of its transform. For more details, see Bracewell (Ref. 7-1).

When $f(x)$ is periodic, its Fourier transform is given by an array of delta functions, i.e.,

$$F(\xi) = \sum_{n=-\infty}^{\infty} c_n \delta(\xi - n\xi_0). \tag{7.112}$$

Equation (7.111) must then be modified as follows:

$$\int_{-\infty}^{\infty} |f(\alpha)|^2 d\alpha = \sum_{n=-\infty}^{\infty} |c_n|^2. \tag{7.113}$$

This expression is sometimes referred to as *Parseval's theorem for Fourier series*.

REFERENCES

7-1 R. Bracewell, *The Fourier Transform and Its Applications*, McGraw-Hill, New York, 1965.

7-2 J. W. Goodman, *Introduction to Fourier Optics*, McGraw-Hill, New York, 1968, p. 7.

7-3 M. J. Lighthill, *Introduction to Fourier Analysis and Generalised Functions*, Cambridge University Press, London, 1958.

7-4 C. R. Wylie, Jr., *Advanced Engineering Mathematics*, McGraw-Hill, New York, 1951, pp. 127–128.

7-5 A. Papoulis, *The Fourier Integral and Its Applications*, McGraw-Hill, New York, 1962, p. 26.

7-6 G. A. Campbell and R. M. Foster, *Fourier Integrals for Practical Applications*, Van Nostrand, New York, 1948.

7-7 A. Papoulis, *Systems and Transforms with Applications in Optics*, McGraw-Hill, New York, 1968.

PROBLEMS

7-1. Given the function $f(x) = \exp\{j\phi(x)\}$, with Fourier transform $F(\xi) = \mathcal{F}\{f(x)\}$, show that:

a. $\mathcal{F}\{\cos\phi(x)\} = 0.5[F(\xi) + F^*(-\xi)]$.

b. $\mathcal{F}\{\sin\phi(x)\} = 0.5j[F^*(-\xi) - F(\xi)]$.

c. $\mathcal{F}\{F(x)\} = e^{j\phi(-\xi)}$.

d. $\mathcal{F}\{F^*(-x)\} = e^{-j\phi(-\xi)}$.

7-2. With $f(x) \xrightarrow{\mathcal{F}} F(\xi)$ an arbitrary Fourier transform pair, and with x_0, ξ_0, and b real constants, show that:

a. $\mathcal{F}\left\{\delta\left(\frac{x - x_0}{b}\right)\right\} = |b| e^{-j2\pi x_0 \xi}$.

b. $\mathcal{F}\left\{f(x) * \delta\left(\frac{x - x_0}{b}\right)\right\} = |b| F(\xi) e^{-j2\pi x_0 \xi}$.

c. $\mathcal{F}\{f(x)\delta(x - x_0)\} = f(x_0) e^{-j2\pi x_0 \xi}$.

d. $f(x) * e^{j2\pi\xi_0 x} = F(\xi_0) e^{j2\pi\xi_0 x}$.

7-3. Find the Fourier transforms of the following functions:

a. $f(x) = \text{rect}\left(\frac{x}{2}\right)$.

b. $g(x) = \text{rect}(x-1)$.

c. $h(x) = 0.5\,\text{rect}\left(\frac{x-1}{2}\right)$.

d. $p(x) = \text{rect}(x-2)e^{j2\pi x}$.

e. $r(x) = 0.5\,\text{rect}(0.5x) + \text{tri}(x)$. Sketch $r(x)$ and $R(\xi)$.

f. $s(x) = \text{rect}(0.5x) - \text{tri}(x)$. Sketch $s(x)$ and $S(\xi)$.

g. $u(x) = 3\,\text{sinc}(3x) - \text{sinc}(x)$. Sketch $u(x)$ and $U(\xi)$.

h. $v(x) = \int_{-\infty}^{x} \text{rect}(\alpha)\,d\alpha$.

7-4. Find the inverse Fourier transforms of the following functions:

a. $F(\xi) = \left(\frac{1}{10}\right)\delta\delta\left(\frac{\xi}{10}\right)$.

b. $G(\xi) = \text{sinc}^2\left(\frac{\xi-1}{2}\right)$.

c. $H(\xi) = \left(\frac{1}{1+j2\pi\xi}\right)e^{-j4\pi\xi}$. Sketch $h(x)$.

d. $P(\xi) = \cos\pi\left[\left(\frac{\xi}{2}\right)^2 - \frac{1}{8}\right]$. Sketch $P(\xi)$ and $p(x)$.

e. $R(\xi) = \text{Gaus}(\xi+5) + \text{Gaus}(\xi-5)$. Sketch $R(\xi)$ and $r(x)$.

f. $S(\xi) = \text{tri}(\xi+1) - \text{tri}(\xi-1)$. Sketch $S(\xi)$ and $s(x)$.

g. $U(\xi) = \text{rect}\left(\frac{\xi}{3}\right) - \text{rect}(\xi)$. Sketch $U(\xi)$ and $u(x)$.

h. $V(\xi) = j2\pi\xi\,\text{Gaus}(\xi)$. Sketch $v(x)$.

7-5. Find the Fourier transforms of the following functions:

a. $f(x) = \text{rect}(x) * \text{rect}(x-1)$.

b. $g(x) = \text{rect}(x) \star \text{rect}(x-1)$.

c. $h(x) = \text{rect}(x-1) \star \text{rect}(x)$.

d. $p(x) = 2\,\text{sinc}(2x)\text{sinc}(x)$.

e. $r(x) = \text{rect}\left(\frac{x}{3}\right)\text{rect}\left(\frac{x+3}{5}\right)$. Sketch $r(x)$.

f. Use the results of part e to show that

$$3\,\text{sinc}(3\xi) * 5\,\text{sinc}(5\xi)e^{j6\pi\xi} = \text{sinc}(\xi)e^{j2\pi\xi}.$$

g. $s(x) = \cos(2\pi x)\text{Gaus}\left(\frac{x}{5}\right)$. Sketch $s(x)$ and $S(\xi)$.

h. $t(x) = \text{rect}(x-1) * \text{rect}(x-1)$.

i. $u(x) = \text{rect}(x-1) \star \text{rect}(x-1)$.

7-6. Given the real constants a and b and the function

$$f(x) = \frac{1}{|a|}\delta\delta\left(\frac{x}{a}\right) * \text{rect}\left(\frac{x}{b}\right),$$

find an expression for the Fourier transform $F(\xi) = \mathcal{F}\{f(x)\}$ when:

a. $a = 10b$. Sketch $f(x)$ and $F(\xi)$.

b. $a = 2b$. Sketch $f(x)$ and $F(\xi)$.

c. $a = b$. Sketch $f(x)$ and $F(\xi)$.

d. $a = 0.5b$. Sketch $f(x)$ and $F(\xi)$.

e. Use the results of part d to show that

$$\cos(\pi b\xi)\text{sinc}(b\xi) = \text{sinc}(2b\xi).$$

7-7. Perform the following convolutions and sketch your results:

a. $f(x) = \text{Gaus}\left(\frac{x}{3}\right) * \text{Gaus}\left(\frac{x}{4}\right)$.

b. $g(x) = \text{sinc}(3x) * \text{sinc}(2x)$.

c. $h(x) = \text{sinc}(x) * \text{sinc}^2(0.5x)$.

d. $p(x) = \dfrac{2}{1+(2\pi x)^2} * \dfrac{2}{1+(2\pi x)^2}$.

7-8. Given the real positive constant ξ_0 and the real constants a and b:

a. For $|b| < (2\xi_0)^{-1}$, show that

$$\frac{1}{|b|}\text{sinc}\left(\frac{x}{b}\right) * \cos(2\pi\xi_0 x) = \cos(2\pi\xi_0 x).$$

b. For $|b| = (2\xi_0)^{-1}$, show that

$$\frac{1}{|b|}\text{sinc}\left(\frac{1}{b}\right) * \sin(2\pi\xi_0 x) = 0.5\sin(2\pi\xi_0 x).$$

c. For $|b| > (2\xi_0)^{-1}$, show that

$$\frac{1}{|b|}\text{sinc}\left(\frac{x}{b}\right) * \cos(2\pi\xi_0 x) = 0.$$

d. For $|b| < |a|$, show that $\text{sinc}\left(\frac{x}{b}\right) * \text{sinc}\left(\frac{x}{a}\right) = |b|\,\text{sinc}\left(\frac{x}{a}\right)$.

e. For $|b|<|0.5a|$, show that $\text{sinc}\left(\frac{x}{b}\right)*\text{sinc}^2\left(\frac{x}{a}\right)=|b|\text{sinc}^2\left(\frac{x}{a}\right)$.

f. For $|b|>|0.5a|$, find an expression for

$$h(x)=\text{sinc}\left(\frac{x}{b}\right)*\text{sinc}^2\left(\frac{x}{a}\right).$$

7-9. Given the real constant b and an arbitrary band-limited function $f(x)$, whose Fourier transform has a total width of W; i.e., $F(\xi)\equiv 0$ for $|\xi|>W/2$.

a. For $|b|<\frac{1}{W}$, show that $\frac{1}{|b|}\text{sinc}\left(\frac{x}{b}\right)*f(x)=f(x)$.

b. Can any such general statement be made if $|b|>\frac{1}{W}$?

7-10. Find the Fourier transforms of the following functions:

a. $f(x)=\frac{1}{5}\text{comb}\left(\frac{x}{5}\right)*\text{rect}(x)$. Sketch $f(x)$ and $F(\xi)$.

b. $g(x)=\frac{1}{5}\text{comb}\left(\frac{x}{5}\right)*\text{rect}\left(\frac{x}{2.5}\right)$. Sketch $g(x)$ and $G(\xi)$.

c. $h(x)=\frac{1}{5}\text{comb}\left(\frac{x}{5}\right)*\text{rect}\left(\frac{x}{4}\right)$. Sketch $h(x)$ and $H(\xi)$.

d. $p(x)=\text{comb}(x)\text{Gaus}(2x)$. Sketch $p(x)$ and $P(\xi)$.

e. $r(x)=\text{comb}(x)\text{Gaus}\left(\frac{x}{5}\right)$. Sketch $r(x)$ and $R(\xi)$.

f. $s(x)=\frac{1}{2}\text{comb}\left(\frac{x}{2}\right)*\text{tri}(x)$. Sketch $s(x)$ and $S(\xi)$.

g. $t(x)=\frac{1}{4}\text{comb}\left(\frac{x}{4}\right)*\text{tri}(x)$. Sketch $t(x)$ and $T(\xi)$.

h. $u(x)=\left[\frac{1}{5}\text{comb}\left(\frac{x}{5}\right)*\text{rect}(x)\right]\cos(60\pi x)$. Sketch $u(x)$ and $U(\xi)$.

7-11. Given the real constants a and b, the positive real integer N, and the function

$$F(\xi)=\frac{1}{|a|}\text{comb}\left(\frac{\xi}{a}\right)\text{rect}\left(\frac{\xi}{b}\right).$$

a. Find a general expression for $f(x)=\mathcal{F}^{-1}\{F(\xi)\}$.

b. How does the $f(x)$ of part a behave as $\left|\frac{b}{a}\right|\to\infty$?

c. For $2(N-1)<\left|\frac{b}{a}\right|<2N$, show that

$$f(x)=1+2\sum_{n=1}^{N-1}\cos(2\pi nax).$$

d. For $\left|\frac{b}{a}\right| = 2N$, show that

$$f(x) = 1 + 2\sum_{n=1}^{N-1} \cos(2\pi nax) + \cos(2\pi Nax).$$

e. For $2N < \left|\frac{b}{a}\right| < 2(N+1)$, show that

$$f(x) = 1 + 2\sum_{n=1}^{N} \cos(2\pi nax).$$

f. Use the results of part c or part e to show that

$$10.1 \sum_{n=-\infty}^{\infty} \text{sinc}\, 10.1(x-n) = 11.9 \sum_{n=-\infty}^{\infty} \text{sinc}\, 11.9(x-n).$$

7-12. Given the real constant b and the function $F(\xi) = \text{comb}(\xi)\text{rect}(\xi/b)$, find the inverse transform $f(x) = \mathcal{F}^{-1}\{F(\xi)\}$, and sketch both $F(\xi)$ and $f(x)$ for the following values of b:

a. $b = 1$.
b. $b = 2$.
c. $b = 3$.
d. $b = 4$.
e. $b = 5$.
f. $b = 10$.
g. $b = 20$.
h. $b = 50$.

7-13. Given the real constant b and the function $F(\xi) = \text{comb}(\xi)\,\text{tri}(\xi/b)$, find the inverse transform $f(x) = \mathcal{F}^{-1}\{F(\xi)\}$, and sketch both $F(\xi)$ and $f(x)$ for the following values of b:

a. $b = 0.5$.
b. $b = 1$.
c. $b = 1.5$.
d. $b = 2$.
e. $b = 2.5$.
f. $b = 5$.
g. $b = 10$.
h. $b = 25$.

7-14. Find the Fourier transforms of the following functions:

a. $f(x) = \left[\text{comb}(x)\text{rect}\left(\frac{x}{5}\right)\right] * \text{rect}(2x)$. Sketch $f(x)$ and $F(\xi)$.

b. $g(x) = \left[\text{comb}(x)\text{rect}\left(\frac{x}{25}\right)\right] * \text{rect}(2x)$. Sketch $g(x)$ and $G(\xi)$.

c. $h(x) = [\text{comb}(x) * \text{rect}(2x)]\text{rect}\left(\frac{x}{25}\right)$. Sketch $h(x)$ and $H(\xi)$.

d. Show that, in general,

$$[\text{comb}(x)s(x)] * t(x) \neq [\text{comb}(x) * t(x)]s(x).$$

Note, however, that an equality results in certain special cases; e.g., $g(x) = h(x)$ in parts b and c above.

e. $p(x)=\left[\text{comb}(x)\text{tri}\left(\frac{x}{10}\right)\right]*\text{rect}(2x)$. Sketch $p(x)$ and $P(\xi)$.

f. $r(x)=\left[\text{comb}(x)*\text{rect}(2x)\right]\text{tri}\left(\frac{x}{10}\right)$. Sketch $r(x)$ and $R(\xi)$.

7-15. Given a linear shift-invariant system with inputs $f_i(x)$, outputs $g_i(x)$, and impulse response $h(x)=7\,\text{sinc}(7x)$. Using a frequency-domain approach and reasonable approximations where appropriate, find the output for each of the following inputs:

a. $f_1(x)=\cos(4\pi x)$.

b. $f_2(x)=\cos(4\pi x)\text{rect}\left(\frac{x}{75}\right)$.

c. $f_3(x)=[1+\cos(8\pi x)]\text{rect}\left(\frac{x}{75}\right)$.

d. $f_4(x)=\text{comb}(x)*\text{rect}(2x)$.

e. $f_5(x)=\left[\text{comb}(x)\text{rect}\left(\frac{x}{75}\right)\right]*\text{rect}(2x)$.

f. $f_6(x)=\left[2\,\text{comb}(2x)\text{rect}\left(\frac{x}{25}\right)\right]*\text{rect}(4x)$.

g. $f_7(x)=\left[4\,\text{comb}(4x)\text{rect}\left(\frac{x}{25}\right)\right]*\text{rect}(8x)$.

7-16. Calculate m_0, $\bar{x}$, and σ^2 for the following functions:

a. $f(x)=\text{rect}(x)$.

b. $g(x)=\text{sinc}(x)$.

c. $h(x)=\text{tri}(x)$.

d. $p(x)=\text{rect}(x)e^{-j2\pi x}$.

e. $r(x)=\text{rect}(x)e^{-j3\pi x}$.

f. $s(x)=\text{Gaus}\left(\frac{x-10}{5}\right)$.

g. $t(x)=\text{rect}\left(\frac{x}{2}\right)*\delta\delta(x)$.

h. $u(x)=-j2\pi x\,\text{Gaus}(x)$.

CHAPTER 8

CHARACTERISTICS AND APPLICATIONS OF LINEAR FILTERS

We have seen that the output of an LSI system may be calculated directly in the domain of the input by convolving the input with the impulse response of the system and that the output may therefore be regarded as a smoothed version of the input. We have also seen that the output may be determined indirectly, and often more easily, by taking a detour through the frequency domain, where a filtering concept is found to be appropriate. In other words the spectrum of the output may be regarded as a filtered version of the input spectrum. In the present chapter we strengthen our understanding of this filtering concept and investigate the characteristics and applications of several types of linear filters. In addition, we introduce the notion of signal sampling and recovery.

8-1 LINEAR SYSTEMS AS FILTERS

Let us consider the LSI system depicted in Fig. 8-1, which is characterized by its impulse response $h(x)$ and transfer function $H(\xi)$. To this system we apply an input signal $f(x)$, with spectrum $F(\xi)$, and we denote the output signal and output spectrum by $g(x)$ and $G(\xi)$, respectively. We know from Eq. (7.87) that the output spectrum is given by

$$G(\xi) = F(\xi)H(\xi), \tag{8.1}$$

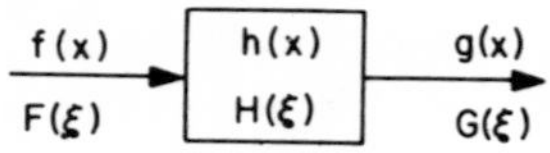

$f(x)$ / $F(\xi)$ → $h(x)$ / $H(\xi)$ → $g(x)$ / $G(\xi)$

Figure 8-1 An LSI system with impulse response $h(x)$ and transfer function $H(\xi)$.

and we wish to determine how $G(\xi)$ is affected by $H(\xi)$. First, we represent the transfer function by the expression

$$H(\xi) = A_H(\xi)e^{-j\Phi_H(\xi)}, \tag{8.2}$$

where both $A_H(\xi)$ and $\Phi_H(\xi)$ are real-valued, bipolar functions. Many authors prefer to restrict $A_H(\xi)$ to nonnegative values only, in which case $A_H(\xi)$ would simply be the modulus of $H(\xi)$, but we feel that certain advantages are gained if $A_H(\xi)$ is allowed to be bipolar. We refer to $A_H(\xi)$ as the *amplitude transfer function* of the system and to $\Phi_H(\xi)$ as its *phase transfer function*.

If we then write the spectra of the input and output signals, respectively, as

$$F(\xi) = A_F(\xi)e^{-j\Phi_F(\xi)}$$

$$G(\xi) = A_G(\xi)e^{-j\Phi_G(\xi)}, \tag{8.3}$$

Eq. (8.1) may be expressed as

$$A_G(\xi)e^{-j\Phi_G(\xi)} = A_F(\xi)A_H(\xi)e^{-j[\Phi_F(\xi)+\Phi_H(\xi)]}. \tag{8.4}$$

Clearly, the amplitude spectrum of the output is just the product of the input amplitude spectrum and the amplitude transfer function, whereas the phase spectrum of the output is simply the sum of the input phase spectrum and the phase transfer function, i.e.,

$$A_G(\xi) = A_F(\xi)A_H(\xi)$$

$$\Phi_G(\xi) = \Phi_F(\xi) + \Phi_H(\xi). \tag{8.5}$$

We may now express the output signal in terms of $A_F(\xi)$, $\Phi_F(\xi)$, $A_H(\xi)$ and $\Phi_H(\xi)$; from Eqs. (8.3) and (8.4) we obtain

$$g(x) = \mathcal{F}^{-1}\{G(\xi)\}$$

$$= \mathcal{F}^{-1}\{A_F(\xi)A_H(\xi)e^{-j[\Phi_F(\xi)+\Phi_H(\xi)]}\}. \tag{8.6}$$

Thus, once we have determined the effects of the amplitude transfer function on the input amplitude spectrum and those of the phase transfer function on the input phase spectrum, we need only to perform the inverse Fourier transform operation of Eq. (8.6) to obtain the output signal.

It will be convenient for us to consider the effects of the amplitude transfer function and those of the phase transfer function separately. If a system is such that $\exp\{-j\Phi_H(\xi)\}=1$ for all ξ, we shall call it an *amplitude filter* because only the amplitude transfer function has an effect on the output. If, on the other hand, $A_H(\xi)=1$ for all ξ, we shall call the system a *phase filter* because only the phase transfer function affects the output. Although most systems encountered in practice are really combinations of these two types of filters, their behavior is often predominantly that of one or the other. We shall first investigate them separately—then in combination.

Perhaps we should address a potentially confusing item before we begin our detailed discussions of these filters: by allowing $A_H(\xi)$ to be bipolar, it is sometimes difficult to decide whether we are dealing with an amplitude filter, a phase filter, or a combination of the two. To illustrate, let us consider a system with a real-valued, bipolar transfer function $H(\xi)$. We may regard such a system to be either an amplitude filter with

$$A_H(\xi)=H(\xi)$$
$$\Phi_H(\xi)=n2\pi, \quad n \text{ an integer} \tag{8.7}$$

or a combination filter with

$$A_H(\xi)=|H(\xi)|$$
$$\Phi_H(\xi)=\begin{cases} n2\pi, & H(\xi)>0 \\ (2n+1)\pi. & H(\xi)<0 \end{cases} \tag{8.8}$$

Furthermore, if $|H(\xi)|=1$, the system could be thought of as a pure phase filter. Any confusion regarding this matter could be avoided by merely requiring that $A_H(\xi)=|H(\xi)|$, but the benefits may not be worth the price of the more complicated forms sometimes required for $A_H(\xi)$ and $\Phi_H(\xi)$.

8-2 AMPLITUDE FILTERS

With the condition that $\exp\{-j\Phi_H(\xi)\}=1$ for amplitude filters, Eq. (8.6) becomes

$$g(x)=\mathcal{F}^{-1}\{A_F(\xi)A_H(\xi)e^{-j\Phi_F(\xi)}\}; \tag{8.9}$$

the amplitude spectrum of the input is modified by $A_H(\xi)$ as it passes through the system, but the phase spectrum is unaltered. The exact nature of the changes produced in the amplitude spectrum depends on the parameters of the system, but in general the strength of the Fourier components in certain regions of the spectrum are attenuated, or amplified, relative to the strength of those in other regions. Because we are interested primarily in passive systems, we will normally assume that the only effect of the amplitude transfer function will be to attenuate the various spectral components of the input to some degree. For simplicity, we assume that $|A_H(\xi)| \leqslant 1$ and $\Phi_H(\xi)=0$.

Let us now consider the class of systems for which $A_H(\xi)$ is a constant, i.e., for which $A_H(\xi)=A$, where $|A| \leqslant 1$. We see from Eq. (8.9) that for this type of system the output becomes $g(x)=Af(x)$: it is simply an attenuated, but undistorted,* version of the input. The explanation for this behavior is straightforward. Not only are all of the spectral components of the input passed by the system, they all undergo exactly the same relative attenuation, which leaves their relative strengths unchanged at the output. As a result, the output has exactly the same shape as the input, differing only in magnitude. We shall refer to such systems as *distortionless amplitude filters*. If $A_H(\xi)$ is not a constant, the system introduces *amplitude distortion* (Ref. 8-1), and the output will no longer have the same form as the input. Later in this section we investigate a number of filters for which this is the case.

The distortionless amplitude filter described above is a type of *all-pass filter*, which derives its name from the fact that it passes all spectral components of the input no matter how high their frequencies. As we learned in Chapter 7, such systems are physically unrealizable because the transfer functions for real-world systems must approach zero for sufficiently large values of $|\xi|$. Of course, if $A=0$, the system is physically realizable but of little interest (and perhaps should be referred to as a *no-pass filter*).

One of the simplest types of amplitude filters is the *binary amplitude filter*, also known as an *ideal filter*. The transfer function of such a filter can assume only values of zero or one, which means that the Fourier components of the input are either passed by the system unaltered, or that they are completely attenuated. As a result, a particular component is either present at the output, with no reduction in strength, or it is absent. True binary filters are usually considered to be physically unrealizable, but, as we shall see in later chapters, they are realizable in certain optical systems. *Continuous amplitude filters* have additional flexibility in that they may cause the various Fourier components of the input to be attenuated to

*One might consider the output to be distorted if A is negative, but we shall not.

any degree, not simply all or none as for the binary filter. Furthermore, continuous amplitude filters are more frequently encountered in practice than binary filters. Typical transfer functions for these two types of amplitude filters are shown in Fig. 8-2(a) and (b).

The above descriptions for binary and continuous amplitude filters may be somewhat restrictive and misleading. For example, it might make sense to use the term binary amplitude filter to describe a system whose transfer function takes on any two values—not necessarily values only of one and zero. The behavior of such a system would, in general, be quite different from that of the binary filter described above, and unless specified differently, the term binary amplitude filter shall refer to the one-zero type. Also, we do not mean to imply that the transfer function of a continuous

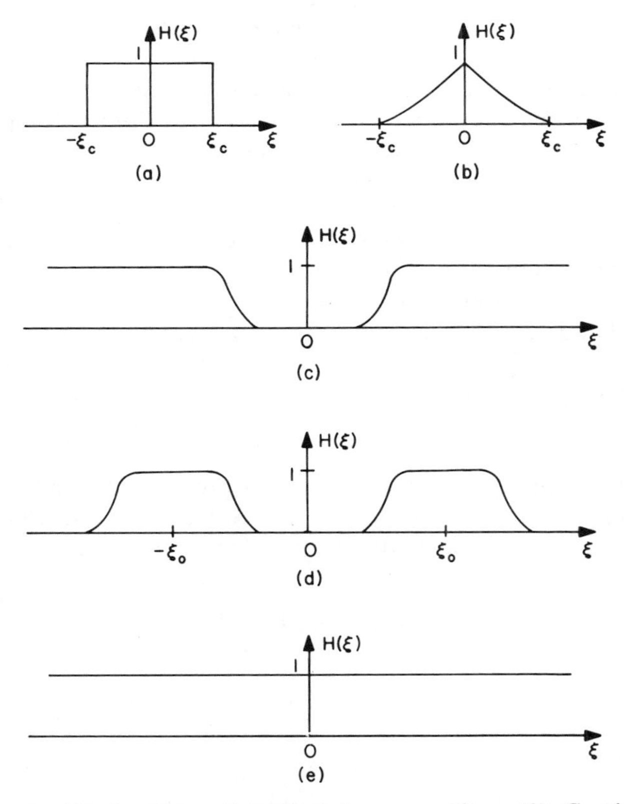

Figure 8-2 Amplitude filters. (*a*) Ideal low-pass filter. (*b*) Continuous-range low-pass filter. (*c*) Continuous-range high-pass filter. (*d*) Continuous-range band-pass filter. (*e*) All-pass filter.

amplitude filter is continuous in the mathematical sense. We only mean that the transfer function has a continuous range on some frequency interval and may take on any value within that range. It might be more appropriate to refer to such a filter as a continuous-range amplitude filter, or a filter whose transfer function is piecewise continuous on some interval. For most of our work here, the transfer functions of continuous amplitude filters shall have values between -1 and 1, inclusive.

The transfer functions of Fig. 8-2(a) and (b) are both illustrative of *low-pass filters*. Such filters pass, with partial or no attenuation, all Fourier components for which $|\xi|$ is less than some *cutoff frequency* ξ_c, while completely attenuating those components for which $|\xi|>\xi_c$. Note that when a band-limited signal is passed through an ideal low-pass filter, no amplitude distortion is introduced as long as the cutoff frequency of the filter is greater than that of the signal spectrum. This statement is related to the result given by Eq. (6.26), in which the quantity $|2b|^{-1}$ corresponds to the cutoff frequency of an ideal low-pass filter.

High-pass filters effectively pass all spectral components for which $|\xi|$ is greater than some specified value, and eliminate those components for which $|\xi|$ is less than this value. As mentioned previously, the transfer function of any real physical system must approach zero for sufficiently large values of $|\xi|$, and therefore a true high-pass filter is physically unrealizable. Nevertheless, it may be helpful to assume the existence of such filters in certain situations.

Filters that pass only the Fourier components lying within specific frequency bands, while blocking those components whose frequencies lie outside these bands, are called *band-pass filters*. Transfer functions for typical high-pass and band-pass filters are also illustrated in Fig. 8-2.

Example

For us to get a better feeling for the effects produced by amplitude filters, we now consider a simple example. For convenience, we choose a low-pass, binary amplitude filter with transfer function of the general form

$$H(\xi)=\text{rect}\left(\frac{\xi}{2\xi_c}\right). \tag{8.10}$$

We specify the input to be the rectangle-wave function

$$f(x)=\text{rect}(x)*\tfrac{1}{2}\,\text{comb}\left(\frac{x}{2}\right) \tag{8.11}$$

with spectrum

$$F(\xi)=\text{sinc}(\xi)\,\text{comb}(2\xi), \tag{8.12}$$

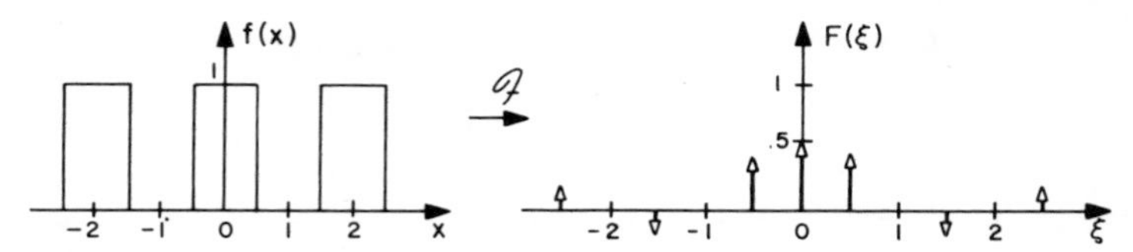

Figure 8-3 Input rectangle-wave function used for amplitude filtering examples and its spectrum.

as shown in Fig. 8-3. The spectrum of the output may then be expressed as

$$G(\xi)=\operatorname{sinc}(\xi)\operatorname{comb}(2\xi)\operatorname{rect}\left(\frac{\xi}{2\xi_c}\right). \tag{8.13}$$

Because the value of the cutoff frequency ξ_c governs the width of the transfer function, and because the width of the transfer function in turn determines how many of the input spectral components will appear at the output, the output spectrum, and hence the output itself, depends greatly on the value of ξ_c.

The impulse response for this system is found to be

$$h(x)=2\xi_c\operatorname{sinc}(2\xi_c x), \tag{8.14}$$

and the convolution form of the output then becomes

$$g(x)=\left[\operatorname{rect}(x)*\tfrac{1}{2}\operatorname{comb}\left(\frac{x}{2}\right)\right]*\left[2\xi_c\operatorname{sinc}(2\xi_c x)\right]. \tag{8.15}$$

As we shall see, Eq. (8.15) is convenient for determining the effects of the system in a qualitative manner, but it is not very useful for obtaining quantitative results. We shall now calculate the output for several values of the cutoff frequency.

1. $\xi_c=\infty$. For $\xi_c=\infty$, the transfer function is unity for all ξ; no filtering action is introduced by the system, and all Fourier components of the input are passed without any attenuation or phase shift. The output spectrum is exactly the same as the input spectrum, with the result that the input and output are identical, i.e.,

$$g(x)=f(x). \tag{8.16}$$

If we approach this problem in the domain of the input rather than in the frequency domain, we obtain the same result. For $\xi_c=\infty$ the impulse response of Eq. (8.14) becomes a unit-area delta function, and from Eq.

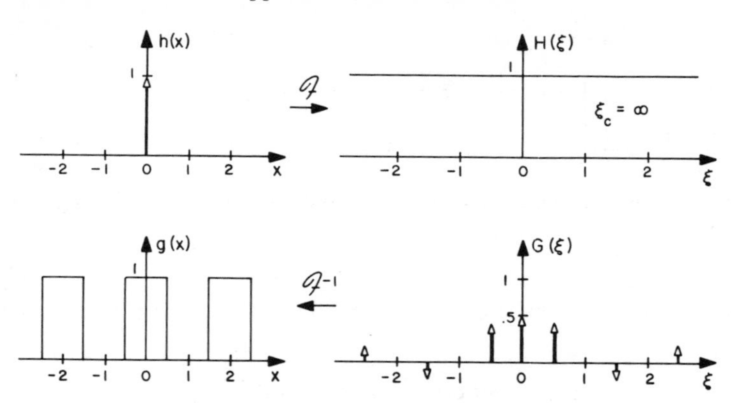

Figure 8-4 Effect of all-pass filter on rectangle-wave input.

(6.21) we know that convolution with a delta function produces no smoothing. Evaluation of Eq. (8.15) then yields $g(x)=f(x)$ (see Fig. 8-4).

2. $\xi_c=2$. For $\xi_c=2$, the transfer function becomes a rectangle function whose width is four units as shown in Fig. 8-5. As a result, the zero-frequency, fundamental, and third-harmonic components of the input are passed unaltered, but all components of higher frequency are filtered out. Thus, the spectrum of the output becomes

$$\begin{aligned}
G(\xi) &= \text{rect}\left(\frac{\xi}{4}\right)\text{sinc}(\xi)\text{comb}(2\xi)\\
&= \text{rect}\left(\frac{\xi}{4}\right)\text{sinc}(\xi)\sum_{n=-\infty}^{\infty}\tfrac{1}{2}\delta\left(\xi-\frac{n}{2}\right)\\
&= \text{rect}\left(\frac{\xi}{4}\right)\sum_{n=-\infty}^{\infty}\tfrac{1}{2}\,\text{sinc}\left(\frac{n}{2}\right)\delta\left(\xi-\frac{n}{2}\right)\\
&= \tfrac{1}{2}\delta(\xi)+\tfrac{1}{2}\left[\frac{2}{\pi}\delta\left(\xi-\tfrac{1}{2}\right)+\frac{2}{\pi}\delta\left(\xi+\tfrac{1}{2}\right)\right]\\
&\quad+\tfrac{1}{2}\left[-\frac{2}{3\pi}\delta\left(\xi-\tfrac{3}{2}\right)-\frac{2}{3\pi}\delta\left(\xi+\tfrac{3}{2}\right)\right]\\
&= \tfrac{1}{2}\delta(\xi)+\frac{2}{\pi}\delta\delta(2\xi)-\frac{2}{9\pi}\delta\delta(2\xi/3).
\end{aligned} \tag{8.17}$$

From Eq. (8.9), the output is found to be

$$g(x)=\tfrac{1}{2}+\frac{2}{\pi}\cos(\pi x)-\frac{2}{3\pi}\cos(3\pi x), \tag{8.18}$$

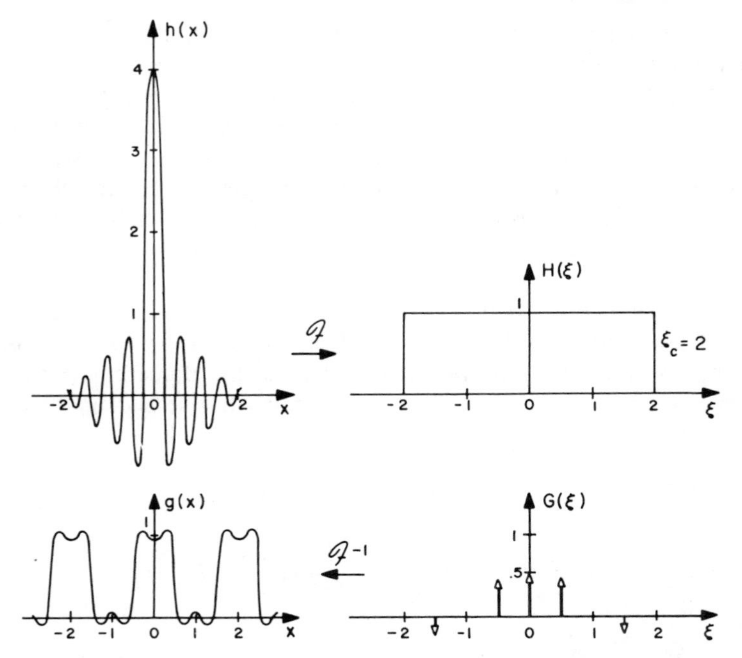

Figure 8-5 Effect of ideal low-pass filter on rectangle-wave input for a cutoff frequency of $\xi_c = 2$.

and is illustrated in Fig. 8-5. The output now exhibits no sharp corners because the high-frequency components required to produce them have been removed by the system.

We can also consider the lack of sharp corners to be a result of smoothing by the impulse response, the width of which is now about one-quarter of the period of $f(x)$ as may be seen in Fig. 8-5.

3. $\xi_c = 1$. The transfer function is now a rectangle function of width two, and the output spectrum is found to be

$$G(\xi) = \text{rect}\left(\frac{\xi}{2}\right) \sum_{n=-\infty}^{\infty} \tfrac{1}{2}\text{sinc}\left(\frac{n}{2}\right)\delta\left(\xi - \frac{n}{2}\right)$$

$$= \tfrac{1}{2}\delta(\xi) + \frac{2}{\pi}\delta\delta(2\xi). \tag{8.19}$$

The inverse transform of this expression is then

$$g(x) = \tfrac{1}{2} + \frac{2}{\pi}\cos(\pi x), \tag{8.20}$$

and we see that the output contains only the zero-frequency and fundamental components of the input. All of the components with frequency higher than the fundamental are filtered out (see Fig. 8-6). It is interesting to note that the output will be exactly as described above as long as $0.5<\xi_c<1.5$. We should point out that if the rectangular transfer function should happen to cut off at exactly the frequency of one of the input Fourier components, this component will be reduced to half strength at the output because the rectangular transfer function has a value of 0.5 at $\xi=\xi_c$.

From the convolution point of view, the impulse response now has a width of approximately half the period of $f(x)$, as shown in Fig. 8-6, and the resultant smoothing is still greater than in the preceding parts of this example. However, we point out once again that it is very difficult to determine the exact form of the output via the convolution method.

4. $\xi_c=0.25$. We now have $H(\xi)=\text{rect}(2\xi)$, so that

$$G(\xi)=\text{rect}(2\xi)\sum_{n=-\infty}^{\infty}\tfrac{1}{2}\text{sinc}\left(\frac{n}{2}\right)\delta\left(\xi-\frac{n}{2}\right)$$

$$=\tfrac{1}{2}\delta(\xi). \tag{8.21}$$

Thus the fundamental Fourier component, as well as all the higher harmonics, has now been eliminated by the system, and the output simply

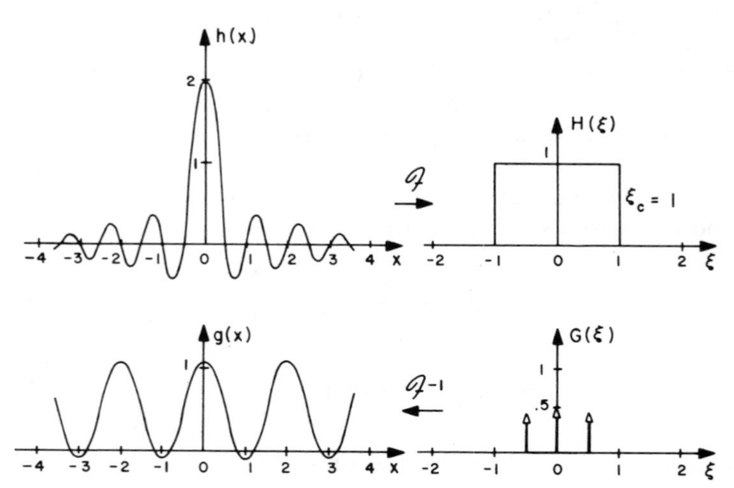

Figure 8-6 Effect of ideal low-pass filter on rectangle-wave input for a cutoff frequency of $\xi_c=1$.

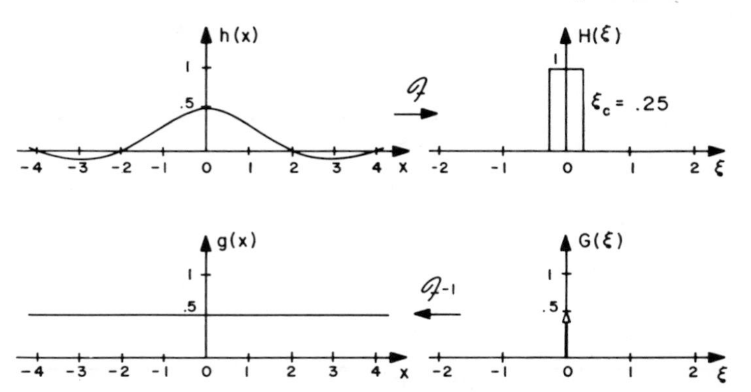

Figure 8-7 Effect of ideal low-pass filter on rectangle-wave input for a cutoff frequency of $\xi_c = 0.25$.

becomes a constant as shown in Fig. 8-7:

$$g(x) = \tfrac{1}{2}. \tag{8.22}$$

Again it is interesting to note that as long as $0 < \xi_c < 0.5$, the output will have this constant value of 0.5.

Viewing the output once more as a convolution, the impulse response is so wide (see Fig. 8-7) that it completely smooths all variations in the input, leaving just the nonvarying part. ■

In this example of a binary amplitude filter, we have seen how the width of the transfer function influences the output of a system: the wider we make the transfer function, the more nearly the output resembles the input, whereas the narrower we make it, the less the resemblance. Mathematically, we can regard this behavior in the following way. The input may be represented by a Fourier series containing an infinite number of terms, and the output may be described by a truncated version of this infinite series. The number of terms included in the output series is directly proportional to the width of the transfer function, and we know from our earlier discussions of Fourier series that the greater this number is allowed to be, the more nearly the truncated series will resemble the infinite series. The overshoot, or ringing, observed in the output can be explained mathematically in terms of Gibb's phenomenon.† For a review of these concepts, see Secs. 4–2 and 4–4.

†Note that the ringing may also be viewed in another fashion: when an oscillatory bipolar function, such as the sinc function, is convolved with a function having discontinuities, such as a step function, rectangle function, rectangle wave, etc., the result is likely to exhibit a distinct ringing in the vicinity of the discontinuities.

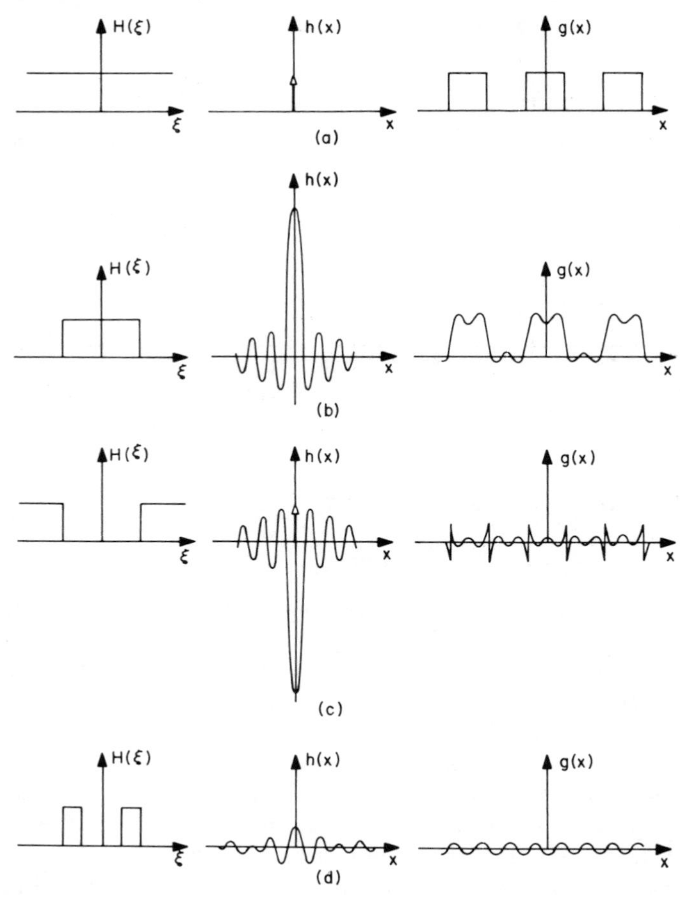

Figure 8-8 Transfer functions, impulse responses, and system response to a rectangle-wave input. (*a*) All-pass filter. (*b*) Ideal low-pass filter. (*c*) Ideal high-pass filter. (*d*) Ideal band-pass filter.

In Fig. 8-8 we illustrate the transfer functions and impulse responses for several different amplitude filters. In addition, we attempt to indicate the behavior of these systems by displaying the output when the input of the previous example [Eq. (8.11) and Fig. 8-3] is applied to each.

8-3 PHASE FILTERS

A *phase filter* is a system that causes the phase spectrum of the input to be altered in some fashion while leaving the amplitude spectrum unchanged. It should be noted, however, that a true phase filter is simply a special kind of all-pass filter, and is therefore physically unrealizable. With the condition that $A_H(\xi)=1$ for phase filters, Eq. (8.6) may be written

$$g(x)=\mathcal{F}^{-1}\{A_F(\xi)e^{-j[\Phi_F(\xi)+\Phi_H(\xi)]}\}$$
$$=\mathcal{F}^{-1}\{F(\xi)e^{-j\Phi_H(\xi)}\}. \tag{8.23}$$

All of the Fourier components of the input are passed by the system without attenuation, but the phase of these components may be shifted. As a result, even though all of the input spectral components are present at the output, their relative positions may have been shifted to such a degree that the output is severely distorted.

We begin our investigation of phase filters by considering the case for which the function $\Phi_H(\xi)$ varies linearly with ξ, i.e., a filter with *linear phase*. Specifically, let us choose

$$\Phi_H(\xi)=2\pi x_0\xi, \tag{8.24}$$

where x_0 is a real constant. We may write the output spectrum as

$$G(\xi)=F(\xi)e^{-j2\pi x_0\xi}$$
$$=A_F(\xi)e^{-j[\Phi_F(\xi)+2\pi x_0\xi]}, \tag{8.25}$$

and it may be seen that while all of the components of the input are present at the output, they have each undergone a phase shift of $2\pi x_0\xi$. The output now becomes

$$g(x)=\mathcal{F}^{-1}\{F(\xi)e^{-j2\pi x_0\xi}\}, \tag{8.26}$$

but from Eq. (7.38) this may be written as

$$g(x)=f(x-x_0), \tag{8.27}$$

and we see that a linear-phase filter simply causes the output to be a shifted version of the input. This effect may be explained as follows. Each Fourier component of the input undergoes a phase shift proportional to its frequency, and because this phase shift, in radians, is determined relative to the period of that individual component, the overall effect is to cause all components to be shifted along the x-axis by the same amount. Thus, when added together at the output, the shifted components combine to produce a function that is identical to the input, except for location. Note that the amount of shift introduced by a linear-phase filter is equal to the slope of $\Phi_H(\xi)$ divided by 2π, i.e.,

$$\begin{aligned}\text{shift} &= \frac{1}{2\pi}\frac{d}{d\xi}(2\pi x_0 \xi)\\ &= x_0.\end{aligned} \tag{8.28}$$

This result follows from the moment theorem of Chapter 7.

The impulse response of this linear-phase filter is given by

$$\begin{aligned}h(x) &= \mathcal{F}^{-1}\{e^{-j2\pi x_0 \xi}\}\\ &= \delta(x - x_0),\end{aligned} \tag{8.29}$$

and from this expression it is apparent why the output is simply a shifted version of the input:

$$\begin{aligned}g(x) &= f(x) * h(x)\\ &= f(x) * \delta(x - x_0)\\ &= f(x - x_0).\end{aligned} \tag{8.30}$$

Because the output is undistorted, albeit shifted, we shall refer to this type of filter as a *distortionless-phase filter*. If the phase transfer function is not a linear function of ξ, the system will normally introduce *phase distortion* (Ref. 8-1); however, for the special case where $\Phi_H(\xi)$ is constant, this statement is not true. For instance, if $\Phi_H(\xi) = B$, where B is a real constant, the output is just the input multiplied by $\exp\{-jB\}$, i.e., $g(x) = f(x)\exp\{-jB\}$. We illustrate this point with three examples. If $\Phi_H(\xi) = 0.5\pi$, $g(x) = -jf(x)$ and we see that a real-valued input is merely converted into an imaginary-valued output, and vice versa, by such a system. Also, we find that $g(x) = f(x)$ whenever $\Phi_H(\xi) = n2\pi$ and that $g(x) = -f(x)$ whenever $\Phi_H(\xi) = (2n+1)\pi$. Note that the latter two filters could be thought of as amplitude filters rather than phase filters, as we discussed in Sec. 8-1.

Example

Let us now consider an example in which phase distortion is encountered. Again we choose the input to be the rectangle-wave function of Eq. (8.11), and the phase transfer function shall be

$$\Phi_H(\xi) = \pi\xi\,\text{rect}\left(\frac{\xi}{2}\right) + \pi\,\text{sgn}(\xi)\left[1 - \text{rect}\left(\frac{\xi}{2}\right)\right], \tag{8.31}$$

as depicted in Fig. 8-9. It is apparent that all spectral components for

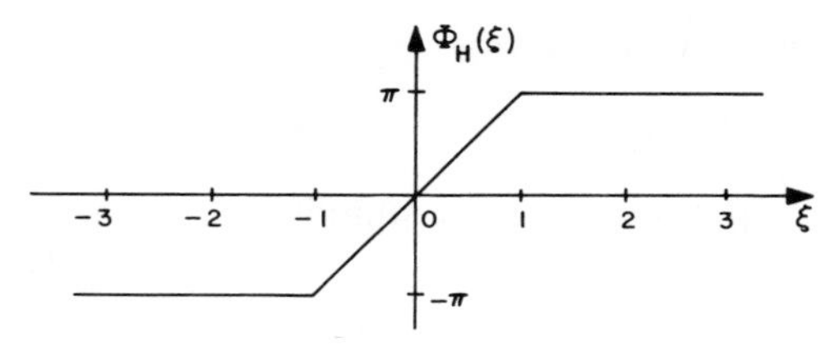

Figure 8-9 Phase transfer function of filter that produces phase distortion.

which $|\xi|>1$ will be shifted in phase by $\pm\pi$ radians, which merely introduces a sign reversal, while those components for which $|\xi|\leqslant 1$ will undergo a phase shift that is proportional to ξ. If we now express the input by its Fourier series expansion,

$$f(x)=\tfrac{1}{2}+\frac{2}{\pi}\cos(\pi x)-\frac{2}{3\pi}\cos(3\pi x)+\frac{2}{5\pi}\cos(5\pi x)-\ldots, \tag{8.32}$$

we see that the effect of $\Phi_H(\xi)$ will be to leave the zero-frequency component unshifted, to shift the fundamental component by $\pi/2$, and to reverse the signs of all components from the third-harmonic on up. Thus, the output may be written as

$$\begin{aligned} g(x)=&\tfrac{1}{2}+\frac{2}{\pi}\cos\pi\left(x-\tfrac{1}{2}\right)+\frac{2}{3\pi}\cos(3\pi x)\\ &-\frac{2}{5\pi}\cos(5\pi x)+\ldots. \end{aligned} \tag{8.33}$$

It will now be convenient to add and subtract the quantity $[\tfrac{1}{2}+(2/\pi)\cos\pi x]$ to the right-hand side of Eq. (8.33), i.e.,

$$\begin{aligned} g(x)=&\tfrac{1}{2}+\frac{2}{\pi}\cos\pi\left(x-\tfrac{1}{2}\right)+\tfrac{1}{2}+\frac{2}{\pi}\cos\pi x\\ &-\tfrac{1}{2}-\frac{2}{\pi}\cos\pi x+\frac{2}{3\pi}\cos(3\pi x)-\ldots, \end{aligned} \tag{8.34}$$

because this allows $g(x)$ to be simplified to

$$\begin{aligned} g(x)&=1-f(x)+\frac{2}{\pi}\left[\cos\pi\left(x-\tfrac{1}{2}\right)+\cos(\pi x)\right]\\ &=1-f(x)+\frac{2\sqrt{2}}{\pi}\cos\pi\left(x-\tfrac{1}{4}\right). \end{aligned} \tag{8.35}$$

It will be instructive to calculate the impulse response for this system.

The transfer function has the form

$$H(\xi)=e^{-j\pi\{\xi\,\mathrm{rect}(\xi/2)+\mathrm{sgn}(\xi)[1-\mathrm{rect}(\xi/2)]\}}, \tag{8.36}$$

and after some manipulation we may simplify this expression to

$$H(\xi)=\mathrm{rect}\left(\frac{\xi}{2}\right)\left[1+e^{-j\pi\xi}\right]-1. \tag{8.37}$$

The impulse response is then

$$\begin{aligned} h(x)&=2\,\mathrm{sinc}(2x)*\left[\delta(x-0.5)+\delta(x)\right]-\delta(x)\\ &=2\,\mathrm{sinc}\,2(x-0.5)+2\,\mathrm{sinc}(2x)-\delta(x). \end{aligned} \tag{8.38}$$

Both $g(x)$ and $h(x)$ are graphed in Fig. 8-10, along with the input, and the severe distortion introduced by the system is readily apparent. ■

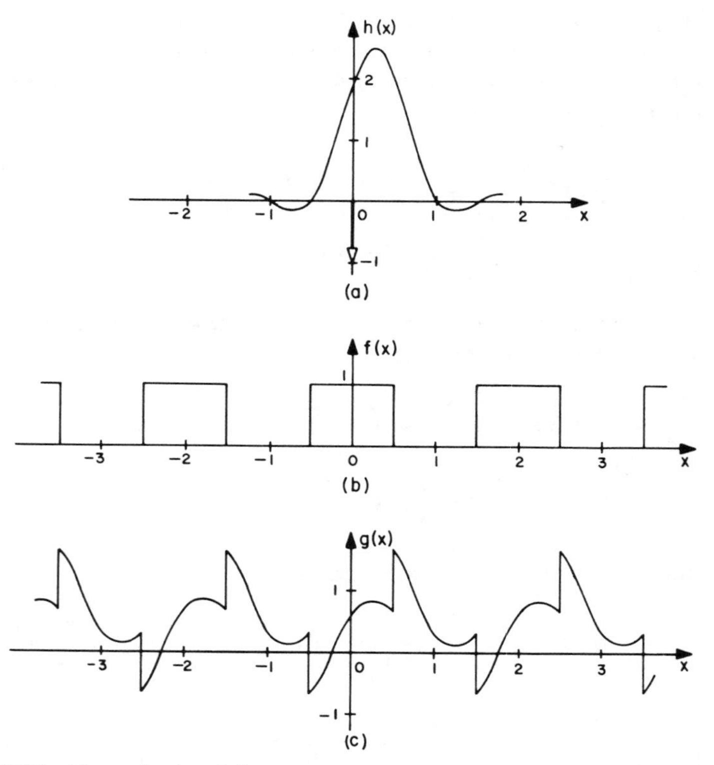

Figure 8-10 Example involving system of Fig. 8-9. (*a*) Impulse response. (*b*) Rectangle-wave input. (*c*) Phase-distorted output.

All of the filters investigated above are *continuous-phase filters* because of the continuous range of values permitted for $\Phi_H(\xi)$. We observe that the transfer functions of these filters are in general complex valued, and may assume any value on the unit circle such that $|H(\xi)|=1$. A *binary-phase filter*, on the other hand, has a phase transfer function that is limited to just two values. Again the transfer function is generally complex valued, with $|H(\xi)|=1$, but it is now limited to only two values on the unit circle.

Example

As an example of a binary-phase filter, let us select a system with phase transfer function

$$\Phi_H(\xi) = -0.5\pi\,\mathrm{rect}(\xi) \tag{8.39}$$

to which we apply the complex-valued input signal

$$f(x) = e^{j0.1\cos(2\pi x)}. \tag{8.40}$$

The input may be approximated by

$$f(x) \cong 1 + j0.1\cos(2\pi x), \tag{8.41}$$

and its spectrum by

$$F(\xi) \cong \delta(\xi) + j0.1(0.5)\delta\delta(\xi). \tag{8.42}$$

The output spectrum then becomes

$$\begin{aligned} G(\xi) &\cong [\delta(\xi) + j0.1(0.5)\delta\delta(\xi)]e^{j0.5\pi\,\mathrm{rect}(\xi)} \\ &\cong j[\delta(\xi) + 0.1(0.5)\delta\delta(\xi)], \end{aligned} \tag{8.43}$$

and finally, the output is found to be

$$g(x) \cong j[1 + 0.1\cos(2\pi x)]. \tag{8.44}$$

For this example, the system causes the phase of the zero-frequency spectral component to be advanced by 0.5π radians such that the zero-frequency and fundamental spectral components of the output are now in phase. This example is related to the phase-contrast microscopy technique discussed by Goodman (Ref. 8-2). ■

It will now be instructive for us to consider a system for which the magnitude of the phase transfer function is small compared with a radian,

i.e.,

$$|\Phi_H(\xi)| \ll 1. \tag{8.45}$$

The transfer function may then be approximated by

$$\begin{aligned} H(\xi) &= e^{-j\Phi_H(\xi)} \\ &\cong 1 - j\Phi_H(\xi), \end{aligned} \tag{8.46}$$

which leads to an impulse response of the form

$$h(x) \cong \delta(x) - j\phi_H(x), \tag{8.47}$$

where $\phi_H(x) = \mathscr{F}^{-1}\{\Phi_H(\xi)\}$.

Example

If we choose $\Phi_H(\xi) = d\sin(2\pi b\xi)$, where $|d| \ll 1$, we obtain

$$\phi_H(x) = -j\frac{d}{2|b|}\delta\delta\left(\frac{x}{b}\right). \tag{8.48}$$

Thus, with

$$g(x) = f(x) * h(x), \tag{8.49}$$

the output is given by

$$\begin{aligned} g(x) &\cong f(x) * \left[\delta(x) - \frac{d}{2|b|}\delta\delta\left(\frac{x}{b}\right)\right] \\ &\cong f(x) - 0.5d\,f(x+b) + 0.5d\,f(x-b). \end{aligned} \tag{8.50}$$

■

We now examine the behavior of a system for which the phase transfer function is of the form

$$\Phi_H(\xi) = \Theta_H(\xi) + 2\pi x_0\xi, \tag{8.51}$$

where $\Theta_H(\xi)$ is a nonlinear, but otherwise arbitrary, function of ξ. The transfer function of the system is then

$$\begin{aligned} H(\xi) &= e^{-j[\Theta_H(\xi) + 2\pi x_0\xi]}. \\ &= e^{-j\Theta_H(\xi)}e^{-j2\pi x_0\xi}, \end{aligned} \tag{8.52}$$

and its impulse response may be written

$$h(x) = \mathcal{F}^{-1}\{e^{-j\Theta_H(\xi)}\} * \delta(x - x_0). \tag{8.53}$$

For such a system, the nonlinear phase term introduces phase distortion in the usual way, whereas the linear term merely produces a shift of the

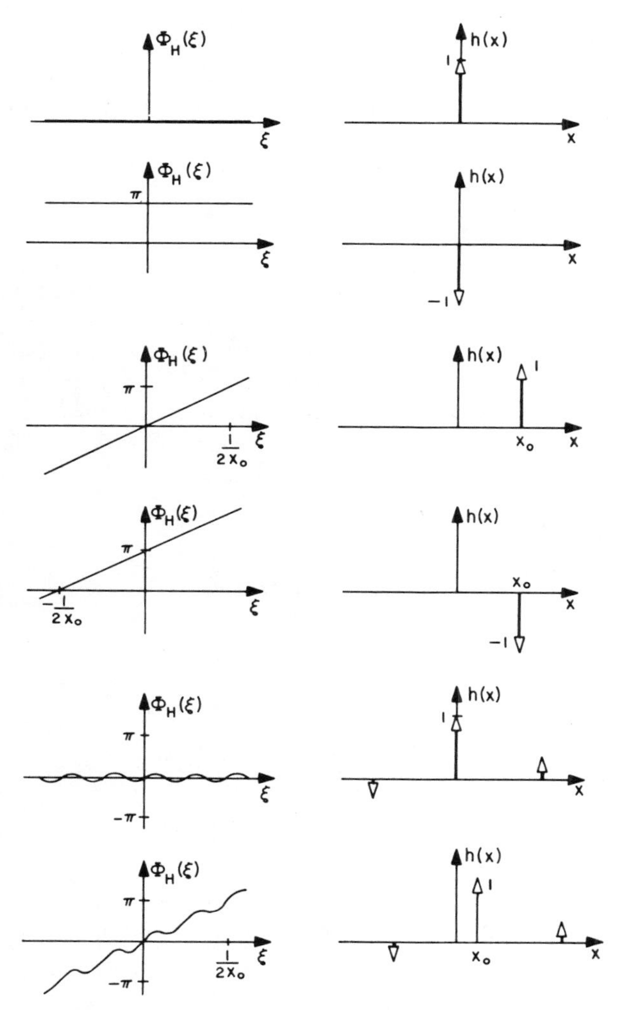

Figure 8-11 Phase transfer functions and impulse response functions of several phase filters.

output. Of course, if $|\Theta_H(\xi)| \ll 1$, the impulse response becomes

$$
\begin{aligned}
h(x) &\cong \mathcal{F}^{-1}\{1 - j\Theta_H(\xi)\} * \delta(x - x_0) \\
&\cong [\delta(x) - j\theta_H(x)] * \delta(x - x_0) \\
&\cong \delta(x - x_0) - j\theta_H(x - x_0),
\end{aligned}
\tag{8.54}
$$

where $\theta_H(x) = \mathcal{F}^{-1}\{\Theta_H(\xi)\}$.

As we did for the amplitude filter, we illustrate the phase transfer functions and impulse responses for several different phase filters in Fig. 8-11.

8-4 CASCADED SYSTEMS

Suppose two LSI systems were arranged in cascade, as depicted in Fig. 8-12, where the subscripts 1 and 2 refer to the various quantities associated with the first and second system, respectively. It is reasonable to inquire about the relationship between the input of the first system, $f_1(x)$, and the output of the second system, $g_2(x)$. Doing so, we may write

$$
g_2(x) = f_2(x) * h_2(x). \tag{8.55}
$$

But $f_2(x) = g_1(x)$, and

$$
g_1(x) = f_1(x) * h_1(x), \tag{8.56}
$$

so that

$$
\begin{aligned}
g_2(x) &= [f_1(x) * h_1(x)] * h_2(x) \\
&= f_1(x) * [h_1(x) * h_2(x)].
\end{aligned}
\tag{8.57}
$$

We may therefore regard these two cascaded systems to be a single

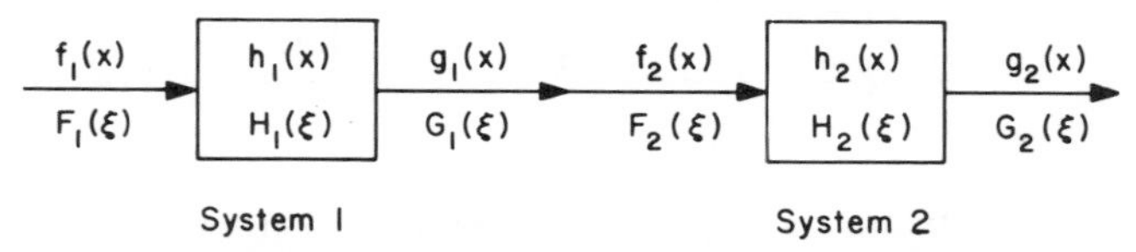

Figure 8-12 Two LSI systems in cascade.

equivalent system with impulse response $h(x)=h_1(x)*h_2(x)$. The transfer function of this equivalent system is simply $H(\xi)=H_1(\xi)H_2(\xi)$, and the output spectrum is given by

$$\begin{aligned} G_2(\xi) &= F_1(\xi)H(\xi) \\ &= F_1(\xi)\left[H_1(\xi)H_2(\xi)\right]. \end{aligned} \tag{8.58}$$

These results may be readily extended to n LSI systems in cascade, as illustrated in Fig. 8-13. The impulse response and transfer function for the equivalent single system (shown dashed in Fig. 8-13) are, respectively,

$$\begin{aligned} h(x) &= h_1(x)*h_2(x)*\cdots*h_n(x) \\ H(\xi) &= H_1(\xi)H_2(\xi)\cdots H_n(\xi). \end{aligned} \tag{8.59}$$

Note that the order in which the above operations are performed is immaterial [see Eq. (6.18)], and hence the order of the systems in the cascade may be rearranged in any convenient manner without changing the outcome.

It is often possible to design a filter having certain desired characteristics by cascading two or more properly selected systems. For example, a band-pass filter can be constructed by cascading a low-pass and high-pass filter with appropriate cutoff frequencies. We shall use this concept a great deal in the sections to follow.

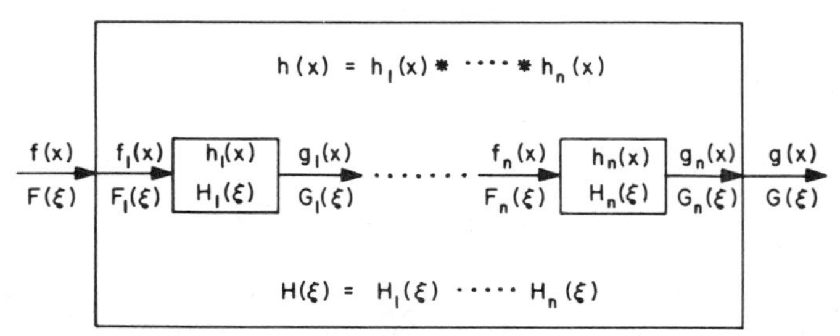

Figure 8-13 A cascade of n LSI systems.

8-5 COMBINATION AMPLITUDE AND PHASE FILTERS

Although certain systems may behave predominantly as either amplitude filters or phase filters, almost all systems encountered in practice are really combinations of the two. Our study of these combination filters will be greatly simplified because we have already investigated the behavior of the component filters. In addition, we will be able to take advantage of our knowledge of systems in cascade.

Let us begin our discussion by writing the expression for the transfer function of a general LSI system, i.e.,

$$H(\xi)=A_H(\xi)e^{-j\Phi_H(\xi)}. \tag{8.60}$$

If we regard this system as a cascaded pair (see Fig. 8-13), with

$$\begin{aligned} H_1(\xi)&=A_H(\xi)\\ H_2(\xi)&=e^{-j\Phi_H(\xi)}, \end{aligned} \tag{8.61}$$

we may determine the behavior of the entire system by successively accounting for the effects of the two filters $H_1(\xi)$ and $H_2(\xi)$. In other words, by first passing the input signal through the filter $H_1(\xi)$, which is an amplitude filter, we may determine the amplitude distortion introduced by the entire system. If we then pass the amplitude-distorted output of $H_1(\xi)$ through the phase filter $H_2(\xi)$, the phase distortion of the system is added. Finally, the output of the filter $H_2(\xi)$ contains all the distortion—both amplitude and phase—introduced by the entire system.

The impulse responses of the two component filters may be written as

$$\begin{aligned} h_1(x)&=a_H(x)\\ h_2(x)&=\mathcal{F}^{-1}\{e^{-j\Phi_H(\xi)}\}, \end{aligned} \tag{8.62}$$

and therefore the output may be expressed as

$$\begin{aligned} g(x)&=f(x)*h_1(x)*h_2(x)\\ &=f(x)*a_H(x)*\mathcal{F}^{-1}\{e^{-j\Phi_H(\xi)}\}. \end{aligned} \tag{8.63}$$

It is often convenient, at least qualitatively, to approach a filtering problem in this manner, because it is easy to understand how the two component filters affect the signal as it passes through them.

This approach may be extended one more step if we assume that the phase transfer function has a linear term in ξ, as in Eq. (8.51). By writing the transfer function for such a system as

$$\begin{aligned} H(\xi)&=A_H(\xi)e^{-j[\Theta_H(\xi)+2\pi x_0\xi]}\\ &=A_H(\xi)e^{-j\Theta_H(\xi)}e^{-j2\pi x_0\xi}, \end{aligned} \tag{8.64}$$

we may regard the system as a cascade of three filters, for which

$$H_1(\xi)=A_H(\xi)$$
$$H_2(\xi)=e^{-j\Theta_H(\xi)}$$
$$H_3(\xi)=e^{-j2\pi x_0\xi}, \tag{8.65}$$

and

$$h_1(x)=a_H(x)$$
$$h_2(x)=\mathcal{F}^{-1}\{e^{-j\Theta_H(\xi)}\}$$
$$h_3(x)=\delta(x-x_0). \tag{8.66}$$

Thus, the output becomes

$$g(x)=f(x)*a_H(x)*\mathcal{F}^{-1}\{e^{-j\Theta_H(\xi)}\}*\delta(x-x_0), \tag{8.67}$$

and we may view the output as a signal that has passed through three filters in cascade; the first causes the signal to be smoothed in the usual way, the second promotes a general deformation of the signal, often severe, and the third introduces an overall shift. This approach to the general filtering problem is illustrated in Fig. 8-14, where we have applied a single rectangular pulse as the input to the system. As indicated in the figure, the amplitude transfer function has a Gaussian shape, and the phase transfer function is composed of a linear-phase term plus a small sinusoidally varying component.

A combination amplitude and phase filter is said to be *distortionless* (Ref. 8-1) if the output can be expressed as

$$g(x)=Af(x-x_0), \tag{8.68}$$

where $f(x)$ is arbitrary and A is a constant (possibly complex-valued).In other words, except for magnitude and location, the input and output signals of a distortionless filter are identical. Clearly, the amplitude and phase transfer functions of a distortionless filter must satisfy the following (or equivalent) conditions: $A_H(\xi)=A$ and $\Phi_H(\xi)=2\pi x_0\xi$. And just as clearly, distortionless filters cannot exist in the real world.

We now investigate the symmetry properties of the transfer functions and impulse responses of combination filters. We begin by putting the

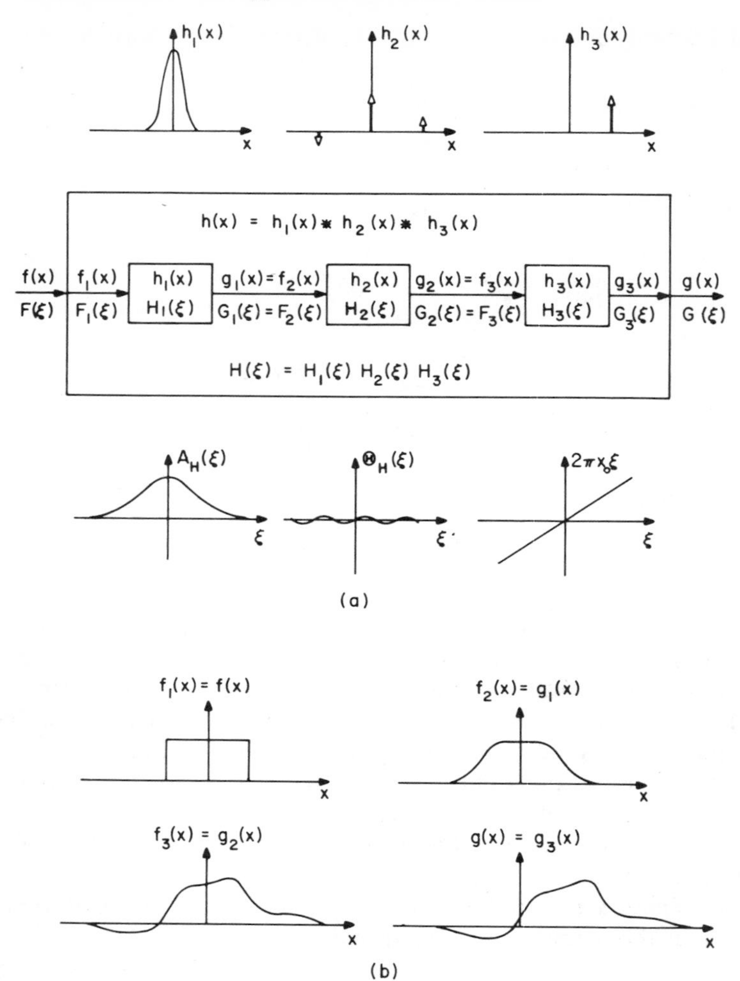

Figure 8-14 A general LSI system. (*a*) Depicted as cascade of amplitude filter, nonlinear phase filter, and linear phase filter. (*b*) Effects of the three filters on a rectangular input pulse.

transfer function in the form

$$H(\xi) = A_H(\xi)e^{-j\Phi_H(\xi)}$$

$$= A_H(\xi)\cos\Phi_H(\xi) - jA_H(\xi)\sin\Phi_H(\xi). \tag{8.69}$$

By assigning various combinations of oddness and evenness to $A_H(\xi)$ and $\Phi_H(\xi)$, we are able to determine whether $H(\xi)$ is odd or even, real or

Table 8-1

Symmetry Properties of the Impulse Response and Transfer Function

	$A_H(\xi)$ Odd	$A_H(\xi)$ Even
$\Phi_H(\xi)$ odd	$H(\xi)$ antihermitian $h(x)$ imaginary, no symmetry	$H(\xi)$ hermitian $h(x)$ real, no symmetry
$\Phi_H(\xi)$ even	$H(\xi)$ complex, odd $h(x)$ complex, odd	$H(\xi)$ complex, even $h(x)$ complex, even
$\Phi_H(\xi)=n\pi$	$H(\xi)$ real, odd $h(x)$ imaginary, odd	$H(\xi)$ real, even $h(x)$ real, even
$\Phi_H(\xi)=(2n+1)\frac{\pi}{2}$	$H(\xi)$ imaginary, odd $h(x)$ real, odd	$H(\xi)$ imaginary, even $h(x)$ imaginary, even

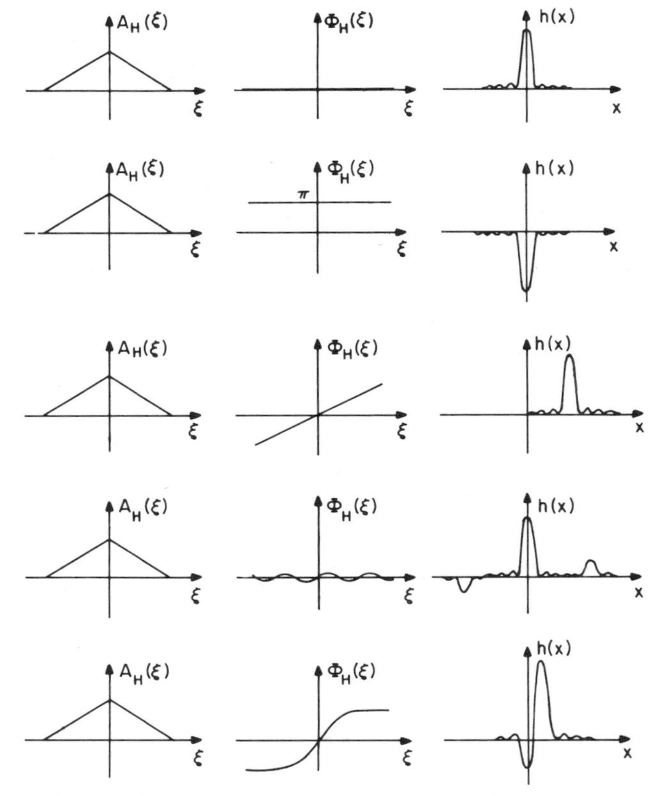

Figure 8-15 Amplitude transfer functions, phase transfer functions, and impulse response functions for a number of combination filters.

complex valued, etc. Finally, from the symmetry properties of the Fourier transform listed in Table 7-1, the same information is determined for the impulse response. The results of this investigation are summarized in Table 8-1.

As we have already done for the amplitude filters and phase filters separately, we show the transfer functions and impulse responses for several typical combination filters in Fig. 8-15. In addition, Fig. 8-16 provides a comparison of the values, on and within the unit circle, allowed for the transfer functions of the various types of filters discussed.

This concludes our brief investigation of general linear filters; for those who are interested in more details, Ref. 8-1 is recommended.

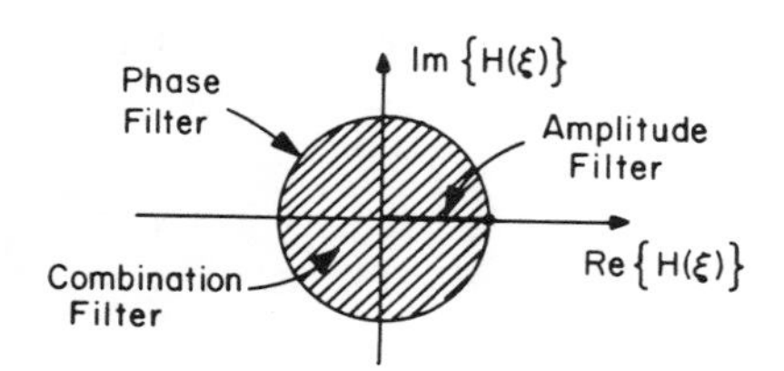

Figure 8-16 Allowed values for transfer functions of various types of filters.

8-6 SIGNAL PROCESSING WITH LINEAR FILTERS

A number of filtering applications are explored in this section to help broaden your understanding and appreciation of the behavior of LSI systems. Before addressing the first of these applications, however, we will

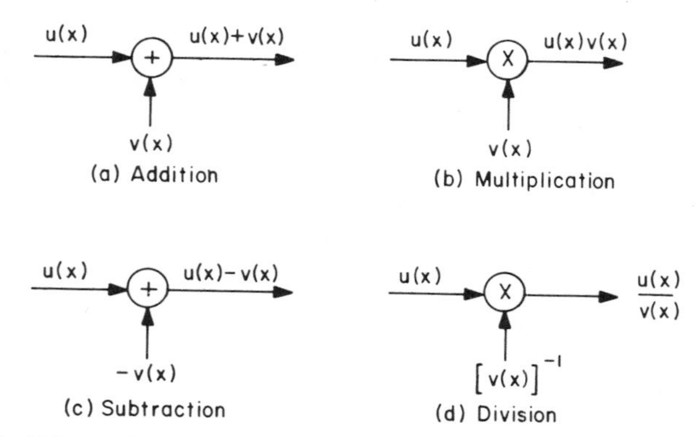

Figure 8-17 Schematic representation of various mathematical operations. (*a*) Addition. (*b*) Multiplication. (*c*) Subtraction. (*d*) Division.

find it helpful to introduce two devices that are familiar to those engaged in electrical engineering or similar disciplines. The first of these devices, which are depicted schematically in Fig. 8-17, causes two signals to be added together, whereas the second causes two signals to be multiplied. Note that subtraction is performed by changing the polarity of the appropriate signal before the addition takes place. Similarly, division is obtained by first inverting the appropriate signal. These various operations are shown in Fig. 8-17.

Next, I wish to inject some editorial comments about the widespread —and often meaningless—use of the word "noise." In other words, I am going to make some noise about the word noise. According to my dictionary (Ref. 8-3), noise is described as sound of any sort, especially that without agreeable musical quality. I suspect this term was initially used in radio engineering to characterize undesired sounds, emanating from a headset or loudspeaker, that interfered with the intelligibility of a desired audio signal. Now, however, it is used by scientists and engineers to denote almost any undesired quantity, irrespective of the nature of that quantity. Thus, optical images with certain undesirable features are termed "noisy," photographically recorded data suffers from "film-grain noise," etc. A little thought will show exactly how inappropriate most of these designations are (have you ever listened to a piece of film and heard the film-grain noise?). However, in spite of my objections, the use of this word has gained such widespread acceptance among the scientific community that a campaign against it would be fruitless, and I, too, shall join the ranks. In this text, the word "noise" shall be used to denote any undesired signal or signal component.

Extraction of Signals from Noise

A frequently encountered filtering problem is one in which a desired signal $s(x)$, corrupted by additive noise $n(x)$, must be restored to its uncorrupted state. We shall assume that both $s(x)$ and $n(x)$ are deterministic functions, with Fourier transforms $S(\xi)$ and $N(\xi)$, respectively, and for this reason our results will be of little practical use in the real world where a statistical approach is generally needed. Nevertheless, this discussion should prove helpful in understanding the nature of the problem.

We apply the signal plus noise as the input to an LSI system, the properties of which have yet to be determined, as shown in Fig. 8-18. The input is given by

$$f(x)=s(x)+n(x), \tag{8.70}$$

and the output becomes

$$g(x)=[s(x)+n(x)]*h(x). \tag{8.71}$$

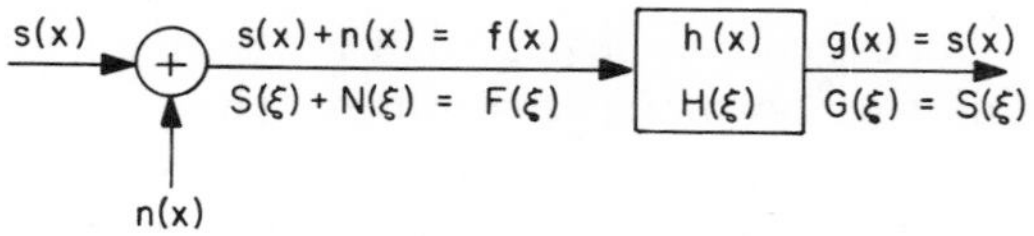

Figure 8-18 Signal and additive noise as the input to an LSI system.

Our problem is to design the filter such that the output is equal to the desired signal alone, i.e., $g(x)=s(x)$. Thus,

$$s(x)=[s(x)+n(x)]*h(x). \tag{8.72}$$

We must therefore find the solution to this equation, which will apparently require some sort of complicated deconvolution operation. However, as is frequently the case, the solution is easily obtained by going to the frequency domain. By Fourier transforming both sides of Eq. (8.72), we obtain

$$S(\xi)=[S(\xi)+N(\xi)]H(\xi), \tag{8.73}$$

which may be rearranged to yield

$$H(\xi)=\frac{S(\xi)}{S(\xi)+N(\xi)}. \tag{8.74}$$

To check our solution we write the output spectrum in terms of the input spectrum and the transfer function:

$$\begin{aligned} G(\xi) &= F(\xi)H(\xi) \\ &= [S(\xi)+N(\xi)]\frac{S(\xi)}{[S(\xi)+N(\xi)]} \\ &= S(\xi), \end{aligned} \tag{8.75}$$

and we see that our solution is correct.

Let us now examine the behavior of the transfer function by dividing the numerator and denominator of Eq. (8.74) by $N(\xi)$, such that

$$H(\xi)=\frac{\dfrac{S(\xi)}{N(\xi)}}{\dfrac{S(\xi)}{N(\xi)}+1}. \tag{8.76}$$

We see, then, that the transfer function should have a value of approximately unity in those spectral regions where the signal spectrum is large relative to the noise spectrum, whereas it should be nearly zero wherever $S(\xi)$ is much smaller than $N(\xi)$. In the regions where the strengths of $S(\xi)$ and $N(\xi)$ are comparable, $H(\xi)$ should have a value of approximately 0.5.

If there is no overlap of the signal and noise spectra in the frequency domain, an ideal (binary) amplitude filter may be used to recover $s(x)$. Examples are shown in Fig. 8-19. On the other hand, the design problem is more difficult when $S(\xi)$ and $N(\xi)$ do overlap, as illustrated in Fig. 8-20. However, it seems impractical to construct a filter in the latter case, because $S(\xi)$ and $N(\xi)$ must already be known in order to do so, and why construct a complicated filter to recover something that is already known?

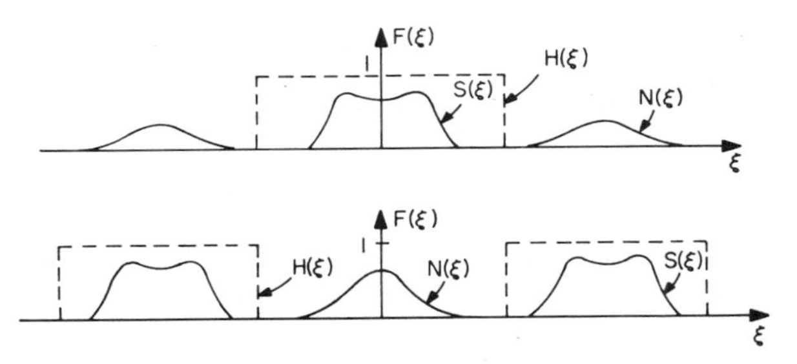

Figure 8-19 Recovery of signal in presence of additive noise when signal and noise spectra do not overlap.

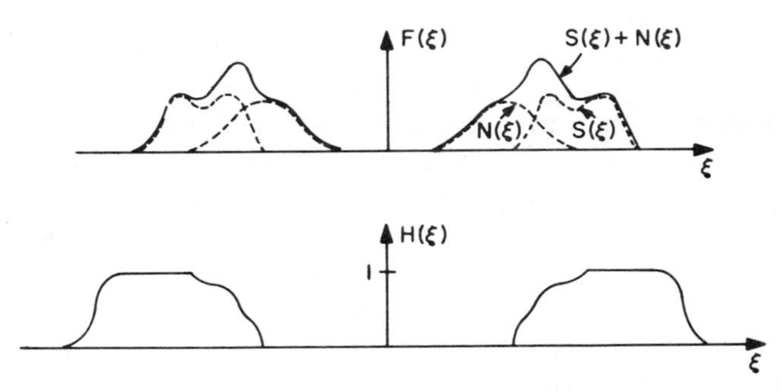

Figure 8-20 Recovery of signal in presence of additive noise when signal and noise spectra overlap.

Equalization

Suppose that an arbitrary signal $s(x)$ has been passed through a filter $H_1(\xi)$, thereby undergoing amplitude and phase distortion, and we wish to restore this distorted signal to its original form. In certain cases this may be done by passing the distorted signal through a second filter $H_2(\xi)$, as shown in Fig. 8-21, whose transfer function is given by

$$H_2(\xi) = \frac{1}{H_1(\xi)}. \tag{8.77}$$

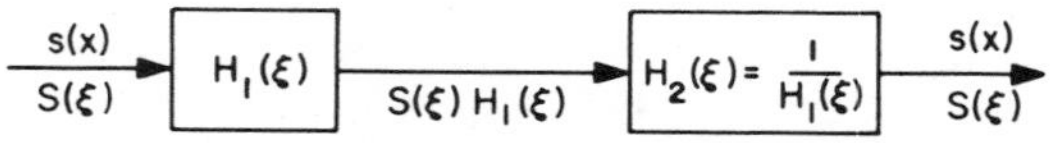

Figure 8-21 Filtering technique of equalization.

The output spectrum of the second filter is then

$$\begin{aligned} G_2(\xi) &= F_2(\xi)H_2(\xi) \\ &= \left[S(\xi)H_1(\xi)\right]\left[\frac{1}{H_1(\xi)}\right] \\ &= S(\xi), \end{aligned} \tag{8.78}$$

and the signal has been restored.

Although equalization is perfectly plausible from a theoretical point of view, it often does not work well in practice. A little thought reveals some of the flaws of this scheme. To begin with, the second system must have gain greater than unity in those spectral regions where $|H_1(\xi)| < 1$, i.e., $|H_2(\xi)| > 1$ wherever $|H_1(\xi)| < 1$, and this excludes any system for which a gain of greater than unity cannot be achieved. Next, $H_1(\xi)$ cannot be allowed to have any zeros, because this would require that $H_2(\xi)$ be infinitely large at those points; clearly this restriction will cause problems, because the transfer functions of all physical systems must go to zero for sufficiently large $|\xi|$. Finally, even more problems are encountered when noise is present, but they are beyond the scope of this book.† In spite of all the difficulties mentioned, equalization techniques have been used with moderate success for certain signal restoration problems.

†Translation: the author doesn't understand them.

Spectral Analysis

In this section we develop a device that performs a Fourier transform operation, i.e., a device whose output is proportional to the spectrum of the input. First, however, it will be convenient to introduce the *quadratic-phase signal*

$$q(x;a)=e^{j\pi a x^2}, \tag{8.79}$$

where a is a real constant. The notation used here for this signal is similar to that suggested by Vander Lugt (Ref. 8-4). Because the phase of $q(x;a)$ varies quadratically with x, its "instantaneous frequency" is a linear function of x; for this reason, $q(x;a)$ is often referred to as a signal with *linear FM* (frequency modulation).

We now list several obvious properties of $q(x;a)$ for reference:

$$\begin{aligned}
q(-x;a)&=q(x;a) & q(x;a_1)q(x;a_2)&=q(x;a_1+a_2),\\
q^*(x;a)&=q(x;-a) & q(x;a_1)q^*(x;a_2)&=q(x;a_1-a_2),\\
q(bx;a)&=q(x;b^2a) & q(x;a)q^*(x;a)&=1,\\
q(x;0)&=1.
\end{aligned} \tag{8.80}$$

The Fourier transform of $q(x;a)$ is found from Eq. (7.79) to be

$$\begin{aligned}
\mathcal{F}\{q(x;a)\}&=Q(\xi;a)\\
&=\frac{\exp\left(j\frac{\pi}{4}\right)}{\sqrt{a}}\exp\left(-j\pi\frac{\xi^2}{a}\right)\\
&=\sqrt{\frac{j}{a}}\,q^*\left(\xi;\frac{1}{a}\right)
\end{aligned} \tag{8.81}$$

and thus

$$\begin{aligned}
\mathcal{F}\{q^*(x;a)\}&=Q^*(-\xi;a)\\
&=\frac{\exp\left(-j\frac{\pi}{4}\right)}{\sqrt{a}}\exp\left(j\pi\frac{\xi^2}{a}\right)\\
&=\frac{1}{\sqrt{ja}}\,q\left(\xi;\frac{1}{a}\right).
\end{aligned} \tag{8.82}$$

The convolution of $q(x;a)$ and $q^*(x;a)$ may be derived as follows:

$$\begin{aligned}\mathcal{F}\{q(x;a)*q^*(x;a)\} &= Q(\xi;a)Q^*(-\xi;a)\\ &= \sqrt{\frac{j}{a}}\, q^*\Big(\xi;\frac{1}{a}\Big)\frac{1}{\sqrt{ja}}\,q\Big(\xi;\frac{1}{a}\Big)\\ &= \frac{1}{|a|},\end{aligned} \tag{8.83}$$

and inverting this result we obtain

$$q(x;a)*q^*(x;a) = \frac{1}{|a|}\delta(x). \tag{8.84}$$

Several other interesting properties that will be useful later are derived below. Given an arbitrary signal $s(x)$, we form the convolution

$$\begin{aligned}s(x)*q(x;a) &= \int_{-\infty}^{\infty} s(\beta)q(x-\beta;a)d\beta\\ &= \int_{-\infty}^{\infty} s(\beta)e^{j\pi a(x^2+\beta^2-2\beta x)}d\beta\\ &= q(x;a)\int_{-\infty}^{\infty} s(\beta)q(\beta;a)e^{-j2\pi a x\beta}d\beta\\ &= q(x;a)\mathcal{F}\{s(x)q(x;a)\}\big|_{\xi=ax}.\end{aligned} \tag{8.85}$$

This form of the result is often useful, but we may use Eqs. (7.44) and (6.29) to obtain

$$\begin{aligned}s(x)*q(x;a) &= q(x;a)\big[S(\xi)*Q(\xi;a)\big]\big|_{\xi=ax}\\ &= q(x;a)a\big[S(ax)*Q(ax;a)\big]\\ &= q(x;a)a\left[S(ax)*\sqrt{\frac{j}{a}}\, q^*\Big(ax;\frac{1}{a}\Big)\right]\\ &= \sqrt{ja}\, q(x;a)\big[S(ax)*q^*(x;a)\big].\end{aligned} \tag{8.86}$$

If we multiply both sides of Eq. (8.86) by $q^*(x;a)$, we may write

$$\big[s(x)*q(x;a)\big]q^*(x;a) = \sqrt{ja}\,\big[S(ax)*q^*(x;a)\big], \tag{8.87}$$

and convolving again with $q(x;a)$ we find that

$$\begin{aligned}\{[s(x)*q(x;a)]q^*(x;a)\}*q(x;a) &= \sqrt{ja}\,[S(ax)*q^*(x;a)]*q(x;a)\\ &= \sqrt{ja}\,S(ax)*[q^*(x;a)*q(x;a)]\\ &= \sqrt{ja}\,S(ax)*\frac{1}{|a|}\delta(x)\\ &= \sqrt{\frac{j}{a}}\,S(ax).\end{aligned} \tag{8.88}$$

Thus, by combining various multiplication and convolution operations involving $q(x;a)$ and $q^*(x;a)$, we are able to produce the spectrum of $s(x)$.

We may obtain a similar result by altering the order in which the multiplication and convolution operations are performed. For example, if we first multiply $s(x)$ by $q^*(x;a)$ and then convolve the product with $q(x;a)$, we have

$$\begin{aligned}[s(x)q^*(x;a)]*q(x;a) &= \int_{-\infty}^{\infty} s(\beta)q^*(\beta;a)q(x-\beta;a)d\beta\\ &= \int_{-\infty}^{\infty} s(\beta)e^{-j\pi a\beta^2}e^{j\pi a(x^2+\beta^2-2x\beta)}\,d\beta\\ &= q(x;a)\int_{-\infty}^{\infty} s(\beta)e^{-j2\pi ax\beta}d\beta\\ &= q(x;a)S(ax),\end{aligned} \tag{8.89}$$

and if we again multiply both sides of this equation by $q^*(x;a)$ we obtain

$$\{[s(x)q^*(x;a)]*q(x;a)\}q^*(x;a) = S(ax). \tag{8.90}$$

Now let us assume that we wish to perform a spectral analysis of an arbitrary signal $f(x)$ by using the filtering techniques we have developed. We apply $f(x)$ as the input to the device depicted in Fig. 8-22, such that the input to the interior LSI system is given by

$$f_1(x) = f(x)q^*(x;a). \tag{8.91}$$

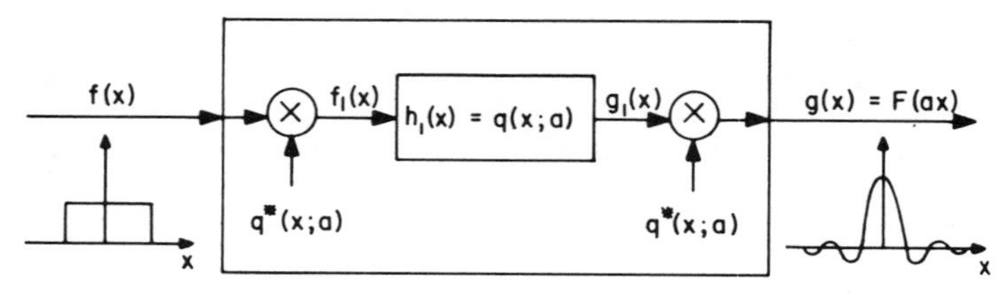

Figure 8-22 System that performs spectral analysis.

Therefore,

$$\begin{aligned} g_1(x) &= f_1(x) * q(x;a) \\ &= [f(x)q^*(x;a)] * q(x;a) \\ &= q(x;a)F(ax), \end{aligned} \tag{8.92}$$

where Eq. (8.89) was used in the last step. With the final multiplication indicated in Fig. 8-22, the output of our device becomes

$$\begin{aligned} g(x) &= g_1(x)q^*(x;a) \\ &= F(ax), \end{aligned} \tag{8.93}$$

which is simply a scaled Fourier transform of the input. Note that even though an LSI system is used as part of our spectrum analyzer, the overall device is not shift invariant. That this is so may easily be determined by shifting the input signal and noting that the output does not shift; e.g., an input of $f(x-x_0)$ leads to an output of $\exp\{-j2\pi a x_0 x\}F(ax)$, which is improper behavior for a shift-invariant system. Such a system is illustrated operationally in Fig. 8-23. The same result can be obtained with the device shown in Fig. 8-24, the details of which are left as an exercise for the reader [hint: use Eq. (8.88)].

In later chapters we encounter similar devices in the form of coherent optical systems; of course, a two-dimensional approach is required for those systems, but the principles of operation are no different. In an optical system, we find that the effect of a lens on an incident wave field can be modeled by a multiplication operation involving a quadratic-phase function. In addition, the propagation of light between lenses can be

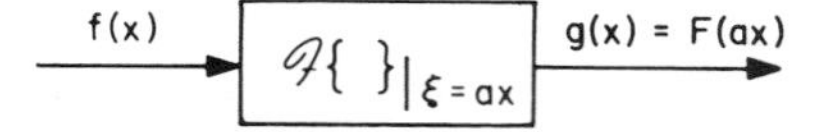

Figure 8-23 Operational representation of a spectrum analyzer.

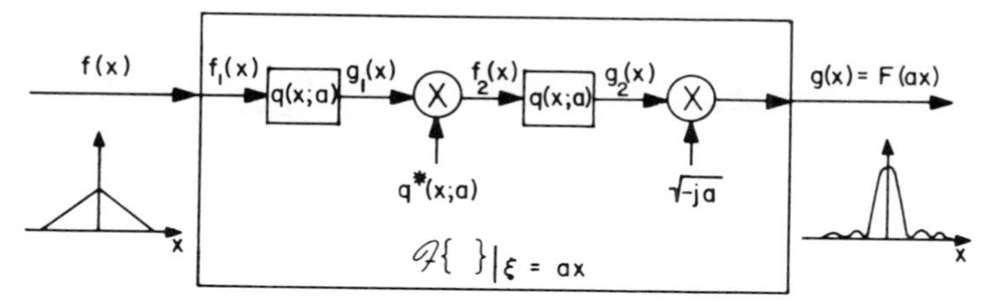

Figure 8-24 Spectrum analyzer consisting of two LSI systems.

accounted for in terms of an LSI system with a quadratic-phase impulse response. Consequently, the results derived in this section will be very useful later on.

Signal Scaling

It is often desirable to reproduce a signal with a change of scale, i.e., a signal that has been expanded or contracted along the axis of the independent variable. A device to perform this scaling can be devised by cascading two of the spectrum analyzers of the previous section, as depicted in Fig. 8-25. These two analyzers employ scaling constants of a_1 and a_2, respectively, and we have added the multiplication by a_1 between the analyzers to obtain a less cumbersome result. If we now apply $f(x)$ as the input to this device, we find

$$g_1(x) = F(a_1 x), \tag{8.94}$$

and with $f_2(x) = a_1 g_1(x)$ and $g(x) = g_2(x)$, we obtain

$$\begin{aligned} g(x) &= \mathcal{F}\{f_2(x)\}|_{\xi = a_2 x} \\ &= \mathcal{F}\{a_1 F(a_1 x)\}|_{\xi = a_2 x} \\ &= f\left(\frac{-a_2 x}{a_1}\right), \end{aligned} \tag{8.95}$$

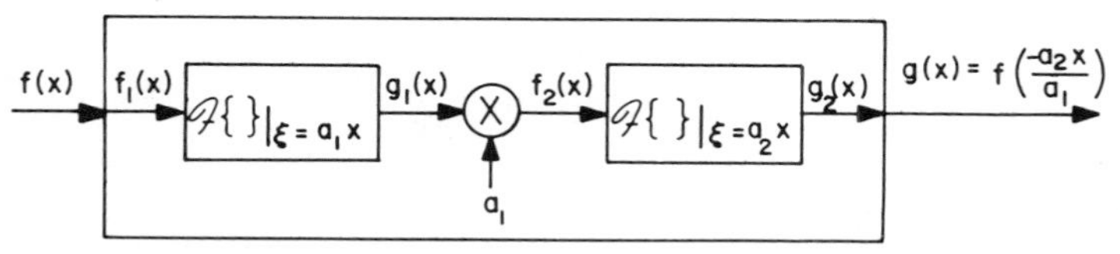

Figure 8-25 Signal scaler as cascade of two spectrum analyzers.

which is the desired result.

If we now designate the ratio $-a_1/a_2$ by m, the output of our signal scaler may be expressed as

$$g(x)=f\left(\frac{x}{m}\right). \tag{8.96}$$

We observe that, in general, this device is not shift invariant: the output is not only expanded or contracted along the x-axis, it is also folded about the origin when m is negative. In addition, when the input is shifted by an amount x_0, the output is shifted by an amount mx_0, as shown in Fig. 8-26, and such behavior is clearly not characteristic of an LSI system.

It should be noted that the above device produces no smoothing of the output, and it is therefore nonphysical. However, a slight modification yields a device that does smooth the output (which, incidentally, is the one-dimensional equivalent of a coherent optical imaging system we will encounter later). The modification is shown in Fig. 8-27, and consists of an additional multiplication operation, by a function $p(x)$, located between the spectrum analyzers of Fig. 8-25. The effect of the function $p(x)$ is to limit the extent of the input to the second analyzer, and in an optical imaging system it corresponds to the transmittance of the exit pupil. With this modification, $f_2(x)=a_1F(a_1x)p(x)$ and

$$\begin{aligned} g(x) &= \mathcal{F}\{a_1F(a_1x)p(x)\}\big|_{\xi=a_2x} \\ &= f\left(\frac{-a_2x}{a_1}\right)*a_2P(a_2x), \end{aligned} \tag{8.97}$$

where $P(\xi)=\mathcal{F}\{p(x)\}$. Again letting $-a_1/a_2=m$, we may write

$$g(x)=f\left(\frac{x}{m}\right)*a_2P(a_2x). \tag{8.98}$$

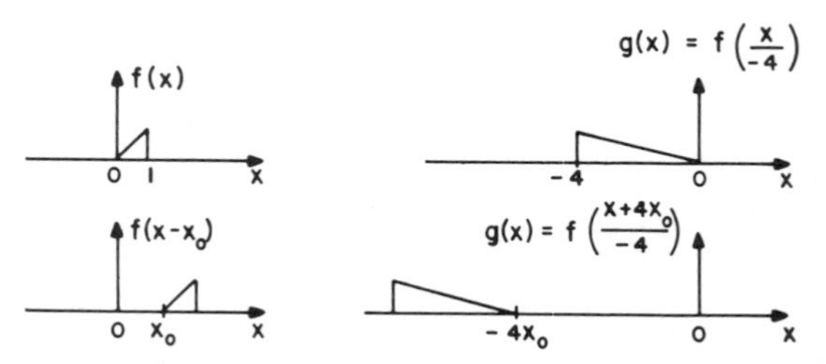

Figure 8-26 Input-output relationships of a signal scaler for which $m=-4$.

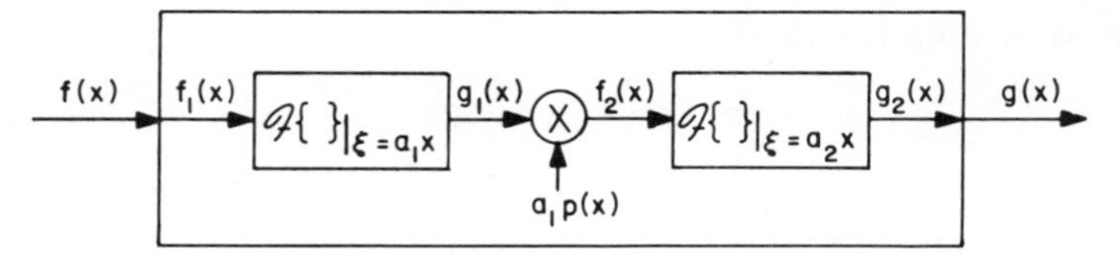

Figure 8-27 Signal scaler with smoothing.

This device is now physically plausible because it introduces smoothing of the output, but it is still not shift invariant. However, we may model it by the cascaded pair of linear systems shown in Fig. 8-28, the first of which is a simple shift-variant scaling device and the second of which is a shift-invariant system. In other words, we may consider the device of Fig. 8-27 to consist of a shift-variant scaling device, which scales and folds the signal without deforming it, followed by a conventional LSI filter that introduces amplitude and/or phase distortion. The impulse response of this filter is seen to be[§]

$$\hat{h}_2(x) = a_2 P(a_2 x), \tag{8.99}$$

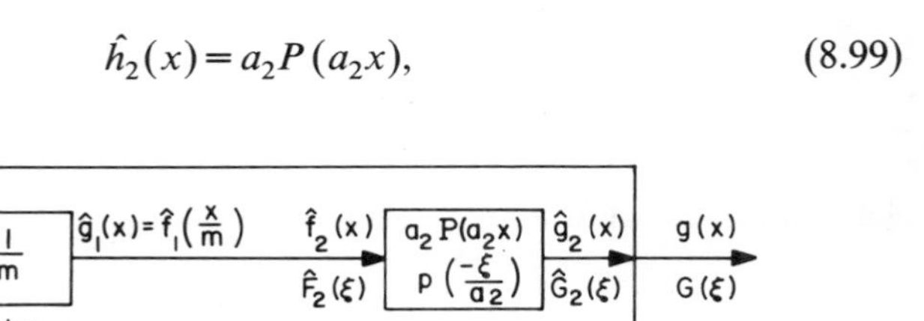

Figure 8-28 System of Fig. 8-27 represented operationally as a cascade of a signal scaler and an LSI filter.

and thus its transfer function is simply

$$\hat{H}_2(\xi) = \mathcal{F}\{\hat{h}_2(x)\}$$

$$= p\left(\frac{-\xi}{a_2}\right). \tag{8.100}$$

By modeling the device of Fig. 8-27 in this fashion, we are able to simplify input-output problems by attacking them in the frequency

[§]The caret notation is used with various functions here to avoid confusion between the internal systems of Fig. 8-28 and those of Fig. 8-27.

domain. Rewriting Eq. (8.98) as

$$g(x)=\hat{f}_2(x)*\hat{h}_2(x), \tag{8.101}$$

where $\hat{f}_2(x)=\hat{g}_1(x)=f(x/m)$, we may express the output spectrum by

$$\begin{aligned} G(\xi) &= \hat{F}_2(\xi)\hat{H}_2(\xi) \\ &= |m|F(m\xi)\,p\left(\frac{-\xi}{a_2}\right). \end{aligned} \tag{8.102}$$

Thus, to calculate the output we need only find the signal spectrum, scale it by m, multiply by $|m|p(-\xi/a_2)$, and then perform an inverse transform of this product.

Example

Let us calculate the output of this device when

$$\begin{aligned} f(x) &= \text{rect}(2x)*\text{comb}(x), \qquad & a_1 &= 2, \\ p(x) &= \text{rect}\left(\frac{x}{4}\right), & a_2 &= 1. \end{aligned} \tag{8.103}$$

The input spectrum and transfer function are

$$\begin{aligned} F(\xi) &= \tfrac{1}{2}\text{sinc}\left(\frac{\xi}{2}\right)\text{comb}(\xi), \\ \hat{H}_2(\xi) &= \text{rect}\left(\frac{-\xi}{4}\right)=\text{rect}\left(\frac{\xi}{4}\right), \end{aligned} \tag{8.104}$$

and with $m=-a_1/a_2=-2$, the output spectrum becomes

$$\begin{aligned} G(\xi) &= |m|F(m\xi)\hat{H}_2(\xi) \\ &= \text{sinc}(\xi)\,\text{comb}(2\xi)\,\text{rect}\left(\frac{\xi}{4}\right) \\ &= \tfrac{1}{2}\delta(\xi)+\frac{2}{\pi}\delta\delta(2\xi)-\frac{2}{9\pi}\delta\delta\left(\frac{2\xi}{3}\right). \end{aligned} \tag{8.105}$$

This result was obtained from Eq. (8.17). Then, from Eq. (8.18), we find the output to be

$$g(x)=\tfrac{1}{2}+\frac{2}{\pi}\cos(\pi x)-\frac{2}{3\pi}\cos(3\pi x). \tag{8.106}$$

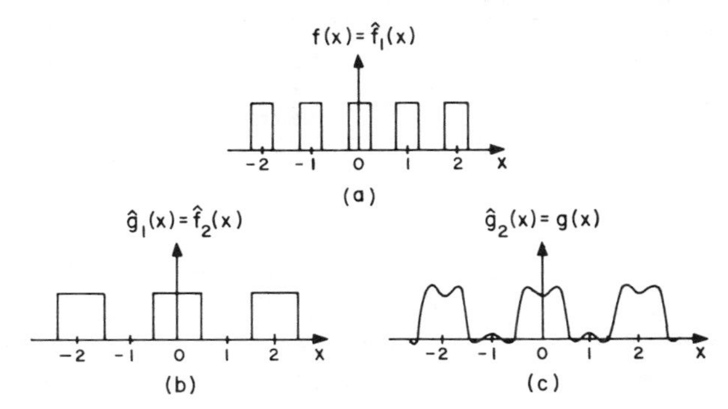

Figure 8-29 Example using system of Fig. 8-28. (*a*) Input signal. (*b*) Output of signal scaler. (*c*) Output signal [see Eq. (8.106)].

This example is illustrated in Fig. 8-29. ■

In the equivalent imaging problem, the input function $f(x)$ represents the light distribution of some object, and the effect of the scaling device is to form a magnified geometrical image of this object, $\hat{g}_1(x)$. Finally, the LSI system introduces the smoothing effects of diffraction and produces the diffraction image $g(x)$. If this brief analogy is confusing, don't worry; there is a lot more discussion of this topic coming up in the later chapters.

Signal Detection—The Matched Filter

We now turn to a filtering problem the purpose of which is not to recover or restore a signal, but merely to *detect the presence* of a signal. More specifically, the intent is to discover when a *particular* signal is present, a determination complicated by the requirement that unwanted signals, or noise, are not allowed to be confused with the desired signal. In other words, any such scheme must be able to distinguish between the desired and the undesired signals. Applications include the reception of radar signals, optical data processing, coded-aperture imaging, etc.

Let us assume we are required to detect the presence of a desired signal, $s(x)$, without detecting any of the ensemble of undesired signals $\{n_i(x)\} = \{n_1(x), n_2(x), \ldots, n_N(x)\}$ that may also be present from time to time. We must devise a scheme that actuates an alarm of some sort whenever $s(x)$ appears at the input, and this scheme must give no indication of detection whenever any of the undesired signals appear at the input. To simplify the problem, we shall require that each of the signals is normalized such that

$$\int_{-\infty}^{\infty} |s(\alpha)|^2 \, d\alpha = \int_{-\infty}^{\infty} |n_i(\alpha)|^2 \, d\alpha$$
$$= K, \tag{8.107}$$

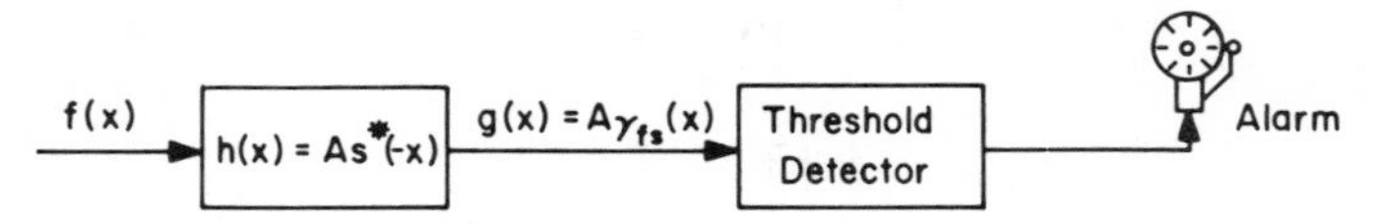

Figure 8-30 Signal detection by matched filtering.

i.e., such that the so-called energy of each is equal to a constant K.

The device to be employed is illustrated in Fig. 8-30, and consists of a special type of LSI system, called a *matched filter*, followed by a threshold detector and alarm. The impulse response of this filter is chosen to be

$$h(x)=As^*(-x), \tag{8.108}$$

where A is a constant and the system is said to be "matched" to the signal $s(x)$. For an arbitrary input $f(x)$, the output may be expressed as

$$\begin{aligned} g(x)&=f(x)*h(x)\\ &=Af(x)*s^*(-x)\\ &=A\gamma_{fs}(x), \end{aligned} \tag{8.109}$$

where $\gamma_{fs}(x)$ is the complex cross-correlation function of $f(x)$ with $s(x)$ [see Eq. (6.52)]. Note, however, that when $f(x)=s(x)$, this cross-correlation function becomes an autocorrelation function, and the output is

$$g(x)=A\gamma_s(x). \tag{8.110}$$

We now exploit the nice properties of the complex autocorrelation function.

We already know from Eq. (6.62) that $\gamma_s(x)$ has a maximum at the origin; it may also be shown, from Eq. (6.63), that for any input signal $f(x)=n_i(x)$ which satisfies Eq. (8.107),

$$\gamma_s(0)\geqslant|\gamma_{fs}(x)|. \tag{8.111}$$

Therefore, the output of the system will reach its maximum value of $A\gamma_s(0)$ when $f(x)=s(x)$, and this maximum will occur at the origin. For any other input signal $f(x)=n_i(x)$, the maximum value attained by the output, whether or not it occurs at the origin, will be less than or equal to $A\gamma_s(0)$. Then, by adjusting the setting of the threshold detector such that a matched-filter output of $A\gamma_s(0)$ will trigger the alarm and any lower value will be rejected, we have solved the problem.

In practice, of course, there are difficulties with this scheme: to be certain of detecting $s(x)$, the threshold must be set at a value slightly less than $A\gamma_s(0)$, say $A\gamma_s(0)-\varepsilon$, where ε is a small number. But, with this lowered threshold setting, there is a possibility that some undesired signal $f(x)=n_i(x)$ will yield an output that exceeds the threshold, thereby producing a false alarm. Consequently, the value of ε selected determines the false alarm rate, which is the ratio of false alarms produced to the total number of undesired signals received. In reality, this is a statistical problem because there are statistical fluctuations in the quantities $\gamma_s(0)$ and $\gamma_{fs}(x)$. As a result, one must be concerned not only with minimizing the false alarm rate, but also with ensuring that an acceptable probability of detection has been established for the desired signal. When a choice is available, these difficulties can be reduced somewhat by selecting desired signals that have sharply peaked autocorrelation functions; e.g., an ideal signal would be one for which $\gamma_s(x)=\delta(x)$.

If we should desire to maximize the output of the matched filter at the point $x=x_0$, rather than at the origin, Eq. (8.108) should be changed to

$$\begin{aligned} h(x) &= As^*(x_0-x) \\ &= As^*\left(\frac{x-x_0}{-1}\right) \\ &= As^*(-x)*\delta(x-x_0). \end{aligned} \tag{8.112}$$

This will yield a general output of

$$\begin{aligned} g(x) &= f(x)*\left[As^*(-x)*\delta(x-x_0)\right] \\ &= A\gamma_{fs}(x)*\delta(x-x_0) \\ &= A\gamma_{fs}(x-x_0), \end{aligned} \tag{8.113}$$

which clearly will have a maximum at $x=x_0$ when $f(x)=s(x)$. We see now that Eq. (8.109) is simply a special case of Eq. (8.113).

We have not yet discussed the transfer function of a matched filter, and it is important that we do so. From Eq. (8.112) we find that

$$\begin{aligned} H(\xi) &= \mathcal{F}\{As^*(-x)*\delta(x-x_0)\} \\ &= AS^*(\xi)e^{-j2\pi x_0\xi}. \end{aligned} \tag{8.114}$$

Thus, for an arbitrary input signal $f(x)$, the output spectrum of the matched filter will be

$$G(\xi)=AF(\xi)S^*(\xi)e^{-j2\pi x_0\xi}, \tag{8.115}$$

and when $f(x)=s(x)$,

$$G(\xi)=A|S(\xi)|^2e^{-j2\pi x_0\xi}. \tag{8.116}$$

We note that when A is real valued, $G(\xi)$ is hermitian and the output $g(x)$ is real valued (with a maximum at $x=x_0$).

Additional details regarding matched filters may be found in Ref. 8-5.

Example

We select the desired signal for this example to be $s(x)=\text{sinc}(x)$, and for simplicity we choose the undesired signals to be of the form $\{n_i(x)\} =\{\sqrt{a_i}\ \text{sinc}(a_i x), a_i\neq 1\}$. Note that $s(x)$ and $\{n_i(x)\}$ all have the same form, with the only difference being the heights and widths. Note also that Eq. (8.107) is satisfied, with $K=1$. Let us assume that $g(x)$ is to be maximized at $x=5$, and that $A=1$. Therefore, we must have

$$\begin{aligned} h(x)&=\text{sinc}(5-x)=\text{sinc}(x-5),\\ H(\xi)&=\text{rect}(\xi)e^{-j10\pi\xi}. \end{aligned} \tag{8.117}$$

The desired signal spectrum is simply

$$S(\xi)=\text{rect}(\xi), \tag{8.118}$$

and when $f(x)=s(x)$, the output spectrum becomes

$$\begin{aligned} G(\xi)&=|\text{rect}(\xi)|^2e^{-j10\pi\xi}\\ &=\text{rect}(\xi)e^{-j10\pi\xi}. \end{aligned} \tag{8.119}$$

Thus, the output $g(x)$ is simply

$$g(x)=\text{sinc}(x-5), \tag{8.120}$$

and we note that the maximum value occurs at $x=5$, as it should. The undesired signal spectra are given by

$$N_i(\xi)=\frac{1}{\sqrt{a_i}}\ \text{rect}\left(\frac{\xi}{a_i}\right), \tag{8.121}$$

such that when $f(x)=n_i(x)$ the output spectrum is given by

$$G(\xi)=\frac{1}{\sqrt{a_i}}\operatorname{rect}\left(\frac{\xi}{a_i}\right)\operatorname{rect}(\xi)e^{-j10\pi\xi}. \tag{8.122}$$

This expression may be reduced to

$$\begin{aligned} G(\xi)&=\frac{1}{\sqrt{a_i}}\operatorname{rect}\left(\frac{\xi}{a_i}\right)e^{-j10\pi\xi}, \quad a_i\leqslant 1 \\ &=\frac{1}{\sqrt{a_i}}\operatorname{rect}(\xi)e^{-j10\pi\xi}, \quad a_i\geqslant 1. \end{aligned} \tag{8.123}$$

Consequently, the output becomes

$$\begin{aligned} g(x)&=\sqrt{a_i}\ \operatorname{sinc} a_i(x-5), \quad a_i\leqslant 1 \\ &=\frac{1}{\sqrt{a_i}}\operatorname{sinc}(x-5), \quad a_i\geqslant 1. \end{aligned} \tag{8.124}$$

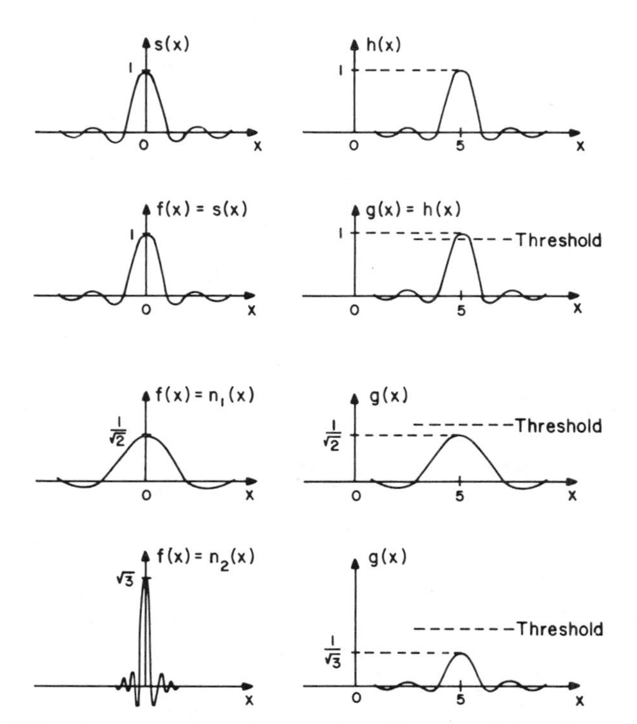

Figure 8-31 Output signals of a matched filter for various input signals.

We now select a few values for a_i and calculate the associated outputs:

$a_1=0.5$	$a_2=3$
$n_1(x)=\frac{1}{\sqrt{2}}\,\mathrm{sinc}\left(\frac{x}{2}\right)$	$n_2(x)=\sqrt{3}\,\mathrm{sinc}(3x)$
$g(x)=\frac{1}{\sqrt{2}}\,\mathrm{sinc}\left(\frac{x-5}{2}\right)$	$g(x)=\frac{1}{\sqrt{3}}\,\mathrm{sinc}(x-5)$

(8.125)

Note that the maximum values of both of these outputs also occur at $x=5$, but both are less than the value of $g(5)$ when the input is $s(x)$. This example is illustrated in Fig. 8-31. ■

8-7 SIGNAL SAMPLING AND RECOVERY

It is often necessary or desirable to describe a function by its values taken on a discrete set of points rather than attempting to describe it everywhere in a continuous fashion. This is particularly true when dealing with the transmission and storage of large quantities of data or when processing such data by digital computer, and the basic reason is related to the limited data-handling capabilities of all physical devices. When sampling is undertaken, the values of the function obtained at the sampling points are referred to as *sampled values*, the separation of the sampling points is called the *sampling interval*, and the inverse of the sampling interval is called the *sampling rate* (or sampling frequency).

As Goodman suggests (Ref. 8-2), it is intuitively obvious that we should be able to adequately describe any function by its sampled values as long as we make the sampling interval small enough; as the sampling interval tends to zero, the function becomes indistinguishable from its sampled values. However, when the sampling interval is nonzero, the following questions naturally arise: How accurately can the function be represented by its sampled values? Are there certain types of functions that can be more accurately described by their samples values than others? Is there an optimum sampling interval, and, if so, how is it determined? We shall try to answer these questions in the paragraphs to follow.

The Sampling Theorem

The so-called *sampling theorem* is generally attributed to Whittaker (Ref. 8-7), and to Shannon (Ref. 8-8), and is sometimes referred to as the Whittaker-Shannon sampling theorem (Ref. 8-2). It is discussed by many others, including Papoulis (Ref. 8-5), Bracewell (Ref. 8-6), and Linden

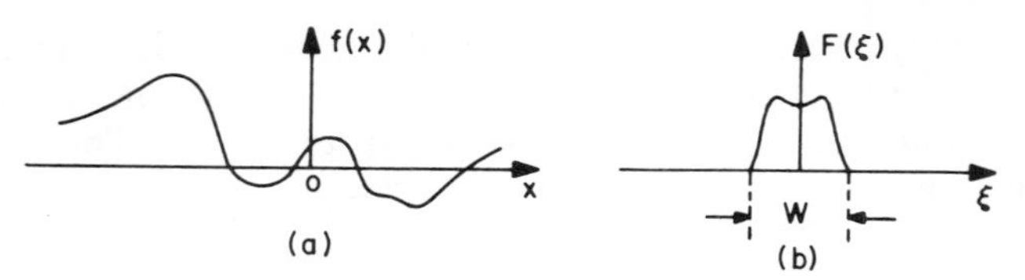

Figure 8-32 A band-limited function.

(Ref. 8-9). There are many versions of the sampling theorem, some more complicated than others, but basically it may be stated as follows: any *band-limited function* can be specified *exactly* by its sampled values, taken at regular intervals, provided that these intervals *do not exceed* some *critical sampling interval*. This statement of a truly remarkable result may, at first glance, appear to be utter nonsense, but in the paragraphs to follow we show that it is not. We stress, however, that it is valid only for band-limited functions; functions that are not band limited cannot, in general, be specified exactly by their sampled values.

Consider the band-limited function $f(x)$, whose Fourier transform $F(\xi) = 0$ for $|\xi| \geqslant W/2$ as shown in Fig. 8-32. We now multiply $f(x)$ by a general sampling function, which we denote by samp(x), to obtain the sampled function $f_s(x)$. This procedure is illustrated operationally in Fig. 8-33. For the present development, we choose the sampling function to be the comb function

$$\text{samp}(x) = \frac{1}{x_s}\,\text{comb}\left(\frac{x}{x_s}\right), \tag{8.126}$$

where the sampling interval x_s is taken to be positive for convenience. The sampled version of $f(x)$ then becomes

$$\begin{aligned} f_s(x) &= f(x)\,\text{samp}(x) \\ &= f(x)\left[\frac{1}{x_s}\,\text{comb}\left(\frac{x}{x_s}\right)\right] \\ &= \sum_{n=-\infty}^{\infty} f(nx_s)\delta(x - nx_s), \end{aligned} \tag{8.127}$$

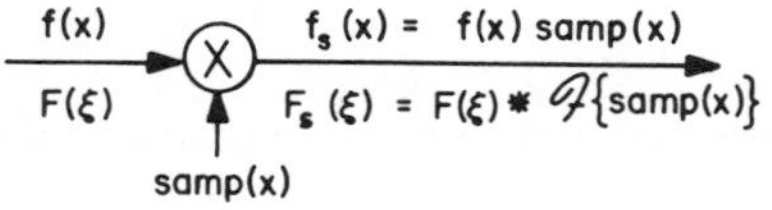

f(x)
F(ξ)
$f_s(x) = f(x)\,\text{samp}(x)$
$F_s(\xi) = F(\xi) * \mathcal{F}\{\text{samp}(x)\}$
samp(x)

Figure 8-33 The sampling operation.

which is simply a weighted set of equally spaced delta functions; the weighting is such that the delta function located at the point $x = nx_s$ has an area equal to the value of the function $f(x)$ at that point. The spectrum of $f_s(x)$ is given by

$$F_s(\xi) = F(\xi) * \text{comb}(x_s\xi)$$

$$= \frac{1}{x_s} \sum_{n=-\infty}^{\infty} F\left(\xi - \frac{n}{x_s}\right), \tag{8.128}$$

or, with the sampling rate defined to be $\xi_s = x_s^{-1}$, we have

$$F_s(\xi) = \xi_s \sum_{n=-\infty}^{\infty} F(\xi - n\xi_s). \tag{8.129}$$

Thus, $F_s(\xi)$ is seen to consist of a set of functions, each having the form of $F(\xi)$, repeated at intervals of ξ_s along the frequency axis as shown in Fig. 8-34. The various replications of $F(\xi)$ are often referred to as the *spectral orders* of $f_s(x)$, with $F(\xi - n\xi_s)$ known as the *nth spectral order*. Thus, $F(\xi)$ is the zero order, $F(\xi + \xi_s)$ is the negative first order, $F(\xi - \xi_s)$ is the positive first order, etc. Note that as long as $\xi_s \geqslant W$, which is the situation depicted in Fig. 8-34, the various spectral orders do not overlap and the use of an ideal low-pass filter to recover the original function $f(x)$ is immediately suggested. By applying the sampled function $f_s(x)$ as the input to a low-pass filter, as shown in Fig. 8-35, and by selecting the cutoff frequency of the filter to satisfy the inequality $W/2 \leqslant \xi_c \leqslant (\xi_s - W/2)$, the spectrum of the filter output will be

$$G(\xi) = F_s(\xi)H(\xi)$$

$$= \left[\xi_s \sum_{n=-\infty}^{\infty} F(\xi - n\xi_s)\right] \text{rect}\left(\frac{\xi}{2\xi_c}\right)$$

$$= \xi_s F(\xi). \tag{8.130}$$

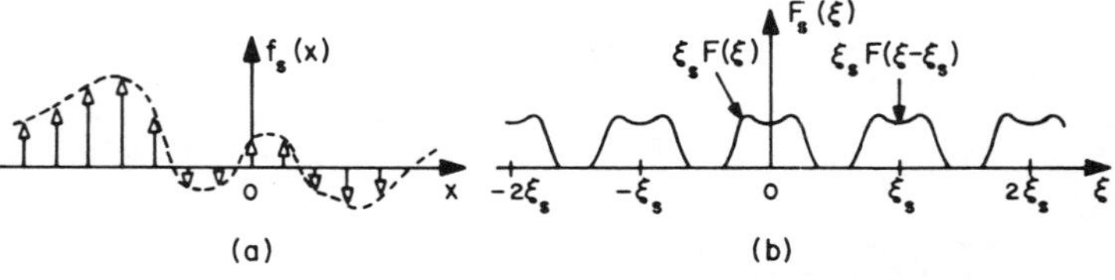

Figure 8-34 Comb-function sampling. (*a*) Sampled function. (*b*) Spectrum of sampled function.

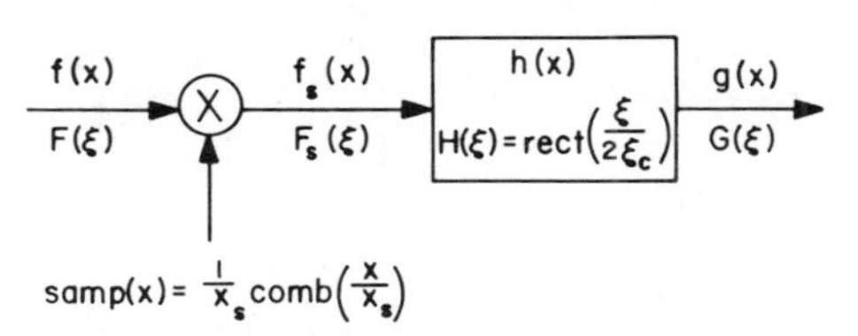

Figure 8-35 Recovery of a band-limited signal from its sampled version.

Consequently, the output itself is

$$g(x) = \xi_s f(x), \tag{8.131}$$

and except for an unimportant multiplicative constant, the original function has been recovered *exactly*.

Let us briefly review the above development. If $f(x)$ is band limited, such that $F(\xi)=0$ for $|\xi| \geqslant W/2$, and if comb-function sampling is used to obtain $f_s(x)$, then $f(x)$ can be recovered from $f_s(x)$ as long as the sampling rate ξ_s is at least as great as the total spectral width W. Thus, the *critical sampling rate* is just W, and the *critical sampling interval* is W^{-1}. We shall also refer to these quantities as the *Nyquist rate* and *Nyquist interval*,¶ respectively, and denote them by

$$\xi_{Ny} = W,$$

$$x_{Ny} = \frac{1}{W}. \tag{8.132}$$

We note that the Nyquist interval depends inversely on the value of W, a dependence that can be explained as follows. It seems reasonable to expect that the required sampling interval should be related to the maximum rate of change of the function and that this relationship should be an inverse one. In other words, the greater the maximum slope of the function, the more closely the sample points should be spaced. Since the highest-frequency spectral component of our band-limited function varies as $\exp\{j\pi Wx\}$, its maximum rate of change is proportional to W; hence, the noted inverse dependence of x_{Ny} on W.

Let us now examine the behavior of $F_s(\xi)$ when the sampling rate is less than the critical value. If $\xi_s < W$, the spectral orders overlap one another and, as a result, we can no longer use an ideal low-pass filter to recover $f(x)$ from $f_s(x)$. The reason for this is apparent from Fig. 8-36. Because of

¶Shannon (Ref. 8-8) called the critical sampling interval the Nyquist interval, after H. Nyquist, and this terminology is now widely used.

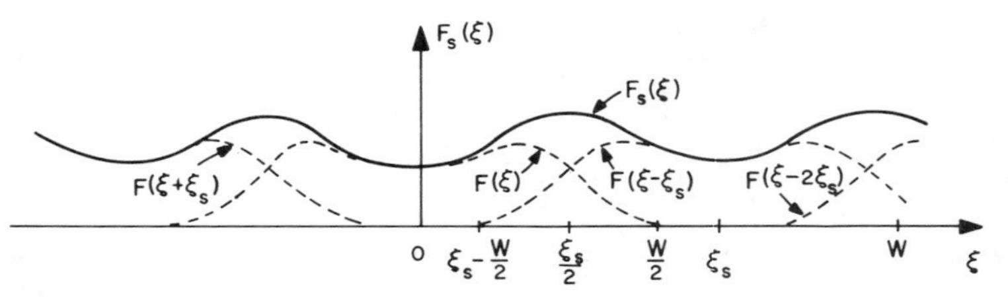

Figure 8-36 Spectrum of a sampled function when the sampling rate is less than the Nyquist rate.

the overlap, $F_s(\xi)$ no longer has the shape of $F(\xi)$ over the entire interval from $-W/2$ to $W/2$. In fact, $F_s(\xi)=F(\xi)$ only where $|\xi|<(\xi_s-W/2)$, and the overlap causes $F_s(\xi)\neq F(\xi)$ outside this region. To illustrate, the original negative-frequency components of $f(x)$ for which $|\xi|>\xi_s/2$ are now masquerading as positive-frequency components with frequencies less than $\xi_s/2$, and the addition of these masquerading components to $F(\xi)$ causes $F_s(\xi)$ to be deformed in the overlapping regions. As a result, any attempt to use an ideal low-pass filter to recover $f(x)$ will fail; if the cutoff frequency of the filter is chosen such that $\xi_c>(\xi_s-W/2)$, the output spectrum will contain the deformed part of $F_s(\xi)$, whereas if $\xi_c<W/2$, the output will be missing any components of $f(x)$ having a frequency higher than ξ_c. This condition, sometimes referred to as *aliasing*, is illustrated in Fig. 8-37, where the shaded areas depict the masquerading components. It is caused by *undersampling* of the function $f(x)$, and generally leads to an output with artifacts and other undesired features. We add that it would be theoretically possible to recover $f(x)$ by using a specially shaped filter, but the design of this filter would require a knowledge of $F(\xi)$ a priori; once again we ask, why build a complicated and expensive filter to recover something that is already known?

There is another way to view the sampling and recovery problem. Let us assume we have sampled the band-limited function $f(x)$ at a rate greater than ξ_{Ny}, and that the sampled function is passed through an ideal

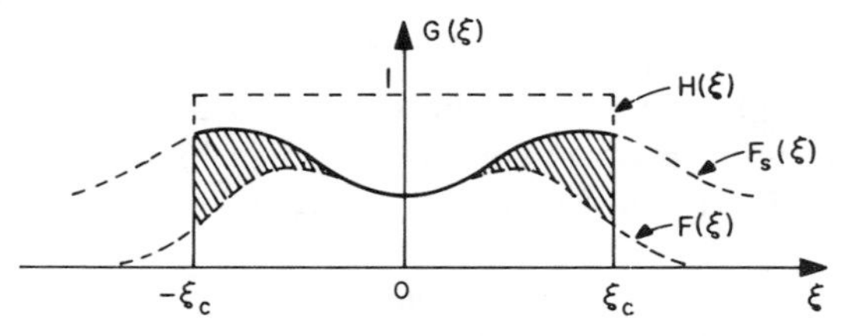

Figure 8-37 Aliasing as a result of undersampling.

low-pass filter with a cutoff frequency satisfying $W/2 \leqslant \xi_c \leqslant \xi_s/2$. We know from Eq. (8.131) that the output of this filter will be simply $\xi_s f(x)$, a result obtained from the frequency-domain approach of Eq. (8.130). However, if we view the output as a convolution of the input $f_s(x)$ with the impulse response $h(x)$, we find that

$$\begin{aligned} g(x) &= f_s(x) * h(x) \\ &= f_s(x) * 2\xi_c \operatorname{sinc}(2\xi_c x), \end{aligned} \tag{8.133}$$

and with $\xi_s = x_s^{-1}$, we combine Eqs. (8.131) and (8.133) to obtain

$$f(x) = f_s(x) * 2\xi_c x_s \operatorname{sinc}(2\xi_c x). \tag{8.134}$$

Thus, $f(x)$ may be described by the convolution of its sampled version with the *interpolating function* $2\xi_c x_s \operatorname{sinc}(2\xi_c x)$; the effect of this convolution is to erect a weighted interpolating function at each of the sample points, i.e.,

$$\begin{aligned} f(x) &= \left[f(x) \frac{1}{x_s} \operatorname{comb}\left(\frac{x}{x_s} \right) \right] * 2\xi_c x_s \operatorname{sinc}(2\xi_c x) \\ &= 2\xi_c x_s \sum_{n=-\infty}^{\infty} f(nx_s) \operatorname{sinc} 2\xi_c (x - nx_s). \end{aligned} \tag{8.135}$$

Note that the weighting of the interpolating function located at the point $x = nx_s$ is proportional to the value of $f(x)$ at that point. Also note that the zeros of the interpolating functions need not occur at the sample points. It seems amazing that the combination of sinc functions specified in Eq. (8.135) could possibly add up to yield $f(x)$, but it does!

The above expression for $f(x)$ is simplified if we choose ξ_c to be just half the sampling frequency; i.e., with $2\xi_c = \xi_s = x_s^{-1}$, we obtain

$$\begin{aligned} f(x) &= f_s(x) * \operatorname{sinc}\left(\frac{x}{x_s} \right) \\ &= \sum_{n=-\infty}^{\infty} f(nx_s) \operatorname{sinc}\left(\frac{x - nx_s}{x_s} \right). \end{aligned} \tag{8.136}$$

The interpolating sinc functions now have zeros at intervals equal to the sampling interval, and the value of their sum at any sampling point is influenced only by the sinc function located at that point. Of course, all of the interpolating functions contribute to the sum between sampling points,

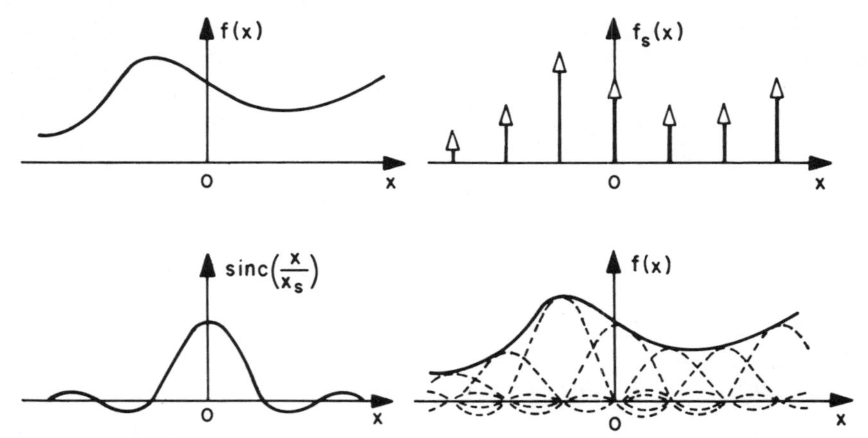

Figure 8-38 Recovery of a band-limited signal from its sampled version by interpolation.

but these contributions are such that their sum is exactly equal to $f(x)$ at every point. This result is depicted in Fig. 8-38.

Alternate Sampling Techniques

Shannon (Ref. 8-8) points out that equally spaced samples are not required for satisfactory recovery of a band-limited function from its sampled version; however, he also points out that the recovery process is more involved and that the sampled values must be known with greater accuracy than for the case with equal sampling intervals. He goes on to state that the sampled values of a function and its derivative, taken at intervals equal to twice the Nyquist interval, are sufficient to specify the function everywhere. Finally, he mentions a recovery technique involving the sampled values of a function and its first two derivatives, taken at intervals equal to three times the Nyquist interval. Basically, the sampling requirements for the unique specification of a band-limited function are as follows: enough independent sampled values of the function and its various derivatives must be obtained to ensure that the *average rate* at which they are taken is at least as great as the Nyquist rate (Ref. 8-6).

The above discussion relates to a quantity encountered frequently in communication theory: the *time-bandwidth product*. The choice of this name implies that the signals under consideration are time-varying signals, but the independent variable need not be time for the quantity to be meaningful. In fact, the two-dimensional equivalent of this quantity, known as the *area-bandwidth product*, is frequently encountered in optical data processing applications. For the purposes of this discussion we consider a function that is band limited, or very nearly band limited, and also effectively zero outside an interval of length X. Therefore, we have

$$f(x) \cong 0, \qquad \begin{cases} x \leqslant x_1 \\ x \geqslant x_1 + X \end{cases}$$

$$F(\xi) \cong 0, \qquad |\xi| \geqslant \frac{W}{2}. \tag{8.137}$$

We use approximation signs in these expressions because it is not possible for both a function and its Fourier transform to have compact support; we simply mean to indicate that they should be small outside the specified intervals. Gabor (Ref. 8-10) has pointed out that, under these conditions, approximately XW independent numbers (sampled values of the function and its derivatives) are sufficient to adequately specify the function everywhere. Note that if we sampled such a function at the Nyquist rate, we would obtain $\xi_{Ny} = W$ sampled values per unit of x. Thus, over an interval of length X, a total of XW sampled values would be collected, which is the number indicated above.

Ordinate and Slope Sampling

We now return to functions that are strictly band limited, i.e., we impose no requirement that the functions themselves be zero outside some interval. If $f(x)$ is band limited such that $F(\xi)=0$ for $|\xi| \geqslant W/2$, it may be recovered from its sampled values and those of its derivative taken at intervals of $x_s = 2/W$, which is twice the Nyquist interval. Bracewell (Ref. 8-6) refers to this as *ordinate and slope sampling*. It is possible to show, albeit with some difficulty, that $f(x)$ can again be described by a linear combination of interpolating functions, i.e.,

$$f(x) = f_s(x) * \mathrm{sinc}^2\left(\frac{Wx}{2}\right) + f_s'(x) * x\,\mathrm{sinc}^2\left(\frac{Wx}{2}\right), \tag{8.138}$$

where

$$f_s'(x) = f'(x)\left[\frac{W}{2}\,\mathrm{comb}\left(\frac{Wx}{2}\right)\right] \tag{8.139}$$

is the sampled derivative of $f(x)$.

In the derivation of Eq. (8.138), we were careful to note that

$$\mathcal{F}\{f_s'(x)\} = [\,j2\pi\xi F(\xi)\,] * \mathrm{comb}\left(\frac{2\xi}{W}\right), \tag{8.140}$$

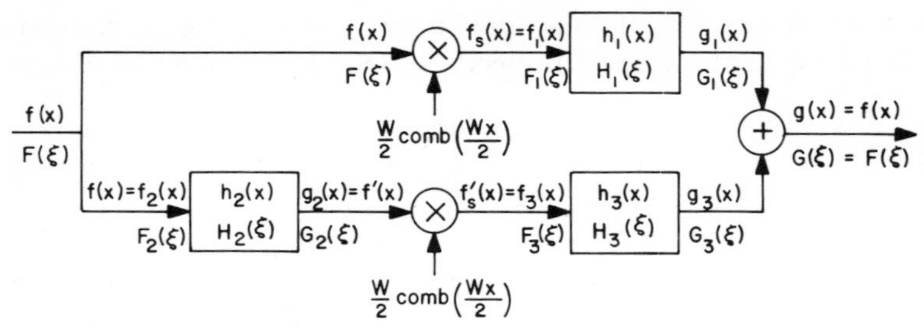

Figure 8-39 Device that performs ordinate and slope sampling.

and that

$$\begin{aligned}\mathcal{F}\{f_s'(x)\} &\neq j2\pi\xi F_s(x)\\ &\neq j2\pi\xi\left[F(\xi)*\mathrm{comb}\left(\frac{2\xi}{W}\right)\right].\end{aligned} \tag{8.141}$$

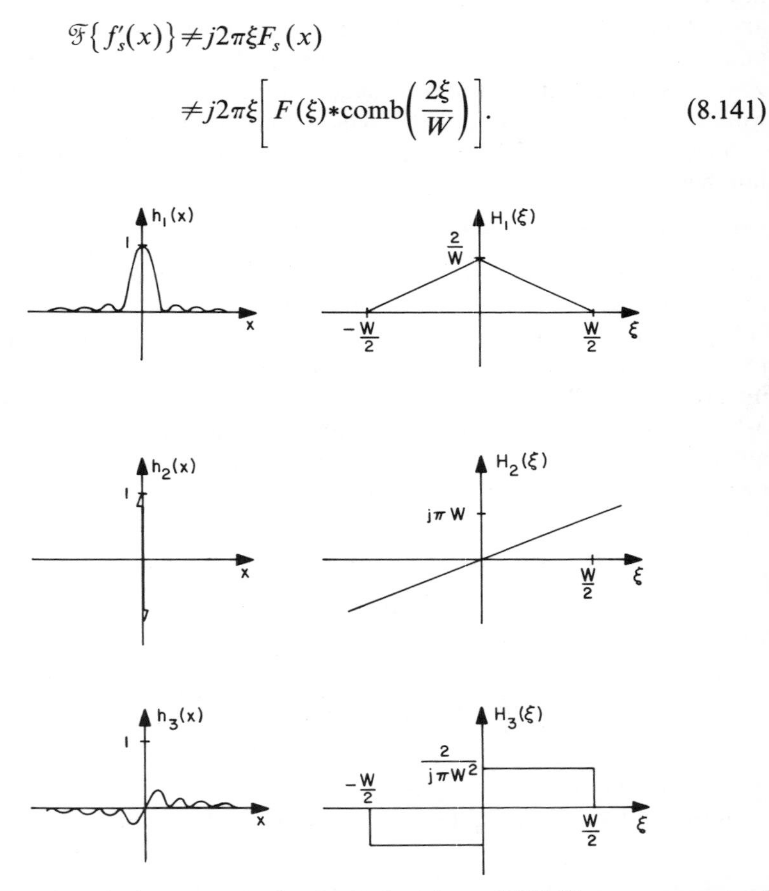

Figure 8-40 Impulse responses and transfer functions of LSI filter components of sampling device shown in Fig. 8-39.

From a systems point of view, we may effect the recovery of $f(x)$ from $f_s(x)$ and $f_s'(x)$ by passing the latter two signals through appropriate filters and combining the outputs of these filters. A device that will do just this is depicted in Fig. 8-39, where

$$h_1(x) = \operatorname{sinc}^2\left(\frac{Wx}{2}\right), \qquad H_1(\xi) = \frac{2}{W}\operatorname{tri}\left(\frac{2\xi}{W}\right),$$

$$h_2(x) = \delta^{(1)}(x), \qquad H_2(\xi) = j2\pi\xi,$$

$$h_3(x) = x\operatorname{sinc}^2\left(\frac{Wx}{2}\right), \qquad H_3(\xi) = \frac{j}{\pi W}\operatorname{tri}'\left(\frac{2\xi}{W}\right). \tag{8.142}$$

These functions are shown in Fig. 8-40, and the spectra at various points in the device are illustrated in Fig. 8-41.

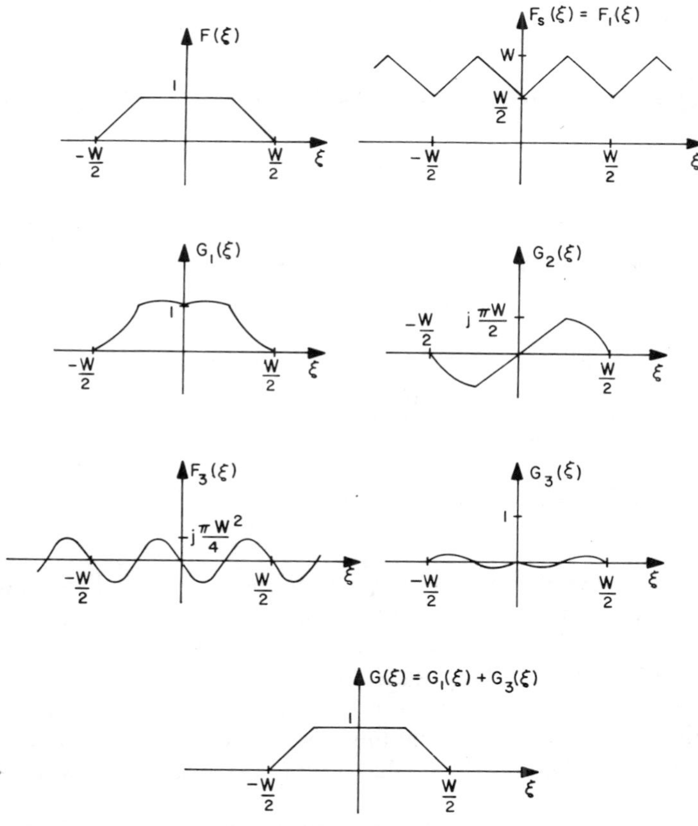

Figure 8-41 Spectra at various points in the device of Fig. 8-39.

Interlaced Sampling

Bracewell also discusses *interlaced sampling*, a technique in which the sampling points are not equally spaced. The function is sampled by two different comb functions, each of which employs a sampling interval x_s equal to twice the Nyquist interval; however, one of these comb functions is shifted slightly to the left, and the other slightly to the right, by a distance that is usually small compared with x_s.† We shall denote these sampling functions by samp$(x+a)$ and samp$(x-a)$, respectively, where samp(x) is given by Eq. (8.126) and a is assumed to be a real constant such that $|a|<x_s/2$. Note that even though every other sampling interval exceeds the Nyquist interval, the average sampling interval satisfies the Nyquist condition.

If we choose $f(x)$ to be the same as in the preceding discussion, i.e., $F(\xi)=0$ for $|\xi|\geqslant W/2$, and if we select the interlacing technique outlined above, it is possible to show, again with some difficulty, that $f(x)$ can be described by

$$\begin{aligned} f(x) &= \left[f(x)\mathrm{samp}(x+a)\right]*m(x)+\left[f(x)\mathrm{samp}(x-a)\right]*m(-x) \\ &= \sum_{n=-\infty}^{\infty} f\left(\frac{2n}{W}-a\right)m\left(x+a-\frac{2n}{W}\right) \\ &\quad + \sum_{n=-\infty}^{\infty} f\left(\frac{2n}{W}+a\right)m\left(-x+a+\frac{2n}{W}\right), \end{aligned} \tag{8.143}$$

where

$$\begin{aligned} \mathrm{samp}(x) &= \frac{W}{2}\mathrm{comb}\left(\frac{Wx}{2}\right), \\ m(x) &= \mathrm{sinc}(Wx)-\frac{\pi W}{2}\cot(\pi Wa)\,x\,\mathrm{sinc}^2\left(\frac{Wx}{2}\right). \end{aligned} \tag{8.144}$$

The functions $m(x)$ and $m(-x)$ are the interpolating functions when this procedure is used, and a graph of $m(x)$ is shown in Fig. 8-42 for $W=2$ and $a=0.1$. In Fig. 8-43 we demonstrate how a combination of these interpolating functions, appropriately weighted and shifted, add up to produce the original function. It is interesting to note that each of the shifted interpolating functions has zeros at all sampling points save the point for which its argument is zero, where it has a value of unity. A system that will perform

†Actually, Bracewell shifted only one of his sampling functions.

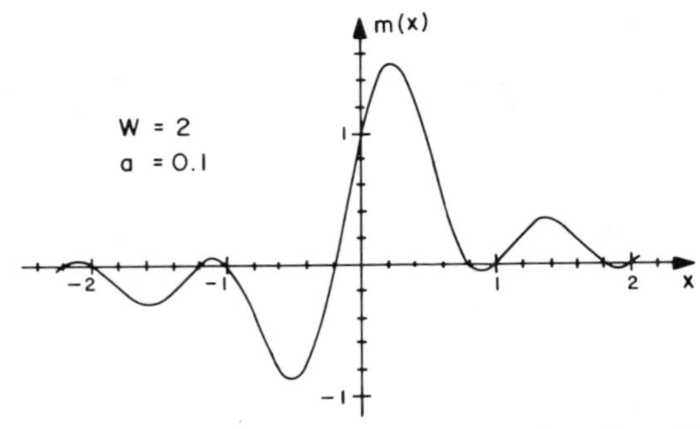

Figure 8-42 Interpolating function for interlaced sampling.

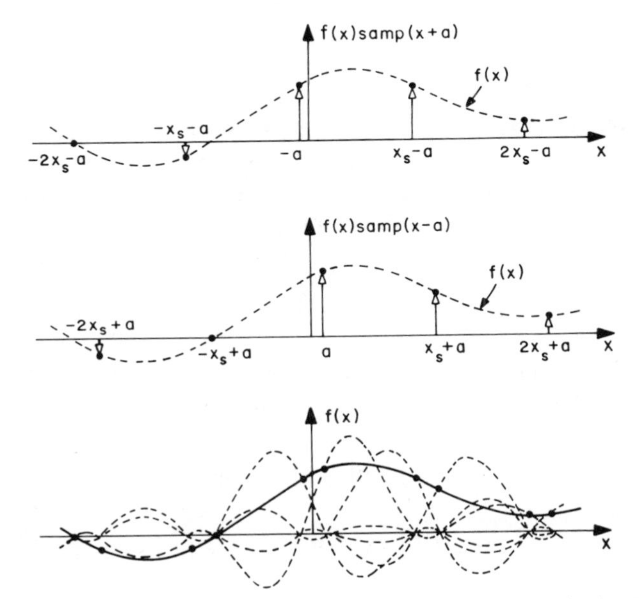

Figure 8-43 Signal recovery from interlaced sampled values by interpolation.

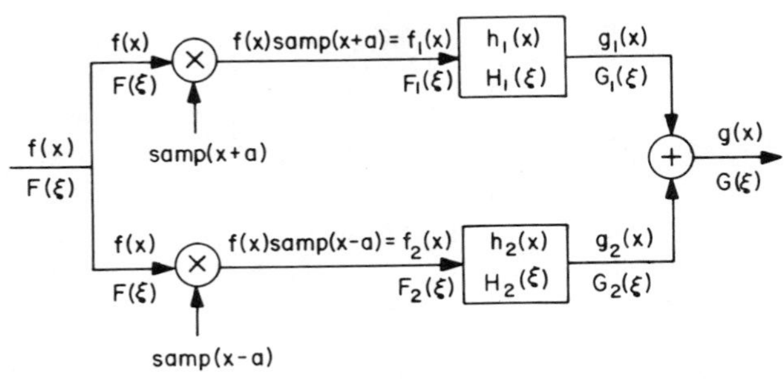

Figure 8-44 System that performs interlaced sampling.

the operations required by Eq. (8.143) is depicted in Fig. 8-44, where

$$h_1(x) = m(x), \qquad H_1(\xi) = \frac{1}{W}\mathrm{rect}\left(\frac{\xi}{W}\right) - \frac{j}{2}\cot(\pi W a)\,\mathrm{tri}'\left(\frac{2\xi}{W}\right),$$

$$h_2(x) = m(-x), \qquad H_2(\xi) = \frac{1}{W}\mathrm{rect}\left(\frac{\xi}{W}\right) + \frac{j}{2}\cot(\pi W a)\,\mathrm{tri}'\left(\frac{2\xi}{W}\right). \tag{8.145}$$

Finite Sampling Arrays

We now assume that we are able to obtain the sampled values of the function $f(x)$ only on a finite interval of length X—the usual situation in the real world. If comb-function sampling is employed, the sampling function of finite extent is expressed as

$$\mathrm{samp}(x) = \mathrm{rect}\left(\frac{x}{X}\right)\left[\frac{1}{x_s}\mathrm{comb}\left(\frac{x}{x_s}\right)\right], \tag{8.146}$$

where we have centered the interval X on the origin for simplicity. Consequently, the sampled function and its spectrum become

$$f_s(x) = f(x)\mathrm{rect}\left(\frac{x}{X}\right)\left[\frac{1}{x_s}\mathrm{comb}\left(\frac{x}{x_s}\right)\right],$$

$$F_s(\xi) = F(\xi) * X\mathrm{sinc}(X\xi) * \mathrm{comb}(x_s\xi). \tag{8.147}$$

This expression may be rewritten as

$$F_s(\xi) = X\mathrm{sinc}(X\xi) * \xi_s \sum_{n=-\infty}^{\infty} F(\xi - n\xi_s), \tag{8.148}$$

which we see is just a convolution of a sinc function with the spectrum that would have been obtained if a sampling function of infinite extent had been used. The net result is that the various spectral orders are smeared out and overlap even when the Nyquist rate is used. Of course, the undesired effects produced by the overlapping can be reduced somewhat by making ξ_s greater than W, but exact recovery of $f(x)$ with an ideal low-pass filter is no longer possible.

Perhaps the reason for this is more easily seen if the effect of the limiting rectangle function is viewed in another manner. Instead of considering $\mathrm{rect}(x/X)$ to be a function that limits the extent of the sampling function, we now regard it as a function that limits the extent of $f(x)$ before the sampling takes place. Hence, even though $f(x)$ is band limited, the product $f(x)\,\mathrm{rect}(x/X)$ is not; consequently, because the sampling theorem holds only for band-limited functions, neither $f(x)\,\mathrm{rect}(x/X)$ nor $f(x)$ can be recovered in the usual way.

Sampling with Other than Comb Functions

In practice, comb-function sampling cannot be used because it is not physically possible to sample a function at discrete points; physical considerations require that the interrogating functions located at each sampling point be of finite extent. Let us now see how this physically imposed condition changes the previous results. Suppose we sample with an array of equally spaced functions $p(x)$, which are usually narrow relative to the sampling interval, such that the sampling function becomes

$$\mathrm{samp}(x) = p(x) * \frac{1}{x_s}\mathrm{comb}\left(\frac{x}{x_s}\right). \tag{8.149}$$

Then the sampled function and its spectrum are given by

$$f_s(x) = f(x)\left[p(x) * \frac{1}{x_s}\mathrm{comb}\left(\frac{x}{x_s}\right)\right],$$

$$F_s(\xi) = F(\xi) * \left[P(\xi)\mathrm{comb}(x_s\xi)\right]$$

$$= \xi_s \sum_{n=-\infty}^{\infty} P(n\xi_s)F(\xi - n\xi_s). \tag{8.150}$$

Note that $P(\xi)$ causes the comb function to be weighted *before* the convolution with $F(\xi)$; thus, only the heights of the various spectral orders are governed by $P(\xi)$, and their shapes are left unaltered.

Example

Let $f(x) = \mathrm{sinc}^2(x)$. We easily calculate that $F(\xi) = \mathrm{tri}(\xi)$ and $W = 2$. If we then select $\xi_s = 2$ and $p(x) = \mathrm{rect}(10x)$, the spectrum of $f_s(x)$ is

$$F_s(\xi) = \tfrac{1}{5} \sum_{n=-\infty}^{\infty} \mathrm{sinc}\left(\frac{n}{5}\right)\mathrm{tri}(\xi - 2n), \tag{8.151}$$

as illustrated in Fig. 8-45. We observe that it would be possible to recover $f(x)$ from $f_s(x)$ in this case because the individual replications of $F(\xi)$ do not overlap and they are not distorted. ■

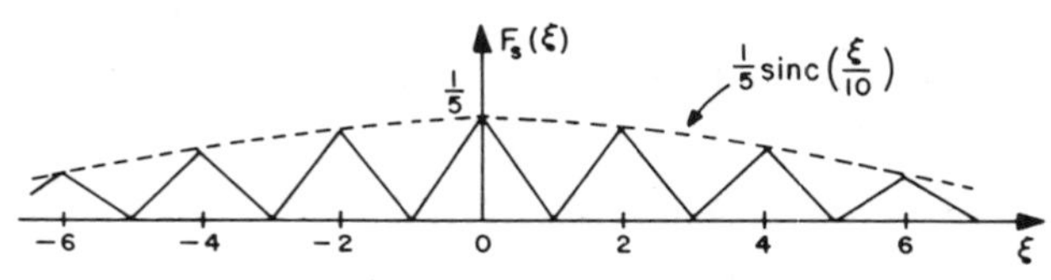

Figure 8-45 Spectrum of sampled function when sampling function of Eq. (8.149) is used to sample $f(x) = \text{sinc}^2(x)$.

An interesting physical example of sampling and recovery is encountered on a daily basis by nearly everyone. The *half-tone pictures* in newspapers and magazines are simply sampled versions of photographs or other objects, and although the sampling is performed in a manner different from that described here, the effects are similar. If you hold such a picture very close to your eye, or use a magnifying glass, you will see a regular array of dots that displays little resemblance to anything in particular. However, if you move the picture to a distance of about 50 cm from your eyes, the dots become indistinguishable and the picture is easily recognized. The following explanation of this phenomenon, although brief and somewhat imprecise, may be helpful in understanding it. The eye acts as a low-pass (spatial) filter for spatially modulated signals, and a half-tone picture is merely a sampled version of a spatially modulated signal. When the picture is held close to the eye, the cutoff frequency of the associated filter is so high relative to the sampling rate that several spectral orders are passed, thereby producing a degradation of the image similar to aliasing. However, by moving the picture away from the eye a sufficient distance, the effective sampling rate is increased to the extent that, for all practical purposes, only the desired zero order is passed by the filter. The result is a recognizable image.

Now let us combine the effects of sampling with finite arrays and those of sampling with functions other than comb functions—the situation usually encountered in the real world. The sampling function may now be written as

$$\text{samp}(x) = p(x) * \left[\text{rect}\left(\frac{x}{X}\right)\frac{1}{x_s}\text{comb}\left(\frac{x}{x_s}\right)\right], \tag{8.152}$$

and the sampled function and its spectrum become

$$f_s(x) = f(x)\left\{p(x)*\left[\text{rect}\left(\frac{x}{X}\right)\frac{1}{x_s}\text{comb}\left(\frac{x}{x_s}\right)\right]\right\},$$

$$F_s(\xi) = F(\xi)*\{P(\xi)[X\text{sinc}(X\xi)*\text{comb}(x_s\xi)]\}$$

$$= F(\xi)*\left\{P(\xi)\xi_s X \sum_{n=-\infty}^{\infty} \text{sinc}[X(\xi - n\xi_s)]\right\}. \quad (8.153)$$

Note that the various spectral orders, including the zero order, are again smeared out by the convolution with the sinc functions, and again it is impossible to effect an exact recovery of $f(x)$. However, with care the undesirable effects can be minimized and sampling can still be a very useful technique. In fact, even with its imperfections, it is indispensable in a number of scientific disciplines.

Sampling of Band-Pass Functions

We now explore the sampling of functions having spectra similar to that depicted in Fig. 8-46(a), where W is the overall width of the spectrum, as before, and W' is the width of each side of the spectrum. If comb function sampling is used with such a function, it is possible to recover $f(x)$ from $f_s(x)$ even though a sampling rate lower than the Nyquist rate is used. In other words, rather than requiring the first spectral order to be shifted by an amount greater than or equal to W, we require this of the nth order. As

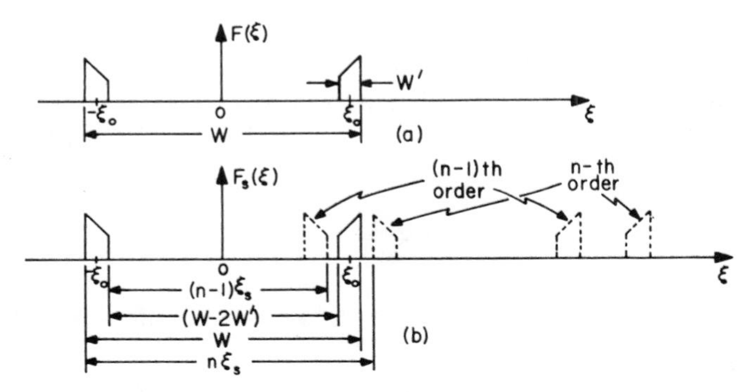

Figure 8-46 Sampling of band-pass signals. (*a*) Spectrum of signal. (*b*) Spectrum of sampled signal.

can be understood with reference to Fig. 8-46(b), the conditions imposed on the nth and $(n-1)$ orders lead to the inequalities

$$n\xi_s \geqslant W,$$
$$(n-1)\xi_s \leqslant W-2W', \tag{8.154}$$

which may be combined to yield

$$\frac{W}{n} \leqslant \frac{W-2W'}{n-1}. \tag{8.155}$$

Solving for n, we find that

$$n \leqslant \frac{W}{2W'}, \tag{8.156}$$

and the required interlacing of the spectral orders will occur if

$$\frac{W}{n} \leqslant \xi_s \leqslant \frac{W-2W'}{n-1}. \tag{8.157}$$

Clearly, the minimum sampling rate that may be used to obtain the desired result will occur when n is the largest integer satisfying Eq. (8.156), i.e.,

$$\xi_{s\min} = \frac{W}{n_{\max}}. \tag{8.158}$$

Of course, the appropriate filter to use for this type of sampling is a band-pass filter.

Example

Let $W=11$ and $W'=1$. Obviously, we could recover $f(x)$ from $f_s(x)$ if we let $\xi_s=11$ and used a low-pass filter of width 11. However, we need not sample at nearly this rate; if we evaluate Eq. (8.156) for the given information, we obtain

$$n \leqslant \frac{11}{2}, \tag{8.159}$$

and we should therefore choose $n=5$. Thus, the minimum sampling frequency in this case is

$$\xi_{s\min} = 2.20, \tag{8.160}$$

which is considerably lower than 11.

Of course, any sampling rate that satisfies Eq. (8.157) is acceptable; with $n=5$

$$2.20 \leqslant \xi_s \leqslant 2.25, \tag{8.161}$$

and with $n=4$

$$2.75 \leqslant \xi_s \leqslant 3.00, \tag{8.162}$$

etc. For any of these sampling frequencies, the band-pass filter $H(\xi) = \text{rect}(\xi+5) + \text{rect}(\xi-5)$ is used to recover $f(x)$. ■

Sampling and Analog-to-Digital Conversion

It is frequently necessary to convert analog data into a form suitable for digital processing, and a commonly used method for accomplishing this involves a special scanning technique. Let us assume that the analog data to be converted is given by $f(x)$, and that the digitization is to take place at intervals of x_s. At any particular point, say $x = nx_s$, we estimate the value of $f(x)$ by the integral

$$\hat{f}(nx_s) = \int_{-\infty}^{\infty} f(\alpha) \frac{1}{|b|} \text{rect}\left(\frac{\alpha - nx_s}{b} \right) d\alpha. \tag{8.163}$$

In other words, we estimate $f(nx_s)$ by computing its average over an interval, of width $|b|$, centered at the point $x = nx_s$. Then, the entire digitized version of $f(x)$ is simply the sum of all the estimates at the digitization points, i.e.,

$$\begin{aligned} f_d(x) &= \sum_{n=-\infty}^{\infty} \left[\int_{-\infty}^{\infty} f(\alpha) \frac{1}{|b|} \text{rect}\left(\frac{\alpha - nx_s}{b} \right) d\alpha \right] \delta(x - nx_s) \\ &= \sum_{n=-\infty}^{\infty} \hat{f}(nx_s) \delta(x - nx_s) \\ &= \hat{f}(x) \sum_{n=-\infty}^{\infty} \delta(x - nx_s) \\ &= \hat{f}(x) \frac{1}{x_s} \text{comb}\left(\frac{x}{x_s} \right) \\ &= \hat{f}_s(x). \end{aligned} \tag{8.164}$$

Thus, $f_d(x)$ is simply a sampled version of the estimate of the original function $f(x)$, and the sampling function is a comb function.

This technique is similar to that used for digitizing analog data stored on photographic film. The film is scanned by an aperture, which performs the averaging operation, but the scanning is done in a stepwise—rather than continuous—fashion. The size of the steps, x_s, is often equal to the aperture width, but this need not be the case. We may express the result of Eq. (8.164) in a different fashion by realizing that

$$\begin{aligned}\hat{f}(x) &= \frac{1}{|b|}\int_{-\infty}^{\infty} f(\alpha)\operatorname{rect}\left(\frac{\alpha - x}{b}\right)d\alpha \\ &= \frac{1}{|b|}\int_{-\infty}^{\infty} f(\alpha)\operatorname{rect}\left(\frac{x-\alpha}{b}\right)d\alpha \\ &= \frac{1}{|b|}\left[f(x)*\operatorname{rect}\left(\frac{x}{b}\right)\right].\end{aligned} \tag{8.165}$$

Thus,

$$\begin{aligned}f_d(x) &= \hat{f}(x)\frac{1}{x_s}\operatorname{comb}\left(\frac{x}{x_s}\right) \\ &= \frac{1}{|b|}\left[f(x)*\operatorname{rect}\left(\frac{x}{b}\right)\right]\frac{1}{x_s}\operatorname{comb}\left(\frac{x}{x_s}\right),\end{aligned} \tag{8.166}$$

and the spectrum of the digitized function is given by

$$F_d(\xi) = \left[F(\xi)\operatorname{sinc}(b\xi)\right]*\operatorname{comb}(x_s\xi). \tag{8.167}$$

Note the difference between this expression and that of Eq. (8.150); even if the sampling rate exceeds the Nyquist rate, we cannot recover the original function with an ideal low-pass filter because of the effects of the sinc function. However, it might be possible to use a low-pass filter with a transfer function of

$$H(\xi) = \frac{1}{\operatorname{sinc}(b\xi)}\operatorname{rect}\left(\frac{\xi}{W}\right), \tag{8.168}$$

where W is the total bandwidth of $F(\xi)$.

If we now consider the effects of a finite sampling array, we find that the same sort of spectral smearing described by Eq. (8.148) occurs again. Thus, a sampling rate much higher than the Nyquist rate may be necessary to reduce the undesirable effects to an acceptable level. We now write the expressions for the digitized function and its spectrum for the situation in which a sampling array of width X is used:

$$f_d(x)=\left[f(x)*\frac{1}{|b|}\operatorname{rect}\left(\frac{x}{b}\right)\right]\left[\frac{1}{x_s}\operatorname{comb}\left(\frac{x}{x_s}\right)\operatorname{rect}\left(\frac{x}{X}\right)\right],$$

$$F_d(\xi)=\left[F(\xi)\operatorname{sinc}(b\xi)\right]*\operatorname{comb}(x_s\xi)*X\operatorname{sinc}(X\xi). \tag{8.169}$$

REFERENCES

8-1 A. Papoulis, *The Fourier Integral and Its Applications*, McGraw-Hill, New York, 1962.

8-2 J. W. Goodman, *Introduction to Fourier Optics*, McGraw-Hill, New York, 1968.

8-3 *Webster's New Collegiate Dictionary*, G.&C. Merriam Co., Springfield, Ill., 1974.

8-4 A. Vander Lugt, "Operational Notation for the Analysis and Synthesis of Optical Data-Processing Systems," *Proc. IEEE* **54**(8): 1055–1063 (1966).

8-5 A. Papoulis, *Systems and Transforms with Applications in Optics*, McGraw-Hill, New York, 1968.

8-6 R. Bracewell, *The Fourier Transform and its Applications*, McGraw-Hill, New York, 1965.

8-7 E. T. Whittaker, "On the Functions which are Represented by the Expansions of the Interpolation Theory," *Proc. Roy. Soc. Edinburgh, Sect. A*, **35**: 181 (1915).

8-8 C. E. Shannon, "Communication in the Presence of Noise," *Proc. IRE* **37**: 10 (1949).

8-9 D. A. Linden, "A Discussion of Sampling Theorems," *Proc. IRE* **47**: 1219 (1959).

8-10 D. Gabor, "Theory of Communication," *J. IEE (London)*, Part 3, **93**(26): 429 (1946).

PROBLEMS

8-1. Given an LSI system, as shown in Fig. 8-1, and a triangle-wave input of finite extent described by

$$f(x)=\left[\tfrac{1}{2}\operatorname{comb}\left(\frac{x}{2}\right)\operatorname{rect}\left(\frac{x}{50}\right)\right]*\operatorname{tri}(x).$$

Using reasonable approximations where appropriate, find the output for each of the following transfer functions. Sketch the transfer function, impulse response, output spectrum, and output in each case.

a. $H(\xi)=\operatorname{rect}\left(\dfrac{\xi}{40}\right)$.

b. $H(\xi)=\operatorname{rect}\left(\dfrac{\xi}{4}\right)$.

c. $H(\xi)=\mathrm{rect}\left(\frac{\xi}{2}\right)$.

d. $H(\xi)=\mathrm{rect}(2\xi)$.

e. $H(\xi)=\mathrm{rect}\left(\frac{\xi}{40}\right)-\mathrm{rect}(2\xi)$.

f. $H(\xi)=\mathrm{rect}\left(\frac{\xi}{4}\right)-\mathrm{rect}\left(\frac{\xi}{2}\right)$.

8-2. Given an LSI system, as shown in Fig. 8-1, and a rectangle-wave input of finite extent described by

$$f(x)=\left[\tfrac{1}{2}\mathrm{comb}\left(\frac{x}{2}\right)\mathrm{rect}\left(\frac{x}{50}\right)\right]*\mathrm{rect}(x).$$

Using reasonable approximations where appropriate, find the output for each of the following transfer functions. Sketch the transfer function, impulse response, output spectrum, and output in each case.

a. $H(\xi)=e^{-j\pi\xi}$.

b. $H(\xi)=e^{-j\pi}$.

c. $H(\xi)=e^{-j\pi\,\mathrm{rect}(2\xi)}$.

d. $H(\xi)=e^{-j\pi[1-\mathrm{rect}(2\xi)]}$.

e. $H(\xi)=e^{-j2\pi\xi\,\mathrm{rect}(0.5\xi)}$.

8-3. Given an LSI system, as shown in Fig. 8-1, and a rectangle-wave input of finite extent described by

$$f(x)=\left[\tfrac{1}{3}\mathrm{comb}\left(\frac{x}{3}\right)\mathrm{rect}\left(\frac{x}{100}\right)\right]*\mathrm{rect}(x).$$

Using reasonable approximations where appropriate, find the output for each of the following system descriptions. Sketch the transfer function, impulse response, output spectrum, and output in each case.

a. $H(\xi)=\left[\mathrm{rect}\left(\frac{\xi}{30}\right)-\mathrm{rect}(6\xi)\right]e^{-j2\pi\xi}$.

b. $h(x)=\mathrm{rect}(x-1)$.

c. $H(\xi)=\mathrm{rect}\left(\frac{\xi}{2}\right)e^{j6\pi\xi^2}$.

d. $H(\xi)=\mathrm{tri}(\xi)e^{j\pi\,\mathrm{rect}(6\xi)}$.

e. $h(x)=4\,\mathrm{sinc}^2(2x)-2\,\mathrm{sinc}^2(x)$.

f. $H(\xi)=\mathrm{rect}\left(\frac{\xi}{4}\right)\mathrm{sgn}(-\xi)$.

8-4. Given an LSI system, as shown in Fig. 8-1, and an input whose amplitude is constant but whose phase is a rectangle-wave function, i.e.,

$$f(x) = e^{-j\Phi(x)},$$

$$\Phi(x) = \pi - \pi\left[\mathrm{comb}(x)\mathrm{rect}\left(\frac{x}{101}\right)\right] * \mathrm{rect}(4x).$$

a. Sketch the phase function $\Phi(x)$.
b. Show that the input can be written as

$$f(x) = 2\left[\mathrm{comb}(x)\mathrm{rect}\left(\frac{x}{101}\right)\right] * \mathrm{rect}(4x) - 1.$$

c. For a transfer function of

$$H(\xi) = \mathrm{rect}\left(\frac{\xi}{51}\right) e^{-j\pi\,\mathrm{rect}(8\xi)},$$

show that the output is given by

$$g(x) \cong 2\left[\mathrm{comb}(x)\mathrm{rect}\left(\frac{x}{101}\right)\right] * \mathrm{rect}(4x).$$

d. Sketch $f(x)$ and $g(x)$.

8-5. A desired signal $s(x)$ is corrupted by additive noise $n(x)$, and you are required to design a filter that can be used to recover the signal for each of the following cases. Find a transfer function $H(\xi)$ such that the output $g(x) \cong s(x)$ when the input is $f(x) = s(x) + n(x)$. Sketch $f(x)$ and $g(x)$.

a. $s(x) = [0.2\,\mathrm{comb}(0.2x) * \mathrm{rect}(x)]\cos(60\pi x)$,
$n(x) = [0.5\,\mathrm{comb}(0.5x) * \mathrm{tri}(x)]\cos(20\pi x)$.

b. $s(x) = \mathrm{Gaus}(0.2x)$,
$n(x) = \mathrm{Gaus}(0.1x)\cos(\pi x)$.

c. $s(x) = \left\{\left[0.25\,\mathrm{comb}(0.25x)\mathrm{rect}\left(\frac{x}{90}\right)\right] * \mathrm{rect}(x)\right\}\cos(40\pi x)$,
$n(x) = 2\,\mathrm{Gaus}\left(\frac{x}{20}\right)\cos(10\pi x)$.

8-6. Given a linear system with impulse response $h(x) = \exp\{j10\pi x^2\}$.

a. Is the system shift invariant?
b. Find the output $g(x)$ when the input is given by

$$f(x) = \mathrm{rect}(0.5x)\left[0.5 + 0.5\cos(5\pi x)\right]e^{-j10\pi x^2}.$$

8-7. Given the spectrum analyzer shown in Fig. 8-24. With $a=0.1$ and $f(x)=\mathrm{Gaus}(0.2x)\cos(2\pi x)$, find the output $g(x)$ and sketch.

8-8. Given the cascaded spectrum analyzers shown in Fig. 8-27, with $a_1=4, a_2=2$, and $p(x)=\mathrm{tri}(x)\exp\{j0.2\pi x\}$:

a. Model this cascade by a combination of a signal scaler and LSI system as in Fig. 8-28.

b. With an input of $f(x)=\mathrm{comb}(x)*\mathrm{rect}(2x-0.5)$, find the output $g(x)$ and sketch.

8-9. A signal $s(x)=\exp\{-x\}\mathrm{step}(x)$ is the input to an LSI system.

a. Find and sketch the output $g(x)$ when $h(x)=s(x)$.

b. Find and sketch the output $g(x)$ when $h(x)=s(-x)$.

c. Describe the impulse response $h(x)$ and transfer function $H(\xi)$ of the matched filter that will maximize the output at $x=2$. Assume that $H(0)=1$, and sketch the output.

8-10. Repeat Problem 8-9 for the signal

$$s(x)=\mathrm{rect}(2x)*[\delta(x)+\delta(x-4)+\delta(x-7)+\delta(x-9)].$$

8-11. Repeat Problem 8-9 for the signal

$$s(x)=\cos(\pi x^2)\mathrm{rect}\left(\frac{x-2.5}{5}\right).$$

8-12. A sampling function $\mathrm{samp}(x)$ is used to sample a signal $f(x)$ as follows: $f_s(x)=f(x)\mathrm{samp}(x)$. The sampled version $f_s(x)$ is then passed through an LSI system with transfer function $H(\xi)$ to produce an output $g(x)$. For

$$f(x)=\mathrm{sinc}^2(100x),$$

$$\mathrm{samp}(x)=\xi_s\,\mathrm{comb}(\xi_s x),$$

$$H(\xi)=\mathrm{rect}\left(\frac{\xi}{\xi_s}\right):$$

a. Find the minimum sampling rate $\xi_s=\xi_{Ny}$ (Nyquist rate) that will permit an exact recovery of $f(x)$ from $f_s(x)$; i.e., such that $g(x)=f(x)$ (to within a multiplicative constant).

b. Find $g(x)$ when $\xi_s=0.75\xi_{Ny}$, and sketch.

c. Find $g(x)$ when $\xi_s=0.5\xi_{Ny}$, and sketch.

8-13. As in Problem 8-12, $f_s(x)=f(x)\text{samp}(x)$. With

$$f(x)=\text{sinc}(10x),$$

$$\text{samp}(x)=\text{rect}(0.2x)\xi_s\,\text{comb}(\xi_s x),$$

$$H(\xi)=\text{rect}\left(\frac{\xi}{\xi_s}\right):$$

a. Find $g(x)$ if ξ_s is chosen to be the usual Nyquist rate associated with a comb sampling function of infinite extent.
b. Find $g(x)$ if ξ_s is chosen to be twice the Nyquist rate determined in part a.
c. What parameter changes will improve signal recovery?
d. Can $f(x)$ ever be recovered exactly in this case?

8-14. Repeat Problem 8-13 when

$$\text{samp}(x)=\left[\text{rect}(0.2x)\xi_s\,\text{comb}(\xi_s x)\right]*\text{rect}(100x).$$

8-15. A function $f(x)$ is digitized in a stepwise fashion to produce

$$f_d(x)=\left[f(x)*p(x)\right]\text{samp}(x),$$

which is then passed through an LSI filter with transfer function $H(\xi)$ to produce an output $g(x)$. For

$$f(x)=\text{sinc}(10x),$$

$$p(x)=100\,\text{rect}(100x),$$

$$\text{samp}(x)=\text{rect}(0.2x)\xi_s\,\text{comb}(\xi_s x),$$

$$H(\xi)=\text{rect}\left(\frac{\xi}{\xi_s}\right):$$

a. Find $g(x)$ if ξ_s is chosen to be the usual Nyquist rate associated with a comb-sampling function of infinite extent.
b. Find $g(x)$ if ξ_s is chosen to be twice the Nyquist rate determined in part a.
c. What parameter changes will improve the signal recovery?
d. Can $f(x)$ ever be recovered exactly in this case?

CHAPTER 9

TWO-DIMENSIONAL CONVOLUTION AND FOURIER TRANSFORMATION

In Chapters 6 and 7 we discussed the one-dimensional convolution and Fourier transform operations, and in Chapter 8 we learned how these operations are useful in solving certain problems involving one-dimensional LSI systems. However, because we will ultimately be dealing with two-dimensional systems, we must extend our capabilities to include convolution and Fourier transformation in two dimensions. This is necessary because a one-dimensional treatment of a two-dimensional problem can produce inaccurate results and lead to incorrect conclusions; although a one-dimensional approach is sometimes useful in determining the behavior of a two-dimensional system in a qualitative manner, a two-dimensional treatment is required to ensure quantitative accuracy.

9-1 CONVOLUTION IN RECTANGULAR COORDINATES

We begin our treatment of two-dimensional convolution with the assumption that the reader has a good understanding of the one-dimensional version, and we omit detailed discussions of such items as existence conditions, elementary properties, etc., because we feel that their extension to the two-dimensional case is straightforward. Given the complex-valued functions $f(x, y)$ and $h(x, y)$, we define their convolution in rectangular coordinates by

$$g(x,y)=\iint_{-\infty}^{\infty} f(\alpha,\beta)h(x-\alpha,y-\beta)\,d\alpha\,d\beta, \tag{9.1}$$

and we denote this integral by $g(x, y)=f(x, y)**h(x, y)$ for convenience.

We may view the convolution operation as one in which we determine the "volume" of the product of the two functions, one having first been folded about each axis, as a function of their relative positions. To help clarify this statement, let us consider the step-by-step procedures involved:

1. We regard $f(x, y)$ as a surface in space, and the height of this surface above the x, y-plane at any point (x, y) corresponds to the value of $f(x, y)$ at that point. We regard $h(x, y)$ similarly.
2. We express these two surfaces as functions of the dummy variables α and β, i.e., $f(\alpha,\beta)$ and $h(\alpha,\beta)$.
3. Next we form a new function $h(-\alpha,-\beta)$ by folding $h(\alpha,\beta)$ about each axis. We may visualize this procedure physically as follows: looking down on the α,β-plane from above, the function $h(-\alpha,-\beta)$ is formed by rotating $h(\alpha,\beta)$ through an angle of $\pm 180°$ about the vertical axis.
4. We then displace $h(-\alpha,-\beta)$ by an amount x in the α-direction and by an amount y in the β-direction to form the function $h(x-\alpha,y-\beta)$.
5. Next the functions $f(\alpha,\beta)$ and $h(x-\alpha,y-\beta)$ are multiplied to obtain a third function $f(\alpha,\beta)h(x-\alpha,y-\beta)$, which we also regard as a surface in space (but one that depends on the displacements x and y).
6. An integration over the entire α,β-plane is then performed to determine the volume of the surface $f(\alpha,\beta)h(x-\alpha,y-\beta)$, which is numerically equal to the value of the convolution $g(x, y)$ for the particular displacements x and y.
7. Finally, the above steps are repeated for all values of x and y to obtain $g(x, y)$ everywhere.

From this discussion we see that the convolution operation maps the two functions $f(x, y)$ and $h(x, y)$ into a third function $g(x, y)$, and we say that $g(x, y)$ is a *functional* of $f(x, y)$ and $h(x, y)$ because its value at any point (x, y) depends on the values of $f(x, y)$ and $h(x, y)$ everywhere.

The convolution operation is simplified somewhat for *one-zero functions*, i.e., functions having values of only one or zero. In this case the volume of the product $f(\alpha,\beta)h(x-\alpha,y-\beta)$ is numerically equal to the overlapping area of $f(\alpha,\beta)$ and $h(x-\alpha,y-\beta)$. The operation is depicted for two such functions in Fig. 9-1, where "top views" of the functions are shown for

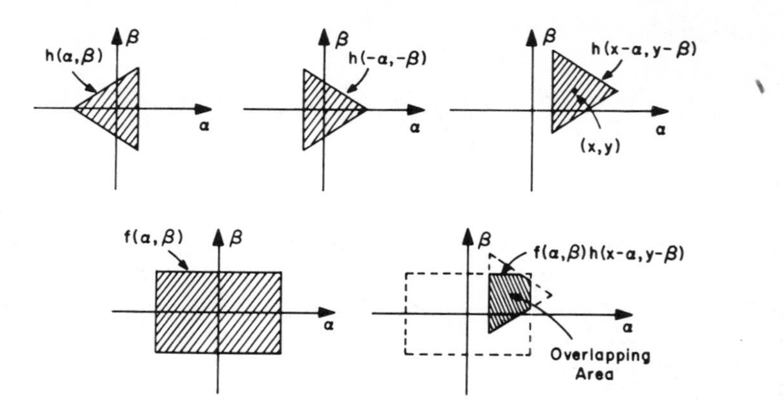

Figure 9-1 Simplified method for convolving one-zero functions.

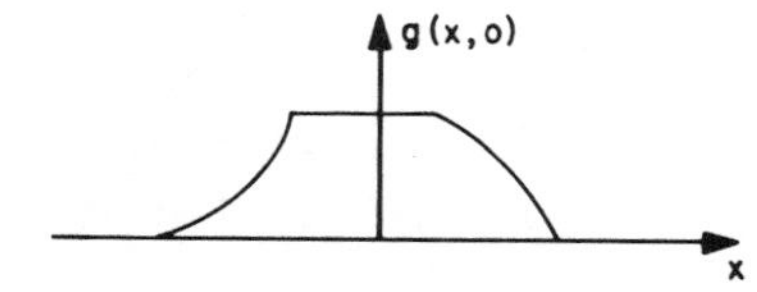

Figure 9-2 An x-axis profile of convolution of one-zero functions shown in Fig. 9-1.

simplicity. An x-profile of the resulting function $g(x, y)$ is shown in Fig. 9-2.

Properties of Convolution in Rectangular Coordinates

If the functions $f(x, y)$ and $h(x, y)$ are separable in rectangular coordinates, their convolution $g(x, y)$ is also separable and is given by the product of one-dimensional convolutions in x and y. To show this we let $f(x, y)=f_1(x)f_2(y)$ and $h(x, y)=h_1(x)h_2(y)$. Therefore,

$$\begin{aligned} g(x,y) &= [f_1(x)f_2(y)]**[h_1(x)h_2(y)] \\ &= \iint_{-\infty}^{\infty} f_1(\alpha)f_2(\beta)h_1(x-\alpha)h_2(y-\beta)\,d\alpha\,d\beta \\ &= \int_{-\infty}^{\infty} f_1(\alpha)h_1(x-\alpha)\,d\alpha \int_{-\infty}^{\infty} f_2(\beta)h_2(y-\beta)\,d\beta \\ &= [f_1(x)*h_1(x)][f_2(y)*h_2(y)] \\ &= g_1(x)\,g_2(y). \end{aligned} \tag{9.2}$$

It follows that if only one of the functions—say $h(x,y)$—is separable, then

$$g(x,y)=f(x,y)**[h_1(x)h_2(y)]$$
$$=f(x,y)*h_1(x)*h_2(y). \tag{9.3}$$

We now list, without derivation, several additional properties that correspond closely to those of the one-dimensional operation. For the sake of brevity, we omit the arguments of the various functions unless their inclusion is required to clarify the expression. In this list A_1 and A_2 are arbitrary constants, x_0 and y_0 are real constants, and b is a nonzero real constant. Also, f, h, u, v, and g are complex-valued functions with $g=f**h$.

$$f**h=h**f, \tag{9.4}$$

$$f**[A_1u+A_2v]=A_1[f**u]+A_2[f**v], \tag{9.5}$$

$$\begin{aligned} f**[u**v]&=[f**u]**v \\ &=f**u**v \\ &=u**f**v \\ &=u**v**f, \text{ etc.,} \end{aligned} \tag{9.6}$$

$$\begin{aligned} f(x-x_0,y-y_0)**h(x,y)&=f(x,y)**h(x-x_0,y-y_0) \\ &=f(x-x_0,y)**h(x,y-y_0) \\ &=f(x,y-y_0)**h(x-x_0,y) \\ &=g(x-x_0,y-y_0), \end{aligned} \tag{9.7}$$

$$f\left(\frac{x}{b},\frac{y}{b}\right)**h\left(\frac{x}{b},\frac{y}{b}\right)=|b|^2g\left(\frac{x}{b},\frac{y}{b}\right). \tag{9.8}$$

In addition, the volume of a convolution is equal to the product of the volumes of the component functions, i.e.,

$$\iint_{-\infty}^{\infty} g(x,y)\,dx\,dy=\iint_{-\infty}^{\infty} f(\alpha,\beta)\,d\alpha\,d\beta\iint_{-\infty}^{\infty} h(\alpha',\beta')\,d\alpha'\,d\beta'. \tag{9.9}$$

In the properties to follow a_i, b_i, c_i, and d_i are real constants and

$$D=a_1b_2-a_2b_1, \qquad x_0=\frac{b_1c_2-b_2c_1}{D}, \qquad y_0=\frac{a_2c_1-a_1c_2}{D},$$

$$g=f**h, \qquad x_{00}=\frac{b_1d_2-b_2d_1}{D}, \qquad y_{00}=\frac{a_2d_1-a_1d_2}{D},$$

$$r(x,y)=f(a_1x+b_1y,\, a_2x+b_2y),$$

$$u(x,y)=f(a_1x+b_1y+c_1,\, a_2x+b_2y+c_2),$$

$$s(x,y)=h(a_1x+b_1y,\, a_2x+b_2y),$$

$$v(x,y)=h(a_1x+b_1y+d_1,\, a_2x+b_2y+d_2),$$

$$t=r**s, \qquad w=u**v. \tag{9.10}$$

It can be shown that

$$\begin{aligned} t(x,y)&=r(x,y)**s(x,y)\\ &=\frac{1}{|D|}\,g(a_1x+b_1y,\, a_2x+b_2y). \end{aligned} \tag{9.11}$$

We know from Eq. (3.84) that

$$\begin{aligned} u(x,y)&=r(x-x_0, y-y_0),\\ v(x,y)&=s(x-x_{00}, y-y_{00}), \end{aligned} \tag{9.12}$$

from which we obtain

$$\begin{aligned} w(x,y)&=u(x,y)**v(x,y)\\ &=t(x-x_0-x_{00}, y-y_0-y_{00})\\ &=\frac{1}{|D|}\,g(a_1x+b_1y+c_1+d_1,\, a_2x+b_2y+c_2+d_2). \end{aligned} \tag{9.13}$$

If $c_1=d_1$ and $c_2=d_2$, this result simplifies to

$$\begin{aligned} w(x,y)&=t(x-2x_0, y-2y_0)\\ &=\frac{1}{|D|}\,g(a_1x+b_1y+2c_1,\, a_2x+b_2y+2c_2). \end{aligned} \tag{9.14}$$

Convolution with Combinations of Delta Functions

It is interesting to note that when a two-dimensional function $f(x, y)$ is convolved with a delta function of the *line-mass type* discussed in Sec. 3-5, the result is effectively an integral of the profile of $f(x, y)$ lying along the line mass. This statement is illustrated by the following equations, which can be easily derived from the results of Chapters 3 and 6:

$$f(x,y)**\delta(x-x_0)=\int_{-\infty}^{\infty} f(x-x_0,\beta)\,d\beta, \tag{9.15}$$

$$f(x,y)**\delta(y-y_0)=\int_{-\infty}^{\infty} f(\alpha,y-y_0)\,d\alpha, \tag{9.16}$$

$$\begin{aligned} f(x,y)**\delta(a_1x+b_1y+c_1)&=\frac{1}{|a_1|}\int_{-\infty}^{\infty} f\left(x+\frac{b_1\beta+c_1}{a_1},\,y-\beta\right)d\beta \\ &=\frac{1}{|b_1|}\int_{-\infty}^{\infty} f\left(x-\alpha,\,y+\frac{a_1\alpha+c_1}{b_1}\right)d\alpha, \end{aligned} \tag{9.17}$$

$$\begin{aligned} f(x,y)**\delta(x-x_0,y-y_0)&=f(x,y)**\left[\delta(x-x_0)\delta(y-y_0)\right] \\ &=f(x,y)*\delta(x-x_0)*\delta(y-y_0) \\ &=f(x-x_0,y-y_0), \end{aligned} \tag{9.18}$$

where

$$f(x,y)*\delta(x-x_0)=f(x-x_0,y)$$

and

$$g(x,y)*\delta(y-y_0)=g(x,y-y_0).$$

Note, however, that

$$f(x,y)**\delta(x-x_0)**\delta(y-y_0)\neq f(x-x_0,y-y_0). \tag{9.19}$$

Also, with D, x_0, and y_0 as defined in Eq. (9.10), we can obtain

$$f(x,y)**\delta(a_1x+b_1y+c_1,\,a_2x+b_2y+c_2)=\frac{1}{|D|}f(x-x_0,y-y_0), \tag{9.20}$$

$$f(x,y)**\operatorname{comb}(a_1x+b_1y+c_1,\,a_2x+b_2y+c_2)$$

$$=\frac{1}{|D|}\sum_{n=-\infty}^{\infty}\sum_{m=-\infty}^{\infty} f\left(x-x_0-\frac{b_2n}{D}+\frac{b_1m}{D},\,y-y_0+\frac{a_2n}{D}-\frac{a_1m}{D}\right). \tag{9.21}$$

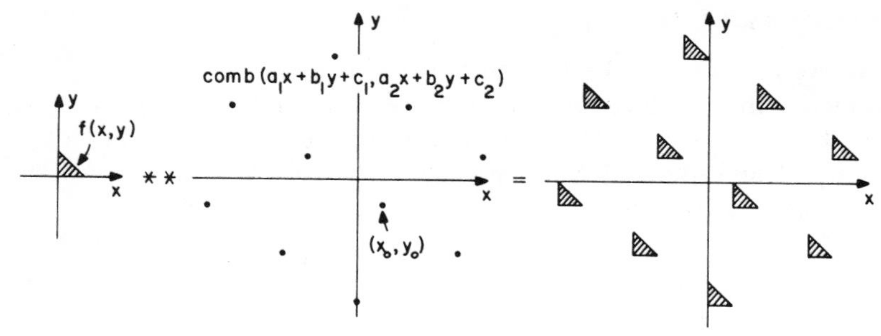

Figure 9-3 Convolution of a function with a skew-periodic comb function.

This result is illustrated in Fig. 9-3.

With k and l nonnegative integers, we define the two-dimensional delta function derivative to be

$$\begin{aligned}\delta^{(k,l)}(x,y)&=\left(\frac{\partial}{\partial x}\right)^k\left(\frac{\partial}{\partial y}\right)^l\delta(x,y)\\&=\delta^{(k)}(x)\delta^{(l)}(y).\end{aligned}\tag{9.22}$$

Thus,

$$\begin{aligned}f(x,y)**\delta^{(k,l)}(x,y)&=f(x,y)*\delta^{(k)}(x)*\delta^{(l)}(y)\\&=\left(\frac{\partial}{\partial x}\right)^k\left(\frac{\partial}{\partial y}\right)^l f(x,y)\\&=f^{(k,l)}(x,y).\end{aligned}\tag{9.23}$$

Convolution by Moment Expansion

We define the joint moments of $f(x, y)$ to be

$$m_{kl}=\iint_{-\infty}^{\infty}\alpha^k\beta^l f(\alpha,\beta)\,d\alpha\,d\beta,\tag{9.24}$$

where $k,l=0,1,2\ldots$. Papoulis (Ref. 9-1) gives the following result for the convolution of $f(x, y)$ and $h(x, y)$ in terms of the moments m_{kl} of $f(x, y)$

and the derivatives $h^{(k,l)}(x, y)$ of $h(x, y)$:

$$\begin{aligned} g(x,y) &= f(x,y)**h(x,y) \\ &= m_{00}h(x,y) - \left[m_{10}h^{(1,0)}(x,y) + m_{01}h^{(0,1)}(x,y)\right] \\ &\quad + \left[m_{20}h^{(2,0)}(x,y) + m_{11}h^{(1,1)}(x,y) \right. \\ &\quad \left. + m_{02}h^{(0,2)}(x,y)\right] - \dots . \end{aligned} \tag{9.25}$$

There is a similar expansion in terms of the moments of $h(x, y)$ and the derivatives of $f(x, y)$.

Cross Correlation and Autocorrelation

As we did in Sec. 6-5 for one-dimensional functions, we now define the (two-dimensional) *cross correlation of* $f(x, y)$ *with* $g(x, y)$ to be

$$\begin{aligned} f(x,y) \star\star g(x,y) &= \iint\limits_{-\infty}^{\infty} f(\alpha,\beta)\, g(\alpha - x, \beta - y)\, d\alpha\, d\beta \\ &= \iint\limits_{-\infty}^{\infty} f(\alpha + x, \beta + y)\, g(\alpha,\beta)\, d\alpha\, d\beta. \end{aligned} \tag{9.26}$$

We note that this operation is similar to the convolution operation, but neither of the functions is folded before the shifting and integration takes place. It can be shown that

$$f(x,y) \star\star g(x,y) = f(x,y)**g(-x, -y), \tag{9.27}$$

and we find that, in general,

$$f(x,y) \star\star g(x,y) \neq g(x,y) \star\star f(x,y). \tag{9.28}$$

When $g(x, y) = f(x, y)$, this operation is called the (two-dimensional) *autocorrelation of* $f(x, y)$.

We also define the (two-dimensional) *complex cross correlation of* $f(x, y)$ *with* $g(x, y)$ to be

$$\begin{aligned} \gamma_{fg}(x,y) &= f(x,y) \star\star g^*(x,y) \\ &= f(x,y)**g^*(-x, -y), \end{aligned} \tag{9.29}$$

and, as for the one-dimensional case [Eqs. (6.54) and (6.55)], we note that

$$\gamma_{fg}(x,y)=\gamma_{gf}^{*}(-x,\ -y),$$
$$\gamma_{gf}(x,y)=\gamma_{fg}^{*}(-x,\ -y). \tag{9.30}$$

When $g(x, y)=f(x, y)$ Eq. (9.29) becomes

$$\begin{aligned}\gamma_{ff}(x,y)&=f(x,y)\star\star f^{*}(x,y)\\&=\gamma_{f}(x,y),\end{aligned} \tag{9.31}$$

which we call the *complex autocorrelation of* $f(x, y)$. Observe that $\gamma_f(x,y)$ is hermitian and that its modulus is maximum at the origin:

$$\gamma_f(x,y)=\gamma_f^{*}(-x,\ -y),$$
$$|\gamma_f(x,y)|\leqslant\gamma_f(0,0). \tag{9.32}$$

The important properties of $\gamma_f(x,y)$ are similar to those listed in Table 6-1 for the one-dimensional complex autocorrelation function.

9-2 CONVOLUTION IN POLAR COORDINATES

In general, convolution in polar coordinates leads to much more cumbersome expressions than convolution in rectangular coordinates. To illustrate, we shall describe the convolution $g(x,y)=f(x,\ y)**h(x,\ y)$ in polar coordinates:

$$\begin{aligned}g(r\cos\theta,\ r\sin\theta)&=f(r\cos\theta,\ r\sin\theta)**h(r\cos\theta,\ r\sin\theta)\\&=\int_0^{\infty}\int_0^{2\pi}f(r'\cos\theta',\ r'\sin\theta')h(r\cos\theta-r'\cos\theta',\ r\sin\theta-r'\sin\theta')r'\,dr'\,d\theta'.\end{aligned} \tag{9.33}$$

If we define

$$u(r,\theta)=f(r\cos\theta,\ r\sin\theta),\qquad v(r,\theta)=h(r\cos\theta,\ r\sin\theta),$$
$$w(r,\theta)=g(r\cos\theta,\ r\sin\theta), \tag{9.34}$$

we may write

$$w(r,\theta)=\int_0^{\infty}\int_0^{2\pi}u(r',\theta')v\left\{\sqrt{r^2+r'^2-2rr'\cos(\theta-\theta')}\,,\right.$$

$$\left.\tan^{-1}\left(\frac{r\sin\theta-r'\sin\theta'}{r\cos\theta-r'\cos\theta'}\right)\right\}r'\,dr'\,d\theta', \tag{9.35}$$

but we must be certain to observe the following conditions in order to avoid ambiguity: with

$$\phi=\tan^{-1}\left(\frac{r\sin\theta-r'\sin\theta'}{r\cos\theta-r'\cos\theta'}\right)=\tan^{-1}(a/b), \tag{9.36}$$

then

$$\begin{aligned} &0\leqslant\phi\leqslant\pi/2 && \text{if} && a>0, && b>0\\ &\pi/2\leqslant\phi\leqslant\pi && \text{if} && a>0, && b<0\\ &\pi\leqslant\phi\leqslant 3\pi/2 && \text{if} && a<0, && b<0\\ &3\pi/2\leqslant\phi\leqslant 2\pi && \text{if} && a<0, && b>0. \end{aligned} \tag{9.37}$$

The tedium associated with keeping track of these values usually renders Eq. (9.35) unsatisfactory for performing convolutions. However, if one of the functions to be convolved is independent of the angle θ, this equation can be greatly simplified. For example, if $h(x,y)=h_R\left(\sqrt{x^2+y^2}\right)$, then $v(r,\theta)=h_R(r)$ and

$$w(r,\theta)=\int_0^{\infty}\int_0^{2\pi}u(r',\theta')h_R\left(\sqrt{r^2+r'^2-2rr'\cos(\theta-\theta')}\right)r'\,dr'\,d\theta'. \tag{9.38}$$

If, in addition, $f(x,y)=f_R\left(\sqrt{x^2+y^2}\right)$, then $u(r,\theta)=f_R(r)$ and

$$w(r,\theta)=\int_0^{\infty}\int_0^{2\pi}f_R(r')h_R\left(\sqrt{r^2+r'^2-2rr'\cos(\theta-\theta')}\right)r'\,dr'\,d\theta', \tag{9.39}$$

which can be shown to be independent of θ. Therefore, without loss of generality, we set $\theta=0$ and $\cos(-\theta')=\cos(\theta')$. Thus,

$$w(r,\theta)=\int_0^{\infty}\int_0^{2\pi}f_R(r')h_R\left(\sqrt{r^2+r'^2-2rr'\cos\theta'}\right)r'\,dr'\,d\theta', \tag{9.40}$$

which depends only on the radial coordinate r. If we then denote $w(r,\theta)= g_R(r)$, we may express the *convolution of circularly-symmetric functions* by

$$g_R(r)=f_R(r)**h_R(r)$$
$$=\int_0^\infty\int_0^{2\pi} f_R(r')h_R\left(\sqrt{r^2+r'^2-2rr'\cos\theta'}\right)r'\,dr'\,d\theta'. \tag{9.41}$$

We retain the double-asterisk notation to avoid confusion with the one-dimensional convolution.

A comparison of the above equations with those of Sec. 9-1 suggests that care must be exercised when performing the convolution operation in polar coordinates; by blindly plugging functions into formulas that are valid only in other coordinate systems, you will likely encounter a great deal of grief.

Properties of Convolution in Polar Coordinates

If we use the notation

$$w(r,\theta)=u(r,\theta)**v(r,\theta) \tag{9.42}$$

to indicate the polar-coordinate convolution operation of Eq. (9.35), then it can be shown that $w(r,\theta)$ is also given by

$$w(r,\theta)=v(r,\theta)**u(r,\theta). \tag{9.43}$$

In addition we have

$$u\left(\frac{r}{d},\theta\right)**v\left(\frac{r}{d},\theta\right)=|d|^2w\left(\frac{r}{d},\theta\right),$$
$$u(r,\theta-\theta_0)**v(r,\theta-\theta_0)=w(r,\theta-\theta_0), \tag{9.44}$$

the first of which follows from Eq. (9.8), while the second is intuitively obvious. Still another property is

$$f_R(r)**h_R(r)=f_R(r)\bigstar\bigstar h_R(r). \tag{9.45}$$

For the sake of completeness it seems in order to include some *nonproperties*:

$$u(r,\theta)**v(r,\theta)\neq\int_0^\infty\int_0^{2\pi}u(r',\theta')v(r-r',\theta-\theta')r'\,dr'\,d\theta',$$

$$f_R(r)**h_R(r)\neq\int_0^\infty f_R(r')h_R(r-r')dr',$$

$$u(r-r_0,\theta)**v(r,\theta)\neq w(r-r_0,\theta),$$

$$u(r-r_0,\theta)**v(r-r_0,\theta)\neq w(r-r_0,\theta). \tag{9.46}$$

Convolution with Delta Functions in Polar Coordinates

It is frequently convenient to use vector notation to describe the convolution operation. With this notation, the convolution of the functions $p(\mathbf{r})$ and $s(\mathbf{r})$ is given by

$$p(\mathbf{r})**s(\mathbf{r})=\iint p(\boldsymbol{\alpha})s(\mathbf{r}-\boldsymbol{\alpha})d\boldsymbol{\alpha}, \tag{9.47}$$

where $\mathbf{r}$ is a position vector in the (r,θ)-plane. With $\delta(\mathbf{r})$ denoting a unit-volume delta function located at the origin, we now show that

$$\delta(\mathbf{r})**s(\mathbf{r})=s(\mathbf{r}). \tag{9.48}$$

We let $s(\mathbf{r})=u(r,\theta)=f(r\cos\theta,\ r\sin\theta)$ and $\delta(\mathbf{r})$ be given by Eq. (3.70). Then Eq. (9.48) becomes

$$\begin{aligned}\delta(\mathbf{r})**s(\mathbf{r})&=\frac{\delta(r)}{\pi r}**u(r,\theta)\\&=\int_0^\infty\int_0^{2\pi}\frac{\delta(r')}{\pi r'}f(r\cos\theta-r'\cos\theta',\ r\sin\theta-r'\sin\theta')r'\,dr'\,d\theta'\\&=\frac{1}{2\pi}\int_0^{2\pi}f(r\cos\theta,\ r\sin\theta)d\theta'\\&=f(r\cos\theta,\ r\sin\theta)\\&=s(\mathbf{r}),\end{aligned} \tag{9.49}$$

as contended. In evaluating the above expression we used

$$\int_0^\infty \delta(\alpha) g(\alpha) d\alpha = \tfrac{1}{2} \int_{-\infty}^\infty \delta(\alpha) g(\alpha) d\alpha$$

$$= \tfrac{1}{2} g(0), \tag{9.50}$$

which stems from the evenness of the delta function.

We now convolve $s(\mathbf{r}) = u(r,\theta) = f(r\cos\theta,\ r\sin\theta)$ with the shifted delta function $\delta(\mathbf{r}-\mathbf{r}_0)$. We use the expression for $\delta(\mathbf{r}-\mathbf{r}_0)$ given by Eq. (3.68) to write

$$\delta(\mathbf{r}-\mathbf{r}_0) ** s(\mathbf{r}) = \frac{1}{r_0}\delta(r-r_0)\delta(\theta-\theta_0) ** u(r,\theta)$$

$$= \int_0^\infty \int_0^{2\pi} \frac{1}{r_0}\delta(r'-r_0)\delta(\theta'-\theta_0)$$

$$\times f(r\cos\theta - r'\cos\theta',\ r\sin\theta - r'\sin\theta') r'\, dr'\, d\theta'$$

$$= f(r\cos\theta - r_0\cos\theta_0,\ r\sin\theta - r_0\sin\theta_0)$$

$$= s(\mathbf{r}-\mathbf{r}_0), \tag{9.51}$$

which corresponds to the result depicted by Eq. (9.18) for rectangular coordinates.

$\mathbf{cyl}(r/d_1) ** \mathbf{cyl}(r/d_2)$

As mentioned in Sec. 9-1, the convolution of two functions is given by the volume of their product as one of them is shifted relative to the other, and for one-zero functions it is numerically equal to their area of overlap as a function of the shift variable. These statements hold in either rectangular coordinates or polar coordinates. We now derive an expression for the convolution of two cylinder functions, a result that will be of significant value in our studies of certain imaging systems. Because the cylinder functions are one-zero functions, we need only calculate the area of overlap as a function of their center-to-center separation. Figure 9-4 will be used to aid with this task.

We denote the diameters of the two functions by d_1 and d_2, and to simplify the development we require that $d_2 \geqslant d_1$. We note that the convolution has a value equal to the area of the smaller function as long as the center-to-center separation is less than $\frac{1}{2}(d_2 - d_1)$, and it is zero for a separation greater than $\frac{1}{2}(d_1 + d_2)$. For intermediate separations, the value of the convolution is numerically equal to the sum of the areas of the

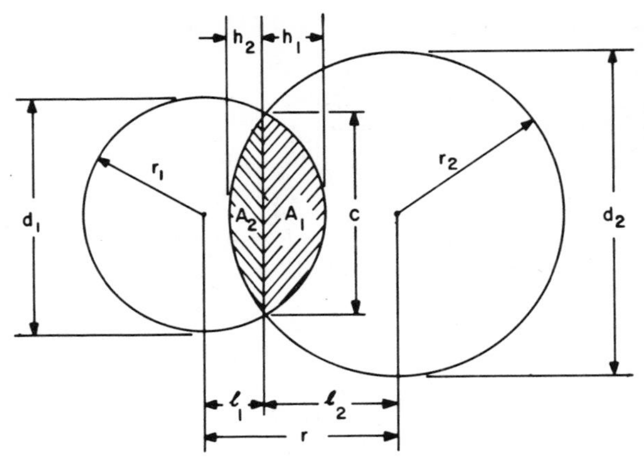

Figure 9-4 Geometry for determining convolution of two cylinder functions.

circular segments denoted by $A_1(r)$ and $A_2(r)$ in Fig. 9-4. The convolution may therefore be written as

$$\text{cyl}\left(\frac{r}{d_1}\right)**\text{cyl}\left(\frac{r}{d_2}\right)=\begin{cases}\dfrac{\pi d_1^2}{4}, & r<\dfrac{d_2-d_1}{2},\\ A_1(r)+A_2(r), & \dfrac{d_2-d_1}{2}\leqslant r\leqslant\dfrac{d_1+d_2}{2},\\ 0, & r>\dfrac{d_1+d_2}{2}.\end{cases} \tag{9.52}$$

With $i=1,2$, the area of the ith segment is given by (Ref. 9-2)

$$A_i(r)=\frac{d_i^2}{4}\cos^{-1}\left[\frac{(d_i/2)-h_i}{d_i/2}\right]-\left(\frac{d_i}{2}-h_i\right)\left(d_ih_i-h_i^2\right)^{1/2} \tag{9.53}$$

Using the relationship $\ell_i=\frac{1}{2}d_i-h_i=\frac{1}{2}(d_i^2-c^2)^{1/2}$, we may solve for ℓ_1 and ℓ_2 in terms of r, d_1, and d_2; the results are

$$\ell_1=\frac{r^2+\frac{1}{4}\left(d_1^2-d_2^2\right)}{2r},\qquad \ell_2=\frac{r^2+\frac{1}{4}\left(d_2^2-d_1^2\right)}{2r}. \tag{9.54}$$

Then, by realizing that

$$\left(d_ih_i-h_i^2\right)^{1/2}=\left(\frac{d_i^2}{4}-\ell_i^2\right)^{1/2}, \tag{9.55}$$

we obtain the following expressions for $A_1(r)$ and $A_2(r)$:

$$A_1(r)=\frac{d_1^2}{4}\cos^{-1}\left[\frac{r^2+\frac{1}{4}(d_1^2-d_2^2)}{rd_1}\right]$$

$$-\frac{d_1^2}{4}\left[\frac{r^2+\frac{1}{4}(d_1^2-d_2^2)}{rd_1}\right]\left\{1-\left[\frac{r^2+\frac{1}{4}(d_1^2-d_2^2)}{rd_1}\right]^2\right\}^{1/2}$$

$$A_2(r)=\frac{d_2^2}{4}\cos^{-1}\left[\frac{r^2+\frac{1}{4}(d_2^2-d_1^2)}{rd_2}\right]$$

$$-\frac{d_2^2}{4}\left[\frac{r^2+\frac{1}{4}(d_2^2-d_1^2)}{rd_2}\right]\left\{1-\left[\frac{r^2+\frac{1}{4}(d_2^2-d_1^2)}{rd_2}\right]^2\right\}^{1/2}. \quad (9.56)$$

Finally, recalling that $d_2 \geqslant d_1$, and with $A_1(r)$ and $A_2(r)$ as given above, Eq. (9.52) may be expressed as

$$\mathrm{cyl}\left(\frac{r}{d_1}\right)**\mathrm{cyl}\left(\frac{r}{d_2}\right)=\frac{\pi d_1{}^2}{4}\mathrm{cyl}\left(\frac{r}{d_2-d_1}\right)$$

$$+\left[A_1(r)+A_2(r)\right]\left[\mathrm{cyl}\left(\frac{r}{d_1+d_2}\right)-\mathrm{cyl}\left(\frac{r}{d_2-d_1}\right)\right]. \quad (9.57)$$

This is an important result, but the expressions for $A_1(r)$ and $A_2(r)$ are obviously quite cumbersome. To simplify things we set

$$d_1=1, \qquad d_2=a, \qquad a\geqslant 1, \qquad b=\tfrac{1}{4}(a^2-1), \quad (9.58)$$

and describe the convolution of Eq. (9.57) in terms of the *normalized cylinder-function cross correlation*

$$\gamma_{\mathrm{cyl}}(r;a)=\frac{4}{\pi}\mathrm{cyl}(r)\star\star\,\mathrm{cyl}\left(\frac{r}{a}\right)$$

$$=\mathrm{cyl}\left(\frac{r}{a-1}\right)+\frac{4}{\pi}\left[A_1(r)+A_2(r)\right]\left[\mathrm{cyl}\left(\frac{r}{a+1}\right)-\mathrm{cyl}\left(\frac{r}{a-1}\right)\right], \quad (9.59)$$

where now

$$A_1(r)=\frac{1}{4}\left\{\cos^{-1}\left(\frac{r^2-b}{r}\right)-\left(\frac{r^2-b}{r}\right)\left[1-\left(\frac{r^2-b}{r}\right)^2\right]^{1/2}\right\},$$

$$A_2(r)=\frac{a^2}{4}\left\{\cos^{-1}\left(\frac{r^2+b}{ar}\right)-\left(\frac{r^2+b}{ar}\right)\left[1-\left(\frac{r^2+b}{ar}\right)^2\right]^{1/2}\right\}. \qquad (9.60)$$

The function $\gamma_{cyl}(r;a)$ has a maximum value of unity and a total width of $a+1$. It is depicted in Fig. 9-5 and tabulated in the Appendix for several values of a.

Once $\gamma_{cyl}(r;a)$ has been determined for a particular value of a, we may use Eqs. (9.59) and (9.44) to obtain

$$\text{cyl}(r)\star\star\,\text{cyl}\left(\frac{r}{a}\right)=\frac{\pi}{4}\gamma_{cyl}(r;a),$$

$$\text{cyl}\left(\frac{r}{d}\right)\star\star\,\text{cyl}\left(\frac{r}{ad}\right)=\frac{\pi d^2}{4}\gamma_{cyl}\left(\frac{r}{d};a\right). \qquad (9.61)$$

Of particular interest is the case for which $a=1$. In this event $A_1(r)=A_2(r)$ and

$$\gamma_{cyl}(r;1)=\frac{2}{\pi}\left[\cos^{-1}(r)-r(1-r^2)^{1/2}\right]\text{cyl}\left(\frac{r}{2}\right). \qquad (9.62)$$

Hufnagel (Ref. 9-3) has given an expression that is a very good approximation to $\gamma_{cyl}(r;1)$ and which can be useful under certain circumstances. In

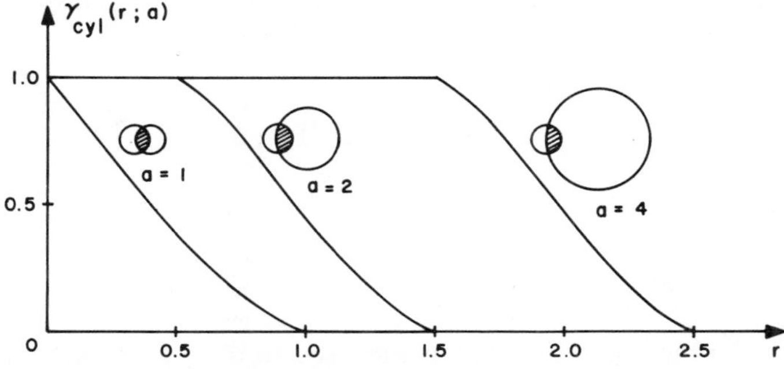

Figure 9-5 Normalized cylinder-function cross correlation for various values of a.

our notation this expression becomes

$$\gamma_{\text{cyl}}(r;1)\cong 0.25\left[5\,\text{tri}(r)-1+r^4\right]\text{cyl}\left(\frac{r}{2}\right). \tag{9.63}$$

9-3 THE FOURIER TRANSFORM IN RECTANGULAR COORDINATES

In this section we investigate the two-dimensional Fourier transform operation in rectangular coordinates, and a brief review of Sec. 4-5 might be in order before proceeding. We omit detailed discussions of such topics as existence conditions, interpretations, etc., which are similar to those for the one-dimensional case and are left as a matter for the reader to pursue. We mention once again, however, that any function that accurately describes a real physical quantity will possess a Fourier transform. In addition, we find that by adopting the notion of generalized Fourier transformation for two-dimensional functions, such nonphysical functions as $\delta(x,y)$, $\text{comb}(x,y)$, etc., also possess Fourier transforms.

Given the function $f(x, y)$, we denote its two-dimensional Fourier transform by

$$F(\xi,\eta)=\iint_{-\infty}^{\infty} f(\alpha,\beta)e^{-j2\pi(\alpha\xi+\beta\eta)}d\alpha\,d\beta, \tag{9.64}$$

where ξ and η are the frequency variables corresponding to the x- and y-directions, respectively. Thus,

$$f(x,y)=\iint_{-\infty}^{\infty} F(\alpha',\beta')e^{j2\pi(\alpha' x+\beta' y)}d\alpha'\,d\beta' \tag{9.65}$$

is the inverse transform of $F(\xi,\eta)$. As in Chap. 7 we use shorthand notation to denote these operations:

$$
\begin{aligned}
F(\xi,\eta)&=\mathcal{F}\mathcal{F}\{f(x,y)\}, & f(x,y)&\overset{\mathcal{F}\mathcal{F}}{\longrightarrow}F(\xi,\eta),\\
f(x,y)&=\mathcal{F}^{-1}\mathcal{F}^{-1}\{F(\xi,\eta)\}, & f(x,y)&\overset{\mathcal{F}^{-1}\mathcal{F}^{-1}}{\longleftarrow}F(\xi,\eta),\\
\tilde{E}(\xi,\eta)&=\mathcal{F}\mathcal{F}\{E(x,y)\}, & E(x,y)&\overset{\mathcal{F}\mathcal{F}}{\longrightarrow}\tilde{E}(\xi,\eta),
\end{aligned} \tag{9.66}
$$

where the double-operator symbols are used to distinguish them from the corresponding one-dimensional operations.

Properties of Two-Dimensional Fourier Transforms

Many properties of two-dimensional Fourier transforms follow easily from the one-dimensional theory, but a few do not and must be derived independently. We note that the kernel of the two-dimensional Fourier transform is separable in x and y and, as a result, the transform of a separable function is itself separable. For example, the transform of the function $f(x,y)=f_1(x)f_2(y)$ may be expressed as

$$\begin{aligned}
F(\xi,\eta) &= \mathcal{F}\mathcal{F}\{f_1(x)f_2(y)\} \\
&= \iint\limits_{-\infty}^{\infty} f_1(\alpha)f_2(\beta)e^{-j2\pi(\alpha\xi+\beta\eta)}d\alpha\, d\beta \\
&= \int_{-\infty}^{\infty} f_1(\alpha)e^{-j2\pi\alpha\xi}d\alpha \int_{-\infty}^{\infty} f_2(\beta)e^{-j2\pi\beta\eta}d\beta \\
&= F_1(\xi)F_2(\eta),
\end{aligned} \tag{9.67}$$

a result that can simplify many problems.

We now discuss an often misunderstood aspect of the two-dimensional transform operation, one that can lead to serious errors for the unsuspecting. Given the two-dimensional transform pair

$$f(x,y)\overset{\mathcal{F}\mathcal{F}}{\rightarrow}F(\xi,\eta),$$

we define the *one-dimensional function* $g(x)$ to be the *profile* of $f(x,y)$ along the y-axis and the *one-dimensional function* $H(\xi)$ to be the *profile* of $F(\xi,\eta)$ along the η-axis, i.e.,

$$g(x)=f(x,0), \qquad H(\xi)=F(\xi,0). \tag{9.68}$$

It is natural to assume that $g(x)$ and $H(\xi)$ form a one-dimensional transform pair, but *in general they do not*! Only if the functions involved are separable will this be so, and a similar statement holds for profiles along the other coordinate axes. In other words, corresponding profiles of a two-dimensional transform pair do not themselves form a one-dimensional transform pair unless the original functions are separable. Mathematically this result may be expressed as follows: if $f(x,y)$ *is not separable*, then

$$\begin{aligned}
\mathcal{F}\{f(x,0)\} &\neq F(\xi,0), \\
\mathcal{F}\{f(0,y)\} &\neq F(0,\eta).
\end{aligned} \tag{9.69}$$

Let us now consider the convolution $f(x, y)**\delta(x)$. From Eq. (9.15) we know

$$f(x,y)**\delta(x)=\int_{-\infty}^{\infty} f(x,\beta)\,d\beta, \tag{9.70}$$

from which we see that $f(x, y)**\delta(x)$ depends only on x and is therefore separable. Denoting this convolution by $u(x)$, we find from Eq. (9.67) that

$$\begin{aligned}\mathfrak{F}\mathfrak{F}\{f(x,y)**\delta(x)\}&=\mathfrak{F}\mathfrak{F}\{u(x)\cdot 1\}\\&=U(\xi)\delta(\eta).\end{aligned} \tag{9.71}$$

However, the convolution theorem, which is similar to that for the one-dimensional case, yields

$$\begin{aligned}\mathfrak{F}\mathfrak{F}\{f(x,y)**\delta(x)\}&=\mathfrak{F}\mathfrak{F}\{f(x,y)\}\mathfrak{F}\mathfrak{F}\{\delta(x)\cdot 1\}\\&=F(\xi,\eta)\delta(\eta)\\&=F(\xi,0)\delta(\eta),\end{aligned} \tag{9.72}$$

and a comparison of Eqs. (9.71) and (9.72) leads to the interesting conclusion that $u(x)$ and $F(\xi,0)$ form a one-dimensional transform pair. There is a corresponding expression for the other coordinate direction, and with $v(y)=f(x, y)**\delta(y)$ we obtain

$$\begin{aligned}u(x)&=\int_{-\infty}^{\infty} f(x,\beta)\,d\beta \overset{\mathfrak{F}}{\longrightarrow} F(\xi,0),\\v(y)&=\int_{-\infty}^{\infty} f(\alpha,y)\,d\alpha \overset{\mathfrak{F}}{\longrightarrow} F(0,\eta).\end{aligned} \tag{9.73}$$

Fourier Transform of $f(a_1x+b_1y+c_1,\ a_2x+b_2y+c_2)$

Given that

$$\begin{aligned}f(x,y)&\overset{\mathfrak{F}\mathfrak{F}}{\longrightarrow} F(\xi,\eta),\\g(x,y)&=f(a_1x+b_1y+c_1,\ a_2x+b_2y+c_2),\end{aligned} \tag{9.74}$$

we shall determine the transform

$$G(\xi,\eta)=\iint_{-\infty}^{\infty} f(a_1\alpha+b_1\beta+c_1,\ a_2\alpha+b_2\beta+c_2)e^{-j2\pi(\alpha\xi+\beta\eta)}\,d\alpha\,d\beta \tag{9.75}$$

in terms of $F(\xi,\eta)$. With the change of variables $\alpha' = a_1\alpha + b_1\beta + c_1$, Eq. (9.75) becomes

$$G(\xi,\eta) = \iint_{-\infty}^{\infty} f\left[\alpha', a_2\left(\frac{\alpha' - b_1\beta - c_1}{a_1}\right) + b_2\beta + c_2\right]$$

$$\times \exp\left\{-j2\pi\left[\left(\frac{\alpha' - b_1\beta - c_1}{a_1}\right)\xi + \beta\eta\right]\right\}\frac{d\alpha'}{|a_1|}d\beta, \quad (9.76)$$

and with another change of variables, e.g.,

$$\beta' = a_2\left(\frac{\alpha' - b_1\beta - c_1}{a_1}\right) + b_2\beta + c_2, \quad (9.77)$$

we obtain

$$G(\xi,\eta) = \frac{1}{|D|}\iint_{-\infty}^{\infty} f(\alpha',\beta')$$

$$\times \exp\left\{-j2\pi\left[\left(\frac{b_2\xi}{D} - \frac{a_2\eta}{D}\right)\alpha' + \left(\frac{-b_1\xi}{D} + \frac{a_1\eta}{D}\right)\beta' + x_0\xi + y_0\eta\right]\right\}d\alpha'\,d\beta',$$

$$(9.78)$$

where

$$D = a_1b_2 - a_2b_1, \qquad x_0 = \frac{b_1c_2 - b_2c_1}{D}, \qquad y_0 = \frac{a_2c_1 - a_1c_2}{D}. \quad (9.79)$$

Finally, we recognize that Eq. (9.78) may be simplified to the form

$$G(\xi,\eta) = \frac{1}{|D|}e^{-j2\pi(x_0\xi + y_0\eta)}F\left(\frac{b_2}{D}\xi - \frac{a_2}{D}\eta, -\frac{b_1}{D}\xi + \frac{a_1}{D}\eta\right), \quad (9.80)$$

which is a fascinating result: if we know the transform of a function $f(x, y)$, we can easily determine the transform of its scaled, shifted, *and skewed* version $g(x, y)$ by simply following the prescription of Eq. (9.80).

Fourier Transforms of Skew-Periodic Functions

Let us consider the skew-periodic function [see Eq. (9.21)]

$$g(x,y)=f(x,y)**|D|\operatorname{comb}(a_1x+b_1y,a_2x+b_2y)$$
$$=\sum_{n=-\infty}^{\infty}\sum_{m=-\infty}^{\infty} f\left(x-\frac{b_2}{D}n+\frac{b_1}{D}m,\,y+\frac{a_2}{D}n-\frac{a_1}{D}m\right), \tag{9.81}$$

where D is as defined above and we have set $c_1=c_2=0$ for simplicity. Application of the convolution theorem and Eq. (9.80) yields

$$G(\xi,\eta)=F(\xi,\eta)\operatorname{comb}\left(\frac{b_2}{D}\xi-\frac{a_2}{D}\eta,\,-\frac{b_1}{D}\xi+\frac{a_1}{D}\eta\right), \tag{9.82}$$

and with the help of Eq. (3.132) we may write

$$G(\xi,\eta)=|D|\sum_{n=-\infty}^{\infty}\sum_{m=-\infty}^{\infty} F(a_1n+a_2m,\,b_1n+b_2m)$$
$$\times\delta(\xi-a_1n-a_2m,\,\eta-b_1n-b_2m). \tag{9.83}$$

Examination of this expression reveals that it consists of a skewed array of delta functions positioned on the reciprocal lattice of $\operatorname{comb}(a_1x+b_1y, a_2x+b_2y)$ and weighted by the value of the transform $F(\xi,\eta)$ at the lattice points. Another useful version of this result is obtained by making the substitutions

$$\alpha_1=\frac{-b_2}{D},\qquad \alpha_2=\frac{b_1}{D},\qquad \beta_1=\frac{a_2}{D},\qquad \beta_2=\frac{-a_1}{D},$$
$$\Delta=\frac{1}{D}=\alpha_1\beta_2-\alpha_2\beta_1,$$
$$h(x,y)=\sum_{n=-\infty}^{\infty}\sum_{m=-\infty}^{\infty} f(x+\alpha_1n+\alpha_2m,\,y+\beta_1n+\beta_2m). \tag{9.84}$$

Then

$$H(\xi,\eta)=\frac{1}{|\Delta|}\sum_{n=-\infty}^{\infty}\sum_{m=-\infty}^{\infty} F\left(-\frac{\beta_2}{\Delta}n+\frac{\beta_1}{\Delta}m,\,\frac{\alpha_2}{\Delta}n-\frac{\alpha_1}{\Delta}m\right)$$
$$\times\delta\left(\xi+\frac{\beta_2}{\Delta}n-\frac{\beta_1}{\Delta}m,\,\eta-\frac{\alpha_2}{\Delta}n+\frac{\alpha_1}{\Delta}m\right). \tag{9.85}$$

Fourier Transforms of One-Zero Functions

Papoulis (Ref. 9-4) gives an interesting method for determining certain profiles of the Fourier transforms of one-zero functions having irregular shapes. Consider the one-zero function

$$\begin{aligned} f(x,y) &= 1, \quad y_1(x) \leqslant y \leqslant y_2(x) \\ &= 0, \quad \text{otherwise} \end{aligned} \tag{9.86}$$

a top view of which is shown in Fig. 9-6(a). The functions $y_1(x)$ and $y_2(x)$ describe the lower and upper boundaries, respectively, and are single-valued functions of x. The Fourier transform of $f(x,y)$ may be expressed as

$$\begin{aligned} F(\xi,\eta) &= \iint_{-\infty}^{\infty} f(\alpha,\beta) e^{-j2\pi(\alpha\xi+\beta\eta)} d\alpha\, d\beta \\ &= \int_{-\infty}^{\infty} e^{-j2\pi\alpha\xi} \left[\int_{y_1(\alpha)}^{y_2(\alpha)} e^{-j2\pi\beta\eta} d\beta \right] d\alpha, \end{aligned} \tag{9.87}$$

and for $\eta = 0$,

$$\begin{aligned} F(\xi,0) &= \int_{-\infty}^{\infty} [y_2(\alpha) - y_1(\alpha)] e^{-j2\pi\alpha\xi} d\alpha \\ &= Y_2(\xi) - Y_1(\xi), \end{aligned} \tag{9.88}$$

where $Y_i(\xi) = \mathcal{F}\{y_i(x)\}$. The profile $F(0,\eta)$ can be found in a similar manner.

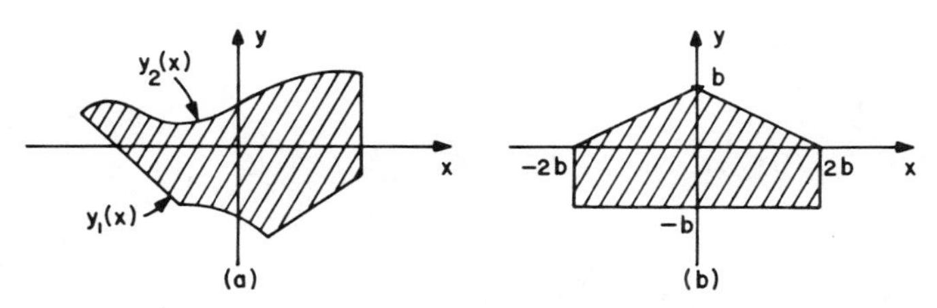

Figure 9-6 Description of one-zero functions. (*a*) General one-zero function. (*b*) One-zero function of Eq. (9.89).

Example

We let

$$y_1(x) = -b\,\mathrm{rect}\left(\frac{x}{2b}\right), \qquad y_2(x) = b\,\mathrm{tri}\left(\frac{x}{b}\right) \tag{9.89}$$

as shown in Fig. 9-6(b). Then from Eq. (9.88)

$$F(\xi,0) = b^2\,\mathrm{sinc}^2(b\xi) + 2b^2\,\mathrm{sinc}(2b\xi), \tag{9.90}$$

and, although this result is only valid along the ξ-axis, it was nevertheless quite easy to determine; it is much more difficult to obtain an expression for $F(\xi,\eta)$ that is valid everywhere when dealing with such irregular functions. Note that if $y_1(x) = -y_2(x) = -b\,\mathrm{tri}(x/b)$ in the present example, the function $f(x, y)$ becomes square in shape with its sides oriented at angles of 45° with respect to the x- and y-axes. In this case,

$$F(\xi,0) = 2b^2\,\mathrm{sinc}^2(b\xi). \tag{9.91}$$

■

Papoulis (Ref. 9-4) has also given a method for obtaining $F(\xi,\eta)$ in the vicinity of the ξ-axis—not just on it—when $y_1(x) = -y_2(x)$. In this event $F(\xi,\eta)$ may be written as

$$\begin{aligned} F(\xi,\eta) &= \int_{-\infty}^{\infty} e^{-j2\pi\alpha\xi}\,d\alpha \int_{-\infty}^{\infty} \mathrm{rect}\left[\frac{\beta}{2y_2(\alpha)}\right] e^{-j2\pi\beta\eta}\,d\beta \\ &= \int_{-\infty}^{\infty} 2y_2(\alpha)\,\mathrm{sinc}\left[2y_2(\alpha)\eta\right] e^{-j2\pi\alpha\xi}\,d\alpha \\ &= \int_{-\infty}^{\infty} \frac{\sin 2\pi y_2(\alpha)\eta}{\pi\eta} e^{-j2\pi\alpha\xi}\,d\alpha, \end{aligned} \tag{9.92}$$

and with the substitution

$$\frac{\sin 2\pi y_2(x)\eta}{\pi\eta} = 2y_2(x) - \frac{4\pi^2}{3}\eta^2\,y_2^{\,3}(x) + \ldots, \tag{9.93}$$

$F(\xi,\eta)$ becomes

$$\begin{aligned} F(\xi,\eta) &= \mathcal{F}\left\{2y_2(x) - \frac{4\pi^2}{3}\eta^2\,y_2^{\,3}(x) + \ldots\right\} \\ &= 2\,Y_2(\xi) - \frac{4\pi^2}{3}\eta^2(j2\pi\xi)^3\,Y_2(\xi) + \ldots \\ &= Y_2(\xi)\left[2 + j\frac{32\pi^5}{3}\eta^2\xi^3 - \ldots\right]. \end{aligned} \tag{9.94}$$

As can be seen, this result is not particularly useful far from the ξ-axis because too many terms of the expansion are required to achieve any sort of accuracy; however, it can be quite helpful in the vicinity of the ξ-axis. There is a corresponding expression that is valid in the vicinity of the η-axis.

Tables of Transform Properties and Pairs

A number of properties of the two-dimensional Fourier transform are summarized in Table 9-1. We have omitted a listing of the symmetry properties of two-dimensional transforms, which are more involved than for the one-dimensional case, but combinations of the first four properties of Table 9-1 allow the various symmetry properties to be determined quickly.

Table 9-1

Properties of Fourier Transforms in Rectangular Coordinates

a_1, a_2, b_1, b_2, c_1, and c_2 real constants

α_1, α_2, β_1, and β_2 real constants — A_1 and A_2 arbitrary constants

x_0, y_0, ξ_0, and η_0 real constants — b and d real nonzero constants

x, y, ξ and η real variables — k and l nonnegative, integers

$g(x,y)=\iint_{-\infty}^{\infty} G(\alpha',\beta')e^{j2\pi(\alpha' x+\beta' y)}d\alpha'\, d\beta'$	$G(\xi,\eta)=\iint_{-\infty}^{\infty} g(\alpha,\beta)e^{-j2\pi(\alpha\xi+\beta\eta)}d\alpha\, d\beta$
$f(\pm x, \pm y)$	$F(\pm\xi, \pm\eta)$
$f^*(\pm x, \pm y)$	$F^*(\mp\xi, \mp\eta)$
$F(\pm x, \pm y)$	$f(\mp\xi, \mp\eta)$
$F^*(\pm x, \pm y)$	$f^*(\pm\xi, \pm\eta)$
$f_1(x)f_2(y)$	$F_1(\xi)F_2(\eta)$
$f_1(x)$	$F_1(\xi)\delta(\eta)$
$f_2(y)$	$\delta(\xi)F_2(\eta)$
$f\left(\frac{x}{b}, \frac{y}{d}\right)$	$\lvert bd\rvert F(b\xi, d\eta)$
$f(x\pm x_0, y\pm y_0)$	$e^{\pm j2\pi x_0\xi}e^{\pm j2\pi y_0\eta}F(\xi,\eta)$
$e^{\pm j2\pi\xi_0 x}e^{\pm j2\pi\eta_0 y}f(x,y)$	$F(\xi\mp\xi_0, \eta\mp\eta_0)$
$e^{\pm j2\pi\xi_0 x}e^{\pm j2\eta_0 y}f\left(\frac{x\pm x_0}{b}, \frac{y\pm y_0}{d}\right)$	$\lvert bd\rvert e^{\pm j2\pi x_0(\xi\mp\xi_0)}e^{\pm j2\pi y_0(\eta\mp\eta_0)}$ $\times F[b(\xi\mp\xi_0), d(\eta\mp\eta_0)]$

Table 9-1 (Continued)

$g(x,y)=\iint_{-\infty}^{\infty} G(\alpha',\beta')e^{j2\pi(\alpha' x+\beta' y)}d\alpha'\,d\beta'$	$G(\xi,\eta)=\iint_{-\infty}^{\infty} g(\alpha,\beta)e^{-j2\pi(\alpha\xi+\beta\eta)}d\alpha\,d\beta$
$f^{(k,l)}(x,y)$	$(j2\pi\xi)^k(j2\pi\eta)^l F(\xi,\eta)$
$(-j2\pi x)^k(-j2\pi y)^l f(x,y)$	$F^{(k,l)}(\xi,\eta)$
$\left(\frac{\partial^2}{\partial x^2}+\frac{\partial^2}{\partial y^2}\right)f(x,y)$	$-4\pi^2(\xi^2+\eta^2)F(\xi,\eta)$
$-4\pi^2(x^2+y^2)f(x,y)$	$\left(\frac{\partial^2}{\partial \xi^2}+\frac{\partial^2}{\partial \eta^2}\right)F(\xi,\eta)$
$\int_{-\infty}^{x}\int_{-\infty}^{\infty} f(\alpha,\beta)d\alpha\,d\beta$	$\frac{1}{2}\left[F(0,0)\delta(\xi)+\frac{1}{j\pi\xi}F(\xi,0)\right]\delta(\eta)$
$\int_{-\infty}^{\infty}\int_{-\infty}^{y} f(\alpha,\beta)d\alpha\,d\beta$	$\frac{1}{2}\delta(\xi)\left[F(0,0)\delta(\eta)+\frac{1}{j\pi\eta}F(0,\eta)\right]$
$\int_{-\infty}^{\infty} f(x,\beta)d\beta$	$F(\xi,0)\delta(\eta)$
$\int_{-\infty}^{\infty} f(\alpha,y)d\alpha$	$\delta(\xi)F(0,\eta)$
$h(x,y)$	$H(\xi,\eta)$
$A_1 f(x,y)+A_2 h(x,y)$	$A_1 F(\xi,\eta)+A_2 H(\xi,\eta)$
$f(x,y)**h(x,y)$	$F(\xi,\eta)H(\xi,\eta)$
$f(x,y)h(x,y)$	$F(\xi,\eta)**H(\xi,\eta)$
$f(x,y)\star\star h(x,y)$	$F(\xi,\eta)H(-\xi,-\eta)$
$f(x,y)h(-x,-y)$	$F(\xi,\eta)\star\star H(\xi,\eta)$
$\gamma_{fh}(x,y)=f(x,y)\star\star h^*(x,y)$	$F(\xi,\eta)H^*(\xi,\eta)$
$f(x,y)h^*(x,y)$	$\gamma_{FH}(\xi,\eta)=F(\xi,\eta)\star\star H^*(\xi,\eta)$
$\gamma_f(x,y)=f(x,y)\star\star f^*(x,y)$	$\lvert F(\xi,\eta)\rvert^2$
$\lvert f(x,y)\rvert^2$	$\gamma_F(\xi,\eta)=F(\xi,\eta)\star\star F^*(\xi,\eta)$

Table 9-1 (Continued)

$g(x,y)=\iint\limits_{-\infty}^{\infty} G(\alpha',\beta')e^{j2\pi(\alpha' x+\beta' y)}d\alpha'\,d\beta'$	$G(\xi,\eta)=\iint\limits_{-\infty}^{\infty} g(\alpha,\beta)e^{-j2\pi(\alpha\xi+\beta\eta)}d\alpha\,d\beta$
$f(a_1x+b_1y+c_1,\ a_2x+b_2y+c_2)$	$\frac{1}{\lvert D\rvert}e^{-j2\pi(x_0\xi+y_0\eta)}F\left(\frac{b_2}{D}\xi-\frac{a_2}{D}\eta,\ -\frac{b_1}{D}\xi+\frac{a_1}{D}\eta\right)$ where: $D=a_1b_2-a_2b_1$, $x_0=\frac{b_1c_2-b_2c_1}{D}$, $y_0=\frac{a_2c_1-a_1c_2}{D}$
$\frac{1}{\lvert D\rvert}e^{j2\pi(\xi_0x+\eta_0y)}f\left(\frac{b_2}{D}x-\frac{a_2}{D}y,\ -\frac{b_1}{D}x+\frac{a_1}{D}y\right)$ where: $D=a_1b_2-a_2b_1$, $\xi_0=\frac{b_1c_2-b_2c_1}{D}$, $\eta_0=\frac{a_2c_1-a_1c_2}{D}$	$F(a_1\xi+b_1\eta+c_1,\ a_2\xi+b_2\eta+c_2)$
$\sum_{n=-\infty}^{\infty}\sum_{m=-\infty}^{\infty} f(x+\alpha_1n+\alpha_2m,\ y+\beta_1n+\beta_2m)$	$\frac{F(\xi,\eta)}{\lvert\Delta\rvert}\sum_{n=-\infty}^{\infty}\sum_{m=-\infty}^{\infty}\delta\left(\xi+\frac{\beta_2}{\Delta}n-\frac{\beta_1}{\Delta}m,\ \eta-\frac{\alpha_2}{\Delta}n+\frac{\alpha_1}{\Delta}m\right)$ where: $\Delta=\alpha_1\beta_2-\alpha_2\beta_1$
$\frac{f(x,y)}{\lvert\Delta\rvert}\sum_{n=-\infty}^{\infty}\sum_{m=-\infty}^{\infty}\delta\left(x+\frac{\beta_2}{\Delta}n-\frac{\beta_1}{\Delta}m,\ y-\frac{\alpha_2}{\Delta}n+\frac{\alpha_1}{\Delta}m\right)$ where: $\Delta=\alpha_1\beta_2-\alpha_2\beta_1$	$\sum_{n=-\infty}^{\infty}\sum_{m=-\infty}^{\infty} F(\xi+\alpha_1n+\alpha_2m,\ \eta+\beta_1n+\beta_2m)$
$\iint\limits_{-\infty}^{\infty} f(\alpha,\beta)d\alpha\,d\beta=F(0,0)$	$f(0,0)=\iint\limits_{-\infty}^{\infty} F(\alpha',\beta')d\alpha'\,d\beta'$

$$\iint\limits_{-\infty}^{\infty} f(\alpha,\beta)h^*(\alpha,\beta)d\alpha\,d\beta=\iint\limits_{-\infty}^{\infty} F(\alpha',\beta')H^*(\alpha',\beta')d\alpha'\,d\beta'$$

$$\iint\limits_{-\infty}^{\infty} \lvert f(\alpha,\beta)\rvert^2d\alpha\,d\beta=\iint\limits_{-\infty}^{\infty} \lvert F(\alpha',\beta')\rvert^2d\alpha'\,d\beta'$$

Table 9-1 (Continued)

If $f(x,y)$ is periodic with $F(\xi,\eta)=\sum_{n=-\infty}^{\infty}\sum_{m=-\infty}^{\infty} c_{nm}\delta(\xi-n\xi_0,\eta-m\eta_0)$

$$\iint_{-\infty}^{\infty} |f(\alpha,\beta)|^2 d\alpha\, d\beta = \sum_{n=-\infty}^{\infty}\sum_{m=-\infty}^{\infty} |c_{nm}|^2$$

$$m_{kl} = \iint_{-\infty}^{\infty} \alpha^k \beta^l f(\alpha,\beta)\, d\alpha\, d\beta = \left.\frac{F^{(k,l)}(\xi,\eta)}{(-j2\pi)^{k+l}}\right|_{\xi=\eta=0} = \frac{F^{(k,l)}(0,0)}{(-j2\pi)^{k+l}}$$

$$m_{00} = \iint_{-\infty}^{\infty} f(\alpha,\beta)\, d\alpha\, d\beta = F^{(0,0)}(\xi,\eta)\big|_{\xi=\eta=0} = F(0,0)$$

$$m_{10} = \frac{F^{(1,0)}(0,0)}{-j2\pi} \qquad \bar{x} = \frac{m_{10}}{m_{00}} = \frac{F^{(1,0)}(0,0)}{-j2\pi F(0,0)}$$

$$m_{01} = \frac{F^{(0,1)}(0,0)}{-j2\pi} \qquad \bar{y} = \frac{m_{01}}{m_{00}} = \frac{F^{(0,1)}(0,0)}{-j2\pi F(0,0)}$$

$$m_{20} = \frac{F^{(2,0)}(0,0)}{-4\pi^2} \qquad \overline{x^2} = \frac{m_{20}}{m_{00}} = \frac{F^{(2,0)}(0,0)}{-4\pi^2 F(0,0)}$$

$$m_{02} = \frac{F^{(0,2)}(0,0)}{-4\pi^2} \qquad \overline{y^2} = \frac{m_{02}}{m_{00}} = \frac{F^{(0,2)}(0,0)}{-4\pi^2 F(0,0)}$$

$$\sigma_x^2 = \overline{x^2} - (\bar{x})^2 \qquad \sigma_y^2 = \overline{y^2} - (\bar{y})^2$$

In Table 9-2 we list a few frequently encountered transform pairs, most of which can be determined immediately with reference to Tables 7-3 and 9-1. You will note that Table 9-2 is not as extensive as Table 7-3 because we felt it was not necessary to list both $\mathfrak{F}\mathfrak{F}\{f(x,y)\}$ and $\mathfrak{F}\mathfrak{F}\{F(x,y)\}$ for most of the functions selected. We do, however, give the transforms of both $f(x,y)$ and $F(x,y)$ when there is a sign reversal that might be difficult to keep track of.

Table 9-2

Elementary Fourier Transform Pairs in Rectangular Coordinates

x_0, y_0, ξ_0 and η_0 real constants — k and l non-negative integers

a and c real constants — x, y, ξ and η real variables

$g(x,y)=\iint_{-\infty}^{\infty} G(\alpha',\beta')e^{j2\pi(\alpha' x+\beta' y)}d\alpha'\,d\beta'$	$G(\xi,\eta)=\iint_{-\infty}^{\infty} g(\alpha,\beta)e^{-j2\pi(\alpha\xi+\beta\eta)}d\alpha\,d\beta$
$\delta(x,y)$	1
$\delta(x\pm x_0, y\pm y_0)$	$e^{\pm j2\pi x_0\xi}e^{\pm j2\pi y_0\eta}$
$e^{\pm j2\pi\xi_0 x}e^{\pm j2\pi\eta_0 y}$	$\delta(\xi\mp\xi_0, \eta\mp\eta_0)$
$\cos(2\pi\xi_0 x)$	$\frac{1}{2\lvert\xi_0\rvert}\delta\delta\left(\frac{\xi}{\xi_0}\right)\delta(\eta)$
$\sin(2\pi\eta_0 y)$	$\frac{j}{2\lvert\eta_0\rvert}\delta(\xi)\delta\delta\left(\frac{\eta}{\eta_0}\right)$
$\mathrm{rect}(x,y)$	$\mathrm{sinc}(\xi,\eta)$
$\mathrm{tri}(x,y)$	$\mathrm{sinc}^2(\xi,\eta)$
$\mathrm{Gaus}(x,y)$	$\mathrm{Gaus}(\xi,\eta)$
$\mathrm{comb}(x,y)$	$\mathrm{comb}(\xi,\eta)$
$x^k y^l$	$\left(\frac{1}{-j2\pi}\right)^{k+l}\delta^{(k,l)}(\xi,\eta)$
$\left(\frac{1}{j2\pi}\right)^{k+l}\delta^{(k,l)}(x,y)$	$\xi^k\eta^l$
$\exp[\pm j\pi(x^2+y^2)]$	$\pm j\exp[\mp j\pi(\xi^2+\eta^2)]$
$\exp\left[-\pi\left(\frac{x^2+y^2}{a+jc}\right)\right], a\geqslant 0, a^2+c^2<\infty$	$(a+jc)\exp\left[-\pi(a+jc)(\xi^2+\eta^2)\right]$

9-4 THE HANKEL TRANSFORM

Functions exhibiting radial symmetry, also called circularly symmetric functions, are frequently encountered in the analysis of optical systems because lenses, aperture stops, etc., are often circular in shape. It will therefore be advantageous for us to investigate the two-dimensional Fourier transform of such functions. Given the circularly symmetric function

$f(x, y) = g\left(\sqrt{x^2+y^2}\right)$, we write its Fourier transform as

$$F(\xi,\eta) = \iint_{-\infty}^{\infty} f(\alpha,\beta)\, e^{-j2\pi(\alpha\xi+\beta\eta)}\, d\alpha\, d\beta$$

$$= \iint_{-\infty}^{\infty} g\left(\sqrt{\alpha^2+\beta^2}\right) e^{-j2\pi(\alpha\xi+\beta\eta)}\, d\alpha\, d\beta. \tag{9.95}$$

We next denote the radial and azimuthal frequency variables by ρ and ϕ, respectively, and with r and ρ nonnegative real variables, we make the following substitutions in Eq. (9.95):

$$x = r\cos\theta, \qquad r = \sqrt{x^2+y^2},$$

$$y = r\sin\theta, \qquad \theta = \tan^{-1}\left(\frac{y}{x}\right),$$

$$\xi = \rho\cos\phi, \qquad \rho = \sqrt{\xi^2+\eta^2},$$

$$\eta = \rho\sin\phi, \qquad \phi = \tan^{-1}\left(\frac{\eta}{\xi}\right). \tag{9.96}$$

The expression for $F(\xi,\eta)$ then becomes

$$F(\rho\cos\phi, \rho\sin\phi) = \int_0^{\infty}\int_0^{2\pi} g(r')\, e^{-j2\pi\rho r'\cos(\theta'-\phi)}\, r'\, dr'\, d\theta', \tag{9.97}$$

where we have used the identity $\cos\theta'\cos\phi + \sin\theta'\sin\phi = \cos(\theta'-\phi)$. Following Goodman (Ref. 9-5), we employ the Bessel-function identity

$$\int_0^{2\pi} e^{-ja\cos(\theta'-\phi)}\, d\theta' = 2\pi J_0(a), \tag{9.98}$$

where $J_0(a)$ is the zero-order Bessel function of the first kind, to obtain

$$F(\rho\cos\phi, \rho\sin\phi) = 2\pi \int_0^{\infty} g(r') J_0(2\pi\rho r')\, r'\, dr'. \tag{9.99}$$

We note that the right-hand side of this equation is independent of ϕ and denote it by some function of ρ alone, say $U(\rho)$. Thus,

$$F(\rho\cos\phi, \rho\sin\phi) = U(\rho). \tag{9.100}$$

Returning to rectangular coordinates momentarily, we find that

$$\begin{aligned} F(\xi,\eta) &= \mathfrak{F}\mathfrak{F}\{f(x,y)\} \\ &= \mathfrak{F}\mathfrak{F}\left\{ g\left(\sqrt{x^2+y^2}\right)\right\} \\ &= U\left(\sqrt{\xi^2+\eta^2}\right), \end{aligned} \tag{9.101}$$

which leads to the observation that the Fourier transform of a circularly symmetric function is also circularly symmetric.

By following a similar procedure, it can be shown that

$$\begin{aligned} f(r\cos\theta, r\sin\theta) &= 2\pi \int_0^\infty U(\rho')J_0(2\pi r\rho')\rho'\,d\rho' \\ &= g(r), \end{aligned} \tag{9.102}$$

which in turn yields

$$\begin{aligned} f(x,y) &= \mathfrak{F}^{-1}\mathfrak{F}^{-1}\{F(\xi,\eta)\} \\ &= \mathfrak{F}^{-1}\mathfrak{F}^{-1}\left\{ U\left(\sqrt{\xi^2+\eta^2}\right)\right\} \\ &= g\left(\sqrt{x^2+y^2}\right). \end{aligned} \tag{9.103}$$

We therefore conclude that $g\left(\sqrt{x^2+y^2}\right)$ and $U\left(\sqrt{\xi^2+\eta^2}\right)$ form a two-dimensional transform pair in the usual way, and we replace U with G to maintain notational consistency, i.e.,

$$g\left(\sqrt{x^2+y^2}\right) \overset{\mathfrak{F}\mathfrak{F}}{\rightarrow} G\left(\sqrt{\xi^2+\eta^2}\right). \tag{9.104}$$

Rewriting Eqs. (9.99) and (9.102) we obtain

$$\begin{aligned} G(\rho) &= 2\pi \int_0^\infty g(r')J_0(2\pi\rho r')r'\,dr', \\ g(r) &= 2\pi \int_0^\infty G(\rho')J_0(2\pi r\rho')\rho'\,d\rho', \end{aligned} \tag{9.105}$$

and in this form $g(r)$ and $G(\rho)$ are known as a *zero-order Hankel transform pair*. The zero-order Hankel transform is simply a special case of the νth order Hankel transform, which we shall now discuss.

The Hankel Transform of Order ν

There are many different forms used to define the Hankel transform (Ref. 9-6); we choose the one that best suits our purposes. Given the function $u(r)$, we define its *Hankel transform of order* ν to be

$$U(\rho;\nu)=2\pi\int_0^{\infty}u(r')J_{\nu}(2\pi\rho r')r'\,dr', \tag{9.106}$$

where J_ν denotes the νth order Bessel function of the first kind. The number ν is unrestricted in general, and may be positive or negative, integer or noninteger, and either real or complex. If Eq. (9.106) is inverted we find

$$u(r)=2\pi\int_0^{\infty}U(\rho';\nu)J_{\nu}(2\pi r\rho')\rho'\,d\rho', \tag{9.107}$$

and it is evident that, unlike the Fourier transform, the Hankel transform is self-reciprocal because the kernels of the transform and inverse transform are identical. For this reason, we refer to both Eqs. (9.106) and (9.107) as Hankel transforms and eliminate references to the inverse operation. The following operational notation will be used to indicate the relationship between $u(r)$ and $U(\rho;\nu)$:

$$\mathcal{H}_{\nu}\{u(r)\}=U(\rho;\nu), \qquad u(r)=\mathcal{H}_{\nu}\{U(\rho;\nu)\},$$

$$u(r)\overset{\mathcal{H}_{\nu}}{\longleftrightarrow}U(\rho;\nu). \tag{9.108}$$

We now present, without derivation, some useful properties and recursion formulas developed from expressions found in Ref. 9-6. Given

$$f(r)\overset{\mathcal{H}_{\nu}}{\longleftrightarrow}F(\rho;\nu), \tag{9.109}$$

we find that

$$f\left(\frac{r}{b}\right)\overset{\mathcal{H}_{\nu}}{\longleftrightarrow}|b|^2F(b\rho;\nu). \tag{9.110}$$

Also, for $k=0,1,2,\ldots,$ we have

$$r^kf(r)\overset{\mathcal{H}_{\nu}}{\longleftrightarrow}\left(\frac{1}{2\pi}\right)^k\frac{1}{\rho^{\nu+k}}\frac{d^k}{d\rho^k}\left[\rho^{\nu+k}F(\rho;\nu+k)\right],$$

$$r^kf(r)\overset{\mathcal{H}_{\nu}}{\longleftrightarrow}\left(\frac{-1}{2\pi}\right)^k\rho^{\nu-k}\frac{d^k}{d\rho^k}\left[\frac{1}{\rho^{\nu-k}}F(\rho;\nu-k)\right]. \tag{9.111}$$

In addition,

$$\frac{f(r)}{r} \overset{\mathcal{H}_\nu}{\longleftrightarrow} 2\pi\rho^{-\nu}\int_0^{2\pi\rho}(\rho')^{\nu}F(\rho';\nu-1)\,d\rho',$$

$$\frac{f(r)}{r} \overset{\mathcal{H}_\nu}{\longleftrightarrow} 2\pi\rho^{\nu}\int_{2\pi\rho}^{\infty}(\rho')^{-\nu}F(\rho';\nu+1)\,d\rho', \tag{9.112}$$

or, for $\nu>0$,

$$\frac{f(r)}{r} \overset{\mathcal{H}_\nu}{\longleftrightarrow} \frac{\pi\rho}{\nu}\left[F(\rho;\nu-1)+F(\rho;\nu+1)\right], \qquad \nu>0. \tag{9.113}$$

Properties of Zero-Order Hankel Transforms

We deal primarily with the Hankel transform of zero order because of its relationship to the Fourier transform, and we therefore list a number of properties associated with it. Most of these properties follow readily from those of the Fourier transform, but a few do not and will be given special attention.

For the sake of brevity, we now alter the previously adopted notation and denote the zero-order Hankel transform of $f(r)$ by simply $F(\rho)$, rather than $F(\rho;0)$, unless the full designation is necessary for clarification. Thus,

$$\mathcal{H}_0\{f(r)\}=F(\rho), \quad \mathcal{H}_0\{F(\rho)\}=f(r), \quad f(r)\overset{\mathcal{H}_0}{\longleftrightarrow}F(\rho). \tag{9.114}$$

Given the transform pair $f(r)$ and $F(\rho)$, we find from Eq. (9.110) that

$$f\left(\frac{r}{b}\right)\overset{\mathcal{H}_0}{\longleftrightarrow}|b|^2F(b\rho), \tag{9.115}$$

or, in terms of the Fourier transform in rectangular coordinates,

$$f\left(\frac{\sqrt{x^2+y^2}}{b}\right)\overset{\mathcal{FF}}{\longrightarrow}|b|^2F\left(b\sqrt{\xi^2+\eta^2}\right). \tag{9.116}$$

If the $x-$ and $y-$coordinates are scaled by different amounts, the resulting function is no longer circularly symmetric; rather, it is elliptically symmetric. As a result, its Fourier transform is elliptically symmetric with the orientation of the major and minor axes interchanged:

$$f\left(\sqrt{\left(\frac{x}{b}\right)^2+\left(\frac{y}{d}\right)^2}\right)\overset{\mathcal{FF}}{\longrightarrow}|bd|F\left(\sqrt{(b\xi)^2+(d\eta)^2}\right). \tag{9.117}$$

When the function $f(r)$ is shifted from the origin, its circular symmetry is destroyed and the Hankel transform cannot be used directly to obtain the Fourier transform. Nevertheless, the shift theorem still holds for the Fourier transform in rectangular coordinates, and we find that

$$f\left(\sqrt{(x\pm x_0)^2+(y\pm y_0)^2}\right)\xrightarrow{\mathcal{F}\mathcal{F}} e^{\pm j2\pi x_0\xi}e^{\pm j2\pi y_0\eta}F\left(\sqrt{\xi^2+\eta^2}\right). \quad (9.118)$$

It is frequently necessary to determine the transform of the product of a function and some power of its argument, and with $\nu=0$, Eqs. (9.111) and (9.112) yield

$$r^k f(r)\overset{\mathcal{H}_0}{\leftrightarrow}\left(\frac{\pm 1}{2\pi\rho}\right)^k\frac{d^k}{d\rho^k}\left[\rho^k F(\rho;\pm k)\right], \qquad k=0,1,2\ldots,$$

$$\frac{f(r)}{r}\overset{\mathcal{H}_0}{\leftrightarrow}2\pi\int_0^{2\pi\rho}F(\rho';-1)\,d\rho',$$

$$\frac{f(r)}{r}\overset{\mathcal{H}_0}{\leftrightarrow}2\pi\int_{2\pi\rho}^{\infty}F(\rho';1)\,d\rho'. \quad (9.119)$$

Also, Eq. (9.113) may be rearranged as follows: with $\nu=1$ and $F(\rho;0)=F(\rho)$,

$$F(\rho)=\frac{1}{\pi\rho}\mathcal{H}_1\left\{\frac{f(r)}{r}\right\}-F(\rho;2). \quad (9.120)$$

We see that in order to use the relationships of Eqs. (9.119) and (9.120), we must know certain higher-order Hankel transforms. A few such transforms are listed in Table 9-5, and an extensive list may be found in Ref. 6-6. In addition, many useful formulas appear in Ref. 9-7.

It can be shown that, with $r=\sqrt{x^2+y^2}$,

$$\left[\frac{\partial^2}{\partial x^2}+\frac{\partial^2}{\partial y^2}\right]f\left(\sqrt{x^2+y^2}\right)=\frac{1}{r}\frac{d}{dr}\left[r\frac{df(r)}{dr}\right]$$

$$=f^{(2)}(r)+\frac{1}{r}f^{(1)}(r). \quad (9.121)$$

Consequently, one of the differentiation properties of the Fourier transform may be extended to the Hankel transform:

$$-4\pi^2 r^2 f(r) \overset{\mathcal{H}_0}{\leftrightarrow} F^{(2)}(\rho) + \frac{1}{\rho} F^{(1)}(\rho)$$

$$f^{(2)}(r) + \frac{1}{r} f^{(1)}(r) \overset{\mathcal{H}_0}{\leftrightarrow} -4\pi^2 \rho^2 F(\rho). \tag{9.122}$$

Moments of f(r)

Many authors define the kth moment of the function $f(r)$ to be

$$m_k = \int_0^\infty r^k f(r) dr, \tag{9.123}$$

and although this is no doubt a useful definition for certain purposes, it can also be very misleading! To illustrate, the m_0 of Eq. (9.123) is merely equal to the area of a one-dimensional function, $f(r)$, for $0 \leqslant r \leqslant \infty$; it does not give the volume of the two-dimensional $f(r)$ over the entire r, θ plane. In other words, the moments of Eq. (9.123) are associated with properties of a one-dimensional curve rather than a two-dimensional surface. Because we are primarily interested in the latter, we shall define the moments of $f(r)$ in a fashion similar to that used in Sec. 9-3, i.e.,

$$m_{kl} = \int_0^\infty \int_0^{2\pi} (r' \cos\theta')^k (r' \sin\theta')^l f(r') r' \, dr' \, d\theta'. \tag{9.124}$$

With this definition we have

$$\begin{aligned} m_{00} &= \int_0^\infty \int_0^{2\pi} f(r') r' \, dr' \, d\theta' \\ &= 2\pi \int_0^\infty f(r') r' \, dr' \\ &= F(0), \end{aligned} \tag{9.125}$$

$$\begin{aligned} m_{10} &= \int_0^\infty \int_0^{2\pi} r' \cos\theta' f(r') r' \, dr' \, d\theta' \\ &= \frac{1}{2\pi} F(0) \int_0^{2\pi} \cos\theta' \, d\theta' \\ &= 0, \end{aligned} \tag{9.126}$$

and

$$m_{20} = \int_0^\infty \int_0^{2\pi} r'^2 \cos^2\theta' f(r') r'\, dr'\, d\theta'$$

$$= \int_0^\infty \int_0^{2\pi} r'^2 \left[\tfrac{1}{2} + \tfrac{1}{2}\cos 2\theta'\right] f(r') r'\, dr'\, d\theta'$$

$$= 2\pi \int_0^\infty \tfrac{1}{2} r'^2 f(r') r'\, dr'$$

$$= \tfrac{1}{2}\mathcal{H}_0\{r^2 f(r)\}\big|_{\rho=0}$$

$$= \frac{-1}{8\pi^2}\left[F^{(2)}(\rho) + \frac{1}{\rho}F^{(1)}(\rho)\right]\Bigg|_{\rho=0}. \tag{9.127}$$

The moment m_{00} clearly has a value equal to the volume of $f(r)$, and the centroid $\bar{r} = m_{10}/m_{00} = m_{01}/m_{00}$ is zero as it should be for a circularly symmetric function. The mean-square radius is given by $\overline{r^2} = m_{20}/m_{00} = m_{02}/m_{00}$, and the variance by $\sigma_r^2 = \overline{r^2} - (\bar{r})^2 = \overline{r^2}$. Thus

$$m_{00} = F(0)$$

$$\bar{r} = 0$$

$$\sigma_r^2 = \frac{-1}{8\pi^2 F(0)}\left[F^{(2)}(\rho) + \frac{1}{\rho}F^{(1)}(\rho)\right]\Bigg|_{\rho=0}. \tag{9.128}$$

In the paragraphs to follow we derive the zero-order Hankel transforms of a number of frequently encountered functions.

$\mathcal{H}_0\{$***somb***$(r)\}$

From the relationship (Ref. 9-6)

$$\frac{J_{\nu+1}(ar)}{r} \overset{\mathcal{H}_\nu}{\leftrightarrow} \left(\frac{2\pi}{a}\right)^{\nu+1} \rho^\nu \operatorname{cyl}\left(\frac{\pi\rho}{a}\right), \qquad a>0, \qquad \operatorname{Re}\{\nu\} > -\tfrac{3}{2} \tag{9.129}$$

and with $\nu = 0$ and $a = \pi$, we find that

$$\frac{J_1(\pi r)}{r} \overset{\mathcal{H}_0}{\leftrightarrow} 2\operatorname{cyl}(\rho). \tag{9.130}$$

Therefore, with somb(r) as defined in Eq. (3.63), we obtain

$$\mathrm{somb}(r) \overset{\mathcal{H}_0}{\leftrightarrow} \frac{4}{\pi}\,\mathrm{cyl}(\rho),$$

$$\mathrm{cyl}(r) \overset{\mathcal{H}_0}{\leftrightarrow} \frac{\pi}{4}\,\mathrm{somb}(\rho). \tag{9.131}$$

In the latter expression, note that the central ordinate of $\mathcal{H}_0\{\mathrm{cyl}(r)\}$ has a value equal to the volume of cyl(r), as it should.

$\mathcal{H}_0\{\delta(\mathbf{r})\}$

With $\delta(\mathbf{r}) = \delta(r)/\pi r$,

$$\begin{aligned}\mathcal{H}_0\{\delta(\mathbf{r})\} &= 2\pi\int_0^\infty \frac{\delta(r')}{\pi r'} J_0(2\pi\rho r')r'\,dr' \\ &= 2\int_0^\infty \delta(r')J_0(2\pi\rho r')dr' \\ &= 1,\end{aligned} \tag{9.132}$$

where we used Eq. (9.50) to evaluate the last integral.

$\mathcal{H}_0\{1/r\}$

From the relationship (Ref. 9-7)

$$\int_0^\infty J_\nu(ar')\,dr' = \frac{1}{a}, \qquad a>0, \qquad \mathrm{Re}\{\nu\} > -1 \tag{9.133}$$

and with $a = 2\pi\rho$ we have

$$\begin{aligned}\mathcal{H}_\nu\left\{\frac{1}{r}\right\} &= 2\pi\int_0^\infty \frac{1}{r'} J_\nu(2\pi\rho r')r'\,dr' \\ &= \frac{1}{\rho}.\end{aligned} \tag{9.134}$$

Thus, as long as $\mathrm{Re}\{\nu\} > -1$, the νth order transform of r^{-1} is simply ρ^{-1}. Included, of course, is the zero-order transform:

$$\frac{1}{r} \overset{\mathcal{H}_0}{\leftrightarrow} \frac{1}{\rho}. \tag{9.135}$$

$\mathcal{H}_0\{J_0(ar)J_0(cr)\}$

For $a>0$, $c>0$ and $\text{Re}\{\nu\}>-1$, we find (Ref. 9-6)

$$\frac{J_\nu(ar)J_\nu(cr)}{r^\nu} \overset{\mathcal{H}_\nu}{\leftrightarrow} \frac{2^{2-4\nu}}{\pi^{\nu-1/2}\Gamma\left(\nu+\frac{1}{2}\right)}$$

$$\times\left[\frac{\left[4\pi^2\rho^2-(a-c)^2\right]^{\nu-1/2}\left[(a+c)^2-4\pi^2\rho^2\right]^{\nu-1/2}}{(ac\rho)^\nu}\right]$$

$$\times\left[\text{cyl}\left(\frac{\pi\rho}{a+c}\right)-\text{cyl}\left(\frac{\pi\rho}{|a-c|}\right)\right], \tag{9.136}$$

where $\Gamma(z)$ is the Gamma function (see Table 9-6 and Ref. 9-7). If we then set $\nu=0$, we obtain

$$J_0(ar)J_0(cr) \overset{\mathcal{H}_0}{\leftrightarrow} 4\left[4\pi^2\rho^2-(a-c)^2\right]^{-1/2}\left[(a+c)^2-4\pi^2\rho^2\right]^{-1/2}$$

$$\times\left[\text{cyl}\left(\frac{\pi\rho}{a+c}\right)-\text{cyl}\left(\frac{\pi\rho}{|a-c|}\right)\right], \tag{9.137}$$

and with $a=c$,

$$J_0^2(ar) \overset{\mathcal{H}_0}{\leftrightarrow} \frac{1}{\pi\rho}(a^2-4\pi^2\rho^2)^{-1/2}\text{cyl}\left(\frac{\pi\rho}{2a}\right). \tag{9.138}$$

$\mathcal{H}_0\{r^{-\mu}J_\mu(ar)\}$

With $a>0$ and $-1<\text{Re}\{\nu\}<\text{Re}\{\mu\}$, and the relationship (Ref. 9-6)

$$r^{\nu-\mu}J_\mu(ar) \overset{\mathcal{H}_\nu}{\leftrightarrow} \frac{\pi^{\nu+1}2^{2\nu+2-\mu}\rho^\nu}{a^\mu\Gamma(\mu-\nu)}(a^2-4\pi^2\rho^2)^{\mu-\nu-1}\text{cyl}\left(\frac{\pi\rho}{a}\right), \tag{9.139}$$

we set $\nu=0$ to obtain

$$\frac{J_\mu(ar)}{r^\mu} \overset{\mathcal{H}_0}{\leftrightarrow} \frac{\pi 2^{2-\mu}}{a^\mu\Gamma(\mu)}(a^2-4\pi^2\rho^2)^{\mu-1}\text{cyl}\left(\frac{\pi\rho}{a}\right). \tag{9.140}$$

Tables of Hankel Transform Properties and Pairs

We list several properties of the zero-order Hankel transform in Table 9-3, and follow this with a number of elementary zero-order Hankel transform pairs in Table 9-4. Then, as an aid in utilizing certain relationships requiring higher-order transforms, Table 9-5 provides a short list of such transforms. In addition, a few elementary properties of Bessel functions and Gamma functions are given in Table 9-6 to help in the evaluation of certain of the transforms.

Table 9-3

Properties of Zero-Order Hankel Transforms

x_0, y_0, ξ_0, and η_0 real constants
b and d nonzero real constants
r and ρ nonnegative real variables

A_1 and A_2 arbitrary constants
k and l nonnegative integers

$g(r)=2\pi\int_0^\infty G(\rho')J_0(2\pi r\rho')\rho'\,d\rho'$	$G(\rho)=2\pi\int_0^\infty g(r')J_0(2\pi\rho r')r'\,dr'$
$f(r)$	$F(\rho)$
$f\left(\frac{r}{b}\right)$	$\lvert b\rvert^2F(b\rho)$
$r^kf(r)$	$\left(\frac{\pm 1}{2\pi\rho}\right)^k\frac{d^k}{d\rho^k}\left[\rho^kF(\rho;\pm k)\right]$
$\frac{f(r)}{r}$	$2\pi\int_0^{2\pi\rho}F(\rho';-1)d\rho'$
$\frac{f(r)}{r}$	$2\pi\int_{2\pi\rho}^\infty F(\rho';1)\,d\rho'$
$f(r)$	$\frac{1}{\pi\rho}\mathcal{H}_1\left\{\frac{f(r)}{r}\right\}-F(\rho;2)$
$-4\pi^2r^2f(r)$	$F^{(2)}(\rho)+\frac{1}{\rho}F^{(1)}(\rho)$
$f^{(2)}(r)+\frac{1}{r}f^{(1)}(r)$	$-4\pi^2\rho^2F(\rho)$
$h(r)$	$H(\rho)$
$A_1f(r)+A_2h(r)$	$A_1F(\rho)+A_2H(\rho)$
$f(r)**h(r)=f(r)\star\star h(r)$	$F(\rho)H(\rho)$
$f(r)h(r)$	$F(\rho)**H(\rho)=F(\rho)\star\star H(\rho)$
$\gamma_{fh}(r)=f(r)\star\star h^*(r)$	$F(\rho)H^*(\rho)$

Table 9-3 (Continued)

$g(r)=2\pi\int_0^\infty G(\rho')J_0(2\pi r\rho')\rho'\,d\rho'$	$G(\rho)=2\pi\int_0^\infty g(r')J_0(2\pi\rho r')r'\,dr'$
$f(r)h^*(r)$	$\gamma_{FH}(\rho)=F(\rho)\star\star H^*(\rho)$
$\gamma_f(r)=f(r)\star\star f^*(r)$	$\lvert F(\rho)\rvert^2$
$\lvert f(r)\rvert^2$	$\gamma_F(\rho)=F(\rho)\star\star F^*(\rho)$
$g\left(\sqrt{x^2+y^2}\right)=\mathcal{F}\mathcal{F}\left\{G\left(\sqrt{\xi^2+\eta^2}\right)\right\}$	$G\left(\sqrt{\xi^2+\eta^2}\right)=\mathcal{F}^{-1}\mathcal{F}^{-1}\left\{f\left(\sqrt{x^2+y^2}\right)\right\}$
$f\left(\sqrt{x^2+y^2}\right)$	$F\left(\sqrt{\xi^2+\eta^2}\right)$
$f\left(\sqrt{\left(\frac{x}{b}\right)^2+\left(\frac{y}{d}\right)^2}\right)$	$\lvert bd\rvert F\left(\sqrt{(b\xi)^2+(d\eta)^2}\right)$
$f\left(\sqrt{(x\pm x_0)^2+(y\pm y_0)^2}\right)$	$e^{\pm j2\pi x_0\xi}e^{\pm j2\pi y_0\eta}F\left(\sqrt{\xi^2+\eta^2}\right)$
$e^{\pm j2\pi\xi_0 x}e^{\pm j2\pi\eta_0 y}f\left(\sqrt{x^2+y^2}\right)$	$F\left(\sqrt{(\xi\mp\xi_0)^2+(\eta\mp\eta_0)^2}\right)$
$\int_{-\infty}^{x}\int_{-\infty}^{\infty}f\left(\sqrt{\alpha^2+\beta^2}\right)d\alpha\,d\beta$	$\frac{1}{2}\left[F(0)\delta(\xi)+\frac{1}{j\pi\xi}F(\xi)\right]\delta(\eta)$
$\int_{-\infty}^{\infty}\int_{-\infty}^{y}f\left(\sqrt{\alpha^2+\beta^2}\right)d\alpha\,d\beta$	$\frac{1}{2}\delta(\xi)\left[F(0)\delta(\eta)+\frac{1}{j\pi\eta}F(\eta)\right]$
$\int_{-\infty}^{\infty}f\left(\sqrt{x^2+\beta^2}\right)d\beta$	$F(\xi)\delta(\eta)$
$\int_{-\infty}^{\infty}f\left(\sqrt{\alpha^2+y^2}\right)d\alpha$	$\delta(\xi)F(\eta)$

$$\int_0^\infty f(r')r'\,dr'=F(0)\qquad f(0)=\int_0^\infty F(\rho')\rho'\,d\rho'$$

$$\int_0^\infty f(r')\,g^*(r')r'\,dr'=\int_0^\infty F(\rho')\,G^*(\rho')\rho'\,d\rho'$$

$$\int_0^\infty \lvert f(r')\rvert^2 r'\,dr'=\int_0^\infty \lvert F(\rho')\rvert^2\rho'\,d\rho'$$

$$m_{kl}=\int_0^\infty\int_0^{2\pi}(r'\cos\theta')^k(r'\sin\theta')^l f(r')r'\,dr'\,d\theta'$$

$$m_{00}=F(0),\qquad m_{10}=m_{01}=0$$

$$m_{20}=m_{02}=\frac{-1}{8\pi^2}\left[F^{(2)}(\rho)+\frac{1}{\rho}F^{(1)}(\rho)\right]\Bigg|_{\rho=0}$$

$$\bar{r}=\frac{m_{10}}{m_{00}}=0,\qquad \sigma_r^2=\overline{r^2}=\frac{m_{20}}{m_{00}}$$

Table 9-4

Elementary Zero-Order Hankel Transform Pairs

a, c, and r_0 real constants | μ—a constant, $\operatorname{Re}\{\mu\}>0$
d—a nonzero real constant | r and ρ nonnegative real variables

$g(r)=2\pi\int_0^\infty G(\rho')J_0(2\pi r\rho')\rho'\,d\rho'$	$G(\rho)=2\pi\int_0^\infty g(r')J_0(2\pi\rho r')r'\,dr'$
$\dfrac{\delta(r)}{\pi r}$	1
$\delta(r-r_0), \quad r_0>0$	$2\pi r_0 J_0(2\pi r_0\rho)$
$\dfrac{1}{r}$	$\dfrac{1}{\rho}$
$\operatorname{cyl}(r)$	$\dfrac{\pi}{4}\operatorname{somb}(\rho)$
e^{-r}	$\dfrac{2\pi}{(4\pi^2\rho^2+1)^{3/2}}$
$\operatorname{Gaus}(r)$	$\operatorname{Gaus}(\rho)$
$\cos(\pi r^2)$	$\sin(\pi\rho^2)$
$e^{\pm j\pi r^2}$	$\pm je^{\mp j\pi\rho^2}$
$\exp\left\{-\pi\left(\dfrac{r^2}{a+jc}\right)\right\}, \quad a\geqslant 0,\ a^2+c^2<\infty$	$(a+jc)e^{-\pi(a+jc)\rho^2}$
$\operatorname{sinc}(r)$	$\dfrac{2\operatorname{cyl}(\rho)}{\pi\sqrt{1-4\rho^2}}$
$\dfrac{\cos(\pi r)}{\pi r}$	$\dfrac{2[1-\operatorname{cyl}(\rho)]}{\pi\sqrt{4\rho^2-1}}$
$\dfrac{\operatorname{cyl}(r)}{r}$	$\pi J_0(\pi\rho)+\dfrac{\pi^2}{2}[J_1(\pi\rho)\mathbf{H}_0(\pi\rho)-J_0(\pi\rho)\mathbf{H}_1(\pi\rho)]^{\dagger}$
$\dfrac{1}{r}[1-\operatorname{cyl}(r)]$	$\dfrac{1}{\rho}-\mathcal{H}_0\left\{\dfrac{\operatorname{cyl}(r)}{r}\right\}$

Table 9-4 (Continued)

$g(r)=2\pi\int_0^\infty G(\rho')J_0(2\pi r\rho')\rho'\,d\rho'$	$G(\rho)=2\pi\int_0^\infty g(r')J_0(2\pi\rho r')r'\,dr'$
$\dfrac{e^{-r}}{r}$	$\dfrac{2\pi}{\sqrt{4\pi^2\rho^2+1}}$
$r^2\,\mathrm{Gaus}(r)$	$\left[\dfrac{1}{\pi}-\rho^2\right]\mathrm{Gaus}(\rho)$
$J_0(ar)J_0(cr), \quad a, c>0$	$4\,[4\pi^2\rho^2-(a-c)^2]^{-1/2}[(a+c)^2-4\pi^2\rho^2]^{-1/2}$ $\times\left[\mathrm{cyl}\left(\dfrac{\pi\rho}{a+c}\right)-\mathrm{cyl}\left(\dfrac{\pi\rho}{\lvert a-c\rvert}\right)\right]$
$J_0^2(ar), \quad a>0$	$\dfrac{1}{\pi\rho\sqrt{a^2-\pi^2\rho^2}}\,\mathrm{cyl}\left(\dfrac{\pi\rho}{2a}\right)$
$\dfrac{J_0(ar)J_1(ar)}{r}, \quad a>0$	$\dfrac{2}{a}\cos^{-1}\left(\dfrac{\pi\rho}{a}\right)\mathrm{cyl}\left(\dfrac{\pi\rho}{2a}\right)$
$\dfrac{J_\mu(ar)}{r^\mu}, \quad a>0,\ \mathrm{Re}\{\mu\}>0$	$\dfrac{\pi 2^{2-\mu}}{a^\mu\Gamma(\mu)}(a^2-4\pi^2\rho^2)^{\mu-1}\,\mathrm{cyl}\left(\dfrac{\pi\rho}{a}\right)$
$\mathrm{cyl}(r)**\mathrm{cyl}(r)$	$\dfrac{\pi^2}{16}\,\mathrm{somb}^2(\rho)$
$\mathrm{cyl}\left(\dfrac{r}{d}\right)**\mathrm{cyl}\left(\dfrac{r}{ad}\right), \quad a\geqslant 1$	$\dfrac{a^2d^4\pi^2}{16}\,\mathrm{somb}(d\rho)\,\mathrm{somb}(ad\rho)$
$\mathrm{somb}(r)**\mathrm{somb}(r)$	$\dfrac{16}{\pi^2}\,\mathrm{cyl}(\rho)$
$\mathrm{somb}\left(\dfrac{r}{d}\right)**\mathrm{somb}\left(\dfrac{r}{ad}\right), \quad a\geqslant 1$	$\dfrac{16a^2d^4}{\pi^2}\,\mathrm{cyl}(ad\rho)$
$\gamma_{\mathrm{cyl}}(r;1)$	$\dfrac{\pi}{4}\,\mathrm{somb}^2(\rho)$
$\gamma_{\mathrm{cyl}}\left(\dfrac{r}{d};1\right)$	$\dfrac{\pi d^2}{4}\,\mathrm{somb}^2(d\rho)$
$\gamma_{\mathrm{cyl}}\left(\dfrac{r}{d};a\right)$	$\dfrac{\pi a^2d^2}{4}\,\mathrm{somb}(d\rho)\,\mathrm{somb}(ad\rho)$

†$\mathbf{H}_0$ and $\mathbf{H}_1$ are Struve functions, see Ref. 9-7.

Table 9-5

Elementary νth Order Hankel Transform Pairs

a, c, and r_0 real constants
n–a positive integer

r and ρ nonnegative real variables
ν–a complex parameter

$g(r)=2\pi\int_0^\infty G(\rho')J_\nu(2\pi r\rho')\rho'\,d\rho'$	$G(\rho)=2\pi\int_0^\infty g(r')J_\nu(2\pi\rho r')r'\,dr'$
$\delta(r-r_0)$	$2\pi r_0 J_\nu(2\pi r_0\rho)$
$\frac{1}{r}$, $\quad \mathrm{Re}\{\nu\}>-1$	$\frac{1}{\rho}$
$r^\nu \,\mathrm{cyl}(r)$, $\quad \mathrm{Re}\{\nu\}>-1$	$\frac{1}{2^{\nu+1}\rho}J_{\nu+1}(\pi\rho)$
$\frac{\mathrm{cyl}(r)}{r^\nu}$	$\frac{\pi^{\nu-1}\rho^{\nu-2}}{\Gamma(\nu)}-\frac{2^{\nu-1}}{\rho}J_{\nu-1}(\pi\rho)$
$r^\nu e^{\pm j\pi r^2}$, $\quad -1<\mathrm{Re}\{\nu\}<\frac{1}{2}$	$\rho^\nu \exp\left[\pm j(\nu+1)\frac{\pi}{2}\right]\exp(\mp j\pi\rho^2)$
$\frac{J_\nu(ar)J_\nu(cr)}{r^\nu}$, $\quad a,c>0$, $\mathrm{Re}\{\nu\}>-1$	$\frac{2^{2-4\nu}\left[4\pi^2\rho^2-(a-c)^2\right]^{\nu-1/2}\left[(a+c)^2-4\pi^2\rho^2\right]^{\nu-1/2}}{\pi^{\nu-1/2}\Gamma\left(\nu+\frac{1}{2}\right)(ac\rho)^\nu}$ $\times\left[\mathrm{cyl}\left(\frac{\pi\rho}{a+b}\right)-\mathrm{cyl}\left(\frac{\pi\rho}{\lvert a-c\rvert}\right)\right]$
${}_{(\nu+n)/2}\left(\frac{ar}{2}\right)J_{(\nu-n)/2}\left(\frac{ar}{2}\right)$, $\quad a>0$, $\mathrm{Re}\{\nu\}>-1$	$\frac{2}{\pi\rho\sqrt{a^2-4\pi^2\rho^2}}\cos\left(n\cos^{-1}\frac{2\pi\rho}{a}\right)\mathrm{cyl}\left(\frac{\pi\rho}{a}\right)$
$\frac{J_{\nu+1}(ar)}{r}$, $\quad a>0$, $\mathrm{Re}\{\nu\}>-\frac{3}{2}$	$\left(\frac{2\pi}{a}\right)^{\nu+1}\rho^\nu\,\mathrm{cyl}\left(\frac{\pi\rho}{a}\right)$

Table 9-6

Properties of Bessel Functions and Gamma Functions

a—a positive real constant	x—a real variable
k—a nonnegative integer	z—a complex variable
n—a positive integer	ν—a complex parameter

$J_0(0)=1$	$\frac{d}{dz}J_0(z)=-J_1(z)$
$J_n(0)=0$	$\frac{d}{dz}zJ_1(z)=zJ_0(z)$
$J_{-1}(z)=-J_1(z)$	$\frac{d}{dz}z^{\nu}J_{\nu}(z)=z^{\nu}J_{\nu-1}(z)$
$J_{-2}(z)=J_2(z)$	$\frac{d}{dz}z^{-\nu}J_{\nu}(z)=-z^{-\nu}J_{\nu+1}(z)$
$J_{-n}(z)=(-1)^nJ_n(z)$	$\left(\frac{d}{zdz}\right)^k z^{\nu}J_{\nu}(z)=z^{\nu-k}J_{\nu-k}(z)$
$\int_0^{\infty}J_{\nu}(ax)\,dx=\frac{1}{a},\quad \operatorname{Re}\{\nu\}>-1$	$\left(\frac{d}{zdz}\right)^k z^{-\nu}J_{\nu}(z)=(-1)^k z^{-\nu-k}J_{\nu+k}(z)$
$\int_0^{a}J_1(x)\,dx=1-J_0(a)$	$z\frac{d}{dz}J_{\nu}(z)+\nu J_{\nu}(z)=zJ_{\nu-1}(z)$
$\int_a^{\infty}J_1(x)\,dx=J_0(a)$	$z\frac{d}{dz}J_{\nu}(z)-\nu J_{\nu}(z)=-zJ_{\nu+1}(z)$
$\Gamma(1)=\Gamma(2)=1$	$\Gamma(n)=(n-1)!$
$\Gamma(0)=\Gamma(-1)=\infty$	$\Gamma(n+\frac{1}{2})=\frac{\sqrt{\pi}}{2^n}[1\cdot 3\cdots(2n-1)]$
$\Gamma(\frac{1}{2})=\sqrt{\pi}$	$\Gamma(n-\frac{1}{2})=\frac{(-1)^n 2^n\sqrt{\pi}}{1\cdot 3\cdots(2n-1)}$
$\Gamma(-\frac{1}{2})=-2\sqrt{\pi}$	$\Gamma(x+1)=x\Gamma(x)$
$\Gamma(\frac{3}{2})=\frac{\sqrt{\pi}}{2}$	$\Gamma(\frac{1}{2}+x)\Gamma(\frac{1}{2}-x)=\frac{\pi}{\cos\pi x}$
$\Gamma(-\frac{3}{2})=\frac{4\sqrt{\pi}}{3}$	$\Gamma(1-x)\Gamma(x)=\frac{\pi}{\sin\pi x}$

It should also be pointed out that a wealth of information regarding Hankel transforms, Bessel functions, etc., is contained in Refs. 9-1, 9-6, 9-7, and 9-8, and you are advised to consult them if the information presented here is inadequate. Also, graphs and tables of numerical values may be found in Refs. 9-9 and 9-10 for a number of the more common Bessel functions.

9-5 DETERMINATION OF TRANSFORMS BY NUMERICAL METHODS

Our previous discussions of Fourier and Hankel transforms have been based on the assumption that all functions of interest can be expressed explicitly by various well-known mathematical formulas; however, this is often not the case. For example, it may be necessary to calculate the Fourier transform of a function defined only by an experimentally acquired data record, and when such a situation is encountered, numerical techniques may be required. Although the subject of numerical transform computation is beyond the scope of this book, we briefly mention a few of the important aspects.

In calculating the Fourier transform of a function by numerical methods, a digital computer performs complex multiplication and addition operations on the sampled values of the function presented to it. The nature of the problem dictates that this function be regarded as a finite segment of some periodic function, and that the length of the segment be equal to the period of the periodic function [see Eq. (4.32)]. Consequently, the resulting Fourier transform consists of a set of Fourier series coefficients, often referred to as a *discrete Fourier transform*. Because of the computer's limited data handling capabilities, both the number of sampled values presented to it and the number of coefficients it calculates must be finite. However, it may not be possible to accurately represent the associated periodic function by a finite number of Fourier components, and the net effect is to truncate the Fourier series before the terms become insignificant. This effect may also be regarded as a consequence of undersampling, or aliasing, and can be minimized with the proper selection of sampling interval, etc.

When the number of sampled values presented to the computer is large, the computation time—and the cost—can also be quite large if direct computation methods are employed. For this reason, a number of algorithms have been developed that significantly reduce both the time and expense of numerical transform calculations (see, for example, Refs. 9-11 and 9-12). One such algorithm, now known as the *fast Fourier transform algorithm*, plays a vital role in current numerical computation schemes. To illustrate, we compare the number of calculations required by the direct method with that required by the fast Fourier transform algorithm in determining the transform of a two-dimensional function represented by an $N \times N$ array of its sampled values. The direct computation method requires N^2 complex multiplications and $N(N-1)$ complex additions, whereas, with $N=2^k$, the fast Fourier transform algorithm requires only $kN/2$ complex multiplications and kN complex additions (Ref. 9-12). If we assume that the computation time, and thus the cost, is proportional to the number of complex multiplications required, we find that the reduction is significant even for modest values of k; e.g., with $N=64$, $k=6$ and the ratio of direct-to-fast Fourier-transform computing time is

$$\frac{N^2}{\left(\frac{kN}{2}\right)} = \frac{2N}{k} = \frac{128}{6} = 21.3. \tag{9.141}$$

This tremendous improvement derives from the fact that the fast Fourier-transform algorithm allows an $N \times N$ matrix to be factored into k matrices, also $N \times N$, in such a fashion that each of the factored matrices requires a minimum number of complex multiplications and additions.

In this brief discussion we have not even scratched the surface of the topic of numerical transform calculation, but a more detailed investigation does not suit the purposes of the present text. For those interested in studying the subject further, Ref. 9-12 is recommended.

9-6 TWO-DIMENSIONAL LINEAR SHIFT-INVARIANT SYSTEMS

The definitions of linearity and shift invariance for two-dimensional systems are simply extensions of those presented previously for one-dimensional systems (see Chapter 5). To illustrate, let us describe a two-dimensional system by the operator $\mathcal{S}\{\ \}$ and denote the output of this system by $g_i(x, y)$ when the input is $f_i(x, y)$, i.e.,

$$\mathcal{S}\{f_i(x,y)\} = g_i(x,y). \tag{9.142}$$

The system is said to be *linear* if, for an input $[A_1 f_1(x, y) + A_2 f_2(x, y)]$, the output is given by

$$\mathcal{S}\{A_1 f_1(x,y) + A_2 f_2(x,y)\} = A_1 g_1(x,y) + A_2 g_2(x,y), \tag{9.143}$$

where A_1 and A_2 are arbitrary constants. We see, then, that the principle of superposition holds for linear systems irrespective of their dimensionality.

The above system is called *shift invariant* (stationary, isoplanatic, etc.) if, for an input of $f_i(x-x_0, y-y_0)$, the output can be written

$$\mathcal{S}\{f_i(x-x_0,y-y_0)\}=g_i(x-x_0,y-y_0), \tag{9.144}$$

where x_0 and y_0 are real constants. In other words, if the input is shifted in the x- and y-directions, the output is shifted by the same amounts in x and y but is otherwise unaltered. As before, we shall use the operator $\mathcal{L}\{\ \}$ to denote systems that are both linear and shift invariant.

We shall assume that most of the two-dimensional systems of interest to us are LSI systems, and, although this is clearly an idealization, it can greatly simplify the treatment of many problems while still yielding surprisingly accurate results. In addition, we shall assume that these systems are noncausal systems with memory (see Chap. 5). Finally, we point out that all of the results of Chap. 8 regarding the characteristics and applications of LSI filters may be extended to include two-dimensional systems.

The Impulse Response

With the operator $\mathcal{L}\{\ \}$ describing an arbitrary two-dimensional LSI system, the impulse response of the system is defined by

$$\mathcal{L}\{\delta(x,y)\}=h(x,y), \tag{9.145}$$

i.e., it is the response of the system to an impulsive input applied at the origin. Because the system is shift invariant, the response to an impulsive input located at the point (x_0, y_0) is simply a shifted version of the impulse response $h(x, y)$; that is,

$$\mathcal{L}\{\delta(x-x_0,y-y_0)\}=h(x-x_0,y-y_0). \tag{9.146}$$

In our studies of diffraction, $h(x-x_0, y-y_0)$ might be used to describe the x and y variations of the wave field produced across some plane by a point source (impulse) of light located at the point (x_0, y_0) in another plane some distance away. When dealing with an imaging system, $h(x-x_0, y-y_0)$ might be used to denote the irradiance distribution at the image plane when the object is an ideal point source located at the point (x_0, y_0). In this latter case, the impulse response is also referred to as the *point spread function* of the imaging system. (As we shall see later, diffraction within the imaging system prevents the image of a point source from being an ideal point image.)

We know that for an arbitrary input signal $f(x, y)$, the output of an LSI system is given by

$$g(x,y)=f(x,y)**h(x,y), \tag{9.147}$$

which may be regarded as a superposition of appropriately weighted and shifted impulse responses (see Sec. 6-4). For an imaging system, we may consider the object to be composed of a weighted set of point sources, each of which produces its own image. As a result, the overall image may be thought of as a superposition of point spread functions, each located and weighted according to the position and intensity of the associated point-source component of the object. This topic will be covered in much greater detail in the chapter on image-forming systems.

In many optical systems, the impulse response is found to be circularly symmetric due to the circular geometry of the optical elements, aperture stops, etc. We denote this type of impulse response either by $h(r)$ or by $h\left(\sqrt{x^2+y^2}\right)$, depending upon the nature of the problem at hand. If the input to such a system is also circularly symmetric, it can be shown that the output, too, possesses circular symmetry, i.e.,

$$f(r)**h(r)=g(r). \tag{9.148}$$

Most of the time, however, the input cannot be expressed solely as a function of r, and this requires that the convolution be put either in the form of Eq. (9.38), or that it be written as

$$g(x,y)=f(x,y)**h\left(\sqrt{x^2+y^2}\right). \tag{9.149}$$

The impulse response completely characterizes a two-dimensional LSI system, just as for the one-dimensional case, and it is therefore often desirable to either calculate it or measure it. Once the impulse response has been determined, the transfer function can be obtained by a Fourier transform operation. When determination of the impulse response is not practical, other quantities such as the line response and step response may be helpful in describing the system. We shall discuss these two quantities later in this section.

The Transfer Function

We have already seen that it is usually quite difficult to calculate the output of a system by the direct evaluation of a convolution integral, and this difficulty is greatly magnified for two-dimensional systems. Consequently, except in special cases, the transfer-function approach is normally applied as a matter of course. With the transfer function defined to be

$$H(\xi,\eta)=\mathcal{F}\mathcal{F}\{h(x,y)\}, \tag{9.150}$$

the output spectrum of an LSI system is given by

$$G(\xi,\eta)=F(\xi,\eta)H(\xi,\eta), \tag{9.151}$$

which is similar to the one-dimensional case.

When a system is characterized by a circularly symmetric impulse response, its transfer function also exhibits circular symmetry and may be written as

$$H(\rho)=\mathcal{H}_0\{h(r)\}. \tag{9.152}$$

If the input to such a system is circularly symmetric, then the output spectrum becomes

$$G(\rho)=F(\rho)H(\rho), \tag{9.153}$$

which is clearly circularly symmetric, too. However, as mentioned in the previous section, the input will not normally possess such symmetry. In that event it is often convenient to describe the various spectra in rectangular coordinates; thus,

$$G(\xi,\eta)=F(\xi,\eta)H\left(\sqrt{\xi^2+\eta^2}\,\right). \tag{9.154}$$

Once the output spectrum has been determined, an inverse transformation will yield the output itself.

Eigenfunctions of Two-Dimensional LSI Systems

The results of Chapter 5 may easily be extended to show that functions of the form

$$\psi(x,y;\xi_0,\eta_0)=e^{j2\pi(\xi_0 x+\eta_0 y)} \tag{9.155}$$

are eigenfunctions of two-dimensional LSI systems. This should not be too surprising, particularly when we recall that the Fourier integral is such a powerful tool in the analysis of LSI systems and that its kernel is also of this form; the Fourier transform operation effectively allows an arbitrary function to be expressed as a linear combination of other functions which just happen to be the eigenfunctions of LSI systems. Thus, we see that it is the nature of the Fourier decomposition that makes this transform so useful when dealing with such systems.

Let us apply one of the eigenfunctions of Eq. (9.155) to the input of an LSI system. The output is given by

$$\begin{aligned} g(x,y) &= \psi(x,y;\xi_0,\eta_0)**h(x,y) \\ &= e^{j2\pi(\xi_0 x+\eta_0 y)}**h(x,y), \end{aligned} \tag{9.156}$$

and the output spectrum may therefore be expressed by

$$\begin{aligned} G(\xi,\eta) &= \delta(\xi-\xi_0,\eta-\eta_0)H(\xi,\eta) \\ &= H(\xi_0,\eta_0)\delta(\xi-\xi_0,\eta-\eta_0). \end{aligned} \tag{9.157}$$

An inverse transform operation yields

$$\begin{aligned} g(x,y) &= H(\xi_0,\eta_0)e^{j2\pi(\xi_0 x+\eta_0 y)} \\ &= H(\xi_0,\eta_0)\psi(x,y;\xi_0,\eta_0), \end{aligned} \tag{9.158}$$

and we see that the functions $\exp\{j2\pi(\xi_0 x+\eta_0 y)\}$ are indeed eigenfunctions. The associated eigenvalues are simply equal to $H(\xi_0,\eta_0)$, the transfer function evaluated at the frequencies ξ_0 and η_0.

If an LSI system has a circularly symmetric transfer function, then functions of the form

$$\psi(r;\rho_0) = J_0(2\pi\rho_0 r) \tag{9.159}$$

are also eigenfunctions. To show this we apply $J_0(2\pi\rho_0 r)$ as the input to a system characterized by an impulse response $h(r)$ and a transfer function $H(\rho)$. The output may be written as

$$g(r) = J_0(2\pi\rho_0 r)**h(r), \tag{9.160}$$

and the output spectrum is therefore equal to

$$\begin{aligned} G(\rho) &= \frac{\delta(\rho-\rho_0)}{2\pi\rho_0}H(\rho) \\ &= H(\rho_0)\frac{\delta(\rho-\rho_0)}{2\pi\rho_0}. \end{aligned} \tag{9.161}$$

An inverse transform operation reveals the output to be

$$g(r) = H(\rho_0)J_0(2\pi\rho_0 r), \tag{9.162}$$

and $J_0(2\pi\rho_0 r)$ is seen to be an eigenfunction as contended. The associated eigenvalue is equal to the value of the transfer function evaluated at ρ_0. Again, we should not be too surprised at this result because the zero-order Hankel transform, which has a kernel of the form $J_0(2\pi\rho_0 r)$, allows a circularly symmetric function to be expressed as a linear combination of these functions. Consequently, the decomposition is once more given in terms of a set of functions that are eigenfunctions of the system.

The Line Response

As should now be clear, the transfer function is nearly always the most useful of the various quantities that can be used to describe the behavior of an LSI system. As such, it would be desirable to be able to make a direct calculation or measurement of this function for any given LSI system; unfortunately, that is often either impractical or impossible. Even when it is possible to calculate a theoretical expression for the transfer function, it may also be necessary to obtain a physical measurement for comparison. Perhaps the most straightforward method for measuring the transfer function would be to apply a large number of eigenfunctions to the input of the system, one after another, and to determine the attenuation (or gain) and phase shift encountered by each of them. The number of measurements required for this determination would depend on the width of the pass band of the system, as well as the spacing (in frequency) that could be tolerated between the various eigenfunctions and could become prohibitively large for some systems. Yet, even when the number of measurements needed is small, this direct method of obtaining the transfer function may be quite difficult and time consuming.

Another method, one used frequently in optics, requires that the impulse response of the system first be calculated or measured. A Fourier-transform operation is then performed, using the numerical techniques discussed in Sec. 9-5, to determine the transfer function. In some cases, however, it is not possible to obtain an accurate representation of the impulse response, and still other methods must be used. We now investigate one of those methods.

We assume that the system under investigation has an impulse response $h(x,y)$ and a transfer function $H(\xi,\eta)$, both of which are unknown. To this system we apply as an input a line mass located along the y-axis, i.e.,

$$f(x,y)=\delta(x). \tag{9.163}$$

We then measure the output, referred to as the *line response* or *line spread function*, which can be expressed in terms of the unknown impulse response as

$$\begin{aligned}\mathfrak{L}\{\delta(x)\} &= \delta(x)**h(x,y)\\ &= \int_{-\infty}^{\infty} h(x,\beta)\,d\beta.\end{aligned} \tag{9.164}$$

We observe that this line response depends only on x, and we denote it by

$$\mathfrak{L}\{\delta(x)\} = l_x(x), \tag{9.165}$$

where the subscript x is used to distinguish this response from those caused by input line masses having different orientations. From Eq. (9.73) we see that

$$\begin{aligned}L_x(\xi) &= \mathfrak{F}\{l_x(x)\}\\ &= \mathfrak{F}\left\{\int_{-\infty}^{\infty} h(x,\beta)\,d\beta\right\}\\ &= H(\xi,0);\end{aligned} \tag{9.166}$$

the *one-dimensional Fourier transform* of the line response $l_x(x)$ is equal to the *ξ-axis profile* of the transfer function. Another way of stating this result is as follows: the line response $l_x(x)$ and the profile $H(\xi,0)$ of the transfer function form a *one-dimensional Fourier-transform pair*.

If we had chosen the input to be $\delta(y)$ rather than $\delta(x)$, the line response would be denoted

$$\mathfrak{L}\{\delta(y)\} = l_y(y), \tag{9.167}$$

and we would then find that $l_y(y)$ and $H(0,\eta)$ form a one-dimensional Fourier-transform pair, i.e.,

$$\begin{aligned}L_y(\eta) &= \mathfrak{F}\{l_y(y)\}\\ &= H(0,\eta).\end{aligned} \tag{9.168}$$

For an input line mass oriented along neither the x-axis nor the y-axis, the line response is most easily described in a coordinate system that has been rotated to align one of the axes with the line mass. For example, with an input of $\delta(a_1x+b_1y)$, where $a_1^2+b_1^2=1$, the notation is simplified if we select a new coordinate system x',y' rotated by an angle θ' such that $x'=a_1x+b_1y$ and $y'=-b_1x+a_1y$ [see Eq. (3.111) and subsequent equations]. In this new coordinate system the impulse response may be described by

$$h_{\theta'}(x',y')=h(a_1x'-b_1y',\,b_1x'+a_1y'), \tag{9.169}$$

and the line response by

$$\begin{aligned} l_{\theta'}(x') &= \mathcal{L}\{\delta(x')\} \\ &= \delta(x')**h_{\theta'}(x',y') \\ &= \int_{-\infty}^{\infty} h_{\theta'}(x',\beta')\,d\beta' \\ &= \int_{-\infty}^{\infty} h(a_1x'-b_1\beta',\,b_1x'+a_1\beta')\,d\beta'. \end{aligned} \tag{9.170}$$

If we also erect a new coordinate system ξ',η' in the frequency domain, where ξ' corresponds to x' and η' to y', the transfer function may be written [see Eq. (9.80)]

$$\begin{aligned} H_{\theta'}(\xi',\eta') &= \mathcal{F}\mathcal{F}\{h_{\theta'}(x',y')\} \\ &= H(a_1\xi'-b_1\eta',\,b_1\xi'+a_1\eta'). \end{aligned} \tag{9.171}$$

We then find that the line response function $l_{\theta'}(x')$ and the profile

$$H_{\theta'}(\xi',0)=H(a_1\xi',b_1\xi') \tag{9.172}$$

form a one-dimensional Fourier-transform pair. In other words

$$\begin{aligned} L_{\theta'}(\xi') &= \mathcal{F}\{l_{\theta'}(x')\} \\ &= H_{\theta'}(\xi',0) \\ &= H(a_1\xi',b_1\xi'). \end{aligned} \tag{9.173}$$

These procedures for determining the line response functions and the associated profiles of the transfer function are illustrated in Figs. 9-7 and 9-8.

By changing the orientation of the input line mass and measuring the resulting line response for each orientation, we can calculate as many profiles of the transfer function as are necessary. Of course, if the impulse response is circularly symmetric, the transfer function is also circularly

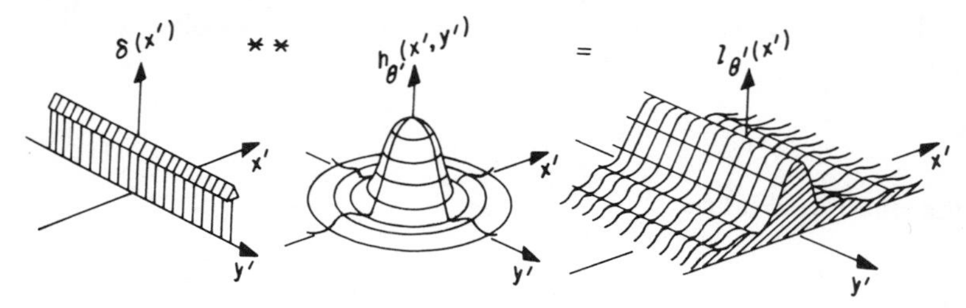

Figure 9-7 Determination of line-response function of LSI system.

symmetric and only one profile is required to specify it completely (actually, only the half of the profile for which $\xi \geqslant 0$ is required). On the other hand, if the impulse response is separable in x and y (or in x' and y'), the transfer function will also be separable, and only the two profiles $H(\xi,0)$ and $H(0,\eta)$ [or $H_{\theta'}(\xi',0)$ and $H_{\theta'}(0,\eta')$] are needed.

It is interesting to note that in the above determination of the transfer function, we did not need to know the impulse response. We only needed to measure one or more of the one-dimensional line response functions associated with the system. This procedure can also be reversed: if the transfer function is known, the line-response functions associated with the system can be calculated by performing inverse transform operations on the appropriate profiles of the transfer function without ever knowing the impulse response.

If the impulse response of an LSI system is both circularly symmetric and known, it may be possible to calculate the line response by using the *Abel transform* operation (see Ref. 9-8). Let us apply the line mass $\delta(x)$ as the input to a system with impulse response $h(r)=h\left(\sqrt{x^2+y^2}\right)$. The output, which is simply the line response $l_x(x)$, may then be expressed

$$\begin{aligned} l_x(x) &= h\left(\sqrt{x^2+y^2}\right) ** \delta(x) \\ &= \int_{-\infty}^{\infty} h\left(\sqrt{x^2+\beta^2}\right) d\beta \\ &= 2\int_{0}^{\infty} h\left(\sqrt{x^2+\beta^2}\right) d\beta. \end{aligned} \tag{9.174}$$

By setting $x^2+\beta^2=r'^2$, we may rewrite Eq. (9.174) as

$$l_x(x) = 2\int_{x}^{\infty} \frac{h(r')r'\,dr'}{\sqrt{r'^2-x^2}}, \tag{9.175}$$

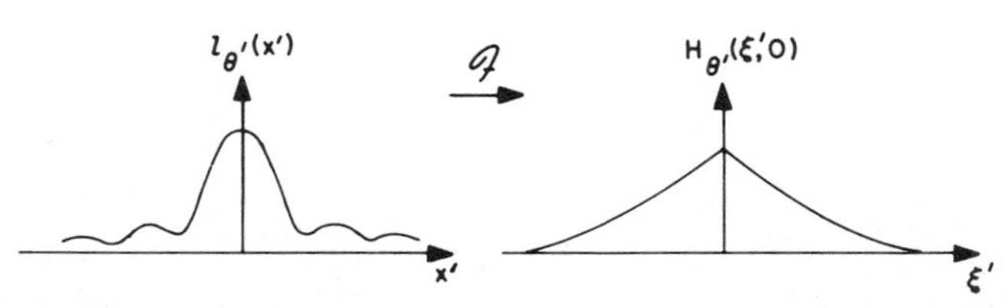

Figure 9-8 Relationship between line-response function and transfer function.

which gives $l_x(x)$ as the Abel transform of $h(r)$. Consequently, it may be possible to evaluate Eq. (9.175) by consulting a table of Abel transforms (e.g., Ref. 9-8). And, if necessary, once $l_x(x)$ has been determined, $H(\rho)$ can be found by a Fourier transform operation as discussed above; i.e.,

$$H(\rho)=L_x(\rho), \qquad \rho\geqslant 0. \tag{9.176}$$

As Bracewell (Ref. 9–8) points out, the Abel transform may be useful whenever it is necessary to obtain a one-dimensional projection of a circularly symmetric two-dimensional quantity. This situation is encountered in optics for any of the following sets of circumstances: when an imaging system with a circularly symmetric impluse response forms an image of a line source; when such a system forms an image of a point source and the image irradiance is scanned with a long narrow slit; when a microdensitometer with a long narrow slit is used to scan a circularly symmetric density distribution on a photographic negative, etc.

It is conceivable that determination of the transfer function by the procedures described here could sometimes be less difficult and less costly than by using the method requiring the direct calculation of the Fourier transform of the impulse response, the reason being that a single two-dimensional transform operation may be more involved than several one-dimensional operations. Each problem should be analyzed to determine the best method for the conditions that exist.

The Edge Response

It sometimes turns out in practice that not only is it impossible to make direct measurements or calculations of the impulse response and transfer function, it is also impossible to make a direct determination of the line response functions. However, not all is lost. To see why, let us apply a step function as the input to the system, i.e.,

$$f(x,y)=\text{step}(a_1x+b_1y). \tag{9.177}$$

The output of the system is then known as the *step response*, or *edge*

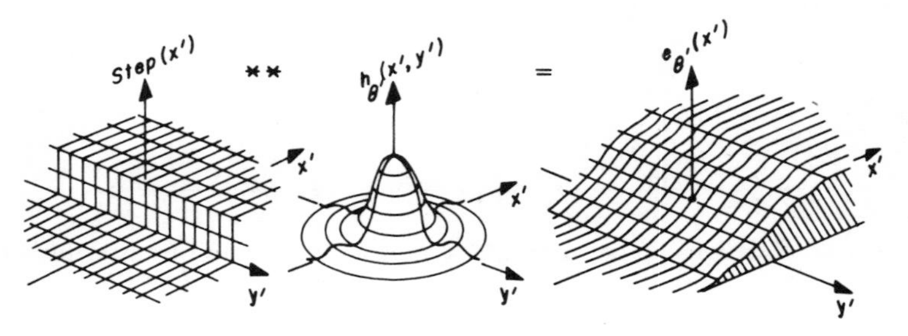

Figure 9-9 Determination of edge-response function of LSI system.

response, and is depicted in Fig. 9-9. We shall again describe the various quantities of interest in the rotated x', y' coordinate system, and when we do the edge response becomes

$$\begin{aligned}
e_{\theta'}(x') &= \mathcal{L}\{\text{step}(x')\} \\
&= h_{\theta'}(x',y') ** \text{step}(x') \\
&= \iint_{-\infty}^{\infty} h_{\theta'}(\alpha',\beta')\,\text{step}(x'-\alpha')\,d\alpha'\,d\beta' \\
&= \int_{-\infty}^{\infty}\left[\int_{-\infty}^{\infty} h_{\theta'}(\alpha',\beta')\,d\beta'\right]\text{step}(x'-\alpha')\,d\alpha' \\
&= \int_{-\infty}^{\infty} l_{\theta'}(\alpha')\,\text{step}(x'-\alpha')\,d\alpha' \\
&= l_{\theta'}(x') * \text{step}(x').
\end{aligned} \tag{9.178}$$

We see from Eqs. (7.51) and (7.53) that Eq. (9.178) may be rewritten as

$$e_{\theta'}(x') = \int_{-\infty}^{x'} l_{\theta'}(\alpha')\,d\alpha', \tag{9.179}$$

and, as a result, we find that the line response $l_{\theta'}(x')$ is equal to the derivative of the edge response $e_{\theta'}(x')$, i.e.,

$$l_{\theta'}(x') = \frac{d}{dx'} e_{\theta'}(x'). \tag{9.180}$$

Therefore, if we can either measure or calculate $e_{\theta'}(x')$, we can determine $l_{\theta'}(x')$ by a simple differentiation operation as illustrated in Fig. 9-10. Of course, once $l_{\theta'}(x')$ has been determined, the corresponding profile of the transfer function can be calculated from Eq. (9.173). The number of

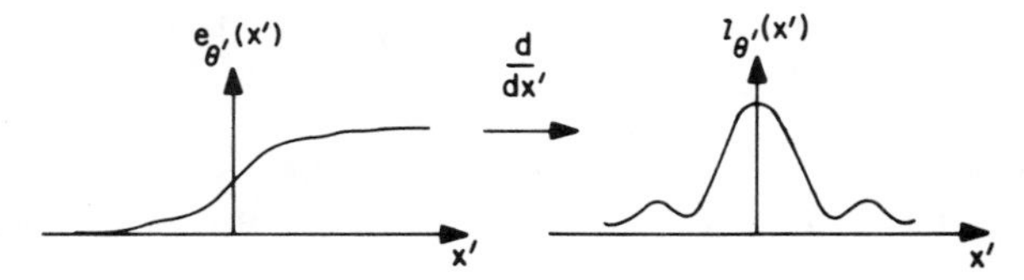

Figure 9-10 Relationship between edge- and line-response functions.

edge-response measurements required is again determined by the symmetry and separability characteristics of the impulse response, just as it was for the line response.

The above result was developed for an arbitrarily oriented step-function input, and the results are completely general. If the input is chosen to be

$$f(x,y)=\text{step}(x),$$

then the results may be expressed in the original x, y-coordinate system:

$$\begin{aligned} e_x(x) &= \mathfrak{L}\{\text{step}(x)\} \\ &= \int_{-\infty}^{x} l_x(\alpha)\,d\alpha, \end{aligned} \tag{9.181}$$

and

$$l_x(x)=\frac{d}{dx}e_x(x). \tag{9.182}$$

A similar result is obtained for an input of step (y).

In an imaging system the edge response is associated with the image of an edge—hence the name. It can often be measured by scanning the image irradiance with a long narrow slit oriented parallel to the edge. The scanning direction is perpendicular to the edge, and a photosensitive device located behind the slit measures the total power passing through it; the resulting variation of measured power as a function of slit position is directly proportional to the edge response. We shall return to this topic in a later chapter.

REFERENCES

9-1 A. Papoulis, *Systems and Transforms with Applications in Optics*, McGraw-Hill, New York, 1968.

9-2 R. S. Burington, *Handbook of Mathematical Tables and Formulas*, Handbook Publishers, Sandusky, Ohio, 1953.

9-3 R. E. Hufnagel, "Simple Approximation to Diffraction-Limited MTF," *Appl. Opt.* **10**(11): 2547–2548 (1971).

9-4 Papoulis, *op. cit.*, p. 95.

9-5 J. W. Goodman, *Introduction to Fourier Optics*, McGraw-Hill, New York, 1968.

9-6 A. Erdélyi, Ed., *Tables of Integral Transforms*, Vol. II, Bateman Manuscript Project, McGraw-Hill, New York, 1954.

9-7 I. S. Gradshteyn and I. M. Ryzhik, *Tables of Integrals, Series, and Products*, 4th edition, prepared by Y. V. Geronimus and M. Y. Tseytlin, translated from the Russian by Scripta Technica, Inc., A. Jeffrey, Ed., Academic Press, New York, 1965.

9-8 R. Bracewell, *The Fourier Transform and its Applications*, McGraw-Hill, New York, 1965.

9-9 E. Jahnke and F. Emde, *Tables of Functions with Formulae and Curves*, 4th edition, Dover Publications, New York, 1945.

9-10 M. Abramowitz and I. A. Stegun, Eds., *Handbook of Mathematical Functions with Formulas, Graphs and Mathematical Tables*, National Bureau of Standards, U.S. Government Printing Office, Washington, D.C., 1967.

9-11 J. W. Cooley and J. W. Tukey, "An Algorithm for the Machine Calculation of Complex Fourier Series," *Math. Comp.*, **19**(90): 297–301 (1965).

9-12 E. O. Brigham, *The Fast Fourier Transform*, Prentice-Hall, Englewood Cliffs, N.J., 1974.

PROBLEMS

9-1. Perform the following convolutions and autocorrelations, and make appropriate sketches of your results; i.e., sketch profiles along the axes and diagonals, sketch "top views," etc.

a. $f(x, y)=\text{rect}(x,y)**\text{rect}(x, y)$.

b. $g(x, y)=\text{rect}(0.5x, y)**\text{rect}(x, 0.5y)$.

c. $h(x,y)=\text{Gaus}\left(\frac{x}{5},\frac{y}{12}\right)**\text{Gaus}\left(\frac{x}{12},\frac{y}{5}\right)$.

d. With $m(x, y)=\text{rect}(0.25x, 0.25y)-\text{rect}(0.5x, 0.5y)$,
find $n(x, y)=m(x, y)\star\star m(x, y)$.

e. With $p(x, y)=\text{rect}(0.1x, y)+\text{rect}(x, 0.1y)-\text{rect}(x, y)$,
find $r(x, y)=p(x, y)**p(x, y)$. [$p(x, y)$describes a Mill's cross.]

f. With $s(x, y)=\delta\delta(x, y)**\text{rect}(x, y)$, where $\delta\delta(x, y)=\delta\delta(x)\delta\delta(y)$,
find $t(x, y)=s(x, y)**s(x, y)$.

g. With $u(x, y)=[\text{rect}(0.5x+1)-\text{rect}(0.5x-1)]\text{rect}(0.1y)$,
find $v(x, y)=u(x, y)**u(x, y)$.

h. Using the $u(x, y)$ given in part g,
find $\gamma_u(x, y)=u(x, y)\star\star u^*(x, y)$.

9-2. Perform the following convolutions and sketch appropriate profiles and "top views" of your results.

a. With $u(x, y)=\text{rect}(\sqrt{2}\,x+\sqrt{2}\,y,\ -\sqrt{2}\,x+\sqrt{2}\,y)$ and

$v(x, y) = \text{rect}(\sqrt{2}\,x + \sqrt{2}\,y - 2,\ -\sqrt{2}\,x + \sqrt{2}\,y + 3)$,
find $w(x, y) = u(x, y) ** v(x, y)$.

b. $f(x, y) = [\text{rect}(x, y)\ \text{sgn}(y)] * \delta(x)$.

c. $g(x, y) = [\text{rect}(x, y)\text{sgn}(y)] ** \delta(x)$.

d. $h(x, y) = \text{sinc}(x, y) ** \delta(x - 4)$.

e. $p(x, y) = \text{rect}(x, y) ** \delta(x - y)$.

f. $r(x, y) = \text{rect}(x, y) ** \delta(3x + 4y - 5)$.

g. $s(x, y) = \text{tri}(10x,\ 10y) ** \text{comb}(x + 2y,\ -x + y)$.

9-3. Perform the following convolutions and autocorrelations, and sketch appropriate profiles and "top views" of your results.

a. $f(r) = \text{cyl}(0.5r) ** \text{cyl}(0.5r)$.

b. With $g(r) = \text{cyl}(0.25r) - \text{cyl}(0.5r)$, find $h(r) = g(r) \star\star g(r)$.

c. $m(r) = \dfrac{\delta(r)}{\pi r} ** \text{Gaus}(r)$.

d. $n(r) = \delta\left(r - 2, \theta - \frac{\pi}{4}\right) ** \text{cyl}(r)$.

e. With $t(x, y) = \text{cyl}\left(\sqrt{x^2 + y^2}\right) + \text{cyl}\left(\sqrt{(x-2)^2 + y^2}\right)$
$+ \text{cyl}\left(\sqrt{x^2 + (y-2)^2}\right)$,
find $u(x, y) = t(x, y) \star\star t(x, y)$.

f. $v(r) = \text{somb}^2(0.5r) ** \text{somb}(r)$.

9-4. Find the Fourier transforms of the following functions and sketch appropriate profiles of your results.

a. $f(x, y) = \text{rect}(0.5x, 2y)$.

b. $g(x, y) = \text{rect}(2x + 2,\ 0.1y) + \text{rect}(2x - 2,\ 0.1y)$.

c. $h(x, y) = \text{cyl}\left(0.5\sqrt{(x+3)^2 + y^2}\right) + \text{cyl}\left(0.5\sqrt{(x-3)^2 + y^2}\right)$.

d. $m(x, y) = 0.2\,\text{comb}(0.2x) * \text{rect}(0.4x)$.

e. $n(x, y) = [\text{comb}(x)\text{rect}(0.01x,\ 0.01y)] * \text{tri}(2x)$.

f. $p(x, y) = \text{rect}(5x - 0.5y,\ 0.25y)$.

g. $s(x, y) = \text{rect}(5x - 0.5y + 2,\ 0.25y) + \text{rect}(5x - 0.5y - 2,\ 0.25y)$.

h. $t(x, y) = \left[0.04\,\text{comb}(0.2x,\ 0.2y)\text{rect}\left(\frac{x}{55}, \frac{y}{55}\right)\right] ** \text{rect}(x, y)$.

i. $u(x, y) = [0.04\,\text{comb}(0.2x,\ 0.2y) ** \text{rect}(x, y)]\text{cyl}\left(\dfrac{\sqrt{x^2 + y^2}}{55}\right)$.

j. $v(x, y) = \text{sinc}^2(0.5x,\ 0.25y) ** \text{sinc}(x,\ 0.5y)$.

k. $w(x, y) = \text{somb}\left(\sqrt{4x^2 + y^2}\right)$.

9-5. Find the zero-order Hankel transforms of the following functions and sketch appropriate profiles of your results.

a. $f(r) = \text{cyl}(2r)$.

b. $g(r) = \text{cyl}(2r) \star\star \text{cyl}(2r)$.

c. $h(r) = \text{cyl}(0.25r) - \text{cyl}(0.5r)$.
d. $s(r) = p(r) \star\star p(r)$, where $p(r) = \text{cyl}(0.25r) - \text{cyl}(0.5r)$.
e. $t(r) = \text{somb}^2(r) \star\star \text{somb}(5r)$.
f. $u(r) = \delta(r-2) ** \delta(r-2)$.
g. $v(r) = -4\pi^2 r^2 \text{Gaus}(2r)$.
h. $w(r) = \exp[-\pi(0.5 - j10)r^2]$.

9-6. Given an LSI system with impulse response $h(x, y) = 25\,\text{sinc}^2(5x, 5y)$, find the output $g_i(x)$ for the following input signals:

a. $f_1(x, y) = 1 + \cos(6\pi x) + \cos(12\pi x)$.
b. $f_2(x, y) = 1 + \cos[6\pi(x+y)] + \cos[12\pi(x+y)]$.
c. $f_3(x, y) = \delta(x)$. [Note that $g_3(x, y)$ is a line response.]
d. $f_4(x, y) = \delta\left(\frac{x}{\sqrt{2}} + \frac{y}{\sqrt{2}}\right)$. [Note that $g_4(x, y)$ is a line response.]

9-7. Given an LSI system with impulse response $h(r) = 25\pi\,\text{somb}(10r)$, find the output $g_i(x)$ for the following input signals:

a. $f_1(x, y) = [1 + \cos(6\pi x) + \cos(12\pi x)]\,\text{rect}\left(\frac{x}{100}, \frac{y}{100}\right)$.
b. $f_2(x, y) = \{1 + \cos[6\pi(x+y)] + \cos[12\pi(x+y)]\}\,\text{rect}\left(\frac{x}{100}, \frac{y}{100}\right)$.
c. $f_3(x, y) = \delta(x)$. [Note that $g_3(x, y)$ is a line response.]
d. $f_4(x, y) = \delta\left(\frac{x}{\sqrt{2}} + \frac{y}{\sqrt{2}}\right)$. [Note that $g_4(x, y)$ is a line response.]
e. $f_5(r) = J_0(6\pi r) + J_0(12\pi r)$.

CHAPTER 10

THE PROPAGATION AND DIFFRACTION OF OPTICAL WAVE FIELDS

We have finally reached the point where we can begin to apply the mathematics of the previous chapters to the solution of problems actually encountered in optics. The first of these to be addressed has to do with the propagation and diffraction of optical wave fields. Although there is a rich history surrounding the development of the various theories associated with these phenomena, our objectives here, which are to gain an understanding of the most relevant theories and to learn problem-solving techniques based on them, do not permit a detailed coverage of the historical aspects. However, for those who are interested, there are a number of excellent works on these subjects that do include discussions of their historical development (e.g., see Refs. 10-1 through 10-9). In fact, serious students of optics are encouraged to consult these references to gain a deeper appreciation of the history of optics and a more thorough understanding of the physics involved.

10-1 MATHEMATICAL DESCRIPTION OF OPTICAL WAVE FIELDS

Before proceeding, please review the part of Sec. 2-3 entitled Monochromatic Light Waves. As stated in that section, we shall restrict our attention almost entirely to monochromatic, linearly polarized wave fields—wave fields that can be adequately described by a real-valued scalar function of position and time. We associate this scalar function with the magnitude of either the electric field vector or the magnetic field vector and denote it by

$$u(x,y,z;t)=a(x,y,z)\cos\left[2\pi\nu_0 t-\phi(x,y,z)\right]. \tag{10.1}$$

This equation is the same as Eq. (2.102) except that the position vector **r** has now been replaced by the rectangular coordinate variables (x, y, z). We define the *complex amplitude* of this wave field to be

$$u(x,y,z)=a(x,y,z)e^{j\phi(x,y,z)}, \tag{10.2}$$

and we find that many problems of interest can be solved using this quantity rather than the real physical wave field description of Eq. (10.1). When necessary, $u(x, y, z; t)$ can be obtained from the complex amplitude by

$$u(x,y,z;t)=\mathrm{Re}\left\{u^*(x,y,z)e^{j2\pi\nu_0 t}\right\}. \tag{10.3}$$

In the optical region of the electromagnetic spectrum, detectors do not respond directly to the electric or magnetic field amplitude; rather, they respond to such quantities as radiant-energy density or the time rate of flow of radiant energy. For example, latent image formation in photographic emulsions is dependent on the total radiant energy per unit area (J/m^2) incident on the emulsion during exposure. Photomultipliers, on the other hand, respond to the rate at which radiant energy impinges on the photodetective surface (J/sec).

The time rate of flow of radiant energy is simply the *power*, or *radiant flux*, associated with a wave field, and is commonly measured in watts. When a wave field impinges on a surface normally, the rate per unit area at which radiant energy arrives at the surface is referred to as the *irradiance* of the wave field at that surface. The irradiance is also known as the *power density*, or *radiant-flux density*, and is measured in watts per square meter. (The radiant-flux density of a wave field leaving a surface in a normal direction is known as the exitance of the wave field at that surface.)

The irradiance usually may be regarded as the time-averaged value of the Poynting vector associated with the wave field,* which is proportional to the square of the electric (or magnetic) field amplitude for a monochromatic wave field. It follows that the irradiance, which we shall denote by

*As pointed out in Ref. 10-1, the physically significant quantity is the integral of the normal component of the Poynting vector rather than the Poynting vector itself. Nevertheless, over small but finite regions of space, the Poynting vector is a measure of the radiant flux density.

$I(x,y,z)$, is given by the squared modulus of the complex amplitude, i.e.,

$$I(x,y,z) = |u(x,y,z)|^2. \tag{10.4}$$

A brief but excellent discussion of the various quantities mentioned above may be found in Ref. 10-10.

Plane Wave Fields

The complex amplitude description of a general plane wave field propagating in a rectangular coordinate system is

$$u(x,y,z) = Ae^{j[k(\gamma_x x + \gamma_y y + \gamma_z z) + \Phi]}, \tag{10.5}$$

where A is the strength, or amplitude, of the wave field, Φ is its phase at the origin, and $(\gamma_x, \gamma_y, \gamma_z)$ are the direction cosines of its propagation vector. As usual, $k = 2\pi/\lambda$, where λ denotes the wavelength of the light, and because we will deal only with monochromatic wave fields, we have dropped the subscript zeros from λ, k, etc. From Eqs. (10.3) and (10.4) we find the real physical field to be

$$u(x,y,z;t) = A\cos\left[2\pi\nu t - k(\gamma_x x + \gamma_y y + \gamma_z z) - \Phi\right], \tag{10.6}$$

and the irradiance to be

$$I(x,y,z) = |A|^2. \tag{10.7}$$

We note that the irradiance of a plane wave field is constant throughout space. We also observe that because irradiance has units of W/m^2, the field amplitude must have units of $(W/m^2)^{1/2}$.

The angles between the positive x-, y-, and z-axes and the propagation vector are $\cos^{-1}(\gamma_x)$, $\cos^{-1}(\gamma_y)$, and $\cos^{-1}(\gamma_z)$, respectively, and with our sign convention the propagation vector has a positive component in any of the coordinate directions whenever the corresponding direction cosine is positive. If a direction cosine is negative, the corresponding component of the propagation vector is negative. In addition, the phase of $u(x,y,z)$ increases with an increase of each coordinate variable when the corresponding direction cosine is positive and vice versa. These statements may be more easily understood with reference to Fig. 10-1, in which we have chosen $\gamma_y = 0$ to simplify the illustration (this causes the propagation vector to lie in the plane of the page). The wavefronts (surfaces of constant phase) associated with this wave field are planes lying perpendicular to the propagation vector, and a number of these wavefronts are depicted in the figure; they are shown separated by a distance of λ for clarity.

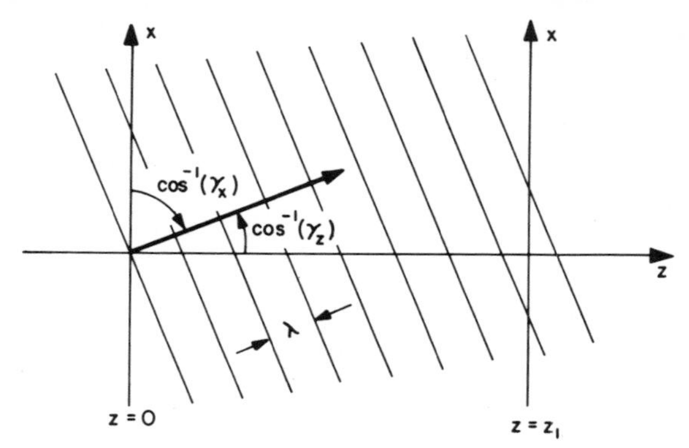

Figure 10-1 Plane-wave field depicted for propagation-vector direction cosines of $(\gamma_x, 0, \gamma_z)$.

From the relationship $\gamma_x^2+\gamma_y^2+\gamma_z^2=1$, we may write

$$\gamma_z=\left[1-\left(\gamma_x^2+\gamma_y^2\right)\right]^{1/2}, \tag{10.8}$$

which is real valued for $(\gamma_x^2+\gamma_y^2)\leqslant 1$ and imaginary valued for $(\gamma_x^2+\gamma_y^2)>1$. When γ_z is real valued, Eqs. (10.5) and (10.6) represent a plane wave field propagating generally in the positive z-direction†; if γ_z is imaginary, these equations describe an evanescent wave that undergoes a rapid exponential attenuation with z (see Ref. 10-3). We will normally be dealing with wave fields for which γ_z is real.

In any plane for which z is constant, say $z=z_1$, Eq. (10.5) may be rewritten as

$$u(x,y,z_1)=A\, e^{j\Phi}\, e^{jk\gamma_z z_1}\, e^{jk(\gamma_x x+\gamma_y y)}. \tag{10.9}$$

This expression describes the x and y dependence of the complex amplitude in the $z=z_1$ plane, and the quantity $\exp\{jk\gamma_z z_1\}$ is simply a complex constant that accounts for the phase difference between any point in the $z=0$ plane and the corresponding point in the plane $z=z_1$. In other words, the difference in phase between the points $(x,y,0)$ and (x,y,z_1) is $k\gamma_z z_1$ for all x and y. We note that the maximum phase difference between these two points occurs when the wave field is propagating parallel with the z-axis, i.e., when $\gamma_z=1$. Conversely, there is no phase difference between the

†It seems to be a universal law of physics that light propagates only in the positive z-direction —i.e., from left to right—and we do not wish to challenge this law here.

points $(x,y,0)$ and (x,y,z_1) when the wave field is propagating perpendicular to the z-axis—a condition for which $\gamma_z=0$.

As long as we are interested in describing a plane wave field only in the $z=z_1$ plane, we may simplify the expression of Eq. (10.9) by an appropriate choice of Φ. Since this choice is entirely arbitrary, we let $\Phi=-k\gamma_z z_1$ and obtain

$$u(x,y,z_1)=A\,e^{jk(\gamma_x x+\gamma_y y)}; \tag{10.10}$$

this corresponds to setting the phase equal to zero at the point $(0,0,z_1)$, which in turn has the effect of setting the constant $\exp\{j(k\gamma_z z_1+\Phi)\}$ equal to unity. However, this procedure should not be employed indiscriminately, because under certain circumstances it is necessary to retain even the constant phase information.

In free space, the nature of a plane wave field does not change as it propagates; a plane wave field at $z=z_1$ is also a plane wave field at any other plane $z=z_i$, and the only difference in the wave field at these two planes is in the phase. Of course, the plane wave fields described here are infinite in extent and are therefore nonphysical. Nevertheless, such descriptions often provide sufficiently accurate representations of real physical wave fields within certain finite regions of space.

Spherical Wave Fields

We next investigate the behavior of spherical wave fields. To do so, we place a point source of light at the point $(\alpha,\beta,0)$ and specify the complex amplitude of the resulting wave field in the region $z>0$. With R and ψ as shown in Fig. 10-2 and D a constant, the usual expression for such a wave

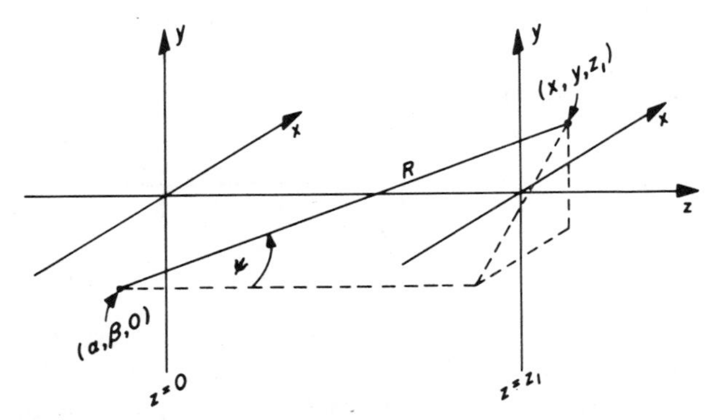

Figure 10-2 Geometry used to describe spherical wave field.

field is (see Ref. 10-3)

$$u(x,y,z) = D\cos\psi \frac{e^{jkR}}{R}. \tag{10.11}$$

The constant D contains information about the nonvarying amplitude and phase of the wave field, and $\cos\psi$ is known as the obliquity factor. This expression is a very good representation of a spherical wave field as long as the observation point is a large number of wavelengths from the source, i.e., $R \gg \lambda$. To depict the complex amplitude as an explicit function of (x, y, z), we note that

$$\cos\psi = \frac{z}{R}$$

$$R = z\left[1 + \left(\frac{x-\alpha}{z}\right)^2 + \left(\frac{y-\beta}{z}\right)^2\right]^{1/2} \tag{10.12}$$

and substitute these expressions into Eq. (10.11) to obtain

$$u(x,y,z) = \frac{Dz}{R^2} e^{jkR}$$

$$= \frac{D\exp\left\{jkz\left[1 + \left(\frac{x-\alpha}{z}\right)^2 + \left(\frac{y-\beta}{z}\right)^2\right]^{1/2}\right\}}{z\left[1 + \left(\frac{x-\alpha}{z}\right)^2 + \left(\frac{y-\beta}{z}\right)^2\right]}. \tag{10.13}$$

When the point source is located at the origin, $\alpha = \beta = 0$ and

$$u(x,y,z) = \frac{D\exp\left\{jkz\left[1 + \left(\frac{x}{z}\right)^2 + \left(\frac{y}{z}\right)^2\right]^{1/2}\right\}}{z\left[1 + \left(\frac{x}{z}\right)^2 + \left(\frac{y}{z}\right)^2\right]}. \tag{10.14}$$

The important things to note about these various representations are the inverse dependence of the modulus on the distance between the source and observation points and the direct dependence of the phase on this distance. As was the case for the plane wave field, the nature of a spherical wave field in free space does not change as it propagates; its amplitude and phase may vary with position, as well as the curvature of its wavefronts, but it remains a spherical wave.

As long as z is positive, Eqs. (10.13) and (10.14) describe spherical wave fields *diverging from* points located in the plane $z=0$. However, when z is negative, these expressions represent spherical wave fields *converging toward* points in that plane. To illustrate, if we specify that $z<0$, Eq. (10.13) may be written as

$$u(x,y,z)=\frac{-D\exp\left\{-jk|z|\left[1+\left(\frac{x-\alpha}{z}\right)^2+\left(\frac{y-\beta}{z}\right)^2\right]^{1/2}\right\}}{|z|\left[1+\left(\frac{x-\alpha}{z}\right)^2+\left(\frac{y-\beta}{z}\right)^2\right]}, \tag{10.15}$$

which describes a converging spherical wave.

We may now generalize the above results: if we allow z to be either positive or negative, the expression

$$u(x,y,z)=\frac{D\exp\left\{jkz\left[1+\left(\frac{x-\alpha}{z}\right)^2+\left(\frac{y-\beta}{z}\right)^2\right]^{1/2}\right\}}{z\left[1+\left(\frac{x-\alpha}{z}\right)^2+\left(\frac{y-\beta}{z}\right)^2\right]} \tag{10.16}$$

denotes either a converging or a diverging spherical wave field. In the region for which $z<0$ it describes a spherical wave converging toward the point $(\alpha,\beta,0)$, which is located a distance $|z|$ to the right of the observation plane. On the other hand, when $z>0$, it represents a spherical wave diverging from the point $(\alpha,\beta,0)$, now located a distance $|z|$ to the left of the observation plane. We see, then, that a negative exponent is associated with a converging wave field, whereas a positive exponent depicts a diverging wave. In either case, the constant D specifies a reference magnitude and phase for the wave field. Note that $u(x,y,z)$ changes sign as the observation distance z changes from negative to positive; physically, this sign change accounts for the phase reversal that the amplitude of the electric (or magnetic) field undergoes when a converging spherical wave passes through a focus and becomes a diverging spherical wave. Of course, the observation distance must be much greater than λ in order for Eq. (10.16) to be accurate.

Simplified Notation for Complex Amplitude

Because we normally describe the wave fields of interest in planes for which z = constant, it is convenient for us to regard the quantity $u(x, y, z)$ as a two-dimensional function of x and y with a parametric dependence on z. Therefore, we denote the complex amplitude in the plane $z = z_i$ by

$$u(x, y, z_i) = u_i(x, y); \tag{10.17}$$

if we leave z unspecified, we use the simplified notation $u(x, y, z) = u(x, y)$.

We also find it convenient to characterize various apertures and semi-transparent objects by their *complex amplitude transmittance* functions. The object of interest is placed in the $z = z_i$ plane and an arbitrary wave field is caused to impinge upon this plane from the left, as shown in Fig. 10-3. If the wave field incident on this plane is denoted by $u(x, y, z_i - \epsilon) = u_i^-(x, y)$ and the wave field leaving it by $u(x, y, z + \epsilon) = u_i^+(x, y)$, where ϵ is a very small distance, then the complex amplitude transmittance of the object is given by

$$t_i(x, y) = \frac{u_i^+(x, y)}{u_i^-(x, y)}. \tag{10.18}$$

For simplicity we usually assume that the object has a thickness of zero in the z-direction, but there are times when we cannot do so. For example, if the object exhibits variations in optical thickness, the amplitude transmittance must contain a phase factor to account for these thickness variations. We note, then, that $t_i(x, y)$ may be complex valued and that, when the object is passive (no gain), the modulus of $t_i(x, y)$ is restricted to values less

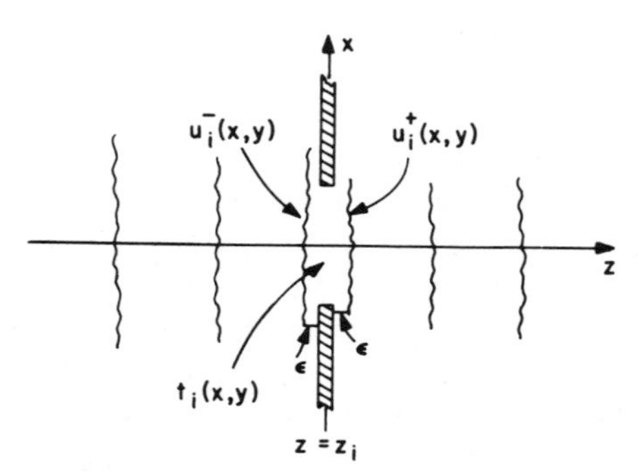

Figure 10-3 Setup for defining complex amplitude transmittance.

than or equal to unity, i.e., $|t_i(x, y)| \leqslant 1$. When appropriate, the amplitude transmittance is referred to as an *aperture function* or *pupil function*.

If we know both the amplitude transmittance for a particular object and the complex amplitude of the incident wave field, the transmitted wave field may be described by

$$u_i^+(x,y) = u_i^-(x,y)t_i(x,y). \tag{10.19}$$

To illustrate, let a clear circular aperture of diameter d be placed in the plane $z = z_1$, and let the plane-wave field

$$u_1^-(x,y) = Ae^{jk\gamma_x x} \tag{10.20}$$

impinge upon it. The transmitted wave field is then simply

$$u_1^+(x,y) = Ae^{jk\gamma_x x}\,\mathrm{cyl}\left(\frac{\sqrt{x^2+y^2}}{d}\right). \tag{10.21}$$

When there exists no aperture or object in a plane, the quantities $u_i^-(x, y)$, $u_i(x, y)$ and $u_i^+(x, y)$ all denote the same complex amplitude.

Plane-Wave Spectrum

We now consider the complex amplitude description of a physically realizable, but otherwise arbitrary, wave field in the plane $z = z_i$. Because this function, $u_i(x, y)$, is associated with a physically realizable wave field, it will possess a Fourier transform $U_i(\xi,\eta) = \mathcal{F}\mathcal{F}\{u_i(x, y)\}$. Consequently, $u_i(x, y)$ may be expressed as the inverse transform of $U_i(\xi,\eta)$:

$$u_i(x,y) = \iint_{-\infty}^{\infty} U_i(\xi,\eta)\,e^{j2\pi(\xi x+\eta y)}\,d\xi d\eta. \tag{10.22}$$

Now recall that a plane wave propagating with direction cosines $(\gamma_x, \gamma_y, \gamma_z)$ may be represented by a function of the form

$$\begin{aligned}\text{Plane wave} &= A\exp[\,jk(\gamma_x x+\gamma_y y)] \\ &= A\exp\left[\,j2\pi\left(\frac{\gamma_x x}{\lambda}+\frac{\gamma_y y}{\lambda}\right)\right],\end{aligned} \tag{10.23}$$

where A is an unimportant complex constant. If we set $\gamma_x = \lambda\xi$ and $\gamma_y = \lambda\eta$, we see that the kernel of the Fourier integral of Eq. (10.22) may be regarded as a unit-amplitude plane wave propagating with direction cosines $\left(\lambda\xi, \lambda\eta, \sqrt{1-\lambda^2(\xi^2+\eta^2)}\right)$. As a result, $u_i(x, y)$ may be regarded as a linear superposition of plane-wave components, each of which is traveling in a direction governed by $\left(\lambda\xi, \lambda\eta, \sqrt{1-\lambda^2(\xi^2+\eta^2)}\right)$ and weighted by $U_i(\xi, \eta)$. More precisely, the component with $\lambda\xi \leqslant \gamma_x \leqslant \gamma\xi + \lambda d\xi$ and $\lambda\eta \leqslant \gamma_y \leqslant \lambda\eta + \lambda d\eta$ has an infinitesimal magnitude given by the volume $U_i(\xi, \eta)d\xi d\eta$. We now see why $U_i(\xi, \eta)$ is often referred to as the *plane-wave spectrum* of $u_i(x, y)$.

Each plane wave component of $u_i(x, y)$ is infinite in extent and propagates as discussed in the section on plane waves. Hence, in some plane $z_l > z_i$, each of these components remains a plane wave, but its phase has been increased by an amount [see Eq. (10.9)]

$$k\gamma_z(z_l - z_i) = k(z_l - z_i)\left[1 - \lambda^2(\xi^2+\eta^2)\right]^{1/2}. \tag{10.24}$$

Because each component undergoes a different phase change in propagating from z_i to z_l, the superposition of these components in the plane $z = z_l$ will, in general, be different from $u_i(x,y)$. In other words,

$$u_l(x, y) = \iint_{-\infty}^{\infty} U_l(\xi, \eta)\, e^{j2\pi(\xi x + \eta y)} d\xi d\eta \tag{10.25}$$

will not have the same form as $u_i(x,y)$, which is given by Eq. (10.22). From a mathematical point of view, then, there must be a difference in the plane wave spectra of the wave fields at these two planes. Reference to Eq. (10.9) and the subsequent discussion leads to the conclusion that $U_l(\xi, \eta)$ and $U_i(\xi, \eta)$ can differ in phase only, and we see that this phase difference is given by Eq. (10.24). Finally we have

$$U_l(\xi, \eta) = U_i(\xi, \eta)\exp\left\{jk(z_l - z_i)\left[1 - \lambda^2(\xi^2+\eta^2)\right]^{1/2}\right\}, \tag{10.26}$$

which is the desired result. We may then regard the quantity

$$\frac{U_l(\xi, \eta)}{U_i(\xi, \eta)} = \exp\left\{jk(z_l - z_i)\left[1 - \lambda^2(\xi^2+\eta^2)\right]^{1/2}\right\} \tag{10.27}$$

as a transfer function for the propagation of a wave field from the plane $z = z_i$ to the plane $z = z_l$. This concept may not make much sense yet, but we will discuss it in greater detail in the sections to follow.

Example

Let us consider the simple plane-wave field of Eq. (10.20):

$$u_1^-(x,y) = Ae^{jk\gamma_x x}. \tag{10.28}$$

If we set $\gamma_x = a$, the plane-wave spectrum of $u_1^-(x,y)$ is

$$\begin{aligned} U_1^-(\xi,\eta) &= \mathcal{F}\mathcal{F}\{Ae^{jkax}\} \\ &= A\mathcal{F}\mathcal{F}\left\{\exp\left[j2\pi\left(\frac{a}{\lambda}\right)x\right]\right\} \\ &= A\delta\left(\xi - \frac{a}{\lambda}, \eta\right), \end{aligned} \tag{10.29}$$

and we see that $u_1^-(x,y)$ consists of a single plane-wave component of amplitude A propagating with direction cosines $(a, 0, \sqrt{1-a^2})$. This agrees with our understanding of Eq. (10.28).

We now calculate the plane-wave spectrum of the transmitted wave field given by Eq. (10.21), and again we let $\gamma_x = a$. We find

$$\begin{aligned} U_1^+(\xi,\eta) &= \mathcal{F}\mathcal{F}\left\{Ae^{jkax}\,\mathrm{cyl}\left(\frac{\sqrt{x^2+y^2}}{d}\right)\right\} \\ &= A\delta\left(\xi - \frac{a}{\lambda}, \eta\right) ** \frac{\pi d^2}{4}\,\mathrm{somb}\left(d\sqrt{\xi^2+\eta^2}\right) \\ &= \frac{\pi A d^2}{4}\,\mathrm{somb}\left(d\sqrt{\left(\xi - \frac{a}{\lambda}\right)^2 + \eta^2}\right). \end{aligned} \tag{10.30}$$

Now, instead of a single plane-wave component, $u_1^+(x,y)$ is composed of an infinite number of such components, each weighted by $U_1^+(\xi,\eta)d\xi d\eta$ according to its direction of propagation. These components are propagating predominately in the direction of the incident wave field $u_1^-(x,y)$, i.e., with direction cosines $(a, 0, \sqrt{1-a^2})$, but there is a spreading in direction as indicated by the function $\mathrm{somb}\left(d\sqrt{(\xi - a/\lambda)^2 + \eta^2}\right)$. Note that this spreading is inversely proportional to the diameter of the aperture: there is very little spreading if d is large, and a great deal of spreading if d is small. Also note that the various plane-wave components of $u_1^+(x,y)$ are all infinite in extent, but when added together they form a finite portion of a single plane-wave field. ■

Definition and Properties of the Function q(x, y; a)

We will find it convenient to define the quadratic-phase signal

$$q(x,y;a)=e^{j\pi a(x^2+y^2)}, \tag{10.31}$$

which is the two-dimensional equivalent of the function given by Eq. (8.79). With a, a_1, a_2, and b real constants, this function has the following properties:

$$q(\pm x,\pm y;a)=q(x,y;a),\quad q(x,y;a_1)q(x,y;a_2)=q(x,y;a_1+a_2),$$

$$q(x,y;-a)=q^*(x,y;a),\quad q(x,y;a_1)q^*(x,y;a_2)=q(x,y;a_1-a_2),$$

$$q(bx,by;a)=q(x,y;b^2a),\quad q(x,y;a_1)**q(x,y;a_2)=\frac{j}{a_1+a_2}q\left(x,y;\frac{a_1a_2}{a_1+a_2}\right),$$

$$q(x,y;0)=1,\quad q(x,y;a)**q^*(x,y;a)=\frac{1}{a^2}\delta(x,y). \tag{10.32}$$

The Fourier transform of this function is

$$\begin{aligned} Q(\xi,\eta;a)&=\mathfrak{F}\mathfrak{F}\{q(x,y;a)\}\\ &=\frac{j}{a}q^*\left(\xi,\eta;\frac{1}{a}\right). \end{aligned} \tag{10.33}$$

With $s(x, y)$ an arbitrary function, we have [similar to Eqs. (8.85)–(8.90)]

$$\begin{aligned} s(x,y)**q(x,y;a)&=q(x,y;a)\mathfrak{F}\mathfrak{F}\{s(x,y)q(x,y;a)\}|_{\xi=ax,\eta=ay}\\ &=jaq(x,y;a)[S(ax,ay)**q^*(x,y;a)]. \end{aligned} \tag{10.34}$$

$$\begin{aligned} [s(x,y)**q(x,y;a)]q^*(x,y;a)&=\mathfrak{F}\mathfrak{F}\{s(x,y)q(x,y;a)\}|_{\xi=ax,\eta=ay}\\ &=ja[S(ax,ay)**q^*(x,y;a)]. \end{aligned} \tag{10.35}$$

$$\{[s(x,y)**q(x,y;a)]q^*(x,y;a)\}**q(x,y;a)=\frac{j}{a}S(ax,ay). \tag{10.36}$$

$$[s(x,y)q^*(x,y;a)]**q(x,y;a)=q(x,y;a)S(ax,ay). \tag{10.37}$$

$$\{[s(x,y)q^*(x,y;a)]**q(x,y;a)\}q^*(x,y;a)=S(ax,ay). \tag{10.38}$$

We will also be interested in the function

$$q(r;a)=e^{j\pi ar^2}, \tag{10.39}$$

where $r^2=x^2+y^2$. This function has the same properties as those given for $q(x,y;a)$ in Eqs. (10.32)–(10.38) above.

10-2 THE SCALAR THEORY OF DIFFRACTION

It is well known that there are many situations in which the behavior of optical wave fields is not adequately described by theories based upon ray optics, or geometric optics. To illustrate, let us consider the experiment depicted in Fig. 10-4: an aperture (with lateral dimensions much greater than a wavelength) is illuminated from the left by a plane-wave field, and the irradiance patterns produced at various planes to the right of the aperture are observed. Ray theory leads to the prediction that the pattern at each of these planes (no matter how far from the aperture) should have exactly the same size and shape as the aperture. However, due to the phenomenon of *diffraction*, this prediction is found to be incorrect. Diffraction causes some of the light passing through the aperture to deviate from its original direction of propagation, which in turn causes the observed irradiance pattern to differ from the aperture in both size and shape. As the distance from the aperture is increased, we find that the diffraction effects generally becomes more pronounced; the observed irradiance

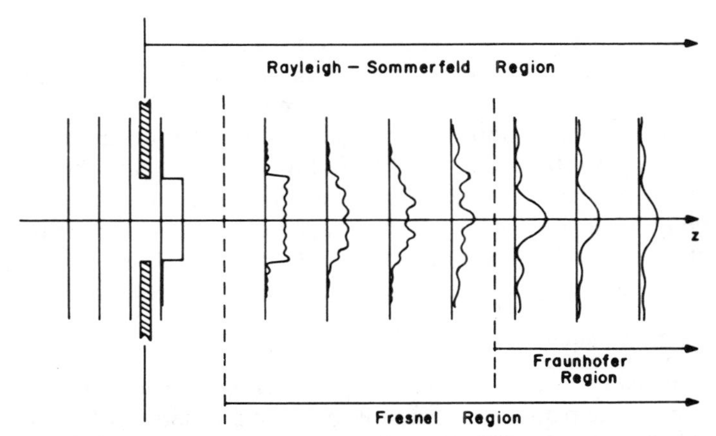

Figure 10-4 Designation of various regions of diffraction associated with an aperture, and nature of the irradiance profiles observed in these regions.

pattern, known as a *diffraction pattern*, undergoes continual and dramatic changes, and the similarity between it and the aperture gradually disappears. However, a point is finally reached beyond which only the size of this pattern, but not its shape, changes with increasing distance.

Although there is some disagreement among authors regarding the designation and extent of the various regions in which diffraction is observed, we shall use the following definitions for our discussions here (refer to Fig. 10-4):

1. *The Rayleigh-Sommerfeld Region.* We define this region to be the entire space to the right of the diffracting aperture. We do so because the general Rayleigh-Sommerfeld diffraction formula, given by Eq. (10.123), is valid throughout this space.
2. *The Fresnel Region.* We define the Fresnel region to be that portion of the Rayleigh-Sommerfeld region within which the Fresnel conditions of Eqs. (10.48) and (10.52) are satisfied. This is the region of validity of the so-called Fresnel approximations, and we note that it extends to infinity.
3. *The Fraunhofer Region.* We define this region to be that portion of the Fresnel region within which the Fraunhofer condition of Eq. (10.89) is satisfied. In other words, this is the region of validity of the Fraunhofer approximation. In the Fraunhofer region the size of the diffraction pattern increases with increasing distance, but its shape is invariant.

The Fraunhofer region is often referred to as the *far field*, which seems appropriate due to the distant location of this region relative to the aperture. However, we disagree with those who consider the Fresnel region and the near field to be synonymous; in fact, because the terms "near field" and "far field" quite naturally seem to imply mutually exclusive regions, and because the Fresnel region contains the Fraunhofer region, we feel that statements to this effect are very misleading. We prefer to regard the near field as that region lying between the diffracting aperture and the Fraunhofer region. In the material to follow we investigate the diffraction phenomenon primarily in the Fresnel and Fraunhofer regions.

A large number of photographs and graphical representations of diffraction patterns may be found in the literature (see Refs. 10-11 through 10-16), and the interested reader is urged to consult these sources.

The Rayleigh-Sommerfeld Diffraction Formula

We now consider the propagation of an arbitrary wave field from the plane $z = z_1$ to the plane $z = z_2$,[‡] and we use the geometry of Fig. 10-5 as an aid. We denote the distance between these two planes by $z_2 - z_1 = z_{12}$ and

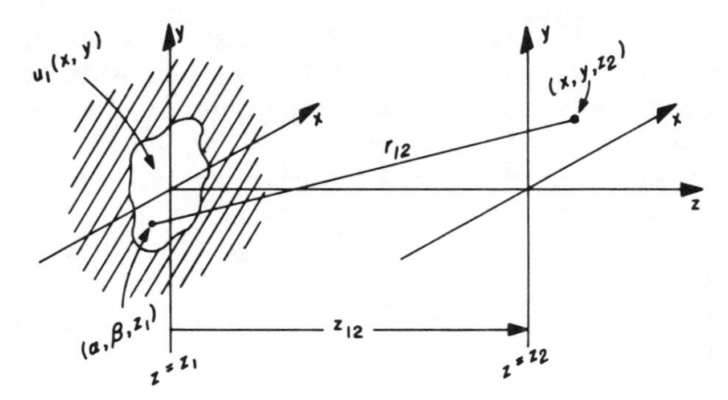

Figure 10-5 Geometry used to describe Rayleigh-Sommerfeld diffraction.

the slant distance between the points (α, β, z_1) and (x, y, z_2) by r_{12}. Thus,

$$r_{12} = \left[(z_2 - z_1)^2 + (x-\alpha)^2 + (y-\beta)^2 \right]^{1/2}$$

$$= z_{12} \left[1 + \left(\frac{x-\alpha}{z_{12}} \right)^2 + \left(\frac{y-\beta}{z_{12}} \right)^2 \right]^{1/2}, \tag{10.40}$$

and for $r_{12} \gg \lambda$, $u_2(x, y)$ may be described by the following approximation of the *Rayleigh-Sommerfeld diffraction formula* (see Ref. 10-9):

$$u_2(x,y) = \iint_{-\infty}^{\infty} u_1(\alpha, \beta) \left(\frac{z_{12}}{j\lambda r_{12}^2} \right) e^{jkr_{12}} d\alpha \, d\beta$$

$$= \iint_{-\infty}^{\infty} u_1(\alpha,\beta) \frac{\exp\left\{ jkz_{12} \left[1 + \left(\frac{x-\alpha}{z_{12}} \right)^2 + \left(\frac{y-\beta}{z_{12}} \right)^2 \right]^{1/2} \right\}}{j\lambda z_{12} \left[1 + \left(\frac{x-\alpha}{z_{12}} \right)^2 + \left(\frac{y-\beta}{z_{12}} \right)^2 \right]} d\alpha \, d\beta. \tag{10.41}$$

‡For the present we choose $z_2 > z_1$, but in later sections we will allow the observation plane to lie to the left of the diffracting object, i.e., $z_2 < z_1$. The development, however, is general and applies to either situation.

Reference to Eq. (10.16) leads to a very interesting interpretation of Eq. (10.41): $u_2(x, y)$ may be regarded as a linear superposition of diverging spherical waves, each of which emanates from a point (α,β,z_1) and is weighted according to its location by $(j\lambda)^{-1}u_1(\alpha,\beta)d\alpha\, d\beta$. This is a mathematical statement of the famed *Huygens-Fresnel principle*, in which each point on a wavefront is considered to be a source of a secondary spherical wave known as a *Huygens' wavelet*; the envelope of all the Huygens' wavelets emanating from a given wavefront at any instant of time is then used to describe that same wavefront at a later instant of time. For more details, see Ref. 10-1.

LSI System Description of Diffraction

Another look at Eq. (10.41) reveals it to be a convolution integral, i.e.,

$$u_2(x,y)=u_1(x,y)**\frac{\exp\left\{jkz_{12}\left[1+\left(\frac{x}{z_{12}}\right)^2+\left(\frac{y}{z_{12}}\right)^2\right]^{1/2}\right\}}{j\lambda z_{12}\left[1+\left(\frac{x}{z_{12}}\right)^2+\left(\frac{y}{z_{12}}\right)^2\right]}. \quad (10.42)$$

As a result, we may consider $u_2(x, y)$ to be the output of an LSI system with input $u_1(x, y)$ and impulse response

$$h_{12}(x,y)=\frac{\exp\left\{jkz_{12}\left[1+\left(\frac{x}{z_{12}}\right)^2+\left(\frac{y}{z_{12}}\right)^2\right]^{1/2}\right\}}{j\lambda z_{12}\left[1+\left(\frac{x}{z_{12}}\right)^2+\left(\frac{y}{z_{12}}\right)^2\right]}. \quad (10.43)$$

Although it is difficult to obtain by direct Fourier transformation, we have already shown the associated transfer function to be [see Eq. (10.27)]

$$H_{12}(\xi,\eta)=\exp\left\{jkz_{12}\left[1-\lambda^2(\xi^2+\eta^2)\right]^{1/2}\right\}. \quad (10.44)$$

Then, with $h_{12}(x, y)$ and $H_{12}(\xi,\eta)$ as given above, we have

$$u_2(x,y)=u_1(x,y)**h_{12}(x,y),$$
$$U_2(\xi,\eta)=U_1(\xi,\eta)H_{12}(\xi,\eta). \quad (10.45)$$

To simplify the terminology, we shall refer to $u_1(x, y)$ and $u_2(x, y)$ as the input and output signals, respectively, and to $U_1(\xi, \eta)$ and $U_2(\xi, \eta)$ as their plane-wave spectra or *spatial-frequency spectra*.

10-3 DIFFRACTION IN THE FRESNEL REGION

Although Eqs. (10.41)–(10.45) are easy to understand from a physical point of view, they are usually difficult to evaluate. However, they may be simplified somewhat by restricting the domains of both the input and output signals to regions lying near the z-axis, the lateral dimensions of which are much smaller than the separation between the input and output planes. To illustrate, let us assume that the input signal has a maximum radial extent of L_1, i.e.,

$$u_1(x, y) = 0, \qquad \sqrt{x^2 + y^2} > L_1, \tag{10.46}$$

and let us agree to confine our observations in the output plane to a region of maximum radial extent L_2. Consequently, the simplified expression we are about to derive for $u_2(x, y)$ will be reasonably accurate as long as $\sqrt{x^2 + y^2} \leqslant L_2$ in the output plane, but outside this region it may be inaccurate. With these restrictions on the extent of both the input signal and the observation region, the maximum value of the quantity $[(x-\alpha)^2 + (y-\beta)^2]$ will be

$$\left[(x-\alpha)^2 + (y-\beta)^2\right]_{\max} = (L_1 + L_2)^2. \tag{10.47}$$

If we then require

$$|z_{12}| \gg L_1 + L_2, \tag{10.48}$$

we may approximate the denominator of the spherical wave in Eq. (10.41) by

$$j\lambda z_{12}\left[1 + \left(\frac{x-\alpha}{z_{12}}\right)^2 + \left(\frac{y-\beta}{z_{12}}\right)^2\right] \cong j\lambda z_{12}. \tag{10.49}$$

More care must be used in simplifying the expression for the phase of the spherical wave because, even though the error is very small relative to the actual value of the phase, it might be a large fraction of a radian; this, in turn, could cause a significant error in the results obtained for $u_2(x, y)$.

We first expand the phase in a binomial series as follows:

$$kr_{12}=kz_{12}+\frac{k}{2z_{12}}\left[(x-\alpha)^2+(y-\beta)^2\right]-\frac{k}{8z_{12}^3}\left[(x-\alpha)^2+(y-\beta)^2\right]^2+\ldots. \tag{10.50}$$

From Eq. (10.47) we know that the maximum absolute value of the third term of this series will be equal to

$$\left|\frac{k}{8z_{12}^3}\left[(x-\alpha)^2+(y-\beta)^2\right]^2\right|_{\max}=\left|\frac{k}{8z_{12}^3}\left[(L_1+L_2)^2\right]^2\right|$$

$$=\left|\frac{\pi(L_1+L_2)^4}{4\lambda z_{12}^3}\right|, \tag{10.51}$$

and we can ensure that the value of this quantity is much less than 1 radian if we require

$$|z_{12}|^3\gg\frac{\pi(L_1+L_2)^4}{4\lambda}. \tag{10.52}$$

With this constraint on z_{12}, we may then approximate the phase by

$$kr_{12}\cong kz_{12}+\frac{\pi}{\lambda z_{12}}\left[(x-\alpha)^2+(y-\beta)^2\right]. \tag{10.53}$$

The constraints of Eqs. (10.48) and (10.52) are sometimes known as the *Fresnel conditions*, and the approximations of Eqs. (10.49) and (10.53) are referred to as the *Fresnel approximations*. We mention that the condition of Eq. (10.52) is *usually* sufficient, but it may not be necessary. Goodman (Ref. 10-3) points out that the higher-order terms of Eq. (10.50) need not be much less than 1 radian as long as they do not change the value of the integral. On the other hand, the value of the integral may be changed by a higher-order term even when that term is much less than a radian. This can happen if certain derivatives of $U_1(\xi,\eta)$ are large enough to offset the small value of the corresponding term of Eq. (10.50) [see Eq. (10.77) and the subsequent discussion].

If we assume the Fresnel approximations to be valid, we may express $u_2(x, y)$ either in integral form

$$u_2(x,y)=\frac{e^{jkz_{12}}}{j\lambda z_{12}}\iint_{-\infty}^{\infty}u_1(\alpha,\beta)$$

$$\times\exp\left\{j\frac{\pi}{\lambda z_{12}}\left[(x-\alpha)^2+(y-\beta)^2\right]\right\}d\alpha\, d\beta, \tag{10.54}$$

or as the output of an LSI system

$$u_2(x,y)=u_1(x,y)**h_{12}(x,y). \tag{10.55}$$

In this latter expression the impulse response now has the form

$$h_{12}(x,y)=\frac{e^{jkz_{12}}}{j\lambda z_{12}}\exp\left[j\frac{\pi}{\lambda z_{12}}(x^2+y^2)\right]. \tag{10.56}$$

In these expressions the wavefronts of the spherical wave fields are described as parabolic surfaces rather than spherical surfaces, but this description is a very good one as long as the Fresnel conditions are satisfied.

The transfer function associated with Fresnel diffraction is found to be

$$\begin{aligned} H_{12}(\xi,\eta) &= \mathscr{F}\mathscr{F}\{h_{12}(x,y)\} \\ &= e^{jkz_{12}}e^{-j\pi\lambda z_{12}(\xi^2+\eta^2)}, \end{aligned} \tag{10.57}$$

and it describes the propagation of plane-wave components from the input plane to the output plane when the Fresnel approximations are valid. For example, if a particular component is propagating with direction cosines $\left(\lambda\xi,\lambda\eta,\sqrt{1-\lambda^2(\xi^2+\eta^2)}\right)$, it undergoes a phase change of $z_{12}[k-\pi\lambda(\xi^2+\eta^2)]$ in traveling from the point (x,y,z_1) to the point (x,y,z_2). Thus, with $U_1(\xi,\eta)$ depicting the distribution of plane wave components in the input plane and $H_{12}(\xi,\eta)$ describing how each of these components propagates to the output plane, the quantity

$$U_2(\xi,\eta)=U_1(\xi,\eta)H_{12}(\xi,\eta) \tag{10.58}$$

represents the distribution of plane-wave components in the output plane.

To make the above expressions still more compact we define

$$B_{il}=\frac{e^{jkz_{il}}}{j\lambda z_{il}}, \tag{10.59}$$

and make use of the function $q(x,y;a)$ [see Eq. (10.31)] to write

$$\begin{aligned} h_{12}(x,y) &= B_{12}q\left(x,y;\frac{1}{\lambda z_{12}}\right), \\ H_{12}(\xi,\eta) &= e^{jkz_{12}}q^*(\xi,\eta;\lambda z_{12}). \end{aligned} \tag{10.60}$$

With this notation we may express the complex amplitude $u_2(x, y)$ as

$$u_2(x,y) = u_1(x,y) ** B_{12} q\left(x, y; \frac{1}{\lambda z_{12}}\right), \tag{10.61}$$

and using Eq. (10.34) we obtain

$$\begin{aligned} u_2(x,y) &= B_{12} q\left(x, y; \frac{1}{\lambda z_{12}}\right) \mathcal{F}\mathcal{F}\left\{ u_1(x,y) q\left(x, y; \frac{1}{\lambda z_{12}}\right)\right\}\bigg|_{\xi = x/\lambda z_{12}, \eta = y/\lambda z_{12}} \\ &= B_{12} q\left(x, y; \frac{1}{\lambda z_{12}}\right) \iint_{-\infty}^{\infty} u_1(\alpha, \beta) q\left(\alpha, \beta; \frac{1}{\lambda z_{12}}\right) \\ &\quad \times \exp\left\{-j2\pi\left(\frac{\alpha x}{\lambda z_{12}} + \frac{\beta y}{\lambda z_{12}}\right)\right\} d\alpha\, d\beta, \end{aligned} \tag{10.62}$$

which is often a useful form in which to write $u_2(x, y)$. Finally, the plane-wave spectrum of the output is

$$U_2(\xi, \eta) = U_1(\xi, \eta) e^{jkz_{12}} q^*(\xi, \eta; \lambda z_{12}). \tag{10.63}$$

We note that $h_{12}(x, y)$ is not only separable in x and y, it is also circularly symmetric. We may therefore express it by

$$\begin{aligned} h_{12}(r) &= B_{12} \exp\left(j \frac{\pi r^2}{\lambda z_{12}}\right) \\ &= B_{12} q\left(r; \frac{1}{\lambda z_{12}}\right), \end{aligned} \tag{10.64}$$

where $q(r; a)$ is defined by Eq. (10.39), and the transfer function by

$$H_{12}(\rho) = e^{jkz_{12}} q^*(\rho; \lambda z_{12}). \tag{10.65}$$

For a circularly symmetric input $u_1(r)$, the output may be written as

$$\begin{aligned} u_2(r) &= u_1(r) ** h_{12}(r) \\ &= u_1(r) ** B_{12} q\left(r; \frac{1}{\lambda z_{12}}\right) \\ &= B_{12} q\left(r; \frac{1}{\lambda z_{12}}\right) \mathcal{H}_0\left\{ u_1(r) q\left(r; \frac{1}{\lambda z_{12}}\right)\right\}\bigg|_{\rho = r/\lambda z_{12}}, \end{aligned} \tag{10.66}$$

and its spectrum as

$$U_2(\rho) = U_1(\rho)H_{12}(\rho)$$
$$= U_1(\rho)e^{jkz_{12}}q^*(\rho;\lambda z_{12}). \tag{10.67}$$

Other Useful Approximations in the Fresnel Region

Let us now assume that we wish to observe the Fresnel diffraction pattern of some object over a region that is much larger than the object. If L_1 and L_2 again denote the maximum radial extent of the object and the observation region, respectively, and if $L_2 \gg L_1$, then Eq. (10.48) may be written simply as

$$|z_{12}| \gg L_2. \tag{10.68}$$

Furthermore, if L_2 is large enough that $(L_1+L_2)^{4/3} \cong L_2^{4/3}$, Eq. (10.52) may be simplified to

$$|z_{12}|^3 \gg \frac{\pi L_2^4}{4\lambda}. \tag{10.69}$$

We now give two examples to illustrate the dependence of the Fresnel conditions on L_1 and L_2.

Example

A circular aperture with a radius of 1 cm is illuminated by a normally incident plane-wave field of wavelength $\lambda = 500$ nm. If we wish to observe the Fresnel diffraction patterns of this aperture only within a region whose radius is also 1 cm, then $L_1 = L_2 = 1$ cm and

$$\frac{\pi(2L_1)^4}{4\lambda} = \frac{4\pi(10^{-2})^4}{5\times 10^{-7}}$$
$$= 0.251 \text{ m}^3. \tag{10.70}$$

From Eq. (10.52) we see that we must have $|z_{12}|^2 \gg 0.251$ m^3, and under the assumption that this condition is satisfied if $|z_{13}|^3$ is at least a factor of 10 larger than the right side of the inequality, we find that

$$|z_{12}| \geqslant 1.36 \text{ m}, \tag{10.71}$$

which also satisfies the requirement of Eq. (10.48). If we now wish to observe these Fresnel diffraction patterns over an area of radius much larger than 1 cm, say 20 cm, we have the situation $L_2 \gg L_1$ just discussed. Thus,

$$\frac{\pi L_2^4}{4\lambda} = \frac{\pi(2\times 10^{-1})^4}{4(5\times 10^{-7})}$$

$$= 2512 \text{ m}^3, \tag{10.72}$$

and from Eq. (10.69) we have the restriction $|z_{12}|^3 \gg 2512$ m^3. If we again use a factor of 10 to satisfy the "much greater than" condition, the requirement on z_{12} becomes

$$|z_{12}| \geqslant 29.3 \text{ m}, \tag{10.73}$$

which also satisfies Eq. (10.68). The actual determination of $u_2(x, y)$ for either of the above conditions is quite complicated, and we choose not to struggle with it here. ■

Although the calculation of virtually all Fresnel diffraction patterns is very difficult, there is at least one more simplifying approximation that can be made: we expand the function $q(x, y; 1/\lambda z_{12})$ in the series

$$q\left(x, y; \frac{1}{\lambda z_{12}}\right) = \exp\left[j\frac{\pi}{\lambda z_{12}}(x^2+y^2)\right]$$

$$= 1 + \frac{j\pi}{\lambda z_{12}}(x^2+y^2) + \frac{1}{2!}\left(\frac{j\pi}{\lambda z_{12}}\right)^2 (x^2+y^2)^2 + \frac{1}{3!}\left(\frac{j\pi}{\lambda z_{12}}\right)^3 (x^2+y^2)^3 + \dots, \tag{10.74}$$

and we may then write Eq. (10.62) as

$$u_2(x,y) = B_{12} q\left(x, y; \frac{1}{\lambda z_{12}}\right) \mathscr{F}\mathscr{F}\Bigg\{ u_1(x,y) + \frac{j\pi}{\lambda z_{12}}(x^2+y^2) u_1(x,y)$$

$$+ \frac{1}{2!}\left(\frac{j\pi}{\lambda z_{12}}\right)^2 (x^2+y^2)^2 u_1(x,y)$$

$$+ \frac{1}{3!}\left(\frac{j\pi}{\lambda z_{12}}\right)^3 (x^2+y^2)^3 u_1(x,y) + \dots \Bigg\}\Bigg|_{\xi = x/\lambda z_{12}, \eta = y/\lambda z_{12}} \tag{10.75}$$

We use the relationship (see Table 9-1)

$$(x^2+y^2)f(x,y) \overset{\mathcal{F}\mathcal{F}}{\rightarrow} -\frac{1}{4\pi^2}\left(\frac{\partial^2}{\partial \xi^2}+\frac{\partial^2}{\partial \eta^2}\right)F(\xi,\eta) \tag{10.76}$$

to evaluate Eq. (10.75), and we obtain

$$\begin{aligned} u_2(x,y) = B_{12}q\left(x,y;\frac{1}{\lambda z_{12}}\right)\Bigg[& U_1\left(\frac{x}{\lambda z_{12}},\frac{y}{\lambda z_{12}}\right) \\ & -\frac{j\lambda z_{12}}{4\pi}\left(\frac{\partial^2}{\partial x^2}+\frac{\partial^2}{\partial y^2}\right)U_1\left(\frac{x}{\lambda z_{12}},\frac{y}{\lambda z_{12}}\right) \\ & -\frac{1}{2}\left(\frac{\lambda z_{12}}{4\pi}\right)^2\left(\frac{\partial^2}{\partial x^2}+\frac{\partial^2}{\partial y^2}\right)^2 U_1\left(\frac{x}{\lambda z_{12}},\frac{y}{\lambda z_{12}}\right) \\ & +\frac{j}{6}\left(\frac{\lambda z_{12}}{4\pi}\right)^3\left(\frac{\partial^2}{\partial x^2}+\frac{\partial^2}{\partial y^2}\right)^3 U_1\left(\frac{x}{\lambda z_{12}},\frac{y}{\lambda z_{12}}\right)+\ldots\ldots\Bigg]. \end{aligned} \tag{10.77}$$

Consequently, $u_2(x,y)$ may be regarded as the sum of the input plane-wave spectrum $U_1(x/\lambda z_{12}, y/\lambda z_{12})$ and various of its derivatives. The number of terms required for this expression to be accurate depends on the extent of $u_1(x, y)$, the distance z_{12}, and the smoothness of $U_1(x/\lambda z_{12}, y/\lambda z_{12})$. In general, $|z_{12}|$ must be considerably larger than the values required either by Eq. (10.52) or by Eq. (10.69) if the number of terms is to be held to just a few, and for this reason Eq. (10.77) is of limited use in many Fresnel diffraction problems.† To see how large $|z_{12}|$ must be, let us return to Eq. (10.74). If we assume once again that the maximum radial extent of $u_1(x, y)$ is L_1, then the maximum absolute value of the nth term of Eq. (10.74) will be

$$|n\text{th term}|_{\max} = \left|\frac{1}{(n-1)!}\left(\frac{\pi L_1^2}{\lambda z_{12}}\right)^{n-1}\right|. \tag{10.78}$$

†Nevertheless, it may be useful in describing the impulse response of a defocused imaging system.

Thus, by choosing

$$|z_{12}| \geqslant \frac{\pi L_1^2}{\lambda} \tag{10.79}$$

we obtain

$$|n\text{th term}| \leqslant \frac{1}{(n-1)!}, \tag{10.80}$$

and the maximum value of the fifth term is only 0.042. As a result, $q(x, y; 1/\lambda z_{12})$ should be adequately described by the first four terms of Eq. (10.74). Finally, if we also assume that the absolute values of the derivatives of $U_1(\xi,\eta)$ do not exceed the absolute value of $U_1(\xi,\eta)$ at any point, i.e.,

$$|U_1^{(k,l)}(\xi,\eta)| \leqslant |U_1(\xi,\eta)|, \tag{10.81}$$

then the first four terms of Eqs. (10.75) and (10.77) are sufficient for describing $u_2(x, y)$.

The last assumption imposes rather severe restrictions on the derivatives of $U_1(\xi,\eta)$ and may further limit the usefulness of these results. However, the effects of the higher order derivatives can be reduced by choosing $|z_{12}|$ to be still larger. For example, by simply doubling the value of $|z_{12}|$ required by Eq. (10.79), the values of the third and fourth terms of Eq. (10.77) are reduced by factors of 4 and 8, respectively.

We are now able to see why the original Fresnel approximation of Eq. (10.53) might not yield valid results even though the third- and higher order terms of Eq. (10.50) are much less than 1 radian: if the values of the derivatives of $U_1(\xi,\eta)$ are sufficiently large, these terms may make significant contributions to the Fresnel integral. Therefore, care must be exercised whenever the Fresnel approximations are used.

Example

Under the assumption that the derivatives of $U_1(\xi,\eta)$ satisfy Eq. (10.81), we now calculate the distance $|z_{12}|$ required by Eq. (10.79) for the case of an object of 1-cm radius and a wavelength of 500 nm. We obtain

$$\begin{aligned} |z_{12}| &\geqslant \frac{\pi(10^{-2})^2}{5\times 10^{-7}} \\ &\geqslant 628 \text{ m}. \end{aligned} \tag{10.82}$$

Then from Eq. (10.69) we find the allowed size of the observation region at a distance of 628 m:

$$
\begin{aligned}
L_2^4 &\ll \frac{4\lambda|z_{12}|^3}{\pi} \\
&\ll \frac{4(5\times 10^{-7})(628)^3}{\pi} \\
&\ll 158\,\mathrm{m}^4 \\
&\leqslant 15.8\,\mathrm{m}^4.
\end{aligned}
\tag{10.83}
$$

Thus,

$$
L_2 \leqslant 1.99\,\mathrm{m}, \tag{10.84}
$$

and the first four terms of Eq. (10.77) will adequately represent $u_2(x, y)$ over a radius of approximately 2 m at a distance of $|z_{12}|=628$ m. It can also be shown that if the distance is doubled, i.e., $|z_{12}|=1256$ m, the first three terms of Eq. (10.77) are sufficient and the region of validity will have a radius of approximately 3.35 m. Finally, if the distance is doubled again, such that $|z_{12}|=2512$ m, the first two terms of Eq. (10.77) should suffice and the radius of the region of validity will be approximately 5.63 m. ■

If the input signal is circularly symmetric we use Eq. (9.121) to evaluate Eq. (10.77), and the result is

$$
\begin{aligned}
u_2(r) = B_{12}q\left(r; \frac{1}{\lambda z_{12}}\right)\Bigg[& U_1\left(\frac{r}{\lambda z_{12}}\right) - \frac{j\lambda z_{12}}{4\pi}\left(\frac{1}{r}\frac{d}{dr}r\frac{d}{dr}\right)U_1\left(\frac{r}{\lambda z_{12}}\right) \\
& - \frac{1}{2}\left(\frac{\lambda z_{12}}{4\pi}\right)^2\left(\frac{1}{r}\frac{d}{dr}r\frac{d}{dr}\right)^2 U_1\left(\frac{r}{\lambda z_{12}}\right) \\
& + \frac{j}{6}\left(\frac{\lambda z_{12}}{4\pi}\right)^3\left(\frac{1}{r}\frac{d}{dr}r\frac{d}{dr}\right)^3 U_1\left(\frac{r}{\lambda z_{12}}\right) + \cdots\cdots \Bigg].
\end{aligned}
\tag{10.85}
$$

The irradiance of the Fresnel diffraction pattern is given by

$$
I_2(x, y) = |u_2(x, y)|^2, \tag{10.86}
$$

and with $u_2(x, y)$ as given by Eq. (10.62) and $|B_{12}q(x, y; 1/\lambda z_{12})|^2 =$

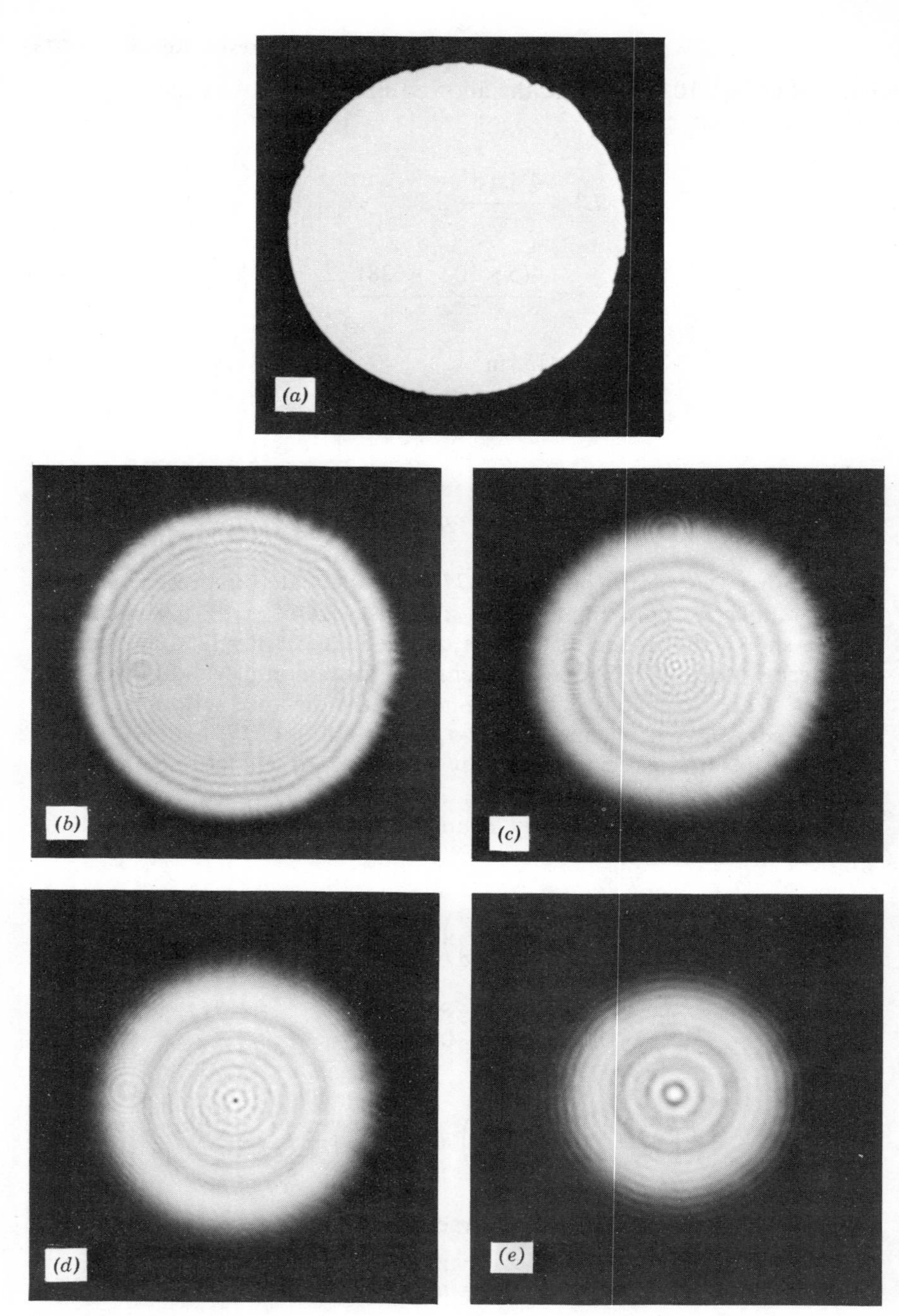

Figure 10-6 Diffraction in the Fresnel region of a circular aperture. (*a*) Circular aperture (almost). (*b–e*) Diffraction patterns at progressively greater distances from aperture. (Note: Converging spherical wave illumination was used, which caused lateral scale of patterns to decrease with increasing distance from aperture.)

$(\lambda z_{12})^{-2}$, it may be expressed as

$$I_2(x,y)=\left(\frac{1}{\lambda z_{12}}\right)^2\left|\iint_{-\infty}^{\infty} u_1(\alpha,\beta)q\left(\alpha,\beta;\frac{1}{\lambda z_{12}}\right)\right.$$

$$\left.\times\exp\left[-j2\pi\left(\frac{\alpha x}{\lambda z_{12}}+\frac{\beta y}{\lambda z_{12}}\right)\right]d\alpha\, d\beta\right|^2. \qquad (10.87)$$

Of course, any appropriate solution may be substituted for the integral in this expression, e.g., Eqs. (10.77) or (10.85). Typical Fresnel diffraction patterns of a clear circular aperture are shown in Fig. 10-6.

10-4 DIFFRACTION IN THE FRAUNHOFER REGION

Let us reconsider the description of Fresnel diffraction given by Eq. (10.75):

$$u_2(x,y)=B_{12}q\left(x,y;\frac{1}{\lambda z_{12}}\right)\mathcal{F}\mathcal{F}\left\{u_1(x,y)+\frac{j\pi}{\lambda z_{12}}(x^2+y^2)u_1(x,y)\right.$$

$$+\frac{1}{2!}\left(\frac{j\pi}{\lambda z_{12}}\right)^2(x^2+y^2)^2u_1(x,y)$$

$$\left.+\frac{1}{3!}\left(\frac{j\pi}{\lambda z_{12}}\right)^3(x^2+y^2)^3u_1(x,y)+\ldots\right\}\Bigg|_{\xi=x/\lambda z_{12},\,\eta=y/\lambda z_{12}} \qquad (10.88)$$

If we now require that

$$|z_{12}|\gg\frac{\pi L_1^2}{\lambda}, \qquad (10.89)$$

where L_1 is again the maximum radial extent of the input signal, then $u_2(x, y)$ is adequately described by the first term of this expression alone, i.e.,

$$u_2(x,y)=B_{12}q\left(x,y;\frac{1}{\lambda z_{12}}\right)U_1\left(\frac{x}{\lambda z_{12}},\frac{y}{\lambda z_{12}}\right). \qquad (10.90)$$

The restriction of Eq. (10.89) is known as the *Fraunhofer* (or *far-field) condition*, and the region for which this condition is satisfied is called the *Fraunhofer region* (or *far field*). Note that the complex amplitude of the diffracted wave field in this region is simply proportional to the Fourier transform of the input signal, evaluated at the spatial frequencies $\xi = x/\lambda z_{12}$ and $\eta = y/\lambda z_{12}$. In other words, the spatial distribution of complex amplitude in the Fraunhofer region describes the spatial-frequency spectrum of the complex amplitude distribution at the input plane; the value of $u_2(x, y)$ at any point (x, y) is proportional to the magnitude of $U_1(\xi, \eta)$ at the spatial frequencies $(x/\lambda z_{12}, y/\lambda z_{12})$. Or, in terms of the plane-wave spectrum of the input, the value of $u_2(x, y)$ at any point (x, y) is proportional to the magnitude of the plane-wave component of $u_1(x, y)$ propagating with direction cosines

$$\left(\frac{x}{z_{12}}, \frac{y}{z_{12}}, \left[1 - \left(\frac{x^2 + y^2}{z_{12}^2} \right) \right]^{1/2} \right).$$

We see, then, that Fraunhofer diffraction produces the same effects on an input signal as those produced by a spatial-frequency spectrum analyzer.

Using Eq. (10.69) to determine the radial extent of the region for which Eq. (10.90) is valid, we obtain

$$L_2^4 \ll \frac{4\lambda |z_{12}|^3}{\pi}. \tag{10.91}$$

If we again use a factor of 10 to satisfy any "much greater than" condition, Eqs. (10.91) and (10.89) may be combined to yield the constraint on L_2, in terms of L_1, at the *minimum distance for which the Fraunhofer condition is satisfied*:

$$\begin{aligned} L_2 &\leqslant \left[\frac{4\lambda |z_{12}|^3}{10\pi} \right]^{1/4} \\ &\leqslant \left[\frac{4\lambda}{10\pi} \left(\frac{10\pi L_1^2}{\lambda} \right)^3 \right]^{1/4} \\ &\leqslant \left(\frac{20\pi L_1^3}{\lambda} \right)^{1/2}. \end{aligned} \tag{10.92}$$

The irradiance of the Fraunhofer pattern is simply

$$I_2(x,y)=|u_2(x,y)|^2$$

$$=\left(\frac{1}{\lambda z_{12}}\right)^2\left|U_1\left(\frac{x}{\lambda z_{12}},\frac{y}{\lambda z_{12}}\right)\right|^2, \tag{10.93}$$

and we note that the multiplicative phase factors of $u_2(x, y)$ vanish when the modulus of $u_2(x, y)$ is squared. We may also express the irradiance as (refer to Table 9-1)

$$I_2(x,y)=\left(\frac{1}{\lambda z_{12}}\right)^2\mathcal{F}\mathcal{F}\{\gamma_{u_1}(x,y)\}\big|_{\xi=x/\lambda z_{12},\eta=y/\lambda z_{12}}$$

$$=\left(\frac{1}{\lambda z_{12}}\right)^2\Gamma_{u_1}\left(\frac{x}{\lambda z_{12}},\frac{y}{\lambda z_{12}}\right), \tag{10.94}$$

where $\gamma_{u_1}(x, y)$ is the complex autocorrelation of $u_1(x, y)$. Thus, the irradiance of the Fraunhofer diffraction pattern is proportional to the Fourier transform of the complex autocorrelation of the input signal; consequently, it may be regarded as a measure of the spatial-frequency spectral distribution of power at the input plane. In communications engineering, this would correspond to the *power spectrum*, or *power spectral density*, of the input, which describes the temporal-frequency spectral distribution of power.

Now consider the situation illustrated in Fig. 10-7, in which a transparency with complex amplitude transmittance $t_1(x,y)$ is illuminated by a

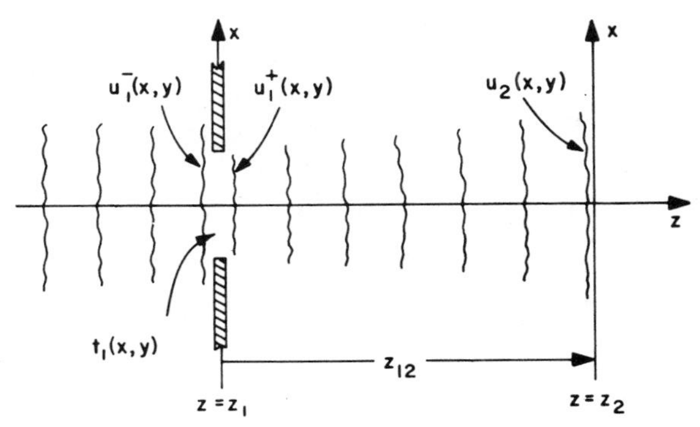

Figure 10-7 General configuration for Fraunhofer diffraction investigation.

wave field $u_1^-(x,y)$. Thus, with

$$u_1^+(x,y)=u_1^-(x,y)t_1(x,y), \tag{10.95}$$

Eq. (10.90) becomes

$$\begin{aligned} u_2(x,y) &= B_{12}q\left(x,y;\frac{1}{\lambda z_{12}}\right)U_1^+\left(\frac{x}{\lambda z_{12}},\frac{y}{\lambda z_{12}}\right) \\ &= B_{12}q\left(x,y;\frac{1}{\lambda z_{12}}\right)\left[U_1^-(\xi,\eta)**T_1(\xi,\eta)\right]\Big|_{\xi=x/\lambda z_{12},\eta=y/\lambda z_{12}} \\ &= B_{12}q\left(x,y;\frac{1}{\lambda z_{12}}\right)\left(\frac{1}{\lambda z_{12}}\right)^2\left[U_1^-\left(\frac{x}{\lambda z_{12}},\frac{y}{\lambda z_{12}}\right)**T_1\left(\frac{x}{\lambda z_{12}},\frac{y}{\lambda z_{12}}\right)\right]. \end{aligned} \tag{10.96}$$

Note that when the illumination is by a normally incident plane-wave field of amplitude A, where A specifies the strength and phase of the field at the plane $z=z_1$, then $u_1^-(x,y)=A$ and we have

$$U_1^-(\xi,\eta)=A\delta(\xi,\eta), \tag{10.97}$$

which results in

$$u_2(x,y)=AB_{12}q\left(x,y;\frac{1}{\lambda z_{12}}\right)T_1\left(\frac{x}{\lambda z_{12}},\frac{y}{\lambda z_{12}}\right), \tag{10.98}$$

and

$$I_2(x,y)=\frac{|A|^2}{(\lambda z_{12})^2}\left|T_1\left(\frac{x}{\lambda z_{12}},\frac{y}{\lambda z_{12}}\right)\right|^2. \tag{10.99}$$

Let us investigate the behavior of this irradiance distribution for different values of the various parameters. To begin with, we define the normalized Fourier transform of $t_1(x,y)$ to be

$$\overline{T}_1(\xi,\eta)=\frac{T_1(\xi,\eta)}{T_1(0,0)}. \tag{10.100}$$

Rearranging this expression we obtain

$$T_1(\xi,\eta)=T_1(0,0)\overline{T}_1(\xi,\eta); \tag{10.101}$$

consequently, we may rewrite Eq. (10.99) as

$$I_2(x,y)=|A|^2\left(\frac{1}{\lambda z_{12}}\right)^2|T_1(0,0)|^2\left|\overline{T}_1\left(\frac{x}{\lambda z_{12}},\frac{y}{\lambda z_{12}}\right)\right|^2. \qquad (10.102)$$

From the central ordinate theorem of Fourier transforms (see Table 9-1) we find that

$$T_1(0,0)=\iint_{-\infty}^{\infty} t_1(\alpha,\beta)\,d\alpha\,d\beta, \qquad (10.103)$$

and we note that the irradiance of the Fraunhofer pattern exhibits the following dependencies: it varies inversely as the square of the wavelength; it varies inversely as the square of the distance to the observation plane; and, it varies directly as the square of the "volume" of the aperture or transparency. In the case of a *clear aperture* [i.e., $t_1(x, y)$ a one-zero function], the volume of the aperture is numerically equal to its area, and the irradiance of the associated Fraunhofer pattern varies as the *square of the aperture area*. The explanation of this behavior is left as a thought exercise for the reader.

We also note that the lateral size, or scale, of the Fraunhofer pattern depends directly on both the wavelength and the distance to the observation plane. In addition, because $\overline{T}_1(\xi,\eta)$ is proportional to the Fourier transform of $t_1(x,y)$, its scale depends inversely on the scale of $t_1(x,y)$. As a result, the lateral scale of the Fraunhofer pattern is inversely proportional to the lateral dimensions of $t_1(x,y)$. We observe, however, that the shape (or form) of the Fraunhofer pattern depends only on the shape of the aperture and is independent of λ, z_{12} and the lateral dimensions of the aperture.

Example

To illustrate the behavior described above, we calculate the Fraunhofer diffraction pattern of a clear rectangular aperture of dimensions a and b. We assume the illumination to be a normally incident plane-wave field of amplitude A, and we also assume that the Fraunhofer conditions are satisfied. With

$$\begin{aligned} u_1^+(x,y) &= u_1^-(x,y)t_1(x,y) \\ &= A\,\mathrm{rect}\left(\frac{x}{a},\frac{y}{b}\right), \end{aligned} \qquad (10.104)$$

we obtain

$$U_1^+(\xi,\eta) = Aab\,\mathrm{sinc}(a\xi, b\eta). \tag{10.105}$$

Finally,

$$I_2(x,y) = |A|^2\left(\frac{ab}{\lambda z_{12}}\right)^2 \mathrm{sinc}^2\left(\frac{ax}{\lambda z_{12}}, \frac{by}{\lambda z_{12}}\right), \tag{10.106}$$

and we see that this pattern exhibits the behavior described above. The area of the aperture is $T_1(0,0) = ab$, and the irradiance does indeed vary as the square of the area. The width of the sinc^2 function in the x- and y-directions depends inversely on the values of a and b, respectively, which are the aperture dimensions. Note that the zeros of the sinc^2 function occur at $x = n\lambda z_{12}/a$ and $y = n\lambda z_{12}/b$, where n is a nonzero integer. Note also that the form of the irradiance is independent of λ, z_{12}, a, and b. The Fraunhofer patterns of two rectangular apertures are shown in Fig. 10-8, first for $b/a = 2$ and then for $b/a = 5$. ■

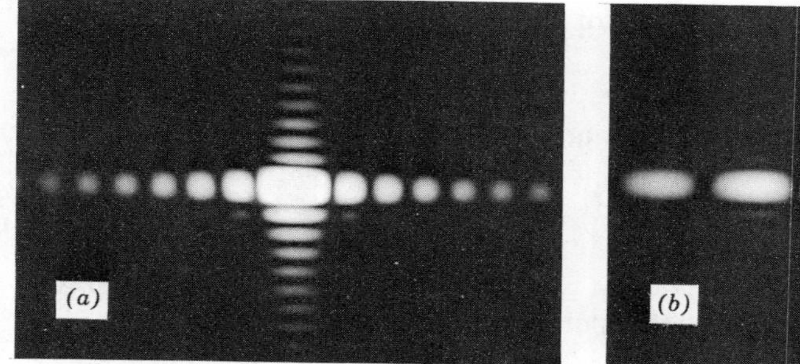

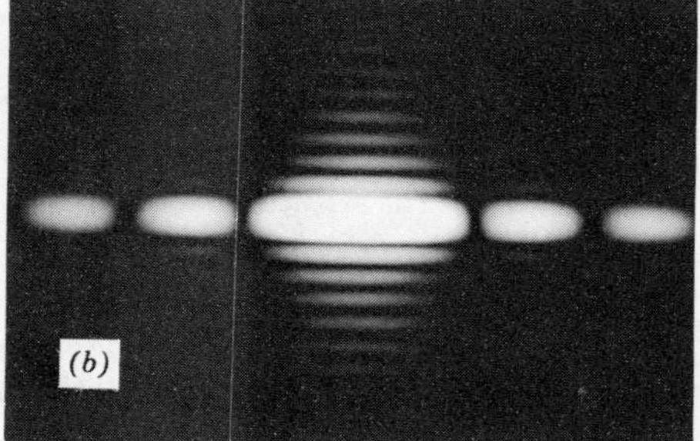

Figure 10-8 Fraunhofer diffraction patterns of rectangular apertures of different height-to-width ratios. (*a*) Height-to-width ratio of 2. (*b*) Height-to-width ratio of 5.

Example

Let us now determine the Fraunhofer diffraction pattern of a clear circular aperture of diameter d. We again assume the illumination to be a normally incident plane-wave field of amplitude A. The wave field transmitted by the aperture is given by

$$\begin{aligned} u_1^+(r) &= u_1^-(r)t_1(r) \\ &= A\,\mathrm{cyl}\left(\frac{r}{d}\right), \end{aligned} \tag{10.107}$$

and its Fourier transform by

$$U_1^+(\rho) = \frac{A\pi d^2}{4} \text{somb}(d\rho). \tag{10.108}$$

Therefore, the irradiance becomes

$$I_2(r) = |A|^2 \left(\frac{\pi d^2}{4\lambda z_{12}} \right)^2 \text{somb}^2\left(\frac{dr}{\lambda z_{12}} \right), \tag{10.109}$$

which again exhibits the dependencies noted above. This expression describes the familiar Airy pattern characterized by a bright central core surrounded by alternating dark and bright rings.

To complete the solution we choose $|A|^2 = 1$ W/cm², $d = 2$ cm, $\lambda = 500$ nm and $z_{12} = 10^4$ m. We first check to see that the Fraunhofer condition is satisfied:

$$\begin{aligned} \frac{\pi L_1^2}{\lambda} &= \frac{\pi (10^{-2})^2}{5 \times 10^{-7}} \\ &= 628 \text{ m}, \end{aligned} \tag{10.110}$$

and with $z_{12} = 10^4$ m $\gg 628$ m, the Fraunhofer condition is indeed satisfied. Next, from Eq. (10.91) we calculate the region of validity for our diffraction pattern results:

$$\begin{aligned} \frac{4\lambda z_{12}^3}{\pi} &= \frac{4(5 \times 10^{-7})(10^4)^3}{\pi} \\ &= 63.7 \times 10^4 \text{m}^4. \end{aligned} \tag{10.111}$$

Therefore, Eq. (10.109) will be valid for $L_2^4 \ll 63.7 \times 10^4 \text{m}^4$, or

$$L_2 \leqslant 15.9 \text{m}. \tag{10.112}$$

Finally, substituting the values given for $|A|^2$, d, λ, and z_{12} into Eq. (10.109) we obtain

$$\begin{aligned} I_2(r) &= (10^4) \left[\frac{\pi (2 \times 10^{-2})^2}{4(5 \times 10^{-7})(10^4)} \right]^2 \text{somb}^2 \left[\frac{2 \times 10^{-2} r}{(5 \times 10^{-7})(10^4)} \right] \\ &= 39.5 \text{ somb}^2(4r) W/\text{m}^2, \qquad r \leqslant 15.9 \text{m}. \end{aligned} \tag{10.113}$$

We note that the radius of the first dark ring of this Airy pattern is equal to 1.22(0.25m)=0.305m. Figure 10-9 shows the Fraunhofer diffraction patterns of two circular apertures, one with a diameter approximately 1.7 times that of the other. ■

Now that we have completed the above examples, it will be instructive to return momentarily to a discussion of the behavior of the Fraunhofer diffraction patterns of clear apertures. If we assume that the aperture in question has a lateral dimension of approximately d, we find that most of the power of the far field pattern has been diffracted into a region of angular extent

$$\theta_F \cong \frac{2\lambda}{d}. \qquad (10.114)$$

Therefore, at a distance of z_{12} the spatial extent of this region is simply $|\theta_F z_{12}|$; its area is then given by

$$\text{Area} \cong \frac{\pi}{4}(\theta_F z_{12})^2$$

$$\cong \frac{\pi\lambda^2 z_{12}^2}{d^2}. \qquad (10.115)$$

The power distributed over this area is just that transmitted by the aperture and is equal to the product of $|A|^2$ and the area of the aperture,

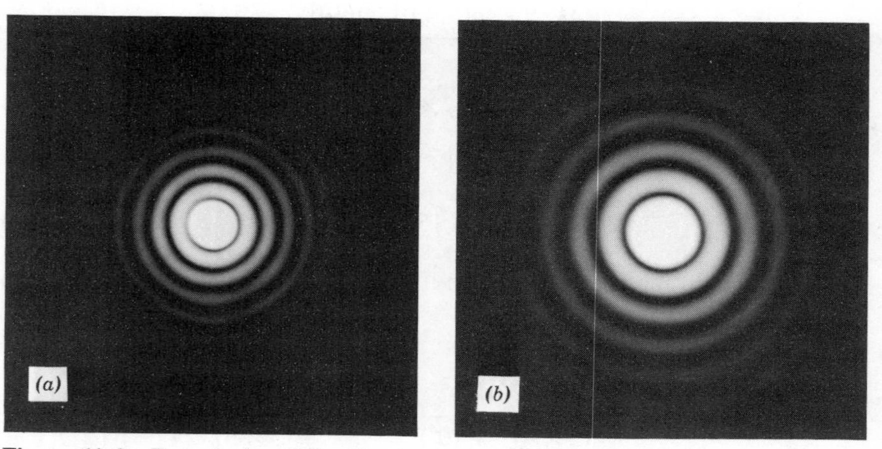

Figure 10-9 Fraunhofer diffraction patterns of circular apertures of different diameters. (*a*) Diameter of d. (*b*) Diameter of $d/1.7$.

i.e.,

$$\text{Power} \cong |A|^2 \frac{\pi d^2}{4}. \qquad (10.116)$$

Finally, an estimate of the average irradiance in the region described by Eq. (10.115) yields

$$\begin{aligned}\text{Avg. Irradiance} &\cong \frac{\text{Power}}{\text{Area}} \\ &\cong |A|^2 \left(\frac{d^2}{2\lambda z_{12}} \right)^2, \end{aligned} \qquad (10.117)$$

and we note how closely this estimate agrees with Eq. (10.109). If we now rewrite Eq. (10.102) as

$$I_2(x,y) = \left[|A|^2 |T_1(0,0)| \right] \left[\left(\frac{1}{\lambda z_{12}} \right)^2 |T_1(0,0)| \right] \left| \overline{T}_1 \left(\frac{x}{\lambda z_{12}}, \frac{y}{\lambda z_{12}} \right) \right|^2,$$

we may identify the first factor as the power transmitted by the aperture, the second as the inverse of the area given by Eq. (10.115), and the third as a form factor that describes the actual spatial distribution of power in the far field. These correspondences may not all be exact, but they provide physical insight for the phenomenon of Fraunhofer diffraction.

Because Fraunhofer diffraction produces the same effects on an input signal as a spectrum analyzer, we no longer consider $u_2(x, y)$ to be the output of a shift-invariant system [see the discussion following Eq. (8.93)]. Physically, of course, it remains shift invariant, but from a mathematical point of view the shift invariance has been destroyed by the Fraunhofer approximation. If the input is shifted by an amount x_0, the Fraunhofer pattern is also shifted by this amount. However, if the Fraunhofer approximation is to remain valid, x_0 must be restricted to a value much less than L_1; as a result, the amount of shift is always very small relative to the lateral extent of the Fraunhofer pattern and thus considered to be negligible. To illustrate, consider the previous example with the aperture shifted by $x_0 = 1$ mm. The first dark ring of the resulting far-field pattern has a radius of 305 mm, which is significantly greater than x_0. With x and y given in meters we obtain

$$\begin{aligned} I_2(x,y) &= 39.5 \,\text{somb}^2\left(4\sqrt{(x-0.001)^2 + y^2}\right) \\ &\cong 39.5 \,\text{somb}^2\left(4\sqrt{x^2+y^2}\right) \frac{W}{m^2}, \end{aligned} \qquad (10.118)$$

which is just Eq. (10.113) described in rectangular coordinates.

Example

Before proceeding to the next topic, we consider one more example: a sinusoidal amplitude grating, with a frequency of ξ_0 in the x-direction, is placed against a circular aperture of diameter d and illuminated with a plane-wave field of amplitude A. Thus,

$$u_1^+(x,y) = A\left[\tfrac{1}{2} + \tfrac{1}{2}\cos(2\pi\xi_0 x)\right]\mathrm{cyl}\left(\frac{\sqrt{x^2+y^2}}{d}\right) \tag{10.119}$$

and

$$\begin{aligned} U_1^+(\xi,\eta) &= \frac{A}{2}\left[\delta(\xi,\eta) + \frac{1}{2|\xi_0|}\delta\delta\left(\frac{\xi}{\xi_0}\right)\delta(\eta)\right] ** \frac{\pi d^2}{4}\,\mathrm{somb}\left(d\sqrt{\xi^2+\eta^2}\right) \\ &= \frac{A\pi d^2}{8}\left[\mathrm{somb}\left(d\sqrt{\xi^2+\eta^2}\right) + \tfrac{1}{2}\mathrm{somb}\left(d\sqrt{(\xi+\xi_0)^2+\eta^2}\right)\right. \\ &\quad \left. + \tfrac{1}{2}\mathrm{somb}\left(d\sqrt{(\xi-\xi_0)^2+\eta^2}\right)\right]. \end{aligned} \tag{10.120}$$

The irradiance is then

$$I_2(x,y) = \left(\frac{1}{\lambda z_{12}}\right)^2 \left|U_1^+\left(\frac{x}{\lambda z_{12}}, \frac{y}{\lambda z_{12}}\right)\right|^2, \tag{10.121}$$

and if $\xi_0 \gg d^{-1}$, the three terms of Eq. (10.120) are well separated (i.e., the cross products are negligible) and we obtain

$$\begin{aligned} I_2(x,y) \cong |A|^2\left(\frac{\pi d^2}{8\lambda z_{12}}\right)^2 &\left[\mathrm{somb}^2\left(\frac{d\sqrt{x^2+y^2}}{\lambda z_{12}}\right)\right. \\ &+ \tfrac{1}{4}\mathrm{somb}^2\left(\frac{d\sqrt{(x+\lambda z_{12}\xi_0)^2+y^2}}{\lambda z_{12}}\right) \\ &\left. + \tfrac{1}{4}\mathrm{somb}^2\left(\frac{d\sqrt{(x-\lambda z_{12}\xi_0)^2+y^2}}{\lambda z_{12}}\right)\right]. \end{aligned} \tag{10.122}$$

We see that the Fraunhofer diffraction pattern in this case consists of a bright Airy pattern centered at the point (0,0), with two weaker Airy patterns centered at the points $(\pm\lambda z_{12}\xi_0, 0)$. Note that the points $x = \pm\lambda z_{12}\xi_0$ in the output plane correspond to input spatial frequency values of $\pm\xi_0$ at the input plane. ■

10-5 A LESS-RESTRICTIVE FORMULATION OF SCALAR DIFFRACTION THEORY

In the previous sections we have concerned ourselves primarily with approximations of the Rayleigh-Sommerfeld diffraction formula that are suitable for describing the propagation of an optical wave field from one plane to another. However, these approximations require that severe limitations be placed upon the lateral extent of both the initial wave field and the observation region and also upon the distance between the two planes. We now briefly discuss a formulation of the theory that is *much less restrictive*; in this formulation, which was developed by Shack and Harvey (Ref. 10-9), results are derived for both plane and converging spherical wave illumination and for both plane and hemispherical observation surfaces.

As usual, we assume that an aperture or transparency with amplitude transmittance $t_1(x, y)$ is placed in the input plane and illuminated by a wave field $u_1^-(x, y)$. The wave field $u_2(x, y)$ may then be expressed in terms of $u_1^-(x, y)$ according to the general *Rayleigh-Sommerfeld diffraction formula* as follows:

$$u_2(x,y) = \int_{-\infty}^{\infty}\int u_1^-(\alpha,\beta)t_1(\alpha,\beta)\left(\frac{1}{kr_{12}} - j\right)\left(\frac{z_{12}}{\lambda r_{12}^2}\right)e^{jkr_{12}}\,d\alpha\,d\beta. \tag{10.123}$$

Shack and Harvey claim that this expression is "valid throughout the entire space in which diffraction occurs—right down to the aperture." There are no limitations on the maximum size of either the aperture or observation region, relative to the observation distance, because no approximations have been made (such limitations are introduced when the Fresnel approximations for the observation distance r_{12} are made). We note that when $r_{12} \gg \lambda$, Eq. (10.123) reduces to Eq. (10.41):

$$u_2(x,y) = \int_{-\infty}^{\infty}\int u_1^-(\alpha,\beta)t_1(\alpha,\beta)\left(\frac{z_{12}}{j\lambda r_{12}^2}\right)e^{jkr_{12}}\,d\alpha\,d\beta. \tag{10.124}$$

We shall work with this form of the diffraction formula for the remainder of the present section.

From the expansion given by Eq. (10.50) for kr_{12}, we obtain

$$kr_{12}=kz_{12}+\frac{k}{2z_{12}}(x^2+\alpha^2+y^2+\beta^2)-\frac{k}{z_{12}}(\alpha x+\beta y)-\dots$$

$$=2\pi\left[\frac{z_{12}}{\lambda}+\frac{1}{\lambda z_{12}}(x^2+\alpha^2+y^2+\beta^2)-\frac{1}{\lambda z_{12}}(\alpha x+\beta y)-\dots\right]. \quad (10.125)$$

Next we express the illuminating wave field by

$$u_1^-(x,y)=a_1^-(x,y)e^{j\phi_1^-(x,y)} \quad (10.126)$$

and rewrite Eq. (10.124) as

$$u_2(x,y)=\int\int_{-\infty}^{\infty} a_1^-(\alpha,\beta)t_1(\alpha,\beta)\left(\frac{z_{12}}{j\lambda r_{12}^2}\right)e^{j[kr_{12}+\phi_1^-(\alpha,\beta)]}d\alpha\, d\beta. \quad (10.127)$$

Then, following Shack and Harvey, we define the phase function

$$W_{12}(\alpha,\beta;x,y)=\frac{r_{12}}{\lambda}+\frac{1}{2\pi}\phi_1^-(\alpha,\beta)+\frac{1}{\lambda z_{12}}(\alpha x+\beta y), \quad (10.128)$$

which includes all of the terms within the square brackets of Eq. (10.125), save the linear term in α and β, plus the phase variations of the incident wave field. Hence,

$$kr_{12}+\phi_1^-(\alpha,\beta)=2\pi W_{12}(\alpha,\beta;x,y)-2\pi\left(\frac{\alpha x}{\lambda z_{12}}+\frac{\beta y}{\lambda z_{12}}\right), \quad (10.129)$$

and with

$$v_1(\alpha,\beta;x,y)=a_1^-(\alpha,\beta)t_1(\alpha,\beta)\left(\frac{z_{12}}{j\lambda r_{12}^2}\right)e^{j2\pi W_{12}(\alpha,\beta;x,y)}$$

$$=\frac{a_1^-(\alpha,\beta)t_1(\alpha,\beta)e^{j2\pi W_{12}(\alpha,\beta;x,y)}}{j\lambda z_{12}\left[1+\left(\frac{x-\alpha}{z_{12}}\right)^2+\left(\frac{y-\beta}{z_{12}}\right)^2\right]} \quad (10.130)$$

defined to be *a generalized pupil function* (see Ref. 10-9), we have

$$u_2(x,y)=\iint_{-\infty}^{\infty} v_1(\alpha,\beta;x,y)\exp\left[-j2\pi\left(\frac{\alpha x}{\lambda z_{12}}+\frac{\beta y}{\lambda z_{12}}\right)\right]d\alpha\, d\beta. \qquad (10.131)$$

Thus, the diffraction formula of Eq. (10.124) has now been expressed as the Fourier transform of a generalized pupil function that includes the amplitude transmittance of the actual aperture, the magnitude and phase of the incident wavefront, the obliquity factor, and all phase terms of the binomial expansion for kr_{12} except the term that was extracted to form the Fourier kernel. This expression, which retains the accuracy of Eq. (10.124), clearly reduces to the Fresnel formula if the appropriate approximations are made.

Without the simplifying Fresnel approximations, Eq. (10.124) is very difficult to evaluate; nevertheless, it is a very useful formula because it allows the diffraction phenomenon to be put into the context of a standard aberration problem of the type encountered in the analysis of imaging systems. To illustrate, let us first assume the illumination to be a normally incident plane-wave field and the observation surface to be the plane $z=z_2$. Then, in an imaging context, $u_2(x, y)$ may be thought of as an aberrated version of $T_1(x/\lambda z_{12}, y/\lambda z_{12})$. In other words, aside from multiplicative constants, $u_2(x, y)$ differs from the desired result $T_1(x/\lambda z_{12}, y/\lambda z_{12})$ because of aberrations that may be associated with the diffraction process. The quantity $W_{12}(\alpha,\beta; x, y)$ may be regarded as an aberration function of the usual sort (i.e., a power series expansion in the pupil and observation space coordinates), and its various terms identified with the standard aberrations of imaging systems. These are piston error, defocus, lateral magnification error, spherical aberration, coma, astigmatism, field curvature, and distortion.

For the case of plane-wave illumination and a plane observation surface, all of the above aberrations are present except for lateral magnification error (which is absent because the power series term associated with it was extracted to form the Fourier kernel). Shack and Harvey present a table (Ref. 10-9) listing the magnitudes of these aberrations as functions of various parameters; from that table it may be seen that the defocus aberration will be significant unless the observation distance is made very large (i.e., unless the Fraunhofer condition is satisfied). In addition, distortion imposes a severe constraint on the size of the observation region within which $u_2(x,y)$ may be regarded as an unaberrated version of $T_1(x/\lambda z_{12}, y/\lambda z_{12})$, a constraint that is related to Eq. (10.69).

If we now illuminate the aperture with a spherical wave field converging to the point $(0,0,z_2)$ as shown in Fig. 10-10, the phase term $\phi_1^-(x,y)$ of the

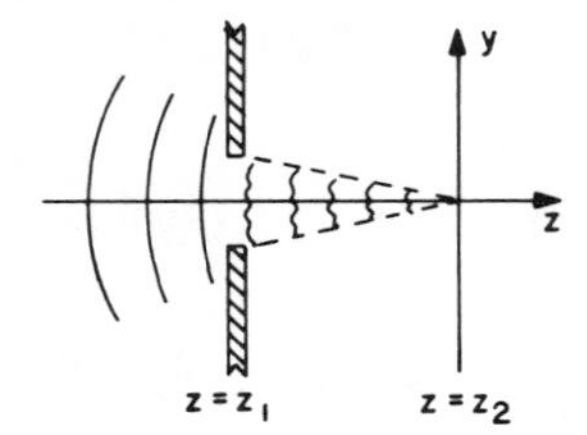

Figure 10-10 Diffraction configuration with converging spherical wave illumination and plane observation surface.

illuminating wave field cancels the terms of the aberration function power series that lead to defocus and spherical aberration; as a result, $u_2(x, y)$ no longer exhibits these two aberrations. However, the other aberrations are not eliminated by this procedure (i.e., the size of the observation region is still severely limited by distortion, etc.).

By describing the diffracted wave field in terms of the direction cosines γ_x and γ_y rather than the spatial variables x and y, and by observing it on a hemispherical surface rather than on a plane (see Fig. 10-11), the results are altered once again. With plane-wave illumination the diffracted wave field observed on the hemisphere is now free of distortion, field curvature, and piston error, but defocus and spherical aberration are present once again.

Finally, by combining the spherical wave illumination with the hemispherical observation surface as shown in Fig. 10-12, all aberrations except coma and astigmatism are eliminated. In other words, with spherical wave illumination, the diffracted wave field $u_2(\gamma_x, \gamma_y)$ observed on the appropriate hemisphere will be free of all aberrations except coma and astigmatism. These two aberrations are generally small when γ_x and γ_y are small (they vanish for $\gamma_x = \gamma_y = 0$), but even for large observation angles they can be minimized by selecting the radius of the hemispherical observation surface to be large relative to the aperture size; when this condition exists, $u_2(\gamma_x, \gamma_y)$ is accurately described by the scaled Fourier transform of $t_1(x, y)$ over the entire hemisphere, not merely over a small region near the z-axis as before.

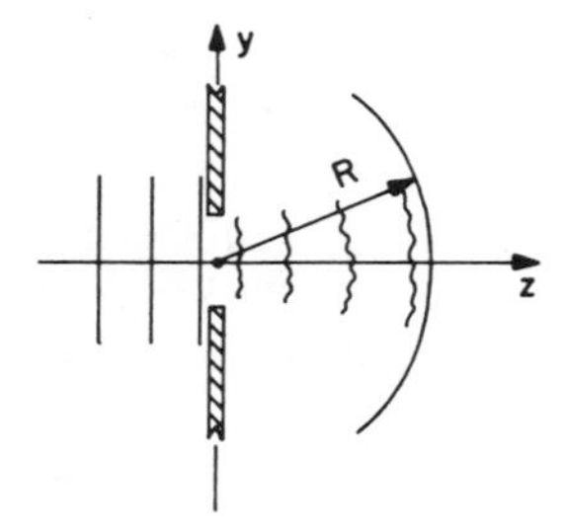

Figure 10-11 Diffraction configuration with plane wave illumination and hemispherical observation surface.

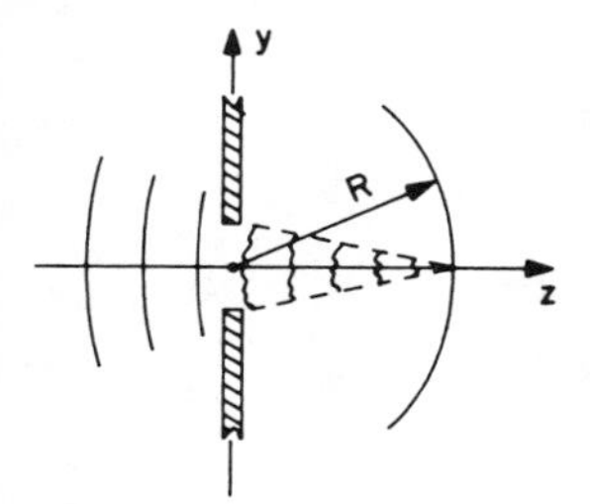

Figure 10-12 Diffraction configuration with converging spherical wave illumination and hemispherical observation surface.

Suppose, now, that the incident spherical wave is caused to converge to a point lying on the hemisphere, but not on the z-axis, and let the (x, y) coordinates of this point be $(0, y_0)$ as illustrated in Fig. 10-13. Shack and Harvey (Ref. 10-9) show that the resulting diffraction pattern, described in direction-cosine space, will be simply a shifted version of the pattern obtained when the incident spherical wave converges to the point $(0,0)$ on the hemisphere. Another way of saying this is as follows: if the pattern observed on the hemisphere is projected back onto the plane of the aperture, the form of the projected pattern is independent of the angle θ_0 even though the pattern on the hemisphere is not.

An interesting consequence of the behavior described above is illustrated by the following example: a sinusoidal amplitude grating $t_1(x, y)=\frac{1}{2}+\frac{1}{2}\cos(2\pi\xi_0 x)$ is illuminated first by a spherical wave converging to the point $(0,0)$ as shown in Fig. 10-12, and then by a spherical wave converging to the off-axis point $(0, y_0)$ as shown in Fig. 10-13. Under the assumption that the three diffracted orders are well separated, the diffraction pattern observed in the first case will consist of three separated Airy-like patterns, the centers of which all lie on the great circle defined by the intersection of the hemisphere and the plane $y=0$. This situation is depicted in Fig. 10-14. Let us now project this pattern back onto the plane of the aperture, and denote the projected pattern by $I_p(x, y)$. In the case of the spherical wave converging to the point $(0, y_0)$, the diffraction pattern observed on the hemisphere still consists of three separated, Airy-like

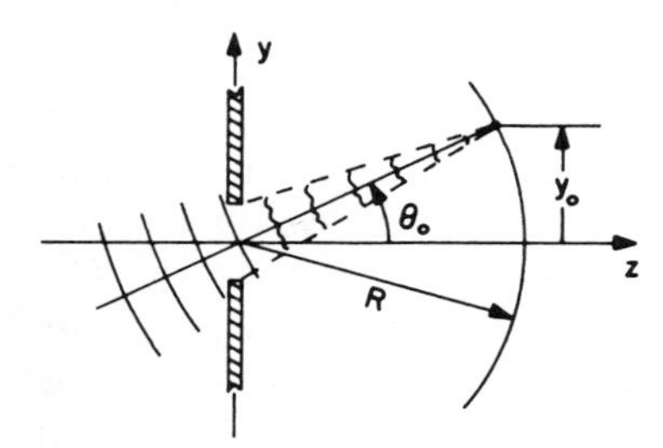

Figure 10-13 Oblique spherical wave illumination.

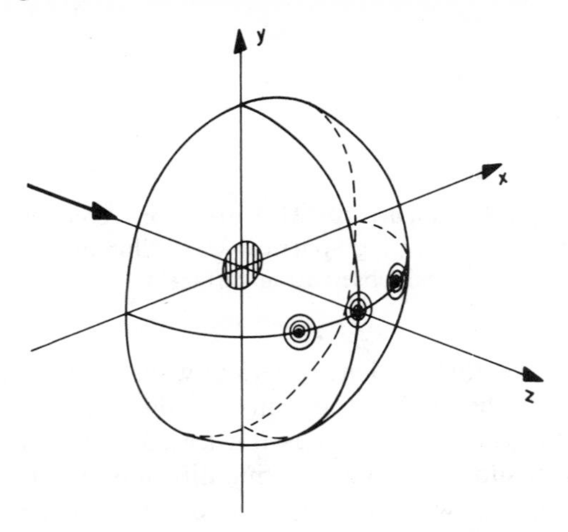

Figure 10-14 Diffraction by a sinusoidal amplitude grating when configuration of Fig. 10-12 is used.

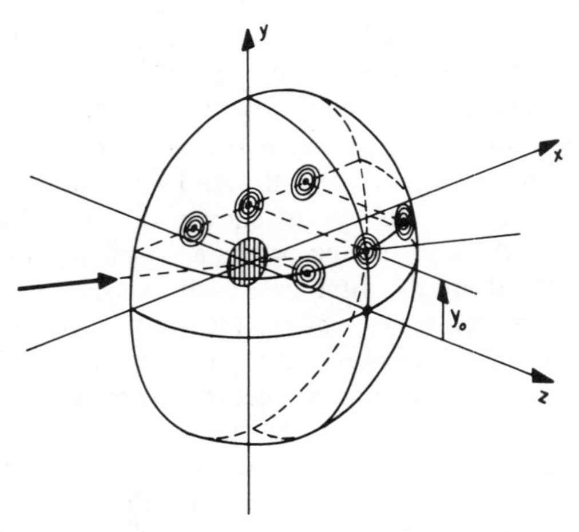

Figure 10-15 Diffracted orders of grating lie on circle of latitude when configuration of Fig. 10-13 is used.

patterns. However, their centers no longer lie on a great circle as might be expected. They now lie on the circle of latitude defined by the intersection of the hemisphere and the plane $y=y_0$, as shown in Fig. 10-15. The projection of this pattern back onto the plane of the aperture yields the same projected pattern as before except that it has been shifted in the y-direction by an amount y_0; i.e., we now have $I_p(x, y-y_0)$. This intriguing result has been verified experimentally by Shack and Harvey (Ref. 10-9).

10-6 EFFECTS OF LENSES ON DIFFRACTION

In this section we explore some of the effects that lenses can have on the diffraction process, and we begin by defining the complex amplitude transmittance functions of various lenses.

Complex Amplitude Transmittance of Lenses

Many authors derive the amplitude transmittance of a lens by first calculating a *thickness function*, which specifies the thickness of the lens as a function of the radial distance from its axis. However, because these calculations are normally carried out for single-element lenses, and because most lenses encountered in practical optical systems consist of more than one element, we shall not follow suit. We shall merely specify the transmittance of a lens in terms of its desired performance.

Most lens elements have spherical surfaces, not necessarily because they give the desired performance but because they are easy to grind and polish relative to other surface shapes. It is usually desired that an incident spherical wave field be converted into another spherical wave field, and this kind of performance generally requires elements with aspheric surfaces (which are difficult and expensive to fabricate). However, by combining several spherical elements having different radii of curvature and different refractive indices, it is possible to construct spherical lenses that exhibit the desired behavior over some range of the governing parameters. Therefore, we shall specify the performance of a spherical lens by requiring that it convert an incident spherical wave into another spherical wave and that it introduce no aberrations into the transmitted wave field.¶

In our development we shall use the quadratic-phase approximation for a spherical wave field introduced in Sec. 10-3. We need not do this, but it greatly simplifies the mathematics and yields valid results as long as the Fresnel conditions are satisfied. Let us place a spherical lens in the plane $z=z_2$, and illuminate it from the left with the spherical wave field

$$\begin{aligned} u_2^-(r) &= A\exp\left(j\frac{\pi r^2}{\lambda z_{12}}\right) \\ &= A\,q\left(r;\frac{1}{\lambda z_{12}}\right), \end{aligned} \tag{10.132}$$

where A describes the magnitude and absolute phase of the incident wave

¶This is clearly an idealization, but aberrations can be minimized by the proper selection of the lens and system configuration.

field and $z_{12}=z_2-z_1$. Consider first the case for which $z_1<z_2$: then z_{12} is positive and $u_2^-(r)$ represents a spherical wave diverging from the point $(r,\theta,z)=(0,0,z_1)$, which lies to the left of the lens. When $z_1>z_2$, however, z_{12} is negative and $u_2^-(r)$ depicts a spherical wave field converging toward the point $(0,0,z_1)$, which lies to the right of the lens. When $z_1=\pm\infty$, such that $z_{12}=\mp\infty$, then $u_2^-(r)$ is a plane-wave field propagating in the positive z-direction.

With $z_{23}=z_3-z_2$, we require that the transmitted complex amplitude be of the form

$$u_2^+(r)=AB_\ell q^*\left(r;\frac{1}{\lambda z_{23}}\right)p_\ell(r), \tag{10.133}$$

which describes a spatially limited spherical wave field that is either converging toward the point $(0,0,z_3)$, if $z_3>z_2$, or diverging from the point $(0,0,z_3)$, if $z_3<z_2$. When $z_3=\pm\infty$, $u_2^+(r)$ represents a plane-wave field as before. In this expression $p_\ell(r)$ is an aperture function that limits the extent of the transmitted wave field, and B_ℓ is a constant that accounts for the optical thickness** of the lens at its center and any losses due to reflection or absorption by the lens. Often a lens is assumed to be lossless, in which case $|B_\ell|=1$.

The complex amplitude transmittance of a spherical lens is then

$$\begin{aligned} t_\ell(r)&=\frac{u_2^+(r)}{u_2^-(r)}\\ &=\frac{AB_\ell q^*\left(r;\dfrac{1}{\lambda z_{23}}\right)p_\ell(r)}{Aq\left(r;\dfrac{1}{\lambda z_{12}}\right)}\\ &=B_\ell q^*\left(r;\frac{1}{\lambda z_{12}}+\frac{1}{\lambda z_{23}}\right)p_\ell(r). \end{aligned} \tag{10.134}$$

We next denote the *effective focal length* (or simply *focal length*) of the lens by f, and require that it satisfy the relationship

$$\frac{1}{f}=\frac{1}{z_{12}}+\frac{1}{z_{23}}. \tag{10.135}$$

**The optical thickness of any element is the product of its physical thickness and its refractive index. We shall assume that this thickness is not large enough to introduce significant errors into our results.

This relationship is known as the *lens law*, and the planes $z=z_1$ and $z=z_3$ are referred to as *conjugate planes*. We may now rewrite Eq. (10.134) as

$$t_\ell(r)=B_\ell q^*\left(r;\frac{1}{\lambda f}\right)p_\ell(r). \tag{10.136}$$

We note that f can be either positive or negative, depending on the magnitudes and polarities of z_{12} and z_{23}. If z_{12} and z_{23} are both positive, f is positive and the lens is referred to as a *positive lens*. If the magnitudes and polarities of z_{12} and z_{23} are such that f is negative, the lens is called a *negative lens*. The lens depicted in Fig. 10-16 is a positive lens, because z_{12} and z_{23} are both positive. Various types of positive and negative lenses are discussed briefly in Appendix A-2.

A positive lens causes a diverging spherical wave field to diverge less rapidly or even to converge as shown in Fig. 10-16; it also causes a converging wave field to converge more rapidly. In addition, such a lens will convert an incident plane-wave field into a spherical wave field converging toward a point in the plane $z_3=z_2+f$ and will convert a spherical wave field diverging from a point in the plane $z_1=z_2-f$ into a plane-wave field.

Conversely, a negative lens causes a diverging spherical wave to diverge more rapidly, and a converging spherical wave to converge less rapidly (or even to diverge). It also converts an incident plane wave into a spherical wave field that appears to diverge from a point in the plane $z_3=z_2-|f|$, and converts an incident spherical wave field, converging toward the point $z_1=z_2+|f|$, into a plane-wave field.

We shall be dealing primarily with positive lenses, for which the plane $z=z_2-f$ is called the *front focal plane* and the plane $z=z_2+f$ is known as the *back focal plane*. The locations of these planes are illustrated in Fig.

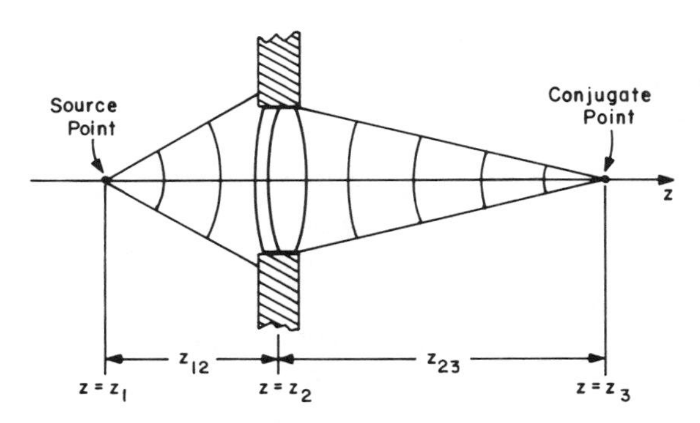

Figure 10-16 Effect of a positive lens on a diverging spherical wave field.

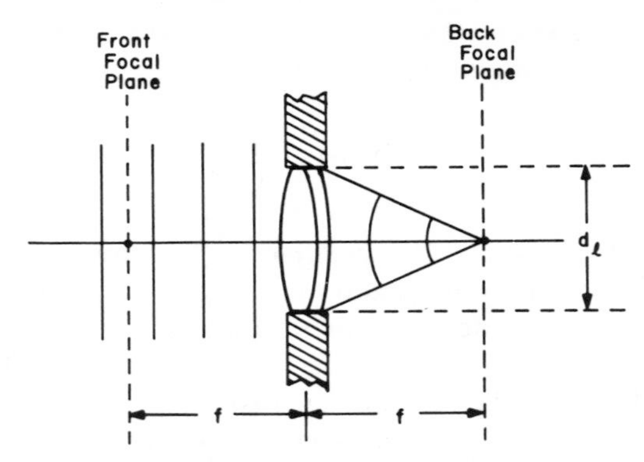

Figure 10-17 A positive lens of focal length f and diameter d_ℓ.

10-17. The *F-number*, abbreviated $F^{\#}$, of a positive lens having a focal length f and diameter d_ℓ is defined by

$$F^{\#} = \frac{f}{d_\ell}, \tag{10.137}$$

and describes the cone angle of the light arriving at the back focal plane when the source is located at $z_1 = -\infty$, i.e., when a plane-wave field is incident on the lens. The relationship is an inverse one, because $F^{\#}$ is small when the cone angle is large, and vice versa, as can be seen in Fig. 10-17. The *effective F-number*, $F^{\#}_{\text{eff}}$, describes the cone angle of the light arriving at a focus located at a plane other than the back focal plane. For example, the cone of light arriving at the plane $z = z_3$ in Fig. 10-16 has an effective F-number given by

$$F^{\#}_{\text{eff}} = \frac{z_{23}}{d_\ell}. \tag{10.138}$$

We see, then, that $F^{\#}$ characterizes a lens while $F^{\#}_{\text{eff}}$ characterizes a particular configuration.

Effects of Converging Spherical Wave Illumination

We now place an aperture or transparency in a converging spherical wave field, as shown in Fig. 10-18, and investigate the resulting diffracted wave field. A positive lens of focal length f is placed at the plane $z = z_2$ and illuminated by the spherical wave field emanating from a point source located at the point $(0, 0, z_1)$. The plane $z = z_1$ lies to the left of the front focal plane, which causes its conjugate plane $z = z_5$ to lie to the right of the lens (in fact, it lies to the right of the back focal plane). The spherical wave

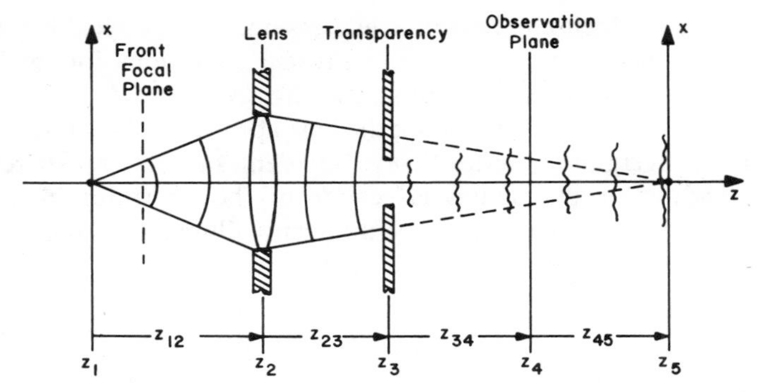

Figure 10-18 Diffraction with converging spherical wave illumination.

field leaving the lens is therefore converging toward the plane $z=z_5$, and the distances z_{12} and z_{25} satisfy the lens law of Eq. (10.135). For the time being we assume the locations of the source, the lens, and the conjugate plane to be fixed, and we describe the wave field leaving the lens by

$$u_2^+(x,y)=A\,B_\ell\,q^*\left(x,y;\frac{1}{\lambda z_{25}}\right)p_\ell(x,y). \tag{10.139}$$

The transparency $t_3(x,y)$ is located between the lens and the plane $z=z_5$, and the observation plane $z=z_4$ is placed to the right of the transparency. The wave field incident on the transparency is given by

$$u_3^-(x,y)=u_2^+(x,y)**B_{23}q\left(x,y;\frac{1}{\lambda z_{23}}\right), \tag{10.140}$$

and the wave field transmitted by it is then

$$u_3^+(x,y)=u_3^-(x,y)t_3(x,y). \tag{10.141}$$

Finally, the complex amplitude in the observation plane may be written as

$$\begin{aligned} u_4(x,y)&=u_3^+(x,y)**B_{34}q\left(x,y;\frac{1}{\lambda z_{34}}\right)\\ &=A\,B_\ell B_{23}B_{34}\left[\left\{\left[q^*\left(x,y;\frac{1}{\lambda z_{25}}\right)p_\ell(x,y)\right]**q\left(x,y;\frac{1}{\lambda z_{23}}\right)\right\}t_3(x,y)\right]\\ &\quad **q\left(x,y;\frac{1}{\lambda z_{34}}\right), \end{aligned} \tag{10.142}$$

which is a complete, albeit cumbersome, expression. Strictly speaking, it is only valid if the distance z_{23} satisfies the Fresnel conditions for the *lens aperture* and if z_{34} satisfies them for the transparency.

We now simplify matters by assuming the lens aperture to be sufficiently large that the converging spherical wave incident on the transparency exhibits no shading or diffraction effects from the aperture; in other words, we assume the transparency to be uniformly illuminated and we set $p_\ell(x, y) = 1$. To simplify the notation we define

$$a_{ij} = \frac{1}{\lambda z_{ij}}, \tag{10.143}$$

and thus $u_3^-(x, y)$ becomes

$$\begin{aligned} u_3^-(x,y) &= AB_\ell B_{23}\left[q^*(x,y;a_{25}) ** q(x,y;a_{23})\right] \\ &= AB_\ell B_{23}\left(\frac{j}{a_{23}-a_{25}}\right)q^*\left(x,y;\frac{a_{23}a_{25}}{a_{23}-a_{25}}\right), \end{aligned} \tag{10.144}$$

where Eq. (10.32) was used to obtain this result. Momentarily returning to the notation in terms of z_{ij}, and with $z_{25} - z_{23} = z_{35}$, we have

$$\begin{aligned} u_3^-(x,y) &= AB_\ell \frac{e^{jkz_{23}}}{j\lambda z_{23}}\left(\frac{j\lambda z_{23}z_{25}}{z_{35}}\right)q^*\left(x,y;\frac{1}{\lambda z_{35}}\right) \\ &= \left(\frac{Az_{25}}{z_{35}}\right)B_\ell e^{jkz_{23}}q^*\left(x,y;\frac{1}{\lambda z_{35}}\right), \end{aligned} \tag{10.145}$$

which is simply a spherical wave converging toward the plane $z = z_5$ and whose amplitude varies with the ratio z_{25}/z_{35}. Equation (10.142) now becomes

$$\begin{aligned} u_4(x,y) &= \left(\frac{Az_{25}}{z_{35}}\right)B_\ell e^{jkz_{23}}B_{34}\left[q^*\left(x,y;\frac{1}{\lambda z_{35}}\right)t_3(x,y)\right] ** q\left(x,y;\frac{1}{\lambda z_{34}}\right) \\ &= \frac{jAa_{35}B_\ell B_{23}B_{34}}{a_{25}a_{23}}\left[q^*(x,y;a_{35})t_3(x,y)\right] ** q(x,y;a_{34}), \end{aligned} \tag{10.146}$$

and with the substitutions

$$\begin{aligned} q^*(x,y;a_{35}) &= q^*(x,y;a_{34})q(x,y;\hat{a}), \\ v(x,y) &= q(x,y;\hat{a})t_3(x,y) \\ \hat{a} &= a_{34} - a_{35}, \end{aligned} \tag{10.147}$$

Eq. (10.37) may be used to write

$$u_4(x,y)=\frac{jAa_{35}B_\ell B_{23}B_{34}}{a_{25}a_{23}}\left[q^*(x,y;a_{34})v(x,y)\right]**q(x,y;a_{34})$$

$$=\frac{jAa_{35}B_\ell B_{23}B_{34}}{a_{25}a_{23}}q(x,y;a_{34})V(a_{34}x,a_{34}y)$$

$$=\frac{jAa_{35}B_\ell B_{23}B_{34}}{a_{25}a_{23}}q(x,y;a_{34})\mathfrak{F}\mathfrak{F}\{t_3(x,y)q(x,y;\hat{a})\}\big|_{\xi=a_{34}x,\eta=a_{34}y}. \tag{10.148}$$

Then, with the definition $\hat{a}=(\lambda\hat{z})^{-1}$, we find

$$\hat{z}=\frac{1}{\lambda\hat{a}}$$

$$=\frac{z_{34}z_{35}}{z_{35}-z_{34}}$$

$$=\frac{z_{34}z_{35}}{z_{45}}, \tag{10.149}$$

and in terms of this new parameter, Eq. (10.148) becomes

$$u_4(x,y)=\left(\frac{Az_{25}}{z_{35}}\right)B_\ell e^{jkz_{23}}B_{34}q\left(x,y;\frac{1}{\lambda z_{34}}\right)$$

$$\times\mathfrak{F}\mathfrak{F}\left\{t_3(x,y)q\left(x,y;\frac{1}{\lambda\hat{z}}\right)\right\}\Bigg|_{\xi=x/\lambda z_{34},\eta=y/\lambda z_{34}}. \tag{10.150}$$

Note that for $\hat{z}>0$, this expression is very similar to the Fresnel diffraction formula of Eq. (10.62); however, there are two important differences. First, it is to be regarded as a Fresnel diffraction formula for $t_3(x, y)$ alone, and not for the entire transmitted wave field $u_3^+(x, y)$. Second, when regarded in this fashion, it is the parameter $\hat{z}$ that determines the nature of the diffraction pattern rather than the distance to the observation plane alone. Consequently, even though $u_4(x, y)$ accurately describes the Fresnel diffraction associated with $t_3(x, y)$, the scaling is different from that obtained with plane-wave illumination. To illustrate, let us denote by $u_d(x, y)$ the Fresnel diffraction pattern of some aperture, at an observation distance of d, when plane-wave illumination is used. A pattern of the same form will occur when converging spherical wave illumination is used, but it will occur at an observation distance for which $\hat{z}=d$. From Eq. (10.149) we find that distance to be

$$z_{34}=\frac{dz_{35}}{d+z_{35}}. \tag{10.151}$$

The pattern observed for this value of z_{34} is then simply a scaled version of $u_d(x,y)$:

$$u_4(x,y)\propto u_d\left(\frac{xd}{z_{34}},\frac{yd}{z_{34}}\right). \tag{10.152}$$

If L_3 denotes the maximum radial extent of the transparency in the plane $z=z_3$, then Eq. (10.150) will be valid within a region of radius L_4 which satisfies

$$(L_3+L_4)^4 \ll \frac{4\lambda|z_{34}|^3}{\pi}. \tag{10.153}$$

If we now locate the observation plane at the conjugate plane of the source we have $z_4=z_5$, $z_{34}=z_{35}$, $z_{45}=0$, $\hat{z}=\infty$, and $\hat{a}=0$. As a result, $q(x,y;\hat{a})=1$ and the Fourier transform operation of Eq. (10.150) is simplified. With $B_{34}=B_{35}$, we obtain

$$u_5(x,y)=\left(\frac{Az_{25}}{z_{35}}\right)B_\ell e^{jkz_{23}}B_{35}q\left(x,y;\frac{1}{\lambda z_{35}}\right)T_3\left(\frac{x}{\lambda z_{35}},\frac{y}{\lambda z_{35}}\right), \tag{10.154}$$

which, aside from some multiplicative constants, has the same form as Eq. (10.98). We see, then, that the complex amplitude in the conjugate plane of the source may be regarded as the *Fraunhofer diffraction pattern of the transparency*, and that converging spherical wave illumination eliminates the need for very large observation distances. We may now draw the following important conclusion: when a transparency $t_3(x,y)$ is illuminated by a spherical wave converging toward the plane $z=z_5$, the *Fresnel diffraction pattern of the total transmitted wave field* $u_3^+(x,y)$, *observed in the plane* $z=z_5$, *is the Fraunhofer diffraction pattern of the transparency alone.* Note that the scale of this Fraunhofer pattern is proportional to the distance z_{35}, and it can therefore be varied by simply changing the location of the transparency: the pattern becomes larger by moving $t_3(x,y)$ toward the lens and smaller by moving it away from the lens. The irradiance of the pattern at the plane $z=z_5$ is given by

$$I_5(x,y)=|AB_\ell|^2\left(\frac{z_{25}}{z_{35}}\right)^2\left(\frac{1}{\lambda z_{35}}\right)^2\left|T_3\left(\frac{x}{\lambda z_{35}},\frac{y}{\lambda z_{35}}\right)\right|^2, \qquad (10.155)$$

and in the future we will frequently refer to that plane as the *Fourier transform plane*, or *Fraunhofer plane*, of the device.

The expressions of Eqs. (10.154) and (10.155) hold when $\hat{z}=\infty$, which occurs when $z_{45}=0$, and they are valid within a region whose maximum radial extent L_5 satisfies

$$(L_3+L_5)^4 \ll \frac{4\lambda|z_{35}|^3}{\pi}. \qquad (10.156)$$

Here L_3 describes the radial extent of $t_3(x, y)$, as before. Outside this region the pattern will suffer from various aberrations; however, as we learned in the previous section, all but coma and astigmatism can be eliminated by choosing the observation surface to be hemispherical rather than plane. If the conditions of Eq. (10.156) are satisfied, we may regard $u_5(x,y)$ as the output of the device depicted in Fig. 10-19 when the input is $t_3(x,y)$: the input is first multiplied by several constants and by $q^*(x,y;a_{35})$, and it is then passed through an LSI system with impulse response $B_{35}q(x, y; a_{35})$. This device is basically a spatial-frequency spectrum analyzer, and as such it is not shift invariant. We mentioned previously that Fraunhofer diffraction is normally considered to be a shift-variant process, but only because of the approximations that are made in order to arrive at the Fraunhofer formula. In the present case, however, no such approximations were made: the device is strictly shift variant. The only effect produced by a lateral shift of the input transparency is the introduction of a linear-phase factor at the output, and the diffraction pattern is not shifted. We note that while the device is not shift invariant, it is a linear device in terms of complex amplitude. (It is not, however, linear in terms of irradiance.)

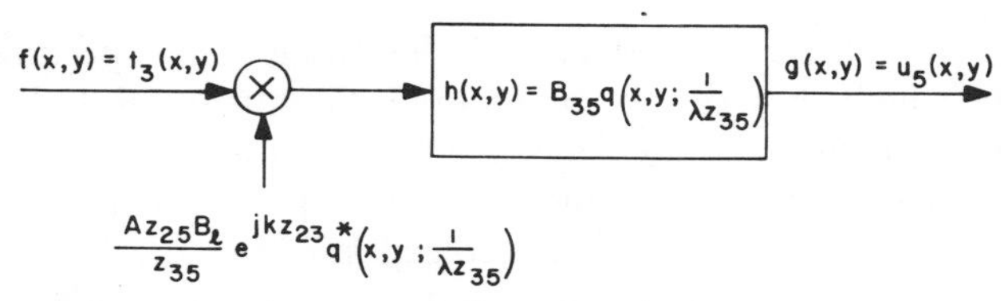

Figure 10-19 Operational representation of optical system shown in Fig. 10-18.

Example

We now calculate the Fraunhofer diffraction pattern for a circular aperture of diameter d which has been placed in the plane $z=z_3$ (see Fig. 10-18). If we choose

$$z_{25}=100 \text{ cm}, \qquad d=8 \text{ mm}, \qquad |AB_\ell|^2=10^{-1} \text{ mW/cm}^2,$$
$$z_{35}=80 \text{ cm}, \qquad \lambda=500 \text{ nm}, \tag{10.157}$$

we find that

$$\begin{aligned} I_5(r) &= |AB_\ell|^2\left(\frac{z_{25}}{z_{35}}\right)^2\left(\frac{\pi d^2}{4\lambda z_{35}}\right)^2 \operatorname{somb}^2\left(\frac{dr}{\lambda z_{35}}\right) \\ &= 2.46 \operatorname{somb}^2\left(\frac{r}{50}\right) \text{W/cm}^2, \qquad r \text{ in } \mu\text{m}. \end{aligned} \tag{10.158}$$

The region of validity is determined as follows:

$$\begin{aligned} \frac{4\lambda|z_{35}|^3}{\pi} &= 3.26\times10^{-7}\text{m}^4, \\ (L_3+L_5)^4 &\ll 3.26\times10^{-7} \\ &\leqslant 3.26\times10^{-8} \\ L_5 &\leqslant (3.26\times10^{-8})^{1/4}-L_3 \\ &\leqslant 1.34\times10^{-2}-0.4\times10^{-2} \\ &\leqslant 9.4 \text{ mm}. \end{aligned} \tag{10.159}$$

Thus, Eq. (10.158) will be accurate within a region whose diameter is approximately 2 cm, and we observe that virtually all of the diffracted light lies within this region. ■

We now investigate the effects of the finite lens aperture for the situations in which they cannot be ignored. Goodman (Ref. 10-3) shows that as long as the plane of the transparency is not located too far out into the Fresnel region, the effect of the finite lens aperture may be accounted for by simply erecting a geometrical projection of the aperture function at the transparency. Thus, if the cone of light incident on the transparency is not sufficiently wide to illuminate it uniformly, we assume the incident wave field to have the form

$$u_3^-(x,y) \cong \left(\frac{Az_{25}}{z_{35}}\right) B_\ell e^{jkz_{23}} q^*\left(x,y;\frac{1}{\lambda z_{35}}\right) p_\ell\left(\frac{z_{25}x}{z_{35}},\frac{z_{25}y}{z_{35}}\right). \qquad (10.160)$$

The transmitted wave field is then

$$u_3^+(x,y) \cong \left(\frac{Az_{25}}{z_{35}}\right) B_\ell e^{jkz_{23}} q^*\left(x,y;\frac{1}{\lambda z_{35}}\right) p_\ell\left(\frac{z_{25}x}{z_{35}},\frac{z_{25}y}{z_{35}}\right) t_3(x,y), \qquad (10.161)$$

and the projected lens aperture function becomes the effective limiting aperture at the plane of the transparency.

If we once again choose the location of the observation plane to be at the Fourier transform plane we have

$$u_5(x,y) = u_3^+(x,y) ** B_{35} q\left(x,y;\frac{1}{\lambda z_{35}}\right), \qquad (10.162)$$

and from Eq. (10.37) we obtain

$$u_5(x,y) \cong \left(\frac{Az_{25}}{z_{35}}\right) B_\ell e^{jkz_{23}} B_{35} q\left(x,y;\frac{1}{\lambda z_{35}}\right)$$

$$\times\left[\left(\frac{1}{\lambda z_{25}}\right)^2 P_\ell\left(\frac{x}{\lambda z_{25}},\frac{y}{\lambda z_{25}}\right) ** T_3\left(\frac{x}{\lambda z_{35}},\frac{y}{\lambda z_{35}}\right)\right]. \qquad (10.163)$$

We see, then, that the effect of the finite lens aperture is manifested in the form of a convolution at the observation plane: the Fourier transform of the aperture function is convolved with the Fourier transform of the transparency, the former having been scaled by z_{25} and the latter by z_{35}. It is interesting to note that even though $t_3(x, y)$ is not entirely illuminated, its entire transform is involved in the convolution; the incomplete illumination of $t_3(x, y)$ is accounted for in the smoothing and spreading of its spectrum by the convolution process. The irradiance associated with

Eq. (10.163) is

$$I_5(x,y) \cong |AB_\ell|^2 \left(\frac{z_{25}}{z_{35}}\right)^2 \left(\frac{1}{\lambda z_{35}}\right)^2$$

$$\times \left| \left(\frac{1}{\lambda z_{25}}\right)^2 P_\ell\left(\frac{x}{\lambda z_{25}}, \frac{y}{\lambda z_{25}}\right) ** T_3\left(\frac{x}{\lambda z_{35}}, \frac{y}{\lambda z_{35}}\right) \right|^2. \tag{10.164}$$

Now let us completely remove the transparency from the device, i.e., let us set $t_3(x, y) = 1$. Then

$$T_3\left(\frac{x}{\lambda z_{35}}, \frac{y}{\lambda z_{35}}\right) = (\lambda z_{35})^2 \delta(x,y), \tag{10.165}$$

and Eqs. (10.163) and (10.164) become, respectively,

$$u_5(x,y) \cong AB_\ell B_{25} q\left(x, y; \frac{1}{\lambda z_{35}}\right) P_\ell\left(\frac{x}{\lambda z_{25}}, \frac{y}{\lambda z_{25}}\right)$$

$$I_5(x,y) \cong |AB_\ell|^2 \left(\frac{1}{\lambda z_{25}}\right)^2 \left| P_\ell\left(\frac{x}{\lambda z_{25}}, \frac{y}{\lambda z_{25}}\right) \right|^2. \tag{10.166}$$

Note that the expression for the complex amplitude is not quite correct: the approximation of Eq. (10.160) causes the quadratic phase factor to be $q(x, y; a_{35})$, and it should really be $q(x, y; a_{25})$. Nevertheless, the expression for the irradiance is correct because the quadratic phase factor vanishes when the modulus of $u_5(x, y)$ is squared.

Let us now review this latest development. When the transparency $t_3(x, y)$ is completely and uniformly illuminated, the complex amplitude at the Fourier-transform plane is proportional to $T_3(a_{35}x, a_{35}y)$ alone. When the transparency is removed from the device, the amplitude at that plane is proportional to $P_\ell(a_{25}x, a_{25}y)$ alone. Finally, when incomplete illumination occurs because the extent of $t_3(x, y)$ is greater than the width of the illuminating cone of light, the amplitude is proportional to the convolution of these two functions.

It should be clear that if we move the point source of Fig. 10-18 closer to the front focal plane of the lens, the illuminating wave field will converge less rapidly, the Fourier-transform plane will be moved farther away from the lens, and the various diffraction patterns will be larger than before. When the point source is located exactly in the front focal plane, we have standard plane-wave illumination of the object and the Fourier-transform plane is shifted to infinity (or some very large distance).

If we now compare the present results with those obtained for plane-wave illumination we see that converging spherical wave illumination has the effect of compressing (and distorting) the z-axis into the region between the diffracting object and the Fourier transform plane. As the observation plane ranges from the object to the Fourier transform plane, the value of the parameter $\hat{z}$ ranges from 0 to ∞ and we observe the standard Fresnel diffraction patterns associated with the object; however, the scaling of these patterns now depends on the value of $\hat{z}$ rather than on that of z_{34} alone. If the observation plane is located at the Fourier transform plane, $\hat{z}$ becomes infinite and the Fraunhofer condition is satisfied for the object (but only for the object). As a result, we observe the standard Fraunhofer diffraction pattern of the object with a scaling factor of z_{35}.

So far, the only difference between plane wave and converging spherical wave illumination is in the location and scaling of the various Fresnel and Fraunhofer patterns; however, if we allow the observation plane to be located to the right of the Fourier-transform plane, i.e., if $z_4 > z_5$, we enter a region that seems to have no counterpart in the case of plane-wave illumination. When $z_4 > z_5$, $\hat{z}$ is negative and the inner quadratic phase factor of Eq. (10.150) describes a converging, rather than diverging, spherical wave field. Consequently, the corresponding region for plane-wave illumination is that region for which the observation distances are negative, i.e., for which the observation plane lies to the left of the transparency. (If this sounds somewhat strange, don't let it worry you—we shall clarify this point in just a bit.) At any rate, the diffraction patterns in this region do resemble the standard Fresnel patterns even though they have a slightly different form. As z_{34} becomes large, $\hat{z}$ approaches a value of $-z_{35}$ and Eq. (10.150) reduces to the Fraunhofer formula for the *product* $t_3(x, y)q^*(x, y; a_{35})$; however, as long as $z_4 > z_5$ there is no location at which the Fraunhofer pattern of $t_3(x, y)$ alone will be observed.

When the observation plane is located very close to, but not at, the Fourier-transform plane, $|\hat{z}|$ is large and an expansion similar to that of Eq. (10.75) may be useful in describing the diffraction pattern. We need only replace z_{12} by $\hat{z}$ *within the Fourier transform brackets* of that expression; the observation distance remains as the scaling factor once the Fourier transform is determined, and in the present case this distance is z_{34}. Under the assumption that the transparency is fully illuminated, i.e., if the lens aperture is large enough that it may be ignored, Eq. (10.150) may be written as

$$u_4(x,y) = \left(\frac{Az_{25}}{z_{35}}\right) B_\ell e^{jkz_{23}} B_{34} q\left(x, y; \frac{1}{\lambda z_{34}}\right) \mathcal{F}\mathcal{F}\left\{ t_3(x,y) + \left(\frac{j\pi}{\lambda \hat{z}}\right)(x^2+y^2) t_3(x,y) + \left(\frac{1}{2!}\right)\left(\frac{j\pi}{\lambda \hat{z}}\right)^2 (x^2+y^2)^2 t_3(x,y) + \left(\frac{1}{3!}\right)\left(\frac{j\pi}{\lambda \hat{z}}\right)^3 (x^2+y^2)^3 t_3(x,y) + \cdots \right\}\Bigg|_{\xi = x/\lambda z_{34},\ \eta = y/\lambda z_{34}} \qquad (10.167)$$

Evaluation of this expression yields a result similar to that of Eq. (10.77):

$$u_4(x,y) = \left(\frac{Az_{25}}{z_{35}}\right) B_\ell e^{jkz_{23}} B_{34} q\left(x, y; \frac{1}{\lambda z_{34}}\right) \left[T_3\left(\frac{x}{\lambda z_{34}}, \frac{y}{\lambda z_{34}}\right) - \frac{j\lambda z_{34}}{4\pi \hat{z}}\left(\frac{\partial^2}{\partial x^2} + \frac{\partial^2}{\partial y^2}\right) T_3\left(\frac{x}{\lambda z_{34}}, \frac{y}{\lambda z_{34}}\right) - \frac{1}{2}\left(\frac{\lambda z_{34}}{4\pi \hat{z}}\right)^2 \left(\frac{\partial^2}{\partial x^2} + \frac{\partial^2}{\partial y^2}\right)^2 T_3\left(\frac{x}{\lambda z_{34}}, \frac{y}{\lambda z_{34}}\right) + \frac{j}{6}\left(\frac{\lambda z_{34}}{4\pi \hat{z}}\right)^3 \left(\frac{\partial^2}{\partial x^2} + \frac{\partial^2}{\partial y^2}\right)^3 T_3\left(\frac{x}{\lambda z_{34}}, \frac{y}{\lambda z_{34}}\right) + \cdots \right]. \qquad (10.168)$$

If the derivatives of $T_3(\xi,\eta)$ satisfy

$$\left|T_3^{(k,l)}(\xi,\eta)\right| \leqslant \left|T_3(\xi,\eta)\right|, \tag{10.169}$$

and if we require

$$|\hat{z}| \geqslant \frac{\pi L_3^2}{\lambda}, \tag{10.170}$$

where L_3 is the maximum radial extent of $t_3(x,y)$, then the nth term of Eq. (10.168) will have a maximum value of

$$|n\text{th term}| \leqslant \frac{1}{(n-1)!}. \tag{10.171}$$

From Eq. (10.149) we have

$$\hat{z} = \frac{z_{34}z_{35}}{z_{45}}, \tag{10.172}$$

and when z_{45} is very small, such that $z_{34} \cong z_{35}$, we have

$$\hat{z} \cong \frac{z_{35}^2}{z_{45}}. \tag{10.173}$$

We may now express the inequality of Eq. (10.170) in terms of z_{45}:

$$|z_{45}| \leqslant \frac{\lambda z_{35}^2}{\pi L_3^2}. \tag{10.174}$$

Thus, with the effective F-number of the system given by

$$F_{\text{eff}}^{\#} = \frac{z_{35}}{2L_3}, \tag{10.175}$$

we obtain

$$|z_{45}| \leqslant \frac{4\lambda\left(F_{\text{eff}}^{\#}\right)^2}{\pi}, \tag{10.176}$$

and we observe that the constraint of Eq. (10.120) is related to the square of effective F-number. As $|z_{45}|$ becomes smaller and smaller, fewer and fewer terms of Eq. (10.168) are required, and when

$$|z_{45}| \ll \frac{4\lambda(F^{\#}_{\text{eff}})^2}{\pi} \leqslant \frac{4\lambda(F^{\#}_{\text{eff}})^2}{10\pi}, \tag{10.177}$$

$u_4(x, y)$ should be adequately described by the first term alone. Consequently, within the range

$$\frac{-4\lambda(F^{\#}_{\text{eff}})^2}{10\pi} \leqslant z_{45} \leqslant \frac{4\lambda(F^{\#}_{\text{eff}})^2}{10\pi}, \tag{10.178}$$

the extent of which is known as the *depth of focus* of the system, we have

$$u_4(x,y) \cong u_5(x,y) \cong \left(\frac{Az_{25}}{z_{35}}\right) B_\ell e^{jkz_{23}} B_{35} q\left(x,y;\frac{1}{\lambda z_{35}}\right) T_3\left(\frac{x}{\lambda z_{35}},\frac{y}{\lambda z_{35}}\right), \tag{10.179}$$

and

$$I_4(x,y) \cong I_5(x,y) \cong |AB_\ell|^2 \left(\frac{z_{25}}{z_{35}}\right)^2 \left(\frac{1}{\lambda z_{35}}\right)^2 \left|T_3\left(\frac{x}{\lambda z_{35}},\frac{y}{\lambda z_{35}}\right)\right|^2. \tag{10.180}$$

The results just obtained can be important in the analysis of an imaging system because the point spread function (impulse response) of such a system is either a Fraunhofer or a Fresnel diffraction pattern of the exit pupil. When such a system is properly focused, the point spread function is the Fraunhofer pattern of the exit pupil; however, when it is defocused, the point spread function is a Fresnel diffraction pattern of the exit pupil. Consequently, more and more terms of the expansion given by Eq. (10.168) are required to describe the point spread function as the focusing error becomes larger and larger. We see, then, that the depth of focus given by Eq. (10.178) can be used to indicate the focusing tolerances of an imaging system.

Effects of Diverging Spherical Wave Illumination

We now investigate the effects of diverging spherical wave field illumination on diffraction, and we use the setup of Fig. 10-20 for the development. A point source located at $(0,0,z_1)$ is used to illuminate a negative lens, and the spherical wave field leaving the lens appears to diverge from the point $(0,0,z_5)$. Thus, the planes $z=z_1$ and $z=z_5$ are conjugate planes as before. The diffracting object is again placed in the plane $z=z_3$, and the observation plane is located at $z=z_4$. If we assume that the lens aperture is large enough to ensure complete illumination of the transparency we have

$$u_3^-(x,y)=\left(\frac{Az_{25}}{z_{35}}\right)B_\ell e^{jkz_{23}}q^*\left(x,y;\frac{1}{\lambda z_{35}}\right), \tag{10.181}$$

where both z_{25} and z_{35} are depicted as being negative in the setup of Fig. 10-20. The transmitted wave field is $u_3^+(x,y)=u_3^-(x,y)t_3(x,y)$, as usual, and the amplitude at the observation plane becomes

$$u_4(x,y)=\left(\frac{Az_{25}}{z_{35}}\right)B_\ell e^{jkz_{23}}B_{34}\left[q^*\left(x,y;\frac{1}{\lambda z_{35}}\right)t_3(x,y)\right]**q\left(x,y;\frac{1}{\lambda z_{34}}\right), \tag{10.182}$$

which is identical to the expression of Eq. (10.146) obtained for the case of converging spherical wave illumination! Consequently, all of the results obtained for converging spherical wave illumination may be applied to the present setup if the proper interpretation is made.

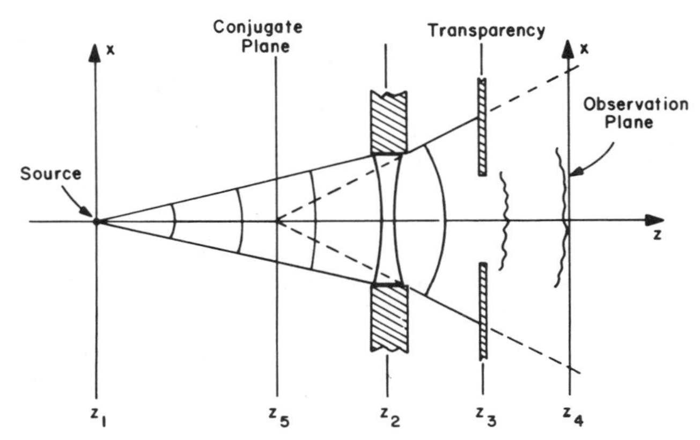

Figure 10-20 Diffraction with diverging spherical wave illumination.

Because the distance z_{35} is now negative, the proper interpretation requires that z_{34} also be negative. This means that the observation plane must be located in front of the transparency, a requirement that may seem physically implausible. However, if we accept the notion of virtual wave fields, virtual diffraction patterns, etc., we may proceed without too much difficulty. A virtual wave field is one that appears to exist at a particular location but which does not exist there physically. (To illustrate, when you look into the mirror in the morning you observe an image of yourself apparently located behind the mirror; clearly, neither you nor the image you see is located behind the mirror physically, and for this reason we call the image a virtual image.) If we now imagine that the observation plane is located to the left of the transparency, the virtual complex amplitude associated with that plane is identical in form to the expression of Eq. (10.150); however, z_{25}, z_{34}, and z_{35} are all negative, and as a result the parameter

$$\hat{z}=\frac{z_{34}z_{35}}{z_{45}} \tag{10.183}$$

is positive when $z_4 < z_5$ and negative when $z_4 > z_5$. Thus, the region to the left of the conjugate plane $z=z_5$ corresponds to the region between the transparency and the Fourier-transform plane in the case of converging spherical wave illumination. Similarly, the region between the conjugate plane and the transparency corresponds to the region to the right of the Fourier transform plane in the previous case. Finally, the conjugate plane $z=z_5$ corresponds to the Fourier transform plane of the previous setup; with $z_4=z_5$, we have $z_{45}=0$ and $\hat{z}=\infty$, and the virtual complex amplitude at this plane becomes

$$\begin{aligned} u_5(x,y) &= \left(\frac{Az_{25}}{z_{35}}\right) B_\ell e^{jkz_{23}} B_{35} q\left(x,y;\frac{1}{\lambda z_{35}}\right) T_3\left(\frac{x}{\lambda z_{35}},\frac{y}{\lambda z_{35}}\right), \\ &= \left(\frac{Az_{25}}{z_{35}}\right) B_\ell e^{jkz_{23}} B_{35} q^*\left(x,y;\frac{1}{|\lambda z_{35}|}\right) T_3\left(\frac{-x}{|\lambda z_{35}|},\frac{-y}{|\lambda z_{35}|}\right). \end{aligned} \tag{10.184}$$

We see that this is simply an inverted (upside down and backward) version of the corresponding amplitude obtained with converging spherical wave illumination, and we therefore regard it as a *virtual Fraunhofer diffraction pattern of* $t_3(x, y)$. The plane $z=z_5$ is referred to as the *virtual Fourier transform plane*, or *virtual Fraunhofer plane*, of the device.

As mentioned above, the wave fields described by Eqs. (10.182) and (10.184) do not really exist in the region $z_4 < z_3$. Nevertheless, if they could somehow be caused to exist in that region, and if the lens and diffracting object were then removed, the resulting wave field in the region $z_4 > z_3$ would be identical to that actually produced by the system. As a result, we may model such a device by the wave field it produces at any plane, and once we have done so we may (mentally) discard the physical hardware that produced that wave field.

In the present development we have assumed that the lens aperture did not introduce any significant diffraction effects or any shading of the transparency. In the event that such effects cannot be neglected, the corresponding results obtained for the case of converging spherical wave illumination may be extended to the present situation.

Because the expression of Eq. (10.182) obtained for diverging illumination has the same mathematical form as that given by Eq. (10.146), we may consider it to be completely general as long as the diffracting object lies to the right of the lens. We need only keep track of the polarities of the distances z_{12}, z_{25}, and z_{34}. Also, Figs. 10-18 and 10-20 may be regarded as two versions of the same general setup: the first was arranged such that z_{25} was positive and the lens emitted a converging spherical wave, whereas the second configuration caused z_{25} to be negative and produced a diverging spherical wave field. Thus, the basic setup is general, and our choice of observation plane location and interpretation of the results depends on the polarity of z_{25}.

Diffracting Object Located in Front of a Lens

We now consider a more general optical system; in this system the diffracting object is placed in front of a lens and illuminated as shown in Fig. 10-21. The source plane and its conjugate are located at $z=z_1$ and $z=z_5$, respectively, the object at $z=z_2$, the lens at $z=z_3$, and the observation plane at $z=z_4$. For the time being we shall assume the setup to be as depicted in the figure: the lens is positive, the source is located to the left of the front focal plane (and its conjugate to the right of the back focal plane), and the observation plane is placed to the right of the lens. Later, however, we see that the results allow for negative lenses, virtual sources, negative observation distances, etc. The only restriction we impose is that the object must lie to the left of the lens and be illuminated from the left.

Because we have already been through similar developments, we omit many of the messy details; however, the key parts of the development will be indicated and the final results given. To begin with, we assume the diverging spherical illumination wave field to be of the form

$$u_2^-(x,y) = \left(\frac{Az_{13}}{z_{12}}\right) q(x,y;a_{12}), \tag{10.185}$$

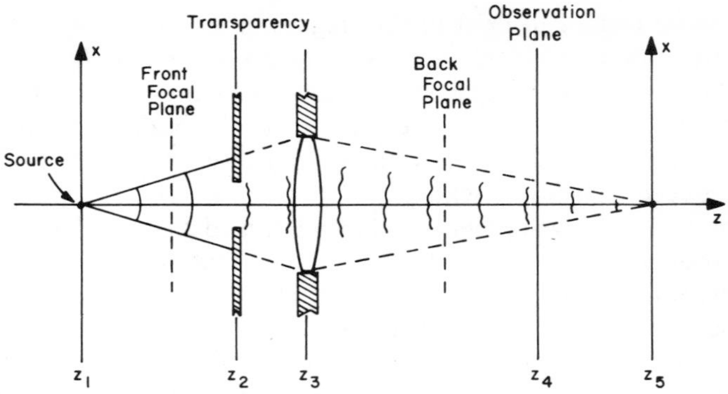

Figure 10-21 Optical system with diffracting object located in front of a lens.

where $a_{12}=(\lambda z_{12})^{-1}$ as usual. Then, with $a_f=(\lambda f)^{-1}$, we have

$$u_2^+(x,y)=\left(\frac{Az_{13}}{z_{12}}\right)q(x,y;a_{12})t_2(x,y), \tag{10.186}$$

$$u_3^-(x,y)=\left(\frac{Az_{13}}{z_{12}}\right)B_{23}\big[q(x,y;a_{12})t_2(x,y)\big]**q(x,y;a_{23}), \tag{10.187}$$

$$u_3^+(x,y)=\left(\frac{Az_{13}}{z_{12}}\right)B_{23}B_\ell\big\{\big[q(x,y;a_{12})t_2(x,y)\big]**q(x,y;a_{23})\big\}$$

$$\times q^*(x,y;a_f)p_\ell(x,y), \tag{10.188}$$

$$u_4(x,y)=\left(\frac{Az_{13}}{z_{12}}\right)B_{23}B_\ell B_{34}\Big[\big\{\big[q(x,y;a_{12})t_2(x,y)\big]**q(x,y;a_{23})\big\}$$

$$\times q^*(x,y;a_f)p_\ell(x,y)\Big]**q(x,y;a_{34}). \tag{10.189}$$

This is the complete result, but it is difficult to evaluate because of all the convolution operations involving quadratic phase factors. One might be tempted to switch to the frequency domain to simplify matters, but such a tactic doesn't work well in the present case because similar convolutions are encountered in that domain as well.

For the moment, we assume that the lens aperture is large enough to pass all of the light diffracted by the transparency, an assumption that may not be valid if the transparency is located too far from the lens. However, when it is valid we may set $p_\ell(x, y) = 1$, and this simplifies the evaluation significantly. With the additional definitions

$$a = a_{12} + a_{23}, \qquad v(x,y) = q(x,y;a)t_2(x,y),$$

$$b = a_{23} + a_{34} - a_f, \qquad w(x,y) = q(x,y;b)V(a_{23}x, a_{23}y),$$

$$\hat{c} = \frac{1}{\lambda\hat{z}} = \left(\frac{a_{34}}{a_{23}}\right)^2\left(a - \frac{a_{23}^2}{b}\right), \qquad g(x,y) = q(x,y;\hat{c})t_2\left(\frac{-a_{34}x}{a_{23}}, \frac{-a_{34}y}{a_{23}}\right),$$

$$K = \left(\frac{Az_{13}}{z_{12}}\right)B_{23}B_\ell B_{34}, \tag{10.190}$$

we obtain

$$\begin{aligned} u_4(x,y) &= K\Big[\big\{\big[q^*(x,y;a_{23})v(x,y)\big]{**}q(x,y;a_{23})\big\} \\ &\qquad \times q^*(x,y;a_f)\Big]{**}q(x,y;a_{34}) \\ &= K\big[\{q(x,y;a_{23})V(a_{23}x,a_{23}y)\}q^*(x,y;a_f)\big]{**}q(x,y;a_{34}) \\ &= K\big[q^*(x,y;a_{34})w(x,y)\big]{**}q(x,y;a_{34}) \\ &= Kq(x,y;a_{34})W(a_{34}x,a_{34}y). \end{aligned} \tag{10.191}$$

From Eq. (10.37) we find that

$$\begin{aligned} W(a_{34}x,a_{34}y) &= \left(\frac{j}{b}\right)\left(\frac{a_{34}}{a_{23}}\right)^2 q\left(x,y;\frac{-a_{34}^2}{b}\right){**}\left[q^*\left(x,y;\frac{-a_{34}^2}{b}\right)g(x,y)\right] \\ &= \left(\frac{j}{b}\right)\left(\frac{a_{34}}{a_{23}}\right)^2 q\left(x,y;\frac{-a_{34}^2}{b}\right)G\left(\frac{-a_{34}^2x}{b}, \frac{-a_{34}^2y}{b}\right), \end{aligned} \tag{10.192}$$

and Eq. (10.191) may therefore be written as

$$\begin{aligned} u_4(x,y) &= \left(\frac{j}{b}\right)\left(\frac{a_{34}}{a_{23}}\right)^2 Kq\left(x,y;a_{34} - \frac{a_{34}^2}{b}\right) \\ &\quad \times \mathfrak{F}\mathfrak{F}\left\{t_2\left(\frac{-a_{34}x}{a_{23}}, \frac{-a_{34}y}{a_{23}}\right)q(x,y;\hat{c})\right\}\Bigg|_{\xi = -a_{34}^2x/b,\ \eta = -a_{34}^2y/b} \end{aligned} \tag{10.193}$$

We note that this expression resembles the Fresnel diffraction formula of Eq. (10.150), but that the parameter governing its behavior is now $\hat{\hat{c}}$. Because we have already thoroughly discussed that formula, we shall not dwell on the subject here; however, we shall determine if and where any Fourier-transform planes are located.

By setting $\hat{\hat{c}}=0$ in Eq. (10.193), which corresponds to setting $\hat{\hat{z}}=\infty$ as was done previously, the inner quadratic phase factor becomes unity and we are left with a Fraunhofer diffraction formula. It can be shown, after considerable manipulation, that the requirement for $\hat{\hat{c}}=0$ is satisfied when $z_{34}=z_{35}$, which means that the observation plane must be located in the plane $z=z_5$. With this condition satisfied, $q(x, y;\hat{\hat{c}})=1$, $b=a_{13}a_{23}/a_{12}$, $a_{34}=a_{35}$ and

$$G\left(\frac{-a_{34}^2 x}{b}, \frac{-a_{34}^2 y}{b}\right)=G\left(\frac{-a_{12}a_{35}^2 x}{a_{13}a_{23}}, \frac{-a_{12}a_{35}^2 y}{a_{13}a_{23}}\right)$$

$$=\left(\frac{a_{23}}{a_{35}}\right)^2 T_2\left(\frac{a_{12}a_{35}x}{a_{13}}, \frac{a_{12}a_{35}y}{a_{13}}\right). \tag{10.194}$$

Finally, we obtain

$$u_5(x,y)=\frac{ja_{12}}{a_{13}a_{23}}Kq\left(x,y;a_{35}\left[1-\frac{a_{12}a_{35}}{a_{13}a_{23}}\right]\right)T_2\left(\frac{a_{12}a_{35}x}{a_{13}}, \frac{a_{12}a_{35}y}{a_{13}}\right)$$

$$=A\left(\frac{z_{13}}{z_{12}}\right)^2 e^{jkz_{23}}B_\ell B_{35}q\left(x,y;\frac{1}{\lambda z_{35}}\left[1-\frac{z_{13}z_{23}}{z_{12}z_{35}}\right]\right)T_2\left(\frac{z_{13}x}{\lambda z_{12}z_{35}}, \frac{z_{13}y}{\lambda z_{12}z_{35}}\right), \tag{10.195}$$

which is simply a Fraunhofer pattern as expected.

It will now be instructive to introduce two new parameters and to put Eq. (10.195) into a slightly different form. With the definitions

$$m'=\frac{f}{f-z_{23}}, \qquad z'_{25}=m'z_{23}+z_{35}, \tag{10.196}$$

it can be shown that

$$\frac{z_{13}}{z_{12}z_{35}}=\frac{m'}{z'_{25}}. \tag{10.197}$$

Therefore, Eq. (10.195) may be rewritten as

$$u_5(x,y)=A\left(\frac{z_{13}}{z_{12}}\right)^2 e^{jkz_{23}}B_\ell B_{35}q\left(x,y;\frac{1}{\lambda z'_{25}}\right)T_2\left(\frac{m'x}{\lambda z'_{25}},\frac{m'y}{\lambda z'_{25}}\right), \quad (10.198)$$

which is simply the result obtained by evaluating the expression

$$u_5(x,y)=A\left(\frac{z_{13}}{z_{12}}\right)^2 e^{jkz_{23}}B_\ell B_{35}q\left(x,y;\frac{1}{\lambda z'_{25}}\right)$$
$$\times\,\mathcal{F}\mathcal{F}\left\{\left(\frac{1}{m'}\right)^2 t_2\left(\frac{x}{m'},\frac{y}{m'}\right)\right\}\Bigg|_{\xi=x/\lambda z'_{25},\,\eta=y/\lambda z'_{25}}. \quad (10.199)$$

If we also define an *effective transparency*

$$t'_2(x,y)=\left(\frac{1}{m'}\right)t_2\left(\frac{x}{m'},\frac{y}{m'}\right), \quad (10.200)$$

Eq. (10.199) may be written as

$$u_5(x,y)=A\left(\frac{z_{13}}{z_{12}}\right)^2 e^{jkz_{23}}B_\ell B_{35}q\left(x,y;\frac{1}{\lambda z'_{25}}\right)$$
$$\times\left(\frac{1}{m'}\right)\mathcal{F}\mathcal{F}\{t'_2(x,y)\}\Big|_{\xi=x/\lambda z'_{25},\,\eta=y/\lambda z'_{25}}$$
$$=A\left(\frac{z_{13}}{z_{12}}\right)^2 e^{jkz_{23}}B_\ell B_{35}q\left(x,y;\frac{1}{\lambda z'_{25}}\right)\left(\frac{1}{m'}\right)T'_2\left(\frac{x}{\lambda z'_{25}},\frac{y}{\lambda z'_{25}}\right), \quad (10.201)$$

where $T'_2(\xi,\eta)$ is the spectrum of $t'_2(x,y)$. Finally, with

$$B'_{25}=\frac{e^{jkz'_{25}}}{j\lambda z'_{25}} \quad (10.202)$$

and m' as given by Eq. (10.197), Eq. (10.201) may be rearranged to yield

$$u_5(x,y)=A\left(\frac{z_{13}}{z_{12}}\right)e^{jkz_{23}(1-m')}B_\ell B'_{25}q\left(x,y;\frac{1}{\lambda z'_{25}}\right)T'_2\left(\frac{x}{\lambda z'_{25}},\frac{y}{\lambda z'_{25}}\right). \quad (10.203)$$

Aside from certain multiplicative constants, this expression may be regarded as the Fraunhofer diffraction pattern associated with the effective transparency at an *effective observation distance of* z'_{25}. In other words, the combination of the point source, the actual transparency $t_2(x, y)$ and the lens may be modeled by an effective transparency and converging spherical wave illumination; the effective transparency is located at the plane $z = z'_2$ and the illuminating wave field is converging toward the point $(0, 0, z_5)$ in the observation plane, as shown in Fig. 10-22. As we shall discover later in the chapter on image-forming systems, the function $t'_2(x, y)$ simply describes the complex amplitude of *the geometrical image of* $t_2(x, y)$, and m' is the *magnification* associated with this image. We note that m' may be either positive or negative, corresponding to an erect or inverted image, respectively. In addition, the plane z'_2 may be located anywhere along the z-axis: if it is located to the left of the lens, the image is virtual, whereas a location to the right of the lens yields a real image.

Aside from the requirement that the diffracting object not be placed too close to the source, which would invalidate the Fresnel approximations and the assumption of uniform illumination, and as long as all the light diffracted by the object is collected by the lens, these results are valid for any configuration similar to that shown in Fig. 10-21 in which the object is placed to the left of the lens and illuminated from the left. If the expressions obtained here are used for a setup in which the object is placed to the right of the lens, the magnitude and scale of the diffraction pattern may be in error even though its general form is correct.

To examine the behavior of $u_5(x, y)$ in greater detail, we now consider specific examples of a few frequently encountered configurations. We shall

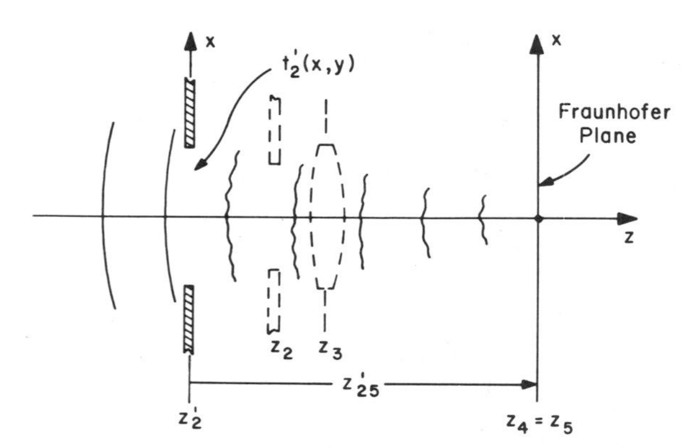

Figure 10-22 Optical system of Fig. 10-21 modeled by an effective transparency $t'_2(x,y)$, converging spherical wave illumination, and an effective observation distance of z'_{25}.

investigate the nature of the geometrical image $t_2'(x, y)$ and determine values for m' and z_{25}' for these configurations; however, in most of these special cases it will be most convenient to express the results in terms of $T_2(\xi,\eta)$ rather than $T_2'(\xi,\eta)$. Other configurations will be left for the reader to evaluate.

1. *Positive lens and point source to left of front focal plane.* A general observation may be made for this set of conditions, for which $z_{13}>f>0$. As may be seen from Eq. (10.195), the lateral size of the diffraction pattern decreases as the transparency is moved away from the lens and increases as the transparency is moved toward the lens. Three specific cases are of interest:

(a) *Object between front focal plane and lens ($f>z_{23}>0$).* For this configuration m' is positive, z_{25}' is positive and $t_2'(x, y)$ describes an erect virtual image located to the left of the object, i.e., $z_2'<z_2$. We may regard $u_5(x, y)$ as the Fraunhofer pattern of $t_2'(x, y)$ viewed at an effective observation distance of $z_{25}'>z_{35}$ as illustrated in Fig. 10-22.

(b) *Object in front focal plane ($z_{23}=f$).* For this situation both m' and z_{25}' are infinite and $t_2'(x, y)$ represents an image located at $z_2'=-\infty$. Manipulation of Eq. (10.198) yields

$$u_5(x,y)=A\left(\frac{z_{13}}{z_{12}}\right)^2 e^{jkf}B_\ell B_{35}T_2\left(\frac{x}{\lambda f},\frac{y}{\lambda f}\right), \tag{10.204}$$

and we see that the quadratic phase factor has now vanished and that the lateral scale of $u_5(x, y)$ is independent of all distances except the focal length f. This situation is depicted in Fig. 10-23.

(c) *Object between source and front focal plane ($z_{13}>z_{23}>f$).* We now find that both m' and z_{25}' are negative, and $t_2'(x, y)$ represents a real

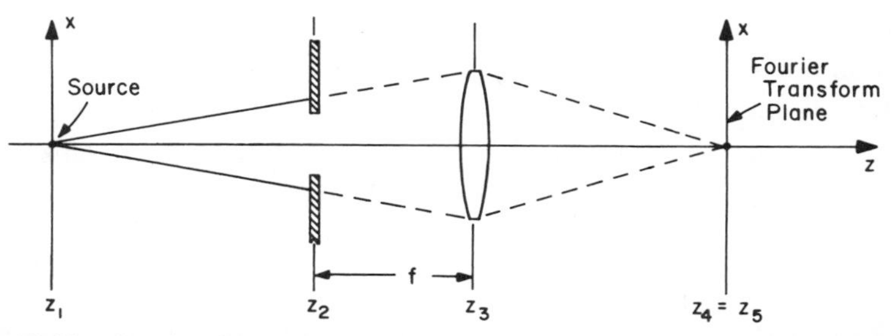

Figure 10-23 Quadratic phase factor vanishes when object is placed in front focal plane.

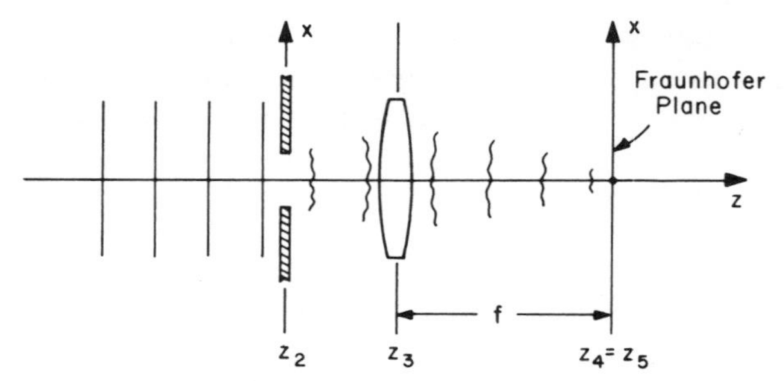

Figure 10-24 Fraunhofer plane is located at back focal plane when plane wave illumination is used.

inverted image located at $z_2' > z_5$. Aside from the fact that $u_5(x, y)$ is not a virtual diffraction pattern in the present case, the relationship between it and $t_2'(x, y)$ is similar to the relationship between $u_5(x, y)$ and $t_3(x, y)$ in Eq. (10.184), for which diverging spherical wave illumination was used. Because z_{25}' is negative, the quadratic phase factor now describes a converging, rather than diverging, spherical wave field.

2. *Positive lens and plane wave field illumination* ($z_{13} = \infty$). With plane wave illumination we have $z_{12} = \infty$, $z_{35} = f$, and $z_{25}' = m'f$. Thus, Eq. (10.198) becomes

$$u_5(x, y) = A \frac{e^{jk(z_{23}+f)}}{j\lambda f} B_\ell q\left(x, y; \frac{1}{\lambda m' f}\right) T_2\left(\frac{x}{\lambda f}, \frac{y}{\lambda f}\right). \qquad (10.205)$$

Note that the lateral scale of this Fraunhofer pattern is independent of z_{23}, being determined only by the focal length f. Also note that the Fraunhofer plane is now located at the back focal plane as shown in Fig. 10-24. Because plane-wave illumination is simply a special case of the more general configuration discussed in paragraph 1 above, the dependence of the Fraunhofer patterns on object position is similar for the two situations. Again we discuss three specific cases:

(a) *Object between front focal plane and lens* ($f > z_{23} > 0$). This case is similar to that discussed in paragraph 1(a) above because m' is positive, $t_2'(x, y)$ is an erect virtual image located at $z_2' < z_2$ and $u_5(x, y)$ may be regarded as the Fraunhofer pattern of $t_2'(x, y)$ viewed at a distance of $z_{25}' > f$. The differences lie in the multiplicative constants and the lateral size of the pattern, which is now determined by f rather than by any other distances.

(b) *Object in front focal plane* ($z_{23}=f$). Again, as for the case of paragraph 1(b) above, both m' and z'_{25} are infinite and $t'_2(x, y)$ describes an image located at $z'_2=-\infty$. Equation (10.205) now becomes

$$u_5(x,y)=A\frac{e^{j2kf}}{j\lambda f}B_\ell T_2\left(\frac{x}{\lambda f},\frac{y}{\lambda f}\right) \tag{10.206}$$

and we see that once again the quadratic phase factor vanishes when the object is located in the front focal plane.

(c) *Object to left of front focal plane* ($z_{23}>f$). This case is similar to that of paragraph 1(c) above in that both m' and z'_{25} are negative and $t'_2(x, y)$ describes a real inverted image located at $z'_2>z_5$. As before, the quadratic phase factor represents a converging spherical wave field for this object location.

3. *Positive or negative lens and object at lens* ($z_{23}=0$). When $z_{23}=0$ we have $z_{12}=z_{13}$, $m'=1$ and $z'_{25}=z_{35}$. Thus Eq. (10.198) becomes

$$u_5(x,y)=AB_\ell B_{35}q\left(x,y;\frac{1}{\lambda z_{35}}\right)T_2\left(\frac{x}{\lambda z_{35}},\frac{y}{\lambda z_{35}}\right). \tag{10.207}$$

Note that in this case $t'_2(x, y)=t_2(x, y)$. Although Eq. (10.154) was derived for a configuration in which the object was located to the right of the lens, it reduces to Eq. (10.207) when the object is located at the lens (in that case we would have $z_{23}=0$ and $z_{25}=z_{35}$). As a result, as long as the object is located very close to the lens, it may be placed on either side and the resulting Fraunhofer pattern will be the same.[††] Two specific cases are of interest:

(a) *Spherical wave field illumination.* Equation (10.207) expresses the result for this type of illumination, but the polarity of z_{35}, and hence the location of the Fraunhofer plane, depends on the magnitudes and polarities of z_{13} and f. Some combinations of z_{13} and f lead to positive values for z_{35} and real Fraunhofer patterns, as shown in Fig. 10-25, whereas others result in negative values of z_{35} and virtual Fraunhofer patterns.

(b) *Plane-wave field illumination.* With plane-wave illumination $z_{35}=f$ and Eq. (10.207) reduces to

$$u_5(x,y)=AB_\ell\frac{e^{jkf}}{j\lambda f}q\left(x,y;\frac{1}{\lambda f}\right)T_2\left(\frac{x}{\lambda f},\frac{y}{\lambda f}\right). \tag{10.208}$$

[††]Strictly speaking, this is not quite true because the aberrations introduced by the lens will be different in the two cases.

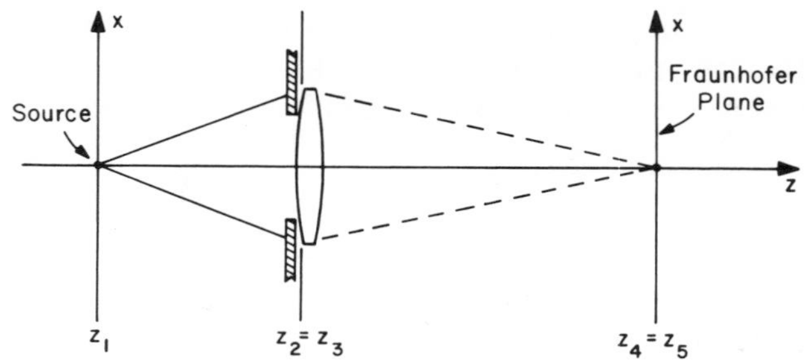

Figure 10-25 Configuration with diffracting object at the lens for which observation distance is positive and Fraunhofer pattern is real.

Now the location of the Fraunhofer plane and the nature of the resulting diffraction pattern are determined by the magnitude and polarity of f alone. ■

We now return to Eq. (10.189) and investigate the effects of the finite lens aperture. As mentioned previously, Eq. (10.189) is very difficult to evaluate exactly; however, certain approximations can be made that greatly simplify the procedure. For example, the propagation from the transparency to the lens may be described in terms of ray theory as long as the distance between them is not too large. We shall assume this to be the case. Thus, following Goodman once again (Ref. 10-3), we erect a geometric projection of the aperture function at the plane of the transparency. In contrast to the previous situation for which this technique was used [e.g., see Eq. (10.161)], the effective center of the projected aperture function is now dependent on the location of the observation point. Appropriate manipulation of Eq. (10.195) yields the following results:

$$u_5(x,y)=A\left(\frac{z_{13}}{z_{12}}\right)^2 e^{jkz_{23}}B_\ell B_{35}q\left(x,y;\frac{1}{\lambda z_{35}}\left[1-\frac{z_{13}z_{23}}{z_{12}z_{35}}\right]\right)$$
$$\times\iint_{-\infty}^{\infty} t_2(\alpha,\beta)\,p_\ell\left(\frac{\alpha+z_{23}x/z_{35}}{z_{12}/z_{13}},\frac{\beta+z_{23}y/z_{35}}{z_{12}/z_{13}}\right)$$
$$\times\exp\left[-j2\pi\left(\frac{\alpha z_{13}x}{\lambda z_{12}z_{35}}+\frac{\beta z_{13}y}{\lambda z_{12}z_{35}}\right)\right]d\alpha\,d\beta. \qquad (10.209)$$

It can be seen from this expression that the light arriving at any point (x,y) in the Fraunhofer plane can arise only from the portion of the transparency contained within the projected aperture function, and that this projected aperture function is effectively caused to scan across the

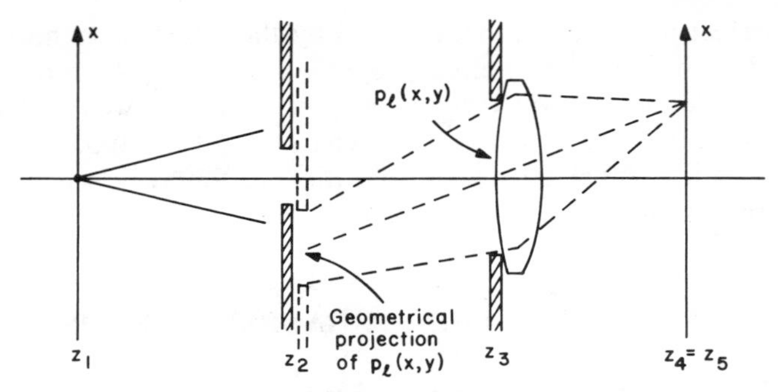

Figure 10-26 Vignetting in an optical system.

transparency as the observation point is moved about the Fraunhofer plane. No light at all will reach those observation points that lie too far from the axis of the system; this occurs when the projected aperture function is shifted so far that it no longer overlaps any part of the transparency, as illustrated in Fig. 10-26. The effect encountered here is known as *vignetting*, and it can be minimized either by placing the transparency close to the lens or by choosing a lens with a sufficiently large aperture.

We may also express Eq. (10.209) in terms of the geometrical image $t_2'(x,y)$ and a geometric projection of $p_\ell(x,y)$ on the plane containing $t_2'(x,y)$. If we now set $\alpha' = m\alpha$ and $\beta' = m\beta$ we obtain

$$u_5(x,y) = A\left(\frac{z_{13}}{z_{12}}\right)^2 e^{jkz_{23}} B_\ell B_{35} q\left(x, y; \frac{1}{\lambda z_{35}}\left[1 - \frac{z_{13}z_{23}}{z_{12}z_{35}}\right]\right)\left(\frac{1}{m'}\right)^2$$
$$\times \iint_{-\infty}^{\infty} t_2\left(\frac{\alpha'}{m'}, \frac{\beta'}{m'}\right) p_\ell\left(\frac{\alpha'/m' + z_{23}x/z_{35}}{z_{12}/z_{13}}, \frac{\beta'/m' + z_{23}y/z_{35}}{z_{12}/z_{13}}\right)$$
$$\times \exp\left[-j2\pi\left(\frac{\alpha' z_{13}x}{\lambda m' z_{12}z_{35}} + \frac{\beta' z_{13}y}{\lambda m' z_{12}z_{35}}\right)\right] d\alpha'\, d\beta'$$
$$= A\left(\frac{z_{13}}{z_{12}}\right) \exp[\,jkz_{23}(1-m')\,] B_\ell B_{25}' q\left(x, y; \frac{1}{\lambda z_{25}'}\right)$$
$$\times \iint_{-\infty}^{\infty} t_2'(\alpha', \beta') p_\ell\left(\frac{\alpha' + m' z_{23}x/z_{35}}{z_{25}'/z_{35}}, \frac{\beta' + m' z_{23}y/z_{35}}{z_{25}'/z_{35}}\right)$$
$$\times \exp\left[-j2\pi\left(\frac{\alpha' x}{\lambda z_{25}'} + \frac{\beta' y}{\lambda z_{25}'}\right)\right] d\alpha'\, d\beta'. \tag{10.210}$$

Thus, the effect of vignetting is described by the effective scanning of the geometrical image by a magnified projection of the lens aperture at the plane $z=z_2'$. We mention once again that this result, as well as that given by Eq. (10.209), is valid as long as the separation of the transparency and the lens is sufficiently small that propagation in this region is adequately described by ray theory.

10-7 PROPAGATION OF GAUSSIAN BEAMS

Because of the importance of *Gaussian beams* in optics, particularly in the area of lasers, we devote the present section to a study of the propagation of these beams. A Gaussian beam, also called a *Gaussian spherical wave field*, is basically a spherical wave field whose modulus in a plane transverse to the propagation direction varies in a Gaussian fashion. As was the case for plane-wave fields and conventional spherical wave fields, *propagation does not alter the fundamental nature of Gaussian beams*; the effective width of a Gaussian beam may change as it propagates, as well as the curvature of its wavefronts, *but it remains a Gaussian beam.*

A great deal has been written about Gaussian beams (see, for example, Refs. 10-17 and 10-18 and the extensive lists of references therein), and here we present only a brief development of the important features of Gaussian beam propagation. Our purpose is to tie the propagation of Gaussian beams in with our general discussion of diffraction and to arrange the results into an easily understood and useful form. For those who desire more details, Siegman's book (Ref. 10-18) is recommended as a starting point.

General Description of a Gaussian Beam

A Gaussian beam of the type we investigate here characterizes the output of a typical laser that is operating in the fundamental, or lowest-order, mode. Such a beam may be described in the plane $z=z_1$ by a complex amplitude distribution of the form

$$u_1(r)=A_1\,\mathrm{Gaus}\left(\frac{r}{b_1}\right)q\left(r;\frac{1}{\lambda R_1}\right), \tag{10.211}$$

where R_1 denotes the radius of curvature of the spherical wavefronts in that plane and b_1 denotes the effective width of the beam.‡‡ We note that

‡‡Care should be used when comparing our results with those of others because our beam-width parameter b_1 is larger, by a factor of $\pi^{1/2}$, than the beam-width parameter found in much of the literature.

the beam is diverging if R_1 is positive and converging if R_1 is negative.

The Gaussian beam of Eq. (10.211) has an irradiance given by

$$\begin{aligned} I_1(r) &= |u_1(r)|^2 \\ &= |A_1|^2 \mathrm{Gaus}^2\left(\frac{r}{b_1}\right) \\ &= |A_1|^2 \mathrm{Gaus}\left(\frac{\sqrt{2}\, r}{b_1}\right), \end{aligned} \tag{10.212}$$

and its total power is simply the integral of $I_1(r)$ over the z_2 plane:

$$\begin{aligned} P_{\mathrm{tot}} &= \int_0^{2\pi}\int_0^{\infty} I_1(r')r'\,dr'\,d\theta' \\ &= 2\pi|A_1|^2 \int_0^{\infty} \mathrm{Gaus}\left(\frac{\sqrt{2}\, r'}{b_1}\right) r'\,dr'. \end{aligned} \tag{10.213}$$

This integral is evaluated easily with the use of the central ordinate theorem, i.e., with

$$\mathrm{Gaus}\left(\frac{\sqrt{2}\, r}{b_1}\right) \overset{\mathcal{H}_0}{\leftrightarrow} \frac{b_1^2}{2}\mathrm{Gaus}\left(\frac{b_1\rho}{\sqrt{2}}\right), \tag{10.214}$$

we use Eq. (9.125) to obtain

$$P_{\mathrm{tot}} = \frac{|A_1|^2 b_1^2}{2}. \tag{10.215}$$

It is sometimes necessary to determine how much of the beam power will be transmitted through a circular aperture of diameter d. Denoting the transmitted power by $P(d/b_1)$, it is not difficult to show that

$$\begin{aligned} P\left(\frac{d}{b_1}\right) &= 2\pi|A_1|^2 \int_0^{\infty} \mathrm{Gaus}\left(\frac{\sqrt{2}\, r'}{b_1}\right) \mathrm{cyl}\left(\frac{r'}{d}\right) r'\,dr' \\ &= 2\pi|A_1|^2 \int_0^{d/2} \mathrm{Gaus}\left(\frac{\sqrt{2}\, r'}{b_1}\right) r'\,dr' \\ &= P_{\mathrm{tot}}\left[1 - \mathrm{Gaus}\left(\frac{d}{\sqrt{2}\, b_1}\right)\right]. \end{aligned} \tag{10.216}$$

Evaluation of this expression reveals that an aperture of diameter $d = b_1$ transmits 79.3% of the total beam power, whereas an aperture of twice this diameter transmits 99.8% of the total power.

Effects of Diffraction on a Gaussian Beam

We now determine how the Gaussian beam of Eq. (10.211) propagates from the plane $z = z_1$ to the plane $z = z_2$. From Eq. (10.66) we may write

$$\begin{aligned} u_2(r) &= B_{12} q\left(r; \frac{1}{\lambda z_{12}}\right) \mathcal{H}_0\left\{ u_1(r) q\left(r; \frac{1}{\lambda z_{12}}\right)\right\}\bigg|_{\rho = r/\lambda z_{12}} \\ &= B_{12} q\left(r; \frac{1}{\lambda z_{12}}\right) \mathcal{H}_0\left\{ A_1 \operatorname{Gaus}\left(\frac{r}{b_1}\right) q\left(r; \frac{1}{\lambda R_1} + \frac{1}{\lambda z_{12}}\right)\right\}\bigg|_{\rho = r/\lambda z_{12}}, \end{aligned} \tag{10.217}$$

and noting that both functions within the curly brackets have a quadratic dependence on r, we combine them into a single function with complex exponent:

$$\begin{aligned} &\operatorname{Gaus}\left(\frac{r}{b_1}\right) q\left(r; \frac{1}{\lambda R_1} + \frac{1}{\lambda z_{12}}\right) \\ &\qquad = \exp\left\{ -\pi r^2 \left[\frac{1}{b_1^2} - \frac{j}{\lambda}\left(\frac{1}{R_1} + \frac{1}{z_{12}}\right)\right]\right\}. \end{aligned} \tag{10.218}$$

The zero-order Hankel transform of this expression is then determined from Table 9-4, and we obtain

$$\begin{aligned} u_2(r) &= B_{12} q\left(r; \frac{1}{\lambda z_{12}}\right) A_1 \left[\frac{1}{b_1^2} - \frac{j}{\lambda}\left(\frac{1}{z_{12}} + \frac{1}{R_1}\right)\right]^{-1} \\ &\quad \times \exp\left\{ -\pi \left(\frac{r}{\lambda z_{12}}\right)^2 \left[\frac{1}{b_1^2} - \frac{j}{\lambda}\left(\frac{1}{z_{12}} + \frac{1}{R_1}\right)\right]^{-1}\right\} \\ &= \frac{A_1 e^{jkz_{12}}}{j\lambda z_{12}\left[\frac{1}{b_1^2} - \frac{j}{\lambda}\left(\frac{1}{z_{12}} + \frac{1}{R_1}\right)\right]} \\ &\quad \times \exp\left\{ -\pi r^2 \left[\left(\frac{1}{\lambda z_{12}}\right)^2 \left[\frac{1}{b_1^2} - \frac{j}{\lambda}\left(\frac{1}{z_{12}} + \frac{1}{R_1}\right)\right]^{-1} - \frac{j}{\lambda z_{12}}\right]\right\}. \end{aligned} \tag{10.219}$$

After considerable manipulation, this expression may be put into the form

$$u_2(r) = A_2 \mathrm{Gaus}\left(\frac{r}{b_2}\right) q\left(r; \frac{1}{\lambda R_2}\right), \tag{10.220}$$

and we see that a Gaussian beam indeed remains a Gaussian beam as it propagates. Here we have identified a new effective beam width b_2, amplitude A_2, and radius of curvature R_2 as follows:

$$b_2 = \left|\frac{\lambda z_{12}}{b_1}\right| \sqrt{1 + \left(\frac{b_1^2}{\lambda}\right)^2 \left(\frac{1}{z_{12}} + \frac{1}{R_1}\right)^2},$$

$$A_2 = A_1 \left(\frac{b_1}{b_2}\right) e^{j(kz_{12} + \Phi_{12})}, \tag{10.221}$$

$$R_2 = z_{12} \left[\frac{1 + \left(\frac{b_1^2}{\lambda}\right)^2 \left(\frac{1}{z_{12}} + \frac{1}{R_1}\right)^2}{1 + \left(\frac{b_1^2}{\lambda}\right)^2 \left(\frac{1}{R_1}\right) \left(\frac{1}{z_{12}} + \frac{1}{R_1}\right)}\right].$$

The quantity Φ_{12} denotes a phase shift given by

$$\Phi_{12} = \tan^{-1}\left[\frac{-\dfrac{\lambda}{b_1^2}}{\dfrac{1}{z_{12}} + \dfrac{1}{R_1}}\right]. \tag{10.222}$$

To gain a better understanding of the behavior of b_2 and R_2 with z_{12}, we now consider the Gaussian beam depicted in Fig. 10-27. We also define the new dimensionless parameter

$$\zeta_1 = \left(\frac{\lambda R_1}{b_1^2}\right)^2 \tag{10.223}$$

to simplify the notation. Thus, the above expressions for b_2 and R_2 may be

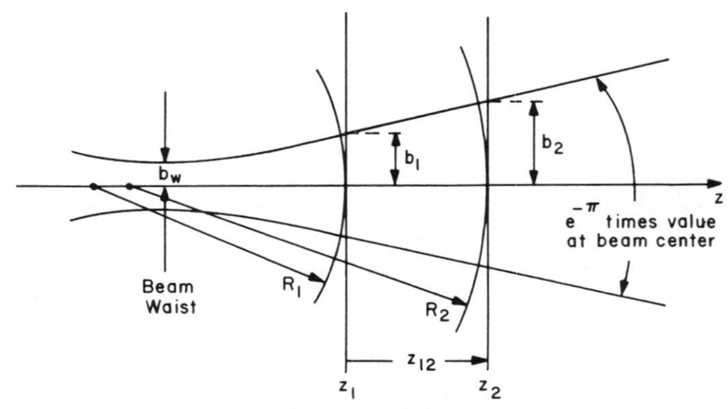

Figure 10-27 A Gaussian beam.

rearranged to obtain

$$b_2 = b_1\sqrt{(\zeta_1+1)\left(\frac{z_{12}}{R_1}\right)^2 + 2\left(\frac{z_{12}}{R_1}\right) + 1}\ ,$$

$$R_2 = R_1\left[\frac{(\zeta_1+1)\left(\frac{z_{12}}{R_1}\right)^2 + 2\left(\frac{z_{12}}{R_1}\right) + 1}{(\zeta_1+1)\left(\frac{z_{12}}{R_1}\right) + 1}\right]. \qquad (10.224)$$

Expressed in this fashion, the dependence of these quantities on ζ_1 and z_{12}/R_1 is easy to see. We may also rewrite the phase shift as

$$\Phi_{12} = \tan^{-1}\left[\frac{-\sqrt{\zeta_1}}{\frac{R_1}{z_{12}} + 1}\right]. \qquad (10.225)$$

For the situation illustrated in Fig. 10-27, in which R_1 is depicted as positive, the beam is expanding at the plane $z = z_1$. Consequently, at any plane to the right of z_1 the beam width b_2 will be greater than the beam width at z_1. Not only that, b_2 is a monotonically increasing function of z_{12}/R_1 in this region. However, the situation is different for observation planes lying to the left of z_1, i.e., for $z_{12} < 0$. As the observation plane is moved to the left of z_1, the beam width decreases at first. However, it reaches a minimum and then begins to increase again as z_2 is moved farther and farther to the left. The location of the minimum beam width is called the *beam waist*, and we denote the plane containing the beam waist by z_w. By setting the derivative of the beam-width expression equal to zero, and with $z_{1w} = z_w - z_1$, we find that the beam waist is located at

$$z_{1w} = \frac{-R_1}{\zeta_1 + 1}. \tag{10.226}$$

Thus, we see that *the beam waist is not found exactly at the point where* $z_{12} = -R_1$ although it may be close to this point. We also note that if R_1 is negative, signifying a converging beam at $z = z_1$, the beam waist will lie to the right of this plane. Substituting the above value of z_{1w} into the expression for b_2, and denoting the beam width at the waist by b_w, we find that

$$\begin{aligned} b_w &= b_1 \sqrt{\frac{\zeta_1}{\zeta_1 + 1}} \\ &= \left| \frac{\lambda R_1}{b_1} \right| \frac{1}{\sqrt{\zeta_1 + 1}}. \end{aligned} \tag{10.227}$$

The general behavior of b_2 as a function of z_{12}/R_1 is shown in Fig. 10-28.

The behavior of R_2 with z_{12}/R_1 is more complicated than that of b_2, as we shall now see. Under the assumption that R_1 is once again positive, as depicted in Fig. 10-27, R_2 will be positive at all observation planes to the

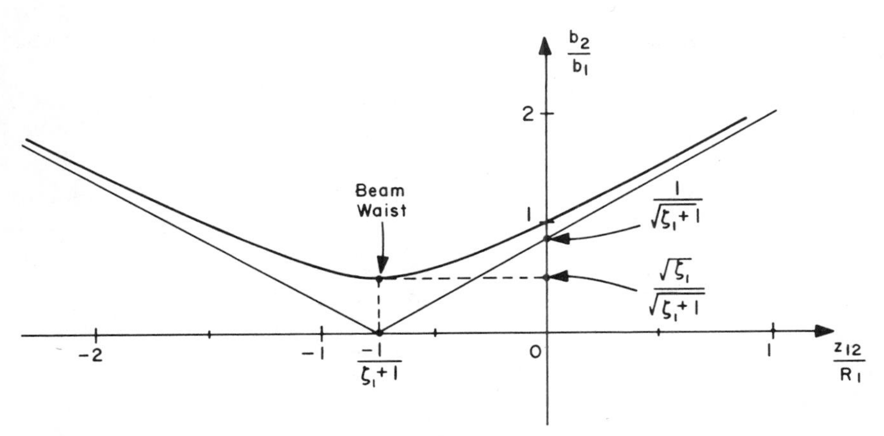

Figure 10-28 Radius of a Gaussian beam as a function of z_{12}/R_1.

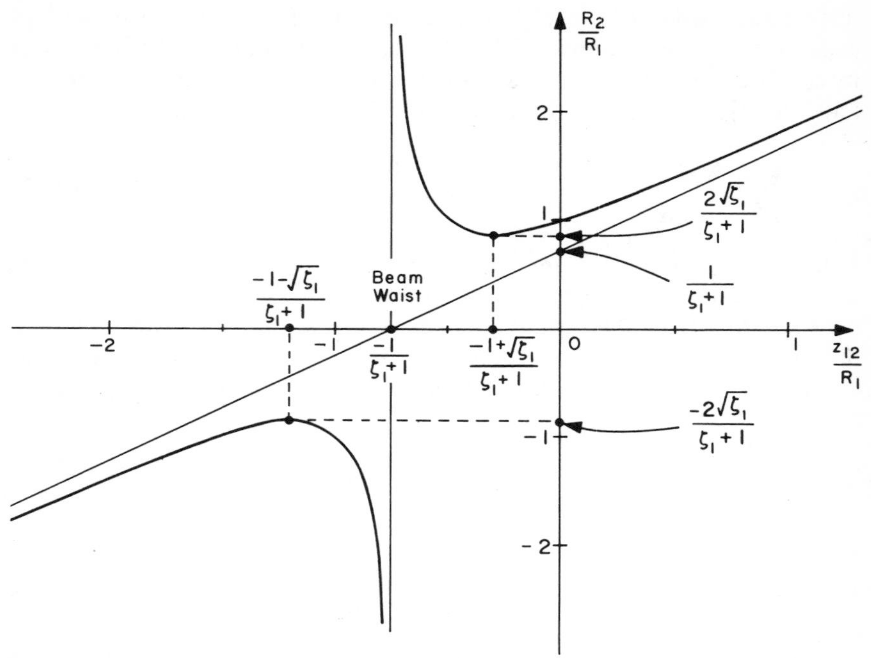

Figure 10-29 Radius of curvature of Gaussian beam wavefronts as a function of z_{12}/R_1.

right of z_1. In fact, it will be positive everywhere to the right of the waist and negative everywhere to the left of the waist. At the waist R_2 has a value of $\pm\infty$, which indicates the wavefronts are plane surfaces at that location. As may be seen in Fig. 10-29, the function describing R_2 exhibits odd symmetry about the beam waist. Thus, starting at the waist, $|R_2|$ decreases at first as the observation plane is moved away from the waist in either direction. At a certain distance from the waist, on either side, $|R_2|$ reaches a minimum value; finally, as the observation plane is moved still farther from the waist, it begins to increase once again. The points at which $|R_2|$ is a minimum, and the value of $|R_2|$ at these points, are found to be

$$(z_{12})_{|R_2|_{\min}} = \frac{-R_1\left(1\mp\sqrt{\zeta_1}\right)}{\zeta_1+1},$$

$$|R_2|_{\min} = \frac{2|R_1|\sqrt{\zeta_1}}{\zeta_1+1}. \tag{10.228}$$

A very curious property of Gaussian beams may be extracted from Eqs. (10.221) or (10.224): whenever the observation distance is equal to the negative of the initial radius of curvature, i.e., whenever $z_{12} = -R_1$, we find that

$$b_2 = \left|\frac{\lambda z_{12}}{b_1}\right| = \left|\frac{\lambda R_1}{b_1}\right|,$$

$$R_2 = z_{12} = -R_1. \tag{10.229}$$

Note that if ζ_1 is very small, i.e., $\zeta_1 \ll 1$, Eq. (10.224) may be written as

$$\left.\begin{aligned} b_2 &\cong b_1\left|\frac{z_{12}}{R_1}+1\right| \\ R_2 &\cong R_1\left(\frac{z_{12}}{R_1}+1\right) \end{aligned}\right\} \quad \zeta_1 \ll 1. \tag{10.230}$$

In addition, Eqs. (10.226)–(10.228) become

$$\left.\begin{aligned} z_{1w} &\cong -R_1 \\ b_w &\cong b_1\sqrt{\zeta_1} \\ (z_{12})_{|R_2|_{\min}} &\cong -(1 \mp \sqrt{\zeta_1}\,)R_1 \\ |R_2|_{\min} &\cong 2|R_1|\sqrt{\zeta_1} \end{aligned}\right\} \quad \zeta_1 \ll 1. \tag{10.231}$$

We have studied the behavior of b_2 and R_2 in some detail in the vicinity of the beam waist, and it will now be instructive to determine their behavior for large values of $|z_{12}/R_1|$. In this event, Eq. (10.224) may be approximated by

$$\left.\begin{aligned} b_2 &\cong b_1\left|\frac{z_{12}}{R_1}\right|\sqrt{\zeta_1+1} \\ R_2 &\cong z_{12} \end{aligned}\right\} \quad \left|\frac{z_{12}}{R_1}\right| \gg 1. \tag{10.232}$$

It is also helpful to determine the behavior of b_2 and R_2 for the situation in which the initial plane $z = z_1$ is located at the beam waist. In this event we have $|R_1| = \infty$ and $b_1 = b_w$, which allows Eq. (10.224) to be altered as follows:

$$\left.\begin{aligned} b_2 &= b_w\sqrt{\left(\frac{\lambda z_{12}}{b_w^2}\right)^2+1} \\ R_2 &= z_{12}\left[\left(\frac{b_w^2}{\lambda z_{12}}\right)^2+1\right] \end{aligned}\right\} \quad z_1 \text{ at waist.} \tag{10.233}$$

If we now define the quantity

$$Z_R = \frac{b_w^2}{\lambda}, \tag{10.234}$$

known as the *Rayleigh range* (see Ref. 10-12), we obtain

$$\left.\begin{aligned} b_2 &= b_w\sqrt{\left(\frac{z_{12}}{Z_R}\right)^2+1} \\ R_2 &= z_{12}\left[\left(\frac{Z_R}{z_{12}}\right)^2+1\right] \end{aligned}\right\} \quad z_1 \text{ at waist.} \tag{10.235}$$

From these expressions we see that, at a distance of $z_{12} = \pm Z_R$ from the beam waist, the beam width has increased by a factor of $\sqrt{2}$ and the radius of curvature has a magnitude equal to just twice the value of Z_R; i.e.,

$$\left.\begin{aligned} b_2 &= \sqrt{2}\, b_w \\ R_2 &= \pm 2Z_R \end{aligned}\right\} \quad z_1 \text{ at waist and } z_{12} = \pm Z_R. \tag{10.236}$$

It is not difficult to show that the minimum value of $|R_2|$ occurs at a distance from the waist equal to the Rayleigh range.

If we again consider the behavior of b_2 and R_2 at large distances from the beam waist, we find that Eq. (10.233) reduces to

$$\left.\begin{aligned} b_2 &\cong b_w\left|\frac{z_{12}}{Z_R}\right| \\ R_2 &\cong z_{12} \end{aligned}\right\} \quad z_1 \text{ at waist and } |z_{12}| \gg Z_R. \tag{10.237}$$

Effects of Lenses on Gaussian Beam Propagation

We now determine the effects of lenses on the propagation of a Gaussian beam, and the configuration shown in Fig. 10-30 is to be used for our discussion. We assume that the beam width b_1 and radius of curvature R_1 are known at the plane $z = z_1$, and that the lens is placed at $z = z_2$. The lens has a focal length of f, which may be either positive or negative, and we wish to determine the beam width b_3 and radius of curvature R_3 at some observation plane $z = z_3$. We start by finding the beam width b_2, the amplitude A_2, and the radius of curvature R_2 just to the left of the lens, and we do so by using the appropriate equations developed above. Thus, we have a wave field of

$$u_2^-(r) = A_2 \operatorname{Gaus}\left(\frac{r}{b_2}\right) q\left(r; \frac{1}{\lambda R_2^-}\right) \tag{10.238}$$

incident on the lens. The complex amplitude of the transmitted wave field is then

$$u_2^+(r) = A_2 B_\ell \operatorname{Gaus}\left(\frac{r}{b_2}\right) q\left(r; \frac{1}{\lambda R_2^-} - \frac{1}{\lambda f}\right) p_\ell(r), \tag{10.239}$$

where B_ℓ and $p_\ell(r)$ are described as following Eq. (10.133).

Let us now assume that the lens aperture is large enough to transmit

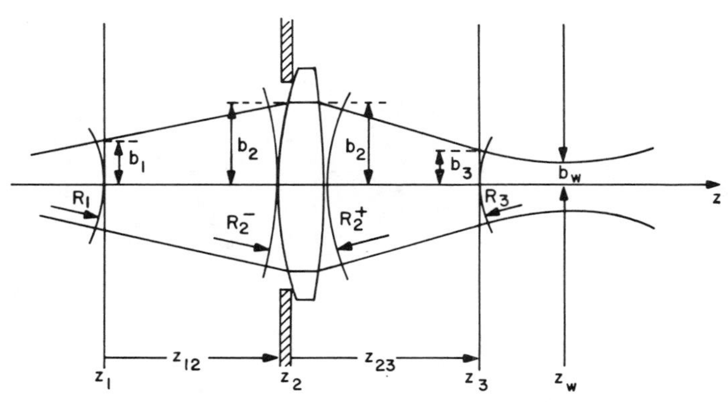

Figure 10-30 Effect of a positive lens on a Gaussian beam.

virtually all of the beam; i.e., if d_ℓ is the diameter of the aperture, let us assume that $d_\ell > 2b_2$. Then, following a procedure similar to that used above, we find that

$$u_3(r) = A_3 \operatorname{Gaus}\left(\frac{r}{b_3}\right) q\left(r; \frac{1}{\lambda R_3}\right), \tag{10.240}$$

where

$$b_3 = \left|\frac{\lambda z_{23}}{b_2}\right| \sqrt{1 + \left(\frac{b_2^2}{\lambda}\right)^2 \left(\frac{1}{z_{23}} + \frac{1}{R_2^-} - \frac{1}{f}\right)^2},$$

$$A_3 = A_2 B_\ell \left(\frac{b_2}{b_3}\right) e^{j(kz_{23} + \Phi_{23})},$$

$$R_3 = z_{23} \left[\frac{1 + \left(\frac{b_2^2}{\lambda}\right)^2 \left(\frac{1}{z_{23}} + \frac{1}{R_2^-} - \frac{1}{f}\right)^2}{1 + \left(\frac{b_2^2}{\lambda}\right)^2 \left(\frac{1}{R_2^-} - \frac{1}{f}\right)\left(\frac{1}{z_{23}} + \frac{1}{R_2^-} - \frac{1}{f}\right)}\right],$$

$$\Phi_{23} = \tan^{-1}\left[\frac{-\dfrac{\lambda}{b_2^2}}{\dfrac{1}{z_{23}} + \dfrac{1}{R_2^-} - \dfrac{1}{f}}\right]. \tag{10.241}$$

If we now define another dimensionless parameter for the z_2^- plane,

$$\zeta_2^- = \left(\frac{\lambda R_2^-}{b_2^2}\right)^2, \tag{10.242}$$

we may rewrite the above expressions to obtain

$$b_3 = b_2 \sqrt{\left[\zeta_2^- + \left(1 - \frac{R_2^-}{f}\right)^2\right]\left(\frac{z_{23}}{R_2^-}\right)^2 + 2\left(1 - \frac{R_2^-}{f}\right)\left(\frac{z_{23}}{R_2^-}\right) + 1},$$

$$R_3 = R_2^- \left\{ \frac{\left[\zeta_2^- + \left(1 - \frac{R_2^-}{f}\right)^2\right]\left(\frac{z_{23}}{R_2^-}\right)^2 + 2\left(1 - \frac{R_2^-}{f}\right)\left(\frac{z_{23}}{R_2^-}\right) + 1}{\left[\zeta_2^- + \left(1 - \frac{R_2^-}{f}\right)^2\right]\left(\frac{z_{23}}{R_2^-}\right) + \left(1 - \frac{R_2^-}{f}\right)} \right\},$$

$$\Phi_{23} = \tan^{-1}\left[\frac{-\sqrt{\zeta_2^-}}{1 + \frac{R_2^-}{z_{23}} - \frac{R_2^-}{f}}\right]. \tag{10.243}$$

Study of these expressions leads to the discovery of another beam waist located at

$$z_{2w} = \frac{-R_2^-\left(1 - \frac{R_2^-}{f}\right)}{\zeta_2^- + \left(1 - \frac{R_2^-}{f}\right)^2}, \tag{10.244}$$

and the width of this waist is given by

$$b_w = b_2 \sqrt{\frac{\zeta_2^-}{\zeta_2^- + \left(1 - \frac{R_2^-}{f}\right)^2}}$$

$$= \left|\frac{\lambda R_2^-}{b_2}\right| \frac{1}{\sqrt{\zeta_2^- + \left(1 - \frac{R_2^-}{f}\right)^2}}. \tag{10.245}$$

Note that if we let $f = R_2^-$, the beam waist is located at the lens and has a width of b_2.

We may simplify these last several expressions by defining

$$\frac{1}{R_2^+} = \frac{1}{R_2^-} - \frac{1}{f},$$

$$\zeta_2^+ = \left(\frac{\lambda R_2^+}{b_2^2} \right)^2, \tag{10.246}$$

where R_2^+ denotes the radius of curvature just to the right of the lens and ζ_2^+ is the parameter associated with R_2^+. Then Eq. (10.239) becomes

$$u_2^+(r) = A_2 B_\ell \operatorname{Gaus}\left(\frac{r}{b_2} \right) q\left(r; \frac{1}{\lambda R_2^+} \right) p_\ell(r), \tag{10.247}$$

and Eq. (10.243) may be written as [note the similarily with Eq. (10.224)]

$$b_3 = b_2 \sqrt{(\zeta_2^+ + 1)\left(\frac{z_{23}}{R_2^+} \right)^2 + 2\left(\frac{z_{23}}{R_2^+} \right) + 1},$$

$$R_3 = R_2^+ \left[\frac{(\zeta_2^+ + 1)\left(\frac{z_{23}}{R_2^+} \right)^2 + 2\left(\frac{z_{23}}{R_2^+} \right) + 1}{(\zeta_2^+ + 1)\left(\frac{z_{23}}{R_2^+} \right) + 1} \right],$$

$$\Phi_{23} = \tan^{-1}\left[\frac{-\sqrt{\zeta_2^+}}{\frac{R_2^+}{z_{23}} + 1} \right]. \tag{10.248}$$

In addition, Eqs. (10.244) and (10.245) reduce to

$$z_{2w} = \frac{-R_2^+}{\zeta_2^+ + 1},$$

$$b_w = b_2 \sqrt{\frac{\zeta_2^+}{\zeta_2^+ + 1}}$$

$$= \left| \frac{\lambda R_2^+}{b_2} \right| \frac{1}{\sqrt{\zeta_2^+ + 1}}. \tag{10.249}$$

We have again assumed that the lens aperture is sufficiently large to transmit all of the incident power.

The expressions above are simplified to some extent if we select the plane $z=z_1$ to coincide with the initial beam waist. To avoid ambiguities in notation we denote the width of the initial beam waist by b_{w1}, the Rayleigh-range parameter associated with this waist by $Z_{R1}=b_{w1}^2/\lambda$, and the distance from this waist to the lens by z_{12}. With the distance to the observation plane again denoted by z_{23}, the observed beam width may be expressed by

$$b_3=b_{w1}\left(\frac{z_{12}}{Z_{R1}}\right)\sqrt{\left(1-\frac{z_{23}}{z_{12}}-\frac{z_{23}}{f}\right)^2+\left(\frac{Z_{R1}}{z_{12}}\right)^2\left(1-\frac{z_{23}}{f}\right)^2}\,,\qquad z_1 \text{ at waist.} \tag{10.249a}$$

This expression is an alternative form of that given by Dickson [Ref. 10-19, Eq. (33)]. Several special cases are of interest, and we list them below:

$$b_3=\begin{cases} b_{w1}\sqrt{\left(\dfrac{z_{23}}{Z_{R1}}\right)^2+\left(1-\dfrac{z_{23}}{f}\right)^2}\,, & z_{12}=0\\[2ex] b_{w1}\sqrt{\left(\dfrac{z_{12}}{Z_{R1}}\right)^2+1}\,, & z_{23}=0\\[2ex] \dfrac{\lambda f}{b_{w1}}, & z_{23}=f. \end{cases} \tag{10.249b}$$

Note that when the observation plane coincides with the back focal plane of the lens, the observed beam width is independent of the waist-to-lens distance z_{12}.

To conclude this part of our studies, we determine the *waist-to-waist properties* of a Gaussian beam that has passed through a lens. We again specify that the initial beam waist lies in the plane $z=z_1$ and has a width of b_{w1}. In addition, we now require the observation plane $z=z_3$ to lie at the waist formed by the lens; this waist is located at a distance of z_{23} from the lens and has a width of b_{w3}. The desired results are obtained by combining

Eqs. (10.235), (10.244), (10.245), and (10.249a). We obtain

$$\left.\begin{aligned} z_{23} &= f + \frac{f^2(z_{12}-f)}{Z_{R1}^2+(z_{12}-f)^2} \\ b_{w3} &= b_{w1}\left[\frac{f}{\sqrt{Z_{R1}^2+(z_{12}-f)^2}}\right] \\ Z_{R3} &= Z_{R1}\left[\frac{f^2}{Z_{R1}^2+(z_{12}-f)^2}\right] \end{aligned}\right\} \text{ waists at } z=z_1 \text{ and } z=z_3, \qquad (10.249c)$$

where Z_{R3} is the Rayleigh-range parameter associated with the waist produced by the lens. Again several special cases are of interest. For example, if the initial waist is placed in the front focal plane of the lens such that $z_{12}=f$, we find that

$$\left.\begin{aligned} z_{23} &= f \\ b_{w3} &= \frac{\lambda f}{b_{w1}} \\ Z_{R3} &= \frac{f^2}{Z_{R1}} \end{aligned}\right\} \text{ initial waist in front focal plane.} \qquad (10.249d)$$

Thus a beam waist in the front focal plane *always* produces a beam waist in the back focal plane. Not only that, Dickson [Ref. 10-19] shows that the width of the waist produced by the lens is *always* a maximum when this waist is located in the back focal plane, and its width decreases when it is moved to either side of the back focal plane. Examination of Eqs. (10.249c) and (10.249d) shows that beam waists are not *imaged* into beam waists in the usual sense. In fact, for certain values of the various parameters, the relationship between the locations of the waists is exactly opposite from the object-to-image relationship of normal imaging. For example, as the initial waist is moved toward the lens, the second waist may also move toward the lens rather than away from it as we might expect from normal imaging considerations.

We now determine the location(s) of the initial beam waist that will produce a second waist of the *same size*. By setting $b_{w3}=b_{w1}$ in Eq.

(10.249c) we find that

$$\left.\begin{aligned} z_{12} &= f \pm \sqrt{f^2 - Z_{R1}^2} \\ z_{23} &= f \pm \sqrt{f^2 - Z_{R1}^2} \end{aligned}\right\} \text{ waists of same size.} \qquad (10.249e)$$

It is not difficult to show that when the initial beam waist is located at either of these two positions, the radius of curvature of the wavefront incident on the lens is simply $R_2^- = 2f$. The lens then converts this wavefront into another of radius $R_2^+ = -2f$, and the symmetrical nature of the situation becomes obvious. Note that if the initial waist is placed in the front focal plane of a lens having a focal length $f = \pm Z_{R1}$, the lens will produce a second waist of the same size in its back focal plane.

Optimum Focusing of Gaussian Beams

We now investigate some applications involving the focusing of Gaussian beams. In the first of these we are to design an optical relay system similar to that shown in Fig. 10-31, and we must select the radius of curvature R_2^+ of the transmitted beam that will produce the minimum beam size at a predetermined distance from the system. We are attempting to maximize the overall beam irradiance at some target, which is located at a fixed distance in the $z = z_3$ plane, and we basically must select the proper focal length of the relay lens. We assume that the width of the beam incident on this lens, b_2, is fixed and that the lens aperture is large enough to transmit all of the incident power. We also assume that we can steer the beam

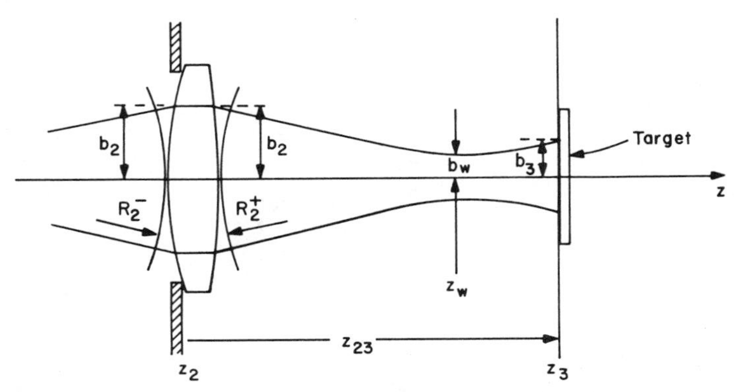

Figure 10-31 Optimum focusing of a Gaussian beam when beam radius is fixed and aperture size is variable.

accurately enough to irradiate the target properly, and we therefore center the target on the z-axis for simplicity.

The optimum radius of curvature R_2^+ is that for which the beam width b_3 at the target is a minimum. At first glance it might appear that the optimum performance would be obtained by selecting a value for R_2^+ that produces a beam waist at the target; however, as we shall now discover, *that is not the case*. To show this, we differentiate the beam-width expression of Eq. (10.248) with respect to R_2^+ and set the derivative equal to zero. In doing so we find that b_3 is minimized if

$$R_2^+ = -z_{23}, \tag{10.250}$$

and the focal length of the lens is then determined from Eq. (10.246). The resulting beam width for this value of R_2^+ is

$$(b_3)_{\min} = \left| \frac{\lambda z_{23}}{b_2} \right|. \tag{10.251}$$

We note that a waist will occur at a distance of

$$\begin{aligned} z_{2w} &= \frac{z_{23}}{\zeta_2^+ + 1} \\ &< z_{23}, \end{aligned} \tag{10.252}$$

where now

$$\zeta_2^+ = \left(\frac{\lambda z_{23}}{b_2^2} \right)^2, \tag{10.253}$$

and that the width of the beam at this waist will be

$$\begin{aligned} b_w &= \left| \frac{\lambda z_{23}}{b_2} \right| \frac{1}{\sqrt{\zeta_2^+ + 1}} \\ &= \frac{(b_3)_{\min}}{\sqrt{\zeta_2^+ + 1}} \\ &< (b_3)_{\min}. \end{aligned} \tag{10.254}$$

However, even though the width of this waist is smaller than $(b_3)_{\min}$, it will not remain smaller if it is relocated at the target.

For the beam waist to occur at the target we must select a new radius of curvature $\hat{R}_2^+$ that satisfies

$$z_{23} = \frac{-\hat{R}_2^+}{\hat{\zeta}_2^+ + 1}. \tag{10.255}$$

Noting that $\hat{\zeta}_2^+ = (\lambda \hat{R}_2^+ / b_2^2)^2$, we may rearrange this expression as follows:

$$\frac{z_{23}\lambda^2}{b_2^4}(\hat{R}_2^+)^2 + \hat{R}_2^+ + z_{23} = 0. \tag{10.256}$$

The solutions of this equation are

$$\hat{R}_2^+ = \frac{-1 \pm \sqrt{1 - 4\left(\frac{\lambda z_{23}}{b_2^2}\right)^2}}{\frac{2\lambda^2 z_{23}}{b_2^4}}$$

$$= \frac{-z_{23}\left(1 \mp \sqrt{1 - 4\zeta_2^+}\right)}{2\zeta_2^+}. \tag{10.257}$$

We see that the solutions for $\hat{R}_2^+$ will be real valued only when

$$\zeta_2^+ \leqslant 0.25, \tag{10.258}$$

and *when* ζ_2^+ *is greater than* 0.25 *the beam waist cannot be located at the target*. Under the assumption that $\zeta_2^+ \leqslant 0.25$, relocating the beam waist at the target will yield a beam width of

$$\hat{b}_3 = \left| \frac{\lambda \hat{R}_2^+}{b_2} \right| \frac{1}{\sqrt{\hat{\zeta}_2^+ + 1}}. \tag{10.259}$$

However, from Eq. (10.255) we know that $\hat{R}_2^+ = -z_{23}(\hat{\zeta}_2^+ + 1)$, and therefore

$$\hat{b}_3 = \left| \frac{\lambda z_{23}}{b_2} \right| \sqrt{\hat{\zeta}_2^+ + 1}$$

$$= (b_3)_{\min} \sqrt{\hat{\zeta}_2^+ + 1}$$

$$> (b_3)_{\min}. \tag{10.260}$$

Consequently, we have shown that, for a target situated a fixed distance from the relay system, *the minimum beam width at the target is not obtained by locating the beam waist at the target*. Rather, we must select a radius of curvature $R_2^+ = -z_{23}$, which causes the beam waist to lie between the relay system and the target.

Example

We now work through an example to demonstrate the above results. We let $z_{23}=50$ cm, $\lambda=500$ nm, and $b_2=1$ mm. By selecting $R_2^+ = -z_{23}$ we find that

$$(b_3)_{\min} = \left| \frac{(5\times 10^{-7})(0.5)}{10^{-3}} \right| = 0.250 \text{ mm},$$

$$\zeta_2^+ = \left[\frac{(5\times 10^{-7})(0.5)}{10^{-6}} \right]^2 = 0.0625,$$

$$z_{2w} = \frac{0.5}{0.0625+1} = 47.1 \text{ cm},$$

$$b_w = \frac{0.250 \text{ mm}}{\sqrt{0.0625+1}} = 0.243 \text{ mm}. \tag{10.261}$$

Thus, for $R_2^+ = -z_{23}$, we see that the beam waist is located slightly in front of the target and that its width is slightly less than $(b_3)_{\min}$. We now select values for $\hat{R}_2^+$ that will move the beam waist out to the target:

$$\hat{R}_2^+ = \frac{-0.5\left[1 \mp \sqrt{1-4(0.0625)}\right]}{2(0.0625)} = \frac{-0.5 \mp 0.433}{0.125}$$

$$= -53.6 \text{ cm}, \ -746 \text{ cm}. \tag{10.262}$$

These values lead to

$$\hat{\zeta}_2^+ = 0.0718, \ 13.9,$$

$$\hat{b}_3 = 0.259 \text{ mm}, \ 0.966 \text{ mm}, \tag{10.263}$$

and even the smaller of these beam waists has a larger width than $(b_3)_{\min} = 0.250$ mm. The difference between $\hat{b}_3$ and $(b_3)_{\min}$ becomes smaller and smaller with decreasing ζ_2^+, as may be easily demonstrated. ■

We now investigate a slightly different version of the preceding application. We again wish to irradiate a target located a fixed distance from the relay system, but there are some important differences. We now wish to maximize only the *peak value* of the irradiance at the target. In addition, not only are we to select the proper radius of curvature R_2^+, we must also choose the optimum value for the beam width b_2 relative to the fixed diameter d_ℓ of the relay lens aperture. In other words, our task boils down to the following: given a Gaussian beam of fixed power, a relay lens aperture of fixed diameter, and a target located at a fixed distance, we must design a beam expander that will maximize the *peak irradiance* at the target. We shall not worry too much about the details of the beam expander itself; however, the desired effects can be obtained by passing the initial beam through an appropriately selected lens that is located ahead of the final relay lens, as shown in Fig. 10-32. The beam incident on the final relay lens will then be of the general form

$$u_2^-(r)=K_2\left(\frac{1}{b_2}\right)\text{Gaus}\left(\frac{r}{b_2}\right)q\left(r;\frac{1}{\lambda R_2^-}\right), \qquad (10.264)$$

where K_2 is a constant that both specifies the phase of the incident beam and is related to the total beam power. The complex amplitude transmitted by the final relay lens is

$$u_2^+(r)=\frac{K_2B_\ell}{b_2}\text{Gaus}\left(\frac{r}{b_2}\right)q\left(r;\frac{1}{\lambda R_2^+}\right)\text{cyl}\left(\frac{r}{d_\ell}\right). \qquad (10.265)$$

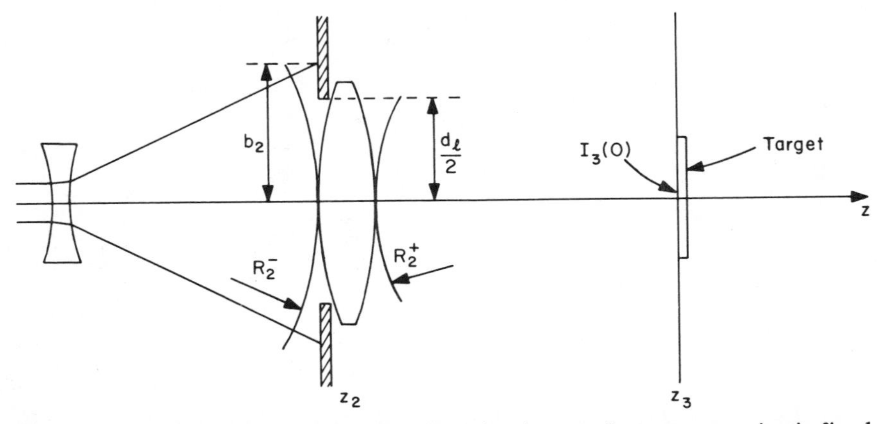

Figure 10-32 Optimum focusing of a Gaussian beam when aperture size is fixed and beam radius is variable.

Because the lens aperture may now truncate the beam, we are unable to use the previous results of this section. However, if we try to use the results of the preceding section [e.g., Eq. (10.164)], we are faced with a convolution operation that is quite difficult to evaluate. We must therefore look for other approaches.

For this particular problem it is reasonable to assume (although difficult to prove) that the peak irradiance at the target will be obtained when $R_2^+ = -z_{23}$ and that it will occur on the z-axis, i.e.,

$$I_3(r)_{\max} = I_3(0)$$
$$= |u_3(0)|^2. \qquad (10.266)$$

Consequently, our task is reduced to one of maximizing $|u_3(0)|$. From Eq. (10.66) we know that

$$u_3(r) = B_{23} q\left(r; \frac{1}{\lambda z_{23}}\right) \mathcal{H}_0\left\{ u_2^+(r') q\left(r'; \frac{1}{\lambda z_{23}}\right)\right\}\Bigg|_{\rho = r/\lambda z_{23}}, \qquad (10.267)$$

and by choosing $R_2^+ = -z_{23}$ and setting $r=0$ we obtain

$$u_3(0) = B_{23} \mathcal{H}_0\left\{ \frac{K_2 B_\ell}{b_2} \operatorname{Gaus}\left(\frac{r'}{b_2}\right) \operatorname{cyl}\left(\frac{r'}{d_\ell}\right)\right\}\Bigg|_{\rho=0}. \qquad (10.268)$$

Then, from the central ordinate theorem [Eq. (9.125)], we have

$$u_3(0) = B_{23} 2\pi \int_0^\infty \frac{K_2 B_\ell}{b_2} \operatorname{Gaus}\left(\frac{r'}{b_2}\right) \operatorname{cyl}\left(\frac{r'}{d_\ell}\right) r'\,dr'$$
$$= K_2 B_\ell B_{23} 2\pi \int_0^{d_\ell/2} \left(\frac{1}{b_2}\right) \operatorname{Gaus}\left(\frac{r'}{b_2}\right) r'\,dr'$$
$$= K_2 B_\ell B_{23} b_2 \left[1 - \operatorname{Gaus}\left(\frac{d_\ell}{2b_2}\right)\right]. \qquad (10.269)$$

Finally, we obtain

$$|u_3(0)| = |K_2 B_\ell B_{23}| b_2 \left[1 - \operatorname{Gaus}\left(\frac{d_\ell}{2b_2}\right)\right], \qquad (10.270)$$

which has a maximum when

$$b_2 = 0.791\, d_\ell . \tag{10.271}$$

We note that for this value of b_2 the power transmitted by the aperture will be somewhat less than the total incident beam power. In fact, we find from Eq. (10.216) that only 91.9% of the incident power passes through the aperture for this value of b_2.

The peak irradiance at the target is given by

$$I_3(0) = |K_2 B_\ell B_{23}|^2 b_2^2 \left[1 - \mathrm{Gaus}\left(\frac{d_\ell}{2b_2} \right) \right]^2$$

$$= |K_2 B_\ell|^2 \left(\frac{0.791\, d_\ell}{\lambda z_{23}} \right)^2 [1 - \mathrm{Gaus}(0.632)]^2$$

$$= 0.320\, |K_2 B_\ell|^2 \left(\frac{d_\ell}{\lambda z_{23}} \right)^2 . \tag{10.272}$$

For comparison, we calculate the peak irradiance for a beam width of $b_2 = 0.5\, d_\ell$, which allows 99.8% of the incident power to be transmitted. We find that

$$I_3(0) = |K_2 B_\ell|^2 \left(\frac{0.5\, d_\ell}{\lambda z_{23}} \right)^2 [1 - \mathrm{Gaus}(1)]^2$$

$$= 0.229\, |K_2 B_\ell|^2 \left(\frac{d_\ell}{\lambda z_{23}} \right)^2 , \tag{10.273}$$

which is only 71.5% of the peak irradiance obtained with $b_2 = 0.791\, d_\ell$. The net result is that even though we lose 8% of the transmitted beam power by overfilling the relay lens aperture, the peak irradiance at the target is increased by nearly 40%.

Results of related problems have been reported by Holmes et al., who determined the on-axis irradiance for Gaussian beams passed through annular apertures (see Ref. 10-20).

Elliptical Gaussian Beams

An elliptical Gaussian beam is one whose cross section is elliptical rather than circular, and such a beam can be produced by an optical element (e.g., a cylindrical lens) that causes the wavefront of the beam to have different radii of curvature in the x- and y-directions. Because the Gaussian function is separable in x and y, the behavior of an elliptical Gaussian beam will be governed separately in each of the coordinate directions. The basic relationships are similar to those given in Eq. (10.221), but now the beam width and wavefront radii of curvature will generally be different in the x- and y-directions at any given plane $z=$constant. It can be shown that a circular cross section will be found at only two points along the beam, and that there is no point along the beam at which the wavefronts are planes. More information on this subject may be found in Yariv (Ref. 10-21).

REFERENCES

10-1 M. Born and E. Wolf, *Principles of Optics*, 3rd ed., Pergamon, New York, 1965.

10-2 J. M. Stone, *Radiation and Optics, An Introduction to the Classical Theory*, McGraw-Hill, New York, 1963.

10-3 J. W. Goodman, *Introduction to Fourier Optics*, McGraw-Hill, New York, 1968.

10-4 B. B. Baker and E. T. Copson, *The Mathematical Theory of Huygen's Principle*, 2nd edition, Clarendon, Oxford, 1949.

10-5 C. J. Bouwkamp, "Diffraction Theory." In A. C. Strickland, Ed., *Reports on Progress in Physics: Vol. XVII*, The Physical Society, London, 1954.

10-6 E. Wolf and E. W. Marchand, "Comparison of the Kirchhoff and the Rayleigh-Sommerfeld Theories of Diffraction at an Aperture," *J. Opt. Soc. Am.* 54:587 (1964).

10-7 A. Sommerfeld, "Optics," *Lectures on Theoretical Physics*, Vol. IV, Academic, New York, 1954.

10-8 E. W. Marchand and E. Wolf, "Diffraction at Small Apertures in Black Screens," *J. Opt. Soc. Am.* 59:79–90 (1969).

10-9 R. V. Shack and J. E. Harvey, "An Investigation of the Distribution of Radiation Scattered by Optical Surfaces," Optical Sciences Center (Univ. of Arizona) Final Report, August 1975.

10-10 E. Hecht and A. Zajac, *Optics*, Addison-Wesley, Reading, Mass., 1974.

10-11 F. S. Harris, Jr., "Light Diffraction Patterns," *Appl. Opt.*, 3(8): 909-913 (1964).

10-12 F. S. Harris, Jr., M. S. Tavenner, and R. L. Mitchell, "Single-Slit Fresnel Diffraction Patterns: Comparison of Experimental and Theoretical Results," *J. Opt. Soc. Am.* 59(3):293–296 (1969).

10-13 M. Cagnet, M. Françon, and J. C. Thrierr, *Atlas of Optical Phenomena*, Springer-Verlag, Berlin, and Prentice-Hall, Englewood Cliffs, N.J., 1962.

10-14 G. B. Parrent and B. J. Thompson, *Physical Optics Notebook*, Society of Photo-Optical Instrumentation Engineers, Redondo Beach, Calif., 1969

10-15 G. Z. Dimitroff and J. G. Baker, *Telescopes and Accessories*, Blakiston, Philadelphia, 1945.

10-16 R. B. Hoover and F. S. Harris, Jr., "Die Beugungserscheinungen: A Tribute to F. M. Schwerd's Monumental Work on Fraunhofer Diffraction," *Appl. Opt.* 8(11): 2161-2164 (1969).

10-17 H. Kogelnik and T. Li, "Laser Beams and Resonators," *Proc. IEEE* 54:1312 (Oct. 1966).

10-18 A. E. Siegman, *An Introduction to Lasers and Masers,* McGraw-Hill, New York, 1971.

10-19 L. D. Dickson, "Characteristics of a Propagating Gaussian Beam," *Appl. Opt.* 9(8): 1854-1861 (1970).

10-20 D. A. Holmes, J. E. Korka, and P. V. Avizonis, "Parametric Study of Apertured Focused Gaussian Beams," *Appl. Opt.* 11(3):565-574 (1972).

10-21 A. Yariv, *Introduction to Optical Electronics*, Holt, Rinehart and Winston, New York, 1976.

PROBLEMS

Use the configuration of Fig. 10-7 for Problems 10-1 and 10-4.

10-1. An aperture is placed in the plane $z=z_1$ and illuminated with a monochromatic, normally incident plane wave field of amplitude A and wavelength λ. With $|A|^2=10^{-2}$ W/cm^2, $\lambda=514.5$ nm, and $d=2$ cm, find an expression for the irradiance along the optical axis in the Fresnel region [i.e., find $I_2(0)=|u_2(0)|^2$ as a function of the observation distance z_{12}] for the following apertures. Sketch $I_2(0)$ in each case.

a. $t_1(r)=\mathrm{cyl}\left(\frac{r}{d}\right)$.

b. $t_1(r)=\mathrm{cyl}\left(\frac{r}{d}\right)-\mathrm{cyl}\left(\frac{2r}{d}\right)$.

10-2. An aperture is placed in the plane $z=z_1$ and illuminated with a monochromatic wave field of amplitude A and wavelength λ. With $|A|^2=1$ W/m^2, $\lambda=514.5$ nm, $b=10$ mm, $d=5$ mm, $\gamma_x=10^{-4}$, $\gamma_y=2\times10^{-4}$, and $z_{12}=10^4$ m, find the irradiance in the observation plane $z=z_2$ for the following illumination wavefields and apertures. Sketch appropriate profiles.

a. $u_1^-(x,y)=A,\ t_1(x,y)=\mathrm{rect}\left(\frac{x}{d},\frac{y}{b}\right)*\frac{1}{b}\delta\delta\left(\frac{x}{b}\right)$.

b. $u_1^-(x,y)=A\exp\left\{-j\frac{2\pi}{\lambda}(\gamma_x x+\gamma_y y)\right\},\ t_1(x,y)=\mathrm{cyl}\left(\frac{\sqrt{x^2+y^2}}{d}\right)$.

c. $u_1^-(x,y)=A,\ t_1(x,y)=\mathrm{rect}\left(\frac{x+y}{\sqrt{2}\,d},\frac{-x+y}{\sqrt{2}\,d}\right)$.

d. $u_1^-(x,y)=A,\ t_1(x,y)=\mathrm{cyl}\left[\sqrt{\left(\frac{x}{b}\right)^2+\left(\frac{y}{d}\right)^2}\right]$.

e. With $u_1^-(x,y)=A$ and $t_1(x,y)=\mathrm{rect}\left(\frac{x}{d},\frac{y}{b}\right)-\mathrm{cyl}\left(\frac{\sqrt{x^2+y^2}}{d}\right)$, find the value of the irradiance at the origin, i.e., find $I_2(0,0)$.

f. Repeat part e with

$$t_1(x,y)=\left[\mathrm{rect}\left(\frac{x}{b},\frac{y}{d}\right)-\mathrm{cyl}\left(\frac{\sqrt{x^2+y^2}}{d}\right)\right]e^{j\pi\,\mathrm{step}(x)}.$$

10-3. Given a transparency with amplitude transmittance

$$t_1(r)=\left[0.5+0.5\cos(\pi a r^2)\right]\mathrm{cyl}\left(\frac{r}{d}\right).$$

This transparency is placed in the plane $z=z_1$ and illuminated with a normally incident plane-wave field at amplitude A and wavelength λ.

a. Show that this transparency behaves as a combination of three lenses simultaneously: a positive lens, a negative lens, and a lens with no power.
b. Find the focal lengths associated with the various lenses, and indicate their dependence on the wavelength of the incident light.
c. With $|A|^2=10^{-2}$ W/m^2, $\lambda=514.5$ nm, $a=10^6$ m^{-2}, and $d=2$ cm, find the amplitude $u_2(r)$ in the vicinity of the optical axis at the plane $z=z_2=z_1+|f|$, where $|f|$ is the magnitude of the finite focal length from part b.

10-4. Given a transparency with amplitude transmittance

$$t(r)=\left\{0.5+0.5\,\mathrm{sgn}\left[\cos(\pi a r^2)\right]\right\}\mathrm{cyl}\left(\frac{r}{d}\right),$$

which is one type of *Fresnel zone plate*:

a. Sketch a profile of $t(r)$.
b. Show that such a zone plate acts as a lens of multiple focal lengths.
c. Find an effective value of the constant $|B_\ell|$ for each focal length.

Use the configuration of Fig. 10-18 for Problems 10-5 through 10-8.

10-5. A transparency is placed at the plane $z=z_3$ and illuminated with a monochromatic wave field, of wavelength λ, emanating from a point source located at $(0,0,z_1)$. This wave field has an amplitude of A just to the left of the lens. Additional information is as follows: $|A|^2=0.1$ mW/cm^2, $\lambda=632.8$ nm, $z_{12}=1$ m, $z_{23}=0.4$ m, $f=0.5$ m, $B_\ell=0.95$, $d=1$ mm, and

$$p_\ell(r)=\text{cyl}\left(\frac{r}{40d}\right),$$

$$t_3(x,y)=\text{cyl}\left(\frac{\sqrt{x^2+y^2}}{2d}\right)*\frac{1}{5d}\delta\delta\left(\frac{x}{5d}\right).$$

With the observation plane located such that $z_4=z_5$:

a. Find an expression for the complex amplitude $u_5(x,y)$.
b. Calculate the irradiance $I_5(x,y)$ and sketch appropriate profiles.
c. Determine the size of the observation region over which the above results are valid.

10-6. Repeat Problem 10-5 with

$$t_3(x,y)=\left[\frac{20}{d}\,\text{comb}\left(\frac{20x}{d}\right)*\text{rect}\left(\frac{40x}{d}\right)\right]\text{rect}\left(\frac{x}{10d},\frac{y}{10d}\right).$$

10-7. Repeat Problem 10-6 with

$$p_\ell(r)=\text{cyl}\left(\frac{r}{10d}\right).$$

10-8. Repeat Problem 10-5 with

$$t_3(x,y)=\left[\left(\frac{10}{d}\right)^2\text{comb}\left(\frac{10x}{d},\frac{10y}{d}\right)**\text{rect}\left(\frac{12.5x}{d},\frac{12.5y}{d}\right)\right]$$
$$\times\,\text{cyl}\left(\frac{\sqrt{x^2+y^2}}{2d}\right).$$

Use the configuration of Fig. 10-21 for Problems 10-9 through 10-13.

10-9. A transparency is placed at the plane $z=z_2$ and illuminated with a monochromatic wave field, of wavelength λ, emanating from a point source located at $(0,0,z_1)$. In the absence of the transparency, this wave field would have an amplitude of A just to the left of the lens. Additional information is as follows: $|A|^2=0.1$

mW/cm^2, $\lambda=632.8$ nm, $z_{13}=1.2$ m, $z_{23}=0.2$ m, $f=0.4$ m, $B_\ell=0.95$, $d=1$ mm, and

$$p_\ell(r)=\text{cyl}\left(\frac{r}{20d}\right),$$

$$t_2(x,y)=\text{rect}\left(\frac{x}{d},\frac{y}{5d}\right).$$

With the observation plane located such that $z_4=z_5$:

a. Find an expression for the complex amplitude $u_5(x,y)$.

b. Calculate the irradiance $I_5(x,y)$ and sketch appropriate profiles.

c. Determine the size of the observation region over which the above results are valid.

10-10. Repeat Problem 10-9 with $z_{23}=0.4$ m (i.e., transparency in front focal plane).

10-11. Repeat Problem 10-9 with $z_{13}=\infty$ (i.e., plane-wave illumination).

10-12. Repeat Problem 10-9 with $z_{23}=0$ and

$$t_2(x,y)=\exp\left[j0.1\cos\left(\frac{20\pi y}{d}\right)\right]\text{rect}\left(\frac{x}{10d},\frac{y}{10d}\right).$$

10-13. Assume that Problem 10-9 is to be repeated, but with

$$t_2(x,y)=\left[\left(\frac{80}{d}\right)^2\text{comb}\left(\frac{80x}{d},\frac{80y}{d}\right)**\text{rect}\left(\frac{100x}{d},\frac{100y}{d}\right)\right]$$

$$\times\,\text{cyl}\left(\frac{\sqrt{x^2+y^2}}{10d}\right).$$

Can the techniques derived in Chapter 10 be used to obtain valid results in this case? Explain.

10-14. Use the configuration of Fig. 10-27 for this problem. The output of an argon laser is described in the plane $z=z_1$ by the expression of Eq. (10.211). With the total beam power $P_{\text{tot}}=100$ mW, $\lambda=488$ nm, $b_1=1$ mm, and $R_1=5$ m:

a. Find the location of the beam waist and the radius of the beam at the waist.

b. Find the locations for which $|R_2|$ is a minimum, find the value of b_2 at these points, and find the value of $|R_2|_{\min}$.

c. If we now let $z_{12}=2$ m, find the description of $u_2(r)$ as given by Eq. (10.220); i.e., find A_2, b_2, and R_2.

10-15. Use the configuration of Fig. 10-30 for this problem. As in Problem 10-14, the output of an argon laser is described in the plane $z=z_1$ by Eq. (10.211), the total beam power is 100 mW, $\lambda=488$ nm, $b_1=1$ mm, and $R_1=5$ m, but now a positive lens of focal length $f=50$ cm and diameter $d_\ell=4$ cm is placed at the plane $z=z_2=z_1+2$ m (i.e., $z_{12}=2$ m).

a. Find the location of the next beam waist, and calculate the radius of the beam at this waist.
b. Find the irradiance $I_3(r)$ at this waist and sketch a radial profile.
c. Find the Rayleigh range associated with this waist.
d. Do the original beam waist (to the left of the lens) and the new beam waist (to the right of the lens) lie in conjugate planes?

10-16. Use the configuration of Fig. 10-31 for this problem. We wish to focus a Gaussian laser beam on a fixed target such that the beam power is concentrated in the smallest possible area. The target is located at a distance of $z_{23}=1000$ m. With $b_2=5$ cm, $R_2^-=0.5$ m, and $\lambda=1$ μm:

a. Find the value of R_2^+ that will yield optimum focusing.
b. Describe a lens that will perform satisfactorily; i.e., choose the focal length f and diameter d_ℓ. Assume that $|B_\ell|=0.98$.
c. Find the radius of the beam, $(b_3)_{\min}$, at the target, and calculate the peak irradiance, $I_3(0)$.
d. Find the location of the beam waist, the radius of the beam waist, and the peak irradiance at the beam waist.
e. Calculate the value(s) of $\hat{R}_2^+$, and select an appropriate lens(es), that will cause a beam waist to be located at the target.
f. Find the radius of the beam waists, $\hat{b}_3$, that result from part e, and calculate the peak irradiance values, $\hat{I}_3(0)$, for each.
g. Show that the "improvement factor" is given by

$$\frac{I_3(0)}{\hat{I}_3(0)}=\left[\frac{\hat{b}_3}{(b_3)_{\min}}\right]^2=\hat{\zeta}_2^+ +1,$$

and calculate it for the above values of $\hat{I}_3(0)$.
h. Determine the largest distance at which the beam waist could be located from the lens, and find the values of $\hat{R}_2^+$ and f required to put it there.

10-17. Here we wish to design the transmitter of a lightweight optical communication system. The modulated output of a small argon laser is passed through an aberration-free beam expander, which is really a Cassegrain telescope operated in reverse, and directed at the detector located some distance away. We model the system by the configuration of Fig. 10-32, and we describe the transmitted wave field by an equation similar to Eq. (10.265); however, we must now include the central obscuration of the Cassegrain telescope. Thus, the transmitted wave field is given by

$$u_2^+(r)=\frac{K_2B_\ell}{b_2}\,\mathrm{Gaus}\left(\frac{r}{b_2}\right)q\left(r;\frac{1}{\lambda R_2^+}\right)\left[\mathrm{cyl}\left(\frac{r}{d_\ell}\right)-\mathrm{cyl}\left(\frac{r}{d_o}\right)\right],$$

where d_o is the diameter of the obscuration. Because of weight considerations, d_ℓ is restricted to a maximum value of 10 cm, and the best 10-cm telescope available has an obscuration of $d_o=0.3d_\ell=3$ cm. We assume that the telescope can be pointed very accurately and that it can be focused satisfactorily. In addition, $\lambda=514.5$ nm, $z_{23}=20$ km, $B_\ell=0.95$, and the total power of the incident beam is 50 mW.

a. Find the value of R_2^+ that will maximize $I_3(0)$.
b. Find the value of b_2 that will maximize $I_3(0)$. [Here it is instructive to determine $I_3(0)$ as a function of b_2/d_ℓ for the given obscuration d_o.]
c. Calculate the ratio of transmitted power to incident power when b_2/d_ℓ has the optimum value.
d. Determine the irradiance $I_3(r)$, sketch a radial profile, and calculate $I_3(0)$ when b_2/d_ℓ has the optimum value.
e. If an unobscured system, as described by Eq. (10.265), were to be used, find the smallest value of d_ℓ that would be required to yield the same value of $I_3(0)$ as obtained in part d. Also find the value of b_2 required for this d_ℓ. [Note that a Cassegrain system can be much more compact and much lighter than a comparable system made with refractive elements.]

CHAPTER 11

IMAGE-FORMING SYSTEMS

In this chapter we apply our understanding of linear shift-invariant systems and our knowledge of diffraction to the analysis of various image-forming systems, first for coherent illumination and then for incoherent illumination. In doing so we introduce the concepts of the subject commonly referred to as *Fourier optics*; however, we again omit the interesting historical background associated with the development of this subject. For those who desire more information, Goodman (Ref. 11-1) presents a brief historical account and provides a list of references.

Although mirrors play important roles in many optical systems, we shall restrict our attention to systems containing only lenses. When necessary, the results obtained for lenses may be extended to include mirrors without great difficulty (we note that mirrors are often modeled as lenses, and vice versa, to simplify the analysis of optical systems). Except when specified to the contrary, we shall assume the lenses in our systems to be free of aberrations to simplify the development; the results obtained for aberration-free lenses must be altered when aberrated lenses are used.

11-1 IMAGE FORMATION WITH COHERENT LIGHT

In this section we derive the basic expessions relating the complex amplitude distributions of object and image for a very general two-lens system. Although the development for such a general system is more complicated than for certain special configurations, the results obtained are much more useful because they are valid for all configurations. The behavior of the special configurations may be assessed by substitution of the appropriate parameter values into the general result. The system to be analyzed is

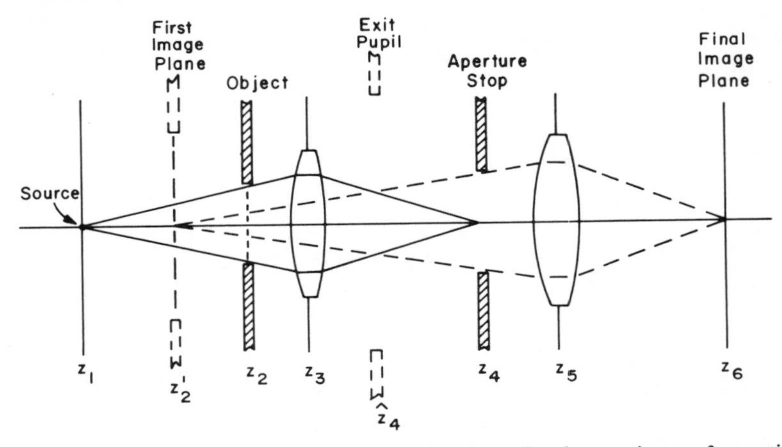

Figure 11-1 Optical system used for investigation of coherent image formation.

similar to the types of systems frequently encountered in coherent optical data processing applications, and is depicted in Fig. 11-1. The object transparency is placed in the plane $z=z_2$ and transilluminated from the left with a spherical wave field of monochromatic light emanating from the point $(0,0,z_1)$.* We place a lens element at the plane $z=z_3$, and we denote by $z=z_4$ the plane that is conjugate with the source. As a result, the first half of this imaging system is effectively the same as the diffracting system of Fig. 10-21 (in that system we designated the conjugate plane by $z=z_5$ rather than by z_4). We may therefore use Eq. (10.198) to obtain the complex amplitude incident on the conjugate plane:

$$u_4^-(x,y)=A\left(\frac{z_{13}}{z_{12}}\right)^2 e^{jkz_{23}}B_{\ell 1}B_{34}q\left(x,y;\frac{1}{\lambda z'_{24}}\right)T_2\left(\frac{m'x}{\lambda z'_{24}},\frac{m'y}{\lambda z'_{24}}\right), \qquad (11.1)$$

where $B_{\ell 1}$ describes the first lens element [see Eq. (10.133)] and m' is the magnification associated with the image of $t_2(x, y)$ produced by this element. From Eq. (10.196) we obtain

$$m'=\frac{f_1}{f_1-z_{23}}, \qquad z'_{24}=m'z_{23}+z_{34}, \qquad (11.2)$$

where f_1 is the focal length of the first element. We now see that the first half of our system is simply a spectrum analyzer that causes the spatial-frequency spectrum of $t_2(x, y)$ to be displayed at the plane $z=z_4$.

*This derivation remains basically valid for objects that are not transilluminated; the magnitude of the image may change, but the form is still specified correctly.

The second half of the system consists of an *aperture stop* located at the plane $z=z_4$, a lens at $z=z_5$, and an image plane at $z=z_6$. Comparison with the first half of the system reveals the second half to be another spectrum analyzer, and the entire system therefore consists of two spectrum analyzers in cascade. However, as we learned in Chap. 8, such a cascade yields a signal scaler, and we conclude that our image-forming system is simply a two-dimensional signal scaler. In fact, aside from some unimportant multiplicative constants, our imaging system is quite similar to the device depicted in Figs. 8-27 and 8-28. Consequently, the concepts developed in Chapter 8 will be useful in the present analysis.

We assume that the aperture stop is the limiting element of the system. In other words, the lens elements are sufficiently large to pass all of the light transmitted by the stop. As will become clear later on, *the aperture stop must be located at a frequency plane of the system in order to yield shift-invariant imaging*. If we denote the complex amplitude transmittance of the stop by $p_4(x, y)$, we may write

$$u_4^+(x,y)=u_4^-(x,y)p_4(x,y)$$

$$=A\left(\frac{z_{13}}{z_{12}}\right)^2 e^{jkz_{23}}B_{\ell 1}B_{34}q\left(x,y;\frac{1}{\lambda z'_{24}}\right)T_2\left(\frac{m'x}{\lambda z'_{24}},\frac{m'y}{\lambda z'_{24}}\right)p_4(x,y). \tag{11.3}$$

If we now define

$$v_4(x,y)=\left(\frac{z'_{24}}{z'_{25}}\right)\left(\frac{z_{13}}{z_{12}}\right)^2 e^{jkz_{23}}B_{\ell 1}B_{34}T_2\left(\frac{m'x}{\lambda z'_{24}},\frac{m'y}{\lambda z'_{24}}\right)p_4(x,y), \tag{11.4}$$

where $z'_{25}=z_5-z'_2=z'_{24}+z_{45}$, we obtain

$$u_4^+(x,y)=A\left(\frac{z'_{25}}{z'_{24}}\right)q\left(x,y;\frac{1}{\lambda z'_{24}}\right)v_4(x,y). \tag{11.5}$$

This expression has exactly the same form as Eq. (10.186), and if we require the distances z'_{25} and z_{56} to satisfy the lens law, i.e.,

$$\frac{1}{z'_{25}}+\frac{1}{z_{56}}=\frac{1}{f_2}, \tag{11.6}$$

we may use Eq. (10.198) once again to obtain the complex amplitude at the image plane $z=z_6$. [Note that when Eq. (11.6) is satisfied, the planes z'_2 and z_6 are conjugate planes.]

The second lens element will produce an image of $v_4(x,y)$ at some plane $z=\hat{z}_4$, and we associate a magnification $\hat{m}$ and distance $\hat{z}_{46}=z_6-\hat{z}_4$ with this image:

$$\hat{m}=\frac{f_2}{f_2-z_{45}},\qquad \hat{z}_{46}=\hat{m}z_{45}+z_{56}. \tag{11.7}$$

We may now write

$$u_6(x,y)=A\left(\frac{z'_{25}}{z'_{24}}\right)^2 e^{jkz_{45}}B_{\ell 2}B_{56}q\left(x,y;\frac{1}{\lambda\hat{z}_{46}}\right)V_4\left(\frac{\hat{m}x}{\lambda\hat{z}_{46}},\frac{\hat{m}y}{\lambda\hat{z}_{46}}\right), \tag{11.8}$$

and with

$$V_4(\xi,\eta)=\left(\frac{z'_{24}}{z'_{25}}\right)\left(\frac{z_{13}}{z_{12}}\right)^2 e^{jkz_{23}}B_{\ell 1}B_{34}$$
$$\times\left[\left(\frac{\lambda z'_{24}}{m'}\right)^2 t_2\left(\frac{-\lambda z'_{24}\xi}{m'},\frac{-\lambda z'_{24}\eta}{m'}\right)**P_4(\xi,\eta)\right], \tag{11.9}$$

we obtain

$$u_6(x,y)=A\left(\frac{z_{13}}{z_{12}}\right)^2\left(\frac{z'_{25}}{z'_{24}}\right)e^{jk(z_{23}+z_{45})}B_{\ell 1}B_{\ell 2}B_{34}B_{56}q\left(x,y;\frac{1}{\lambda\hat{z}_{46}}\right)$$
$$\times\left[\left(\frac{\hat{m}z'_{24}}{m'\hat{z}_{46}}\right)^2 t_2\left(\frac{-\hat{m}z'_{24}x}{m'\hat{z}_{46}},\frac{-\hat{m}z'_{24}y}{m'\hat{z}_{46}}\right)**P_4\left(\frac{\hat{m}x}{\lambda\hat{z}_{46}},\frac{\hat{m}y}{\lambda\hat{z}_{46}}\right)\right]$$
$$=A\left(\frac{z_{13}}{z_{12}}\right)e^{jk(z_{23}+z_{34}+z_{45}+z_{56})}B_{\ell 1}B_{\ell 2}$$
$$\times\left(\frac{z_{13}z'_{25}\hat{z}_{46}}{z_{12}m'\hat{m}}\right)\left(\frac{-1}{\lambda^2 z_{34}z_{56}}\right)q\left(x,y;\frac{1}{\lambda\hat{z}_{46}}\right)$$
$$\times\left[\left(\frac{\hat{m}z'_{24}}{m'\hat{z}_{46}}\right)t_2\left(\frac{-\hat{m}z'_{24}x}{m'\hat{z}_{46}},\frac{-\hat{m}z'_{24}y}{m'\hat{z}_{46}}\right)**\left(\frac{\hat{m}}{\hat{z}_{46}}\right)^2 P_4\left(\frac{\hat{m}x}{\lambda\hat{z}_{46}},\frac{\hat{m}y}{\lambda\hat{z}_{46}}\right)\right]$$
$$=A\left(\frac{z_{13}}{z_{12}}\right)e^{jkz_{26}}B_{\ell 1}B_{\ell 2}\left(\frac{z_{13}z'_{25}\hat{z}_{46}}{z_{12}z_{34}z_{56}m'\hat{m}}\right)q\left(x,y;\frac{1}{\lambda\hat{z}_{46}}\right)$$
$$\times\left[\left(\frac{-\hat{m}z'_{24}}{m'\hat{z}_{46}}\right)t_2\left(\frac{-\hat{m}z'_{24}x}{m'\hat{z}_{46}},\frac{-\hat{m}z'_{24}y}{m'\hat{z}_{46}}\right)**\left(\frac{\hat{m}}{\lambda\hat{z}_{46}}\right)^2 P_4\left(\frac{\hat{m}x}{\lambda\hat{z}_{46}},\frac{\hat{m}y}{\lambda\hat{z}_{46}}\right)\right]. \tag{11.10}$$

However, we know that

$$\frac{z_{13}}{z_{12}z_{34}}=\frac{m'}{z'_{24}}, \qquad \frac{z'_{25}}{z'_{24}z_{56}}=\frac{\hat{m}}{\hat{z}_{46}}, \tag{11.11}$$

which yields

$$\frac{z_{13}z'_{25}\hat{z}_{46}}{z_{12}z_{34}z_{56}m'\hat{m}}=1. \tag{11.12}$$

We now define the *overall magnification of the entire system* to be

$$m''=\frac{-m'\hat{z}_{46}}{\hat{m}z'_{24}}, \tag{11.13}$$

and we choose the function

$$t''_2(x,y)=\frac{1}{m''}t_2\left(\frac{x}{m''},\frac{y}{m''}\right) \tag{11.14}$$

to denote *the geometrical image of $t_2(x, y)$ formed by both lenses*. In addition, we regard

$$\hat{p}_4(x,y)=\frac{1}{\hat{m}}p_4\left(\frac{x}{\hat{m}},\frac{y}{\hat{m}}\right) \tag{11.15}$$

to be the geometrical image of the aperture stop produced by the second lens alone, and we shall refer to it as the *exit pupil* of the system. Thus, with

$$\left(\frac{\hat{m}}{\lambda\hat{z}_{46}}\right)^2 P_4\left(\frac{\hat{m}x}{\lambda\hat{z}_{46}},\frac{\hat{m}y}{\lambda\hat{z}_{46}}\right)=\hat{m}\left(\frac{1}{\lambda\hat{z}_{46}}\right)^2\hat{P}_4\left(\frac{x}{\lambda\hat{z}_{46}},\frac{y}{\lambda\hat{z}_{46}}\right), \tag{11.16}$$

where $\hat{P}_4(\xi,\eta)=\mathfrak{F}\mathfrak{F}\{\hat{p}_4(x,y)\}$, the image amplitude becomes

$$u_6(x,y)=A\left(\frac{z_{13}}{z_{12}}\right)e^{jkz_{26}}B_{\ell 1}B_{\ell 2}q\left(x,y;\frac{1}{\lambda\hat{z}_{46}}\right)$$
$$\times\left[t''_2(x,y)**\hat{m}\left(\frac{1}{\lambda\hat{z}_{46}}\right)^2\hat{P}_4\left(\frac{x}{\lambda\hat{z}_{46}},\frac{y}{\lambda\hat{z}_{46}}\right)\right]. \tag{11.17}$$

Aside from the constants and quadratic phase factor, the image amplitude is simply given by the convolution of the geometrical image $t''_2(x, y)$ with the Fraunhofer diffraction pattern of the exit pupil. We see, then, that image formation with coherent light is linear in complex amplitude, and we shall use this result in the next section to cast the imaging process into the form of a linear filtering operation.

We point out that the expression of Eq. (11.17) is valid as long as the following conditions are satisfied: the aperture stop must be the limiting element of the system and must be located at a frequency plane; and the Fresnel conditions must be satisfied for each segment of the system. The latter condition places a restriction on the field of view over which the impulse response is invariant.

11-2 LINEAR FILTER INTERPRETATION OF COHERENT IMAGING

The convolution of Eq. (11.17) suggests that the image might be regarded as the output of a linear shift-invariant system, and indeed it may. In general, however, the shift-invariant property holds only for the geometrical image $t_2''(x, y)$ and not for the object itself. To demonstrate the correctness of this statement, we consider our imaging device to be a linear system $\mathcal{L}\{\ \}$ with input $t_2(x, y)$ and output $u_6(x, y)$. It is not difficult to show that a shifted input $t_2(x-x_0, y-y_0)$ will lead to an output of

$$\mathcal{L}\{t_2(x-x_0, y-y_0)\} = u_6(x-m''x_0, y-m''y_0), \tag{11.18}$$

which exhibits a shift of the same magnitude and direction only when $m=1$. Consequently, shift invariance is not a general characteristic of this system in terms of $t_2(x, y)$. Nevertheless, the system may be regarded as shift invariant with respect to $t_2''(x, y)$, and we take advantage of this fact in the paragraphs to follow.

Coherent Impulse Response

We assume that our imaging system consists of the three separate components illustrated in Fig. 11-2: a geometric scaler, a linear shift-invariant system, and a multiplier. The input $t_2(x, y)$ is first passed through the geometric scaler to produce a geometrical image of the proper magnification, and this geometrical image is then passed through the LSI system to account for the smoothing effects of diffraction. Finally, the magnified and smoothed image is multiplied by several constants and a quadratic phase factor to provide the correct magnitude and phase for the final image.

If we denote the *coherent impulse response* of the system by†

$$h_{26}(x, y) = \hat{m}\left(\frac{1}{\lambda \hat{z}_{46}}\right)^2 \hat{P}_4\left(\frac{x}{\lambda \hat{z}_{46}}, \frac{y}{\lambda \hat{z}_{46}}\right), \tag{11.19}$$

†We use the subscripts to indicate that the transformation is from the plane $z = z_2$ to the plane $z = z_6$.

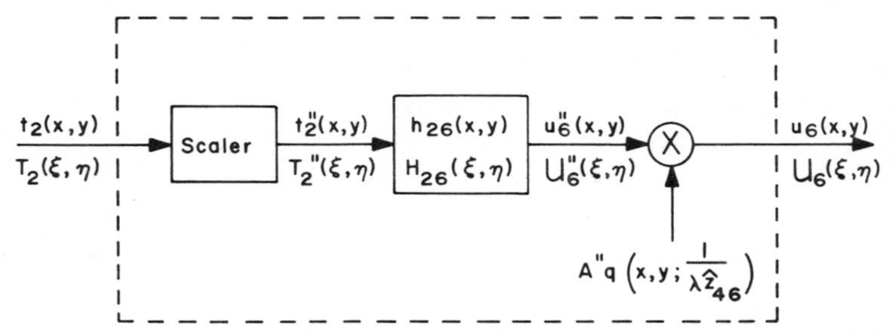

Figure 11-2 Operational representation of imaging system shown in Fig. 11-1.

and if we define

$$A'' = A\left(\frac{z_{13}}{z_{12}}\right)e^{jkz_{26}}B_{\ell 1}B_{\ell 2}, \tag{11.20}$$

we may rewrite Eq. (11.17) as

$$u_6(x,y) = A''q\left(x,y;\frac{1}{\lambda\hat{z}_{46}}\right)\left[t_2''(x,y)**h_{26}(x,y)\right]. \tag{11.21}$$

The smoothing of this image is clearly described by the convolution operation alone, and may be accounted for entirely in terms of a *normalized diffraction image*

$$u_6''(x,y) = t_2''(x,y)**h_{26}(x,y), \tag{11.22}$$

which contains neither the phase factor $q(x, y; 1/\lambda\hat{z}_{46})$ nor the constant A''. Thus, we obtain the following concise expression for the final image amplitude:

$$u_6(x,y) = A''q\left(x,y;\frac{1}{\lambda\hat{z}_{46}}\right)u_6''(x,y). \tag{11.23}$$

We note that $h_{26}(x,y)$ is the amplitude response of the imaging system to a delta-function input, i.e., it is the image of a point source. However, because it is the Fraunhofer diffraction pattern of a finite aperture, $h_{26}(x,y)$ cannot be a delta function itself; thus, we see that the system is incapable of mapping a point object into a point image. Consequently, for an arbitrary input $t_2(x,y)$, the image is simply a superposition of appropriately weighted and shifted versions of $h_{26}(x,y)$ that overlap to produce a smoothing effect. Put another way, the system maps a collection of points into a collection of overlapping diffraction patterns, and much of the fine detail of the object tends to be washed out in the image.

The irradiance of the image is given by

$$
\begin{aligned}
I_6(x,y) &= |u_6(x,y)|^2 \\
&= |A'' u_6''(x,y)|^2,
\end{aligned}
\tag{11.24}
$$

and we see that the quadratic phase factor has now vanished. When the image irradiance is the quantity of importance, as it often is, the phase factor is of no consequence and might just as well be dropped at an early stage of the development. It must be retained, however, if $u_6(x, y)$ is to be reimaged by another system because the locations of frequency planes and image planes in succeeding systems are dependent on its behavior.

Coherent Transfer Function

Because the diffraction image $u_6''(x, y)$ may be regarded as the output of a linear filter, we may use the frequency-domain approach developed in Chapters 7–9 to determine the smoothing effects of diffraction. The *coherent transfer function* of this linear filter is found to be

$$
\begin{aligned}
H_{26}(\xi,\eta) &= \mathfrak{F}\mathfrak{F}\{h_{26}(x,y)\} \\
&= \mathfrak{F}\mathfrak{F}\left\{ \hat{m}\left(\frac{1}{\lambda \hat{z}_{46}}\right)^2 \hat{P}_4\left(\frac{x}{\lambda \hat{z}_{46}}, \frac{y}{\lambda \hat{z}_{46}}\right)\right\} \\
&= \hat{m}\hat{p}_4(-\lambda \hat{z}_{46}\xi, -\lambda \hat{z}_{46}\eta),
\end{aligned}
\tag{11.25}
$$

which is simply a scaled version of the exit pupil $\hat{p}_4(x, y)$. This curious result may be explained as follows. We recall from our earlier discussions of LSI systems that it is the multiplicative nature of the transfer function from which the advantages of such systems are derived. The effect of placing the aperture stop at the frequency plane $z = z_4$ is to cause the spectrum of the input to be multiplied by an appropriately scaled version of the aperture stop transmittance function, which may therefore be regarded as a transfer function. We note that the system will no longer be shift invariant if the stop is located at a plane other than the frequency plane, the reason being that the spectrum of the input is displayed only at this plane. At all other planes the complex amplitude does not describe the spectrum of the input, and multiplication of the pupil function and the input spectrum cannot occur except at the plane $z = z_4$. Consequently, shift invariance is lost if the stop is located improperly. A helpful discussion of this characteristic may be found in Ref. 11-2.

The spectrum of the diffraction image is given by

$$\begin{aligned} U_6''(\xi,\eta) &= T_2''(\xi,\eta)H_{26}(\xi,\eta) \\ &= m''T_2(m''\xi, m''\eta)\hat{m}\hat{p}_4(-\lambda\hat{z}_{46}\xi, -\lambda\hat{z}_{46}\eta). \end{aligned} \tag{11.26}$$

Once the image spectrum has been calculated, the actual image is determined from:

$$\begin{aligned} u_6(x,y) = A''q\left(x,y;\frac{1}{\lambda\hat{z}_{46}}\right)\mathcal{F}^{-1}\mathcal{F}^{-1}\{m''T_2(m''\xi, m''\eta) \\ \times\hat{m}\hat{p}_4(-\lambda\hat{z}_{46}\xi, -\lambda\hat{z}_{46}\eta)\}. \end{aligned} \tag{11.27}$$

For a clear aperture, the transfer function is a zero-one function and the spectral components of the input are either passed without attenuation or completely eliminated, depending on whether they lie inside or outside the passband of $H_{26}(\xi,\eta)$. For a more general aperture, i.e., one that exhibits variations in complex amplitude transmittance within the pupil, the various spectral components undergo different amounts of attenuation and phase shift as they pass through the system. In either case, the nature of the aperture controls the degree of smoothing and/or distortion of the image.

The *cutoff frequency* of an imaging system is the value of the frequency variable beyond which the transfer function is zero. For circularly symmetric exit pupils, this quantity is characterized by a single number. However, for systems with exit pupils that are not circularly symmetric, the cutoff frequency is different for signals of different orientations. We now calculate the cutoff frequencies of circular and rectangular apertures to illustrate this difference. For a circular aperture of diameter d we have the following relationships:

$$\begin{aligned} p_4(r) &= \text{cyl}\left(\frac{r}{d}\right), \\ \hat{p}_4(r) &= \frac{1}{\hat{m}}\,\text{cyl}\left(\frac{r}{\hat{m}d}\right), \\ H_{26}(\rho) &= \hat{m}\hat{p}_4(\lambda\hat{z}_{46}\rho) \\ &= \text{cyl}\left(\frac{\lambda\hat{z}_{46}\rho}{\hat{m}d}\right). \end{aligned} \tag{11.28}$$

The transfer function has a value of zero whenever the argument of the cylinder function is greater than one half; thus, the cutoff frequency ρ_c is found to be

$$\rho_c = \left| \frac{\hat{m}d}{2\lambda \hat{z}_{46}} \right|. \tag{11.29}$$

On the other hand, for a rectangular aperture of dimensions a and b, we have

$$p_4(x, y) = \operatorname{rect}\left(\frac{x}{a}, \frac{y}{b} \right),$$

$$\hat{p}_4(x, y) = \frac{1}{\hat{m}} \operatorname{rect}\left(\frac{x}{\hat{m}a}, \frac{y}{\hat{m}b} \right),$$

$$\begin{aligned} H_{26}(\xi, \eta) &= \hat{m}\hat{p}_4(-\lambda \hat{z}_{46}\xi, -\lambda \hat{z}_{46}\eta) \\ &= \operatorname{rect}\left(\frac{\lambda \hat{z}_{46}\xi}{\hat{m}a}, \frac{\lambda \hat{z}_{46}\eta}{\hat{m}b} \right). \end{aligned} \tag{11.30}$$

The cutoff frequencies in the ξ- and η-directions, respectively, are now

$$\xi_c = \left| \frac{\hat{m}a}{2\lambda \hat{z}_{46}} \right|, \qquad \eta_c = \left| \frac{\hat{m}b}{2\lambda \hat{z}_{46}} \right|. \tag{11.31}$$

In this latter case, however, we note that the absolute cutoff frequency of the system is greater than either ξ_c or η_c. In other words, a signal with a frequency higher than either ξ_c or η_c would be passed by the system if it were oriented properly. The absolute cutoff frequency for a rectangular aperture is

$$(\rho_c)_{\text{abs}} = \sqrt{\xi_c^2 + \eta_c^2} = \frac{\hat{m}}{2\lambda \hat{z}_{46}} \sqrt{a^2 + b^2}\,. \tag{11.32}$$

Example

Consider an object with transmittance

$$t_2(x, y) = 0.5\left[1 + \cos(2\pi \xi_0 x) \right] \tag{11.33}$$

and geometrical image

$$t_2''(x, y) = \frac{0.5}{m''}\left[1 + \cos\left(\frac{2\pi \xi_0 x}{m''} \right) \right]. \tag{11.34}$$

Thus, the spectrum of the geometrical image is found to be

$$T_2''(\xi,\eta) = \frac{0.5}{m''}\left[\delta(\xi,\eta) + \frac{m''}{2\xi_0}\delta\delta\left(\frac{m''\xi}{\xi_0}\right)\right], \tag{11.35}$$

and the spectrum of the normalized image is then determined from Eq. (11.26). For a system with a circular exit pupil, all of the input spectral components will be passed as long as

$$\rho_c > \frac{\xi_0}{m''}; \tag{11.36}$$

if the exit pupil is rectangular, these components will be passed if

$$\xi_c > \frac{\xi_0}{m''}. \tag{11.37}$$

When either Eq. (11.36) or (11.37) holds, we have

$$U_6''(\xi,\eta) = T_2''(\xi,\eta) \tag{11.38}$$

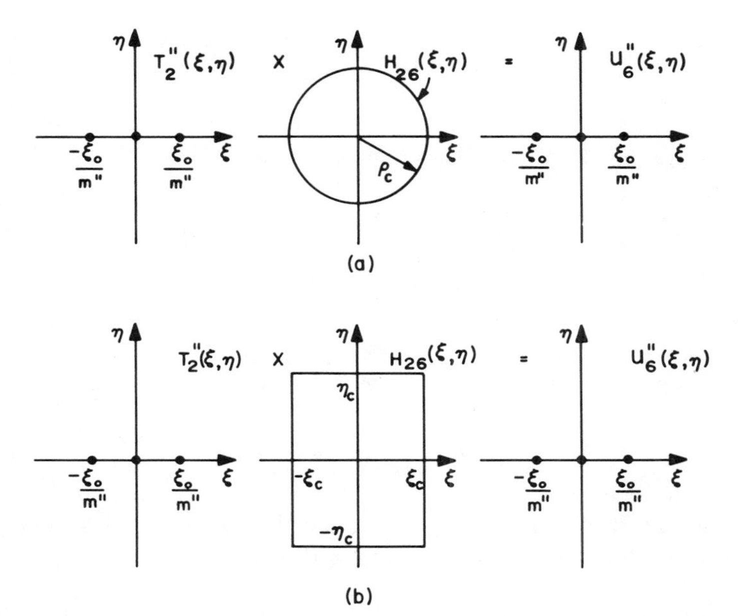

Figure 11-3 Depiction of coherent transfer function in the frequency domain. (*a*) Clear circular aperture. (*b*) Clear rectangular aperture.

as shown in Fig. 11-3. Consequently,

$$u_6''(x,y) = t_2''(x,y), \tag{11.39}$$

and the final image becomes

$$\begin{aligned} u_6(x,y) &= A'' q\left(x,y;\frac{1}{\lambda \hat{z}_{46}}\right) u_6''(x,y) \\ &= \frac{0.5A''}{m''} q\left(x,y;\frac{1}{\lambda \hat{z}_{46}}\right)\left[1+\cos\left(\frac{2\pi\xi_0 x}{m''}\right)\right]. \end{aligned} \tag{11.40}$$

■

Example

Let us now consider an object

$$t_2(x,y) = 0.5\left[1+\cos 2\pi(\xi_0 x + \eta_0 y)\right], \tag{11.41}$$

with geometrical image

$$t_2''(x,y) = \frac{0.5}{m''}\left[1+\cos 2\pi\left(\frac{\xi_0 x}{m''}+\frac{\eta_0 y}{m''}\right)\right]. \tag{11.42}$$

The spectrum of this geometrical image is then

$$T_2''(\xi,\eta) = \frac{0.5}{m''}\left[\delta(\xi,\eta) + \tfrac{1}{2}\delta\left(\xi+\frac{\xi_0}{m''},\eta+\frac{\eta_0}{m''}\right) + \tfrac{1}{2}\delta\left(\xi-\frac{\xi_0}{m''},\eta-\frac{\eta_0}{m''}\right)\right]. \tag{11.43}$$

For a system with a circular exit pupil, we must have

$$\rho_c > \frac{1}{m''}\sqrt{\xi_0^2+\eta_0^2} \tag{11.44}$$

for the entire input spectrum to be passed. However, for the rectangular-pupil system, the entire spectrum is passed if both

$$\xi_c > \frac{\xi_0}{m''} \quad \text{and} \quad \eta_c > \frac{\eta_0}{m''}. \tag{11.45}$$

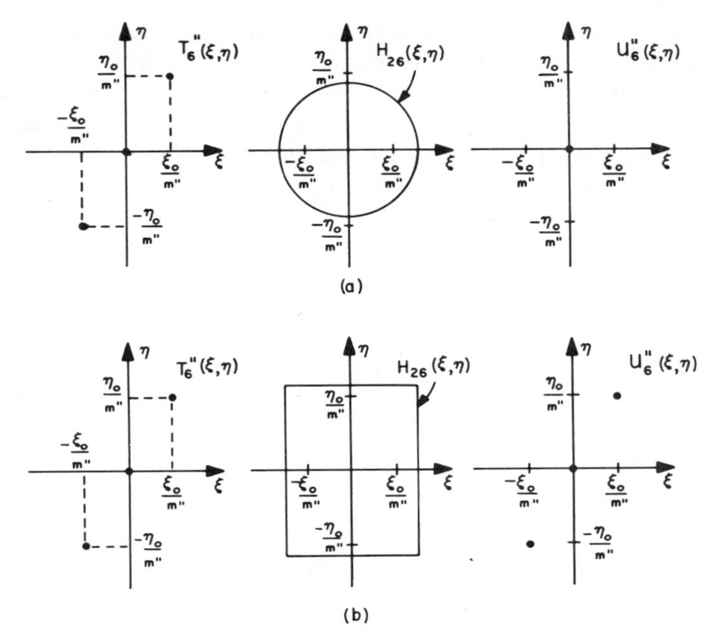

Figure 11-4 Dependence of cutoff frequency on aperture shape and signal orientation. (*a*) Clear circular aperture. (*b*) Clear rectangular aperture.

This will be true even though

$$\xi_c, \eta_c < \frac{1}{m''}\sqrt{\xi_0^2 + \eta_0^2}\ , \tag{11.46}$$

as can be seen in Fig. 11-4. In each of these examples we have assumed the object to be of infinite extent so that we might concentrate on the effects of the exit pupil; in the future we shall limit the extent of the object. ■

Review of Coherent-Image Formation

We now review the previous developments and collect all of the important equations for the reader's convenience. For the results to be valid, the imaging system must conform to the basic configuration depicted in Fig. 11-5. The object is placed to the left of a lens and transilluminated from the left. A second lens is then located to the right of the first, and an aperture stop is placed at the frequency plane of the system; this latter statement implies that the Fraunhofer pattern of the object must be displayed at some physically accessible location, i.e., it must be a real (as opposed to virtual) diffraction pattern. As long as we conform to the configuration just described, the results are quite general: the distances between the various planes may take on either positive or negative values; the lenses may be either positive or negative; the intermediate image planes z_2' and $\hat{z}_4$ may lie almost anywhere along the z-axis, as may the final image plane z_6; the final image may be either real or virtual, and it may be either inverted or erect; etc.

We now list the important relationships for this general imaging configuration:

$$u_6(x,y)=A''q\left(x,y;\frac{1}{\lambda\hat{z}_{46}}\right)u_6''(x,y),$$

$$A''=A\left(\frac{z_{13}}{z_{12}}\right)e^{jkz_{26}}B_{\ell 1}B_{\ell 2},$$

$$q\left(x,y;\frac{1}{\lambda\hat{z}_{46}}\right)=\exp\left[j\frac{\pi}{\lambda\hat{z}_{46}}(x^2+y^2)\right],$$

$$u_6''(x,y)=t_2''(x,y)**h_{26}(x,y),$$

$$t_2''(x,y)=\frac{1}{m''}t_2\left(\frac{x}{m''},\frac{y}{m''}\right),$$

$$h_{26}(x,y)=\hat{m}\left(\frac{1}{\lambda\hat{z}_{46}}\right)^2\hat{P}_4\left(\frac{x}{\lambda\hat{z}_{46}},\frac{y}{\lambda\hat{z}_{46}}\right)$$

$$=\left(\frac{\hat{m}}{\lambda\hat{z}_{46}}\right)^2P_4\left(\frac{\hat{m}x}{\lambda\hat{z}_{46}},\frac{\hat{m}y}{\lambda\hat{z}_{46}}\right),$$

$$U_6''(\xi,\eta)=T_2''(\xi,\eta)H_{26}(\xi,\eta),$$

$$T_2''(\xi,\eta)=m''T_2(m''\xi,m''\eta),$$

$$H_{26}(\xi,\eta)=\hat{m}\hat{p}_4(-\lambda\hat{z}_{46}\xi,-\lambda\hat{z}_{46}\eta)$$

$$=p_4\left(\frac{-\lambda\hat{z}_{46}\xi}{\hat{m}},\frac{-\lambda\hat{z}_{46}\eta}{\hat{m}}\right). \tag{11.47}$$

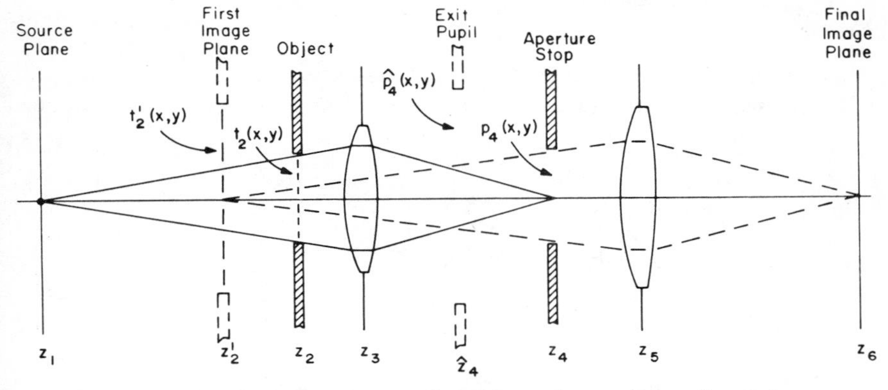

Figure 11-5 Optical system analyzed for coherent image formation.

Helpful auxiliary expressions are:

$$z'_{2i} = z_i - z'_2, \qquad \hat{z}_{4i} = z_i - \hat{z}_4,$$

$$z'_{24} = z'_{23} + z_{34}, \qquad \hat{z}_{46} = \hat{z}_{45} + z_{56},$$

$$z'_{23} = m' z_{23}, \qquad \hat{z}_{45} = \hat{m} z_{45},$$

$$z_{34} = \frac{z_{13} f_1}{z_{13} - f_1}, \qquad z_{56} = \frac{z'_{25} f_2}{z'_{25} - f_2},$$

$$m' = \frac{f_1}{f_1 - z_{23}}, \qquad \hat{m} = \frac{f_2}{f_2 - z_{45}},$$

$$\frac{m'}{z'_{24}} = \frac{z_{13}}{z_{12} z_{34}}, \qquad \frac{\hat{m}}{\hat{z}_{46}} = \frac{z'_{25}}{z'_{24} z_{56}},$$

$$z'_{25} = z'_{24} + z_{45}, \qquad z_{35} = z_{34} + z_{45},$$

$$m'' = \frac{-m' \hat{z}_{46}}{\hat{m} z'_{24}} = \left[\left(\frac{z_{35}}{f_2} - 1 \right) \left(\frac{z_{23}}{f_1} - 1 \right) - \frac{z_{23}}{f_2} \right]^{-1}. \tag{11.48}$$

Example

We now determine the image amplitude and irradiance for the system of Fig. 11-5 when the input is a Ronchi ruling with amplitude transmittance

$$t_2(x,y)=\left[\operatorname{rect}\left(\frac{x}{a}\right)*\left(\frac{1}{2a}\right)\operatorname{comb}\left(\frac{x}{2a}\right)\right]\operatorname{rect}\left(\frac{x}{b},\frac{y}{b}\right) \qquad (11.49)$$

and the system parameters are as follows:

$$\begin{aligned}
&z_{13}=100\text{ cm}, && z_{23}=10\text{ cm}, && z_{35}=50\text{ cm},\\
&f_1=20\text{ cm}, && f_2=30\text{ cm}, && \lambda=514.5\text{ nm},\\
&|B_{\ell 1}|=|B_{\ell 2}|=0.95, && && |A|^2=10\text{ mW/cm}^2,\\
&a=20\ \mu\text{m}, && b=1\text{ cm}, && \\
&p_4(r)=\operatorname{cyl}\left(\frac{r}{d}\right), && d=1.2\text{ cm}. &&
\end{aligned} \qquad (11.50)$$

From Eqs. (11.47) and (11.48) we obtain

$$\begin{aligned}
&z_{34}=\frac{100(20)}{100-20}=25\text{ cm}, && z_{45}=50-25=25\text{ cm},\\
&m'=\frac{20}{20-10}=2, && z'_{24}=2(10)+25=45\text{ cm},\\
&z'_{25}=45+25=70\text{ cm}, && z_{56}=\frac{70(30)}{70-30}=52.5\text{ cm},\\
&\hat{m}=\frac{30}{30-25}=6, && \hat{z}_{46}=6(25)+52.5=202.5\text{ cm},\\
&m''=\frac{-2(202.5)}{6(45)}=-1.5. &&
\end{aligned} \qquad (11.51)$$

The cutoff frequency of this system is found to be

$$\rho_c=\frac{6(1.2\times10^{-2})}{2(5.145\times10^{-7})(2.025)}=3.46\times10^{4}\text{ m}^{-1}=34.6\text{ mm}^{-1}, \qquad (11.52)$$

and the transfer function then becomes

$$H_{26}(\rho)=\mathrm{cyl}\left(\frac{\rho}{69.2}\right), \qquad \rho \text{ in mm}^{-1}. \tag{11.53}$$

The object spectrum is given by

$$T_2(\xi,\eta)=\left[a\,\mathrm{sinc}(a\xi)\,\mathrm{comb}(2a\xi)\delta(\eta)\right]**b^2\,\mathrm{sinc}(b\xi,b\eta), \tag{11.54}$$

and the spectrum of the geometrical image by

$$\begin{aligned} T_2''(\xi,\eta)&=(m'')^3\left[a\,\mathrm{sinc}(m''a\xi)\,\mathrm{comb}(2m''a\xi)\delta(m''\eta)\right]\\ &\quad **b^2\,\mathrm{sinc}(m''b\xi,\, m''b\eta)\\ &=\left[a\,\mathrm{sinc}(m''a\xi)\,\mathrm{comb}(2m''a\xi)m''\,\delta(m''\eta)\right]\\ &\quad **(m''b)^2\,\mathrm{sinc}(m''b\xi,\, m''b\eta)\\ &=\frac{1}{2m''}\left[\mathrm{sinc}(m''a\xi)2m''a\,\mathrm{comb}(2m''a\xi)\right]\\ &\quad *(m''b)^2\,\mathrm{sinc}(m''b\xi,\, m''b\eta) \end{aligned} \tag{11.55}$$

where we used Eq. (9.8) to retain the correct magnitude. Then, with

$$\begin{aligned} 2m''&=-3, & m''a&=-0.03 \text{ mm},\\ 2m''a&=-0.06 \text{ mm}, & m''b&=15 \text{ mm}, \end{aligned} \tag{11.56}$$

we obtain

$$T_2''(\xi,\eta)=-\tfrac{1}{3}\left[\mathrm{sinc}(0.03\xi)0.06\,\mathrm{comb}(0.06\xi)\right]*(15)^2\,\mathrm{sinc}(15\xi,15\eta), \tag{11.57}$$

where ξ and η are in mm^{-1}. This quantity consists of an array of very narrow sinc functions, spaced at intervals of $(0.06)^{-1}=16.7\ \text{mm}^{-1}$ in frequency space, and weighted by a broad sinc-function envelope as shown in Fig. 11-6. Multiplication by $H_{26}(\rho)=H_{26}(\sqrt{\xi^2+\eta^2}\,)$ yields

$$\begin{aligned} U_6''(\xi,\eta)&=T_2''(\xi,\eta)H_{26}\left(\sqrt{\xi^2+\eta^2}\,\right)\\ &\cong-\tfrac{1}{3}\left[\delta(\xi)+\frac{2}{\pi}(0.06)\delta\delta(0.06\xi)\right]*(15)^2\,\mathrm{sinc}(15\xi,15\eta), \end{aligned} \tag{11.58}$$

and an inverse transform produces

$$u_6''(x,y)\cong-\tfrac{1}{3}\left[1+\frac{4}{\pi}\cos 2\pi(16.7x)\right]\mathrm{rect}\left(\frac{x}{15},\frac{y}{15}\right) \tag{11.59}$$

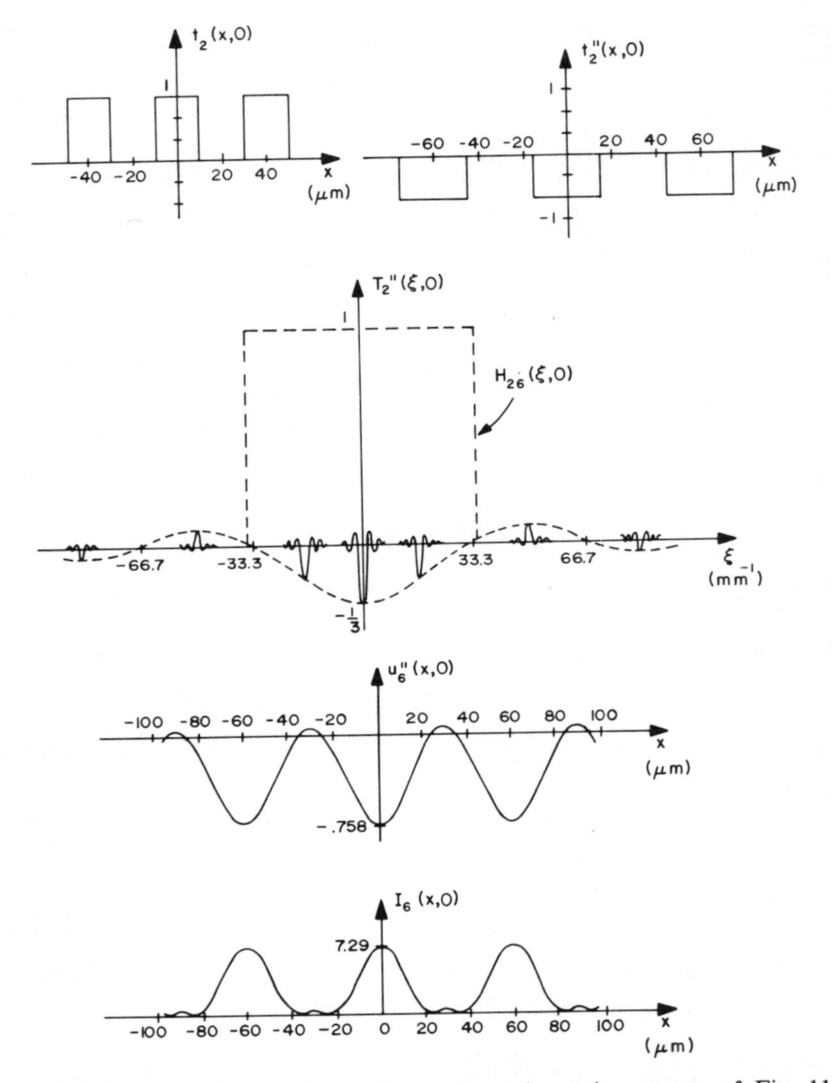

Figure 11-6 Example of coherent image formation using system of Fig. 11-5, object of Eq. (11.49), and parameter values of Eq. (11.50).

where x and y are in millimeters. In obtaining this result, which should bring back memories of Chapter 8, we have assumed that the sinc-function components of the spectrum are so narrow relative to their spacing that they are either passed by the system without attenuation or completely eliminated. In other words, we have assumed that

$$\operatorname{sinc}[15(\xi - n16.7), 15\eta]\operatorname{cyl}\left(\frac{\sqrt{\xi^2+\eta^2}}{\rho_c}\right)$$
$$\cong \begin{cases} \operatorname{sinc}[15(\xi - n16.7), 15\eta], & n16.7 < \rho_c \\ 0, & n16.7 > \rho_c \end{cases} \tag{11.60}$$

where n is an integer that denotes the spectral order. As a consequence of our assumption, Eq. (11.59) indicates that the interior portions of the image exhibit severe smoothing, whereas the edges of the limiting rectangle function are not smoothed at all. This is physically untenable, of course, and we can rest assured that the edges of the image will indeed undergo the same degree of smoothing as the rest of the image. The advantage of this assumption is that it allowed us to obtain a reasonably accurate description of the interior portions of the image, and to do so quite easily. The final image amplitude can be obtained from Eq. (11.47) if desired.

Recognizing that $\operatorname{rect}^2(x, y) = \operatorname{rect}(x, y)$, the final image irradiance is given by

$$I_6(x,y) \cong |A'' u_6''(x,y)|^2$$
$$\cong |A|^2\left(\frac{z_{13}}{z_{12}}\right)^2 |B_{\ell 1} B_{\ell 2}|^2 \left(\frac{-1}{3}\right)^2 \left[1 + \frac{4}{\pi}\cos 2\pi(16.7x)\right]^2$$
$$\times \operatorname{rect}\left(\frac{x}{15}, \frac{y}{15}\right)$$
$$\cong 10\left(\frac{100}{80}\right)^2 |0.95|^4 \left(\frac{1}{9}\right)\left[1 + \frac{8}{\pi}\cos 2\pi(16.7x)\right.$$
$$\left. + \frac{16}{\pi^2}\cos^2 2\pi(16.7x)\right]\operatorname{rect}\left(\frac{x}{15}, \frac{y}{15}\right)$$
$$\cong 1.41\left[1 + \frac{8}{\pi}\cos 2\pi(16.7x) + \frac{8}{\pi^2} + \frac{8}{\pi^2}\cos 2\pi(33.3x)\right]$$
$$\times \operatorname{rect}\left(\frac{x}{15}, \frac{y}{15}\right)$$
$$\cong 2.55[1 + 1.41\cos 2\pi(16.7x) + 0.45\cos 2\pi(33.3x)]$$
$$\times \operatorname{rect}\left(\frac{x}{15}, \frac{y}{15}\right), \tag{11.61}$$

where x and y are in mm and $I_6(x, y)$ is in mW/cm^2. This irradiance distribution is illustrated in Fig. 11-6. ■

Coherent Line and Edge Response Functions

We now determine the relationships among the line response, edge response, and transfer function for an imaging system that employs coherent illumination. For simplicity, we shall treat only the configuration for which the line source or edge is situated along the y-axis; for other configurations, the results of Sec. 9-6 may be used. The *coherent line response* is given by [see Eq. (9.166)]

$$\begin{aligned} l_x(x) &= \mathcal{F}^{-1}\{H_{26}(\xi, 0)\} \\ &= \mathcal{F}^{-1}\{\hat{m}\hat{p}_4(-\lambda\hat{z}_{46}\xi, 0)\}, \end{aligned} \tag{11.62}$$

and is simply the one-dimensional inverse Fourier transform of the ξ-axis profile of the coherent transfer function. Note that it is not, in general, a profile of the two-dimensional inverse transform of $H_{26}(\xi,\eta)$, which would simply be a profile of the impulse response. It is interesting to note that, *for any system with a clear aperture $p_4(x, y)$, the coherent line response will exhibit a sinc-function behavior*, irrespective of the shape of the aperture. To illustrate, consider an aperture of the form

$$p_4(x,y) = \text{cyl}\left(\frac{\sqrt{x^2+y^2}}{d}\right). \tag{11.63}$$

Then

$$\begin{aligned} \hat{m}\hat{p}_4(-\lambda\hat{z}_{46}\xi, 0) &= \text{cyl}\left(\frac{|\lambda\hat{z}_{46}\xi|}{\hat{m}d}\right) \\ &= \text{rect}\left(\frac{\lambda\hat{z}_{46}\xi}{\hat{m}d}\right), \end{aligned} \tag{11.64}$$

which leads to

$$\begin{aligned} l_x(x) &= \mathcal{F}^{-1}\left\{\text{rect}\left(\frac{\lambda\hat{z}_{46}\xi}{\hat{m}d}\right)\right\} \\ &= \frac{\hat{m}d}{\lambda\hat{z}_{46}}\,\text{sinc}\left(\frac{\hat{m}dx}{\lambda\hat{z}_{46}}\right). \end{aligned} \tag{11.65}$$

We observe that this is not simply a profile of the coherent impulse response, which is a sombrero function for the present system. Even when the aperture stop is not centered on the optical axis, the fundamental form of the line response is that of a sinc function; the only effect produced by a decentered aperture will be the introduction of a linear phase factor into the line response.

From Eq. (9.181) we know that the *coherent edge response* is related to the coherent line response by

$$e_x(x)=\int_{-\infty}^{x} l_x(\alpha)\,d\alpha. \tag{11.66}$$

If the system has a clear aperture that is centered on the axis and has a width d in the x-direction, then the line response is given by Eq. (11.65) and

$$e_x(x)=\int_{-\infty}^{x}\frac{\hat{m}d}{\lambda\hat{z}_{46}}\operatorname{sinc}\left(\frac{\hat{m}d\alpha}{\lambda\hat{z}_{46}}\right)d\alpha. \tag{11.67}$$

We now let $c=(\hat{m}d/\lambda\hat{z}_{46})$ and use the relationship

$$\int_{-\infty}^{x} c\operatorname{sinc}(c\alpha)\,d\alpha=0.5+\int_{0}^{x} c\operatorname{sinc}(c\alpha)\,d\alpha \tag{11.68}$$

to obtain

$$\begin{aligned}
e_x(x)&=0.5+\int_0^x c\operatorname{sinc}(c\alpha)\,d\alpha\\
&=0.5+\frac{1}{\pi}\int_0^x\frac{\sin(\pi c\alpha)}{\alpha}\,d\alpha\\
&=0.5+\frac{1}{\pi}\left[\pi cx-\frac{(\pi cx)^3}{3\cdot 3!}+\frac{(\pi cx)^5}{5\cdot 5!}-\cdots\right]\\
&=0.5+\frac{1}{\pi}\left[\frac{\pi\hat{m}dx}{\lambda\hat{z}_{46}}-\frac{1}{18}\left(\frac{\pi\hat{m}dx}{\lambda\hat{z}_{46}}\right)^3+\frac{1}{600}\left(\frac{\pi\hat{m}dx}{\lambda\hat{z}_{46}}\right)^5-\cdots\right].
\end{aligned} \tag{11.69}$$

This function, along with the associated line response, is illustrated in Fig. 11-7; note the oscillatory, or ringing, behavior of the edge response, which can be associated with the Gibb's phenomenon discussed in Chapter 4.

It is instructive to calculate the squared modulus of the coherent line and edge response functions because these quantities govern the nature of

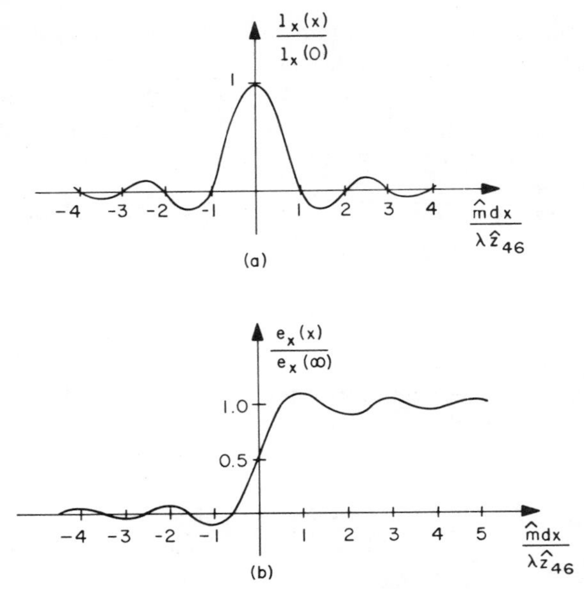

Figure 11-7 Normalized coherent line and edge response functions. (*a*) Line response. (*b*) Edge response.

the image irradiance for line-source or edge-type objects, respectively. The squared modulus of the line response is simply

$$|l_x(x)|^2 = \left|\frac{\hat{m}d}{\lambda\hat{z}_{46}}\right|^2 \operatorname{sinc}^2\left(\frac{\hat{m}dx}{\lambda\hat{z}_{46}}\right). \tag{11.70}$$

However, the squared modulus of the edge response is more difficult to

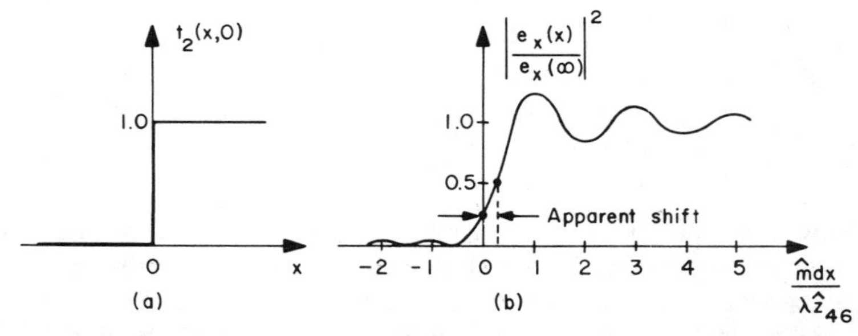

Figure 11-8 Irradiance associated with normalized coherent edge response.

calculate; in fact, we shall only observe its behavior graphically. The result is depicted in Fig. 11-8, along with the edge-type object that produced it. Note that there has been *an apparent shift of the edge toward the region of higher transmittance*. In addition, the irradiance variations (fringes) on the dark side of the edge occur with twice the frequency of those on the light side, and the magnitude of these oscillations is smaller on the dark side.

11-3 SPECIAL CONFIGURATIONS FOR COHERENT IMAGING

We now investigate several special imaging configurations, pointing out the salient features of each. In doing so we make the following assumptions: all lenses are positive; the source lies outside the front focal plane (FFP) of the first lens; and the object transparency is placed at or inside the FFP of the first lens. These assumptions are not really necessary, but they help to reduce the number of possible variations by a substantial amount and to maintain consistency between the mathematical results and the illustrations.

Object in FFP of First Lens

With $z_{23}=f_1$, it may be shown that

$$m' = z'_{24} = \infty, \qquad z_{56} = f_2,$$

$$m'' = \frac{-f_2}{f_1}, \qquad \hat{z}_{46} = \frac{f_2^2}{f_2 - z_{45}}. \tag{11.71}$$

Thus, irrespective of the other parameter values, *the overall magnification is simply equal to the negative of the focal length ratio*, and *the image plane is always found in the back focal plane (BFP) of the second lens*. This configuration is illustrated in Fig. 11-9.

Aperture Stop in FFP of Second Lens

Here we have $z_{45}=f_2$, which leads to

$$\hat{m} = \hat{z}_{46} = \infty, \qquad q\left(x, y; \frac{1}{\lambda \hat{z}_{46}}\right) = 1,$$

$$\frac{\hat{z}_{46}}{\hat{m}} = f_2, \qquad H_{26}(\xi, \eta) = p_4(-\lambda f_2 \xi, -\lambda f_2 \eta). \tag{11.72}$$

Therefore, whenever the frequency plane (and thus the stop) lies in the

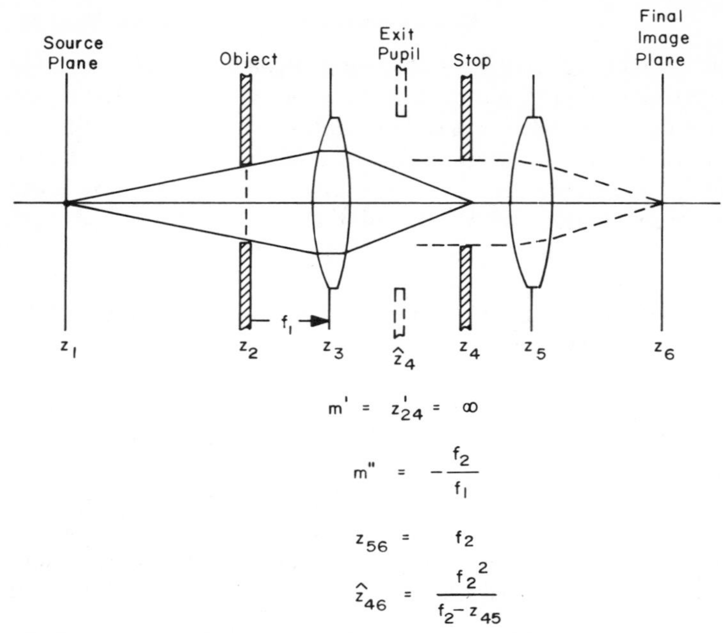

Figure 11-9 Imaging configuration with object in FFP of first lens element.

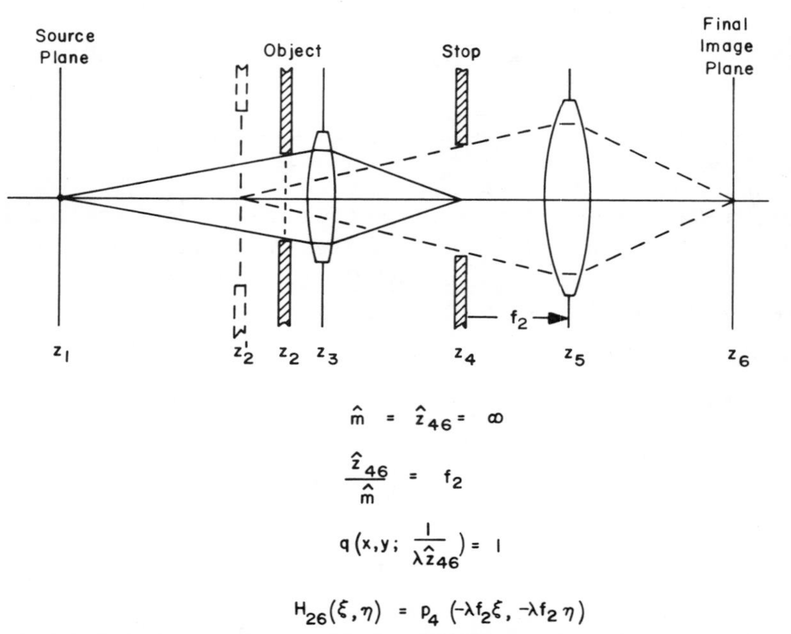

Figure 11-10 Aperture stop in FFP of second lens element.

FFP of the second lens, *the quadratic phase factor vanishes at the image plane* and the scaling factor for the transfer function is simply equal to f_2. This system is shown in Fig. 11-10. Note that if this configuration is combined with the preceding one, the special features of each are obtained.

Lens Elements Separated by the Sum of their Focal Lengths

When $z_{35}=f_1+f_2$, as shown in Fig. 11-11, we find that

$$m''=\frac{-f_2}{f_1},\qquad z_{26}=\frac{(f_1+f_2)^2}{f_1}+z_{23}\left(1-\frac{f_2^2}{f_1^2}\right),\tag{11.73}$$

and the magnification is again equal to the (negative) ratio of the focal lengths of the lens elements; however, *it is now independent of object position.* In addition, the distance between the object and image planes can be made independent of object position by setting $f_1=f_2$. *With $f_1=f_2=f$, the object and image planes are always separated by a distance of $z_{26}=4f$.*

A system for which $z_{35}=f_1+f_2$ is referred to as an *afocal system* because it has no finite focal points; i.e., rays entering the system parallel with the optical axis also exit parallel with the axis. As a consequence, afocal systems exhibit the following interesting behavior: light entering the system as a collimated (parallel) beam leaves as a collimated beam. The cross section and direction of propagation may have changed, but the beam is still collimated. Beam expanders used in certain laser applications are often of this design, which is basically that of a Keplerian telescope. The

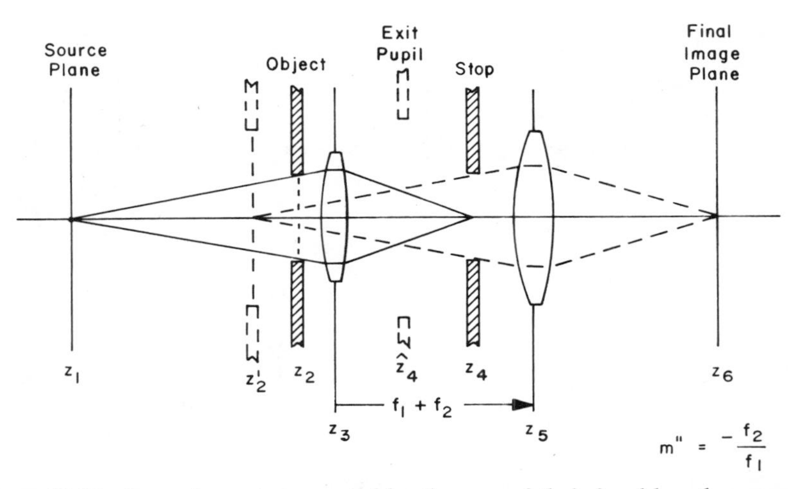

Figure 11-11 Lens elements separated by the sum of their focal lengths: *an afocal system.*

overall length of such a beam expander can be reduced if the first lens element is negative, as shown in Fig. 10-32 (that system is effectively a reversed Galilean telescope).

Plane-Wave Illumination with Object in FFP of First Lens and Stop in FFP of Second Lens

This configuration, which is commonly found in textbooks because of its simplicity, is basically a combination of the three preceding configurations; one additional condition is the specification of an infinite source distance. Thus, we have $z_{13}=z_{12}=\infty$, $z_{23}=f_1$, and $z_{45}=f_2$, which leads to

$$m'=\hat{m}=\infty, \qquad z'_{24}=\hat{z}_{46}=\infty,$$

$$z_{34}=f_1, \qquad z_{56}=f_2,$$

$$\frac{\hat{z}_{46}}{\hat{m}}=f_2, \qquad m''=\frac{-f_2}{f_1},$$

$$q\left(x,y;\frac{1}{\lambda\hat{z}_{46}}\right)=1, \qquad H_{26}(\xi,\eta)=p_4(-\lambda f_2\xi,-\lambda f_2\eta). \tag{11.74}$$

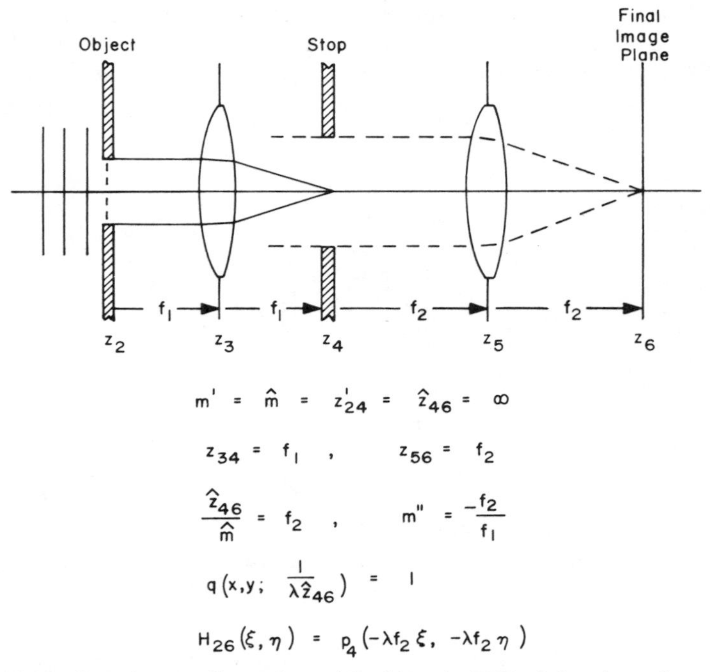

Figure 11-12 Imaging configuration with object in FFP of first lens element, stop in FFP of second lens element, and plane-wave illumination.

This configuration is illustrated in Fig. 11-12. It may be identified as a cascade of two ideal spectrum analyzers, each of which has an output that is free of quadratic phase factors. Consequently, the final image may be regarded as being proportional to the Fourier transform of the (modified) Fourier transform of the object, as in the one-dimensional system of Fig. 8-27. Although this configuration is quite easy to analyze, it has the disadvantages that the object and frequency planes lie quite far from the lenses, which could result in vignetting unless special care is taken. In addition, it is less versatile than other configurations and may even be undesirable when the effects of lens aberrations are taken into account. As mentioned earlier, the existence of a quadratic phase factor at the image plane in no way influences the quality of the image. Hence, in using a system that eliminates this phase factor, the penalty paid in the laboratory may be greater than the advantages gained in the analysis.

Object Located at the First Lens

When $z_{23}=0$, we find that

$$m'=1, \qquad z'_{25}=z_{35},$$

$$m''=\frac{f_2}{f_2-z_{35}}, \qquad z_{56}=-m''z_{35}. \tag{11.75}$$

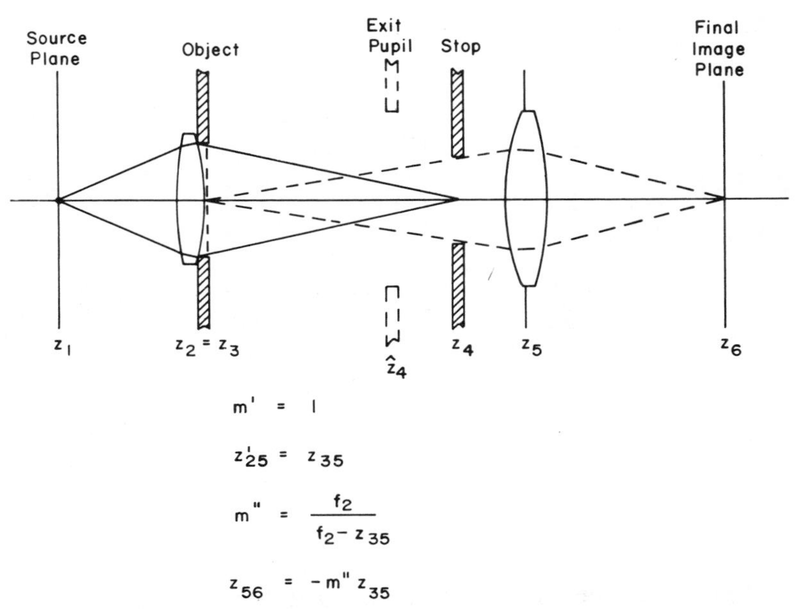

Figure 11-13 Object located at first lens element.

We see that the overall magnification is determined only by the focal length and position of the second lens, as is the location of the final image plane. The same result is obtained if the object is moved to the other side of the first lens as shown in Fig. 11-13. This is a departure from our original conditions, but it has the advantage that the first lens now needs to be corrected for spherical aberration only; because it is "working" only on axis, the off-axis aberrations of coma, astigmatism, etc., will not be a factor in selecting this lens. This system is often considered to be a single-lens system because the imaging is done by the second lens alone. The first lens merely launches a converging spherical wave that causes the object spectrum to be displayed at the frequency plane. We note that the object need not even be located at the lens in order for the basic results to be valid, and such was the case for the system analyzed in Ref. 11-2. However, care must be taken to adjust the magnitude of the illuminating wave field and to make certain that correct values are used for the various distances.

Second Lens Located at the Frequency Plane

For this configuration we have $z_{45}=0$, from which we obtain

$$\hat{m}=1, \qquad \hat{z}_{46}=z_{56},$$

$$H_{26}(\xi,\eta)=p_4(-\lambda z_{56}\xi, -\lambda z_{56}\eta). \tag{11.76}$$

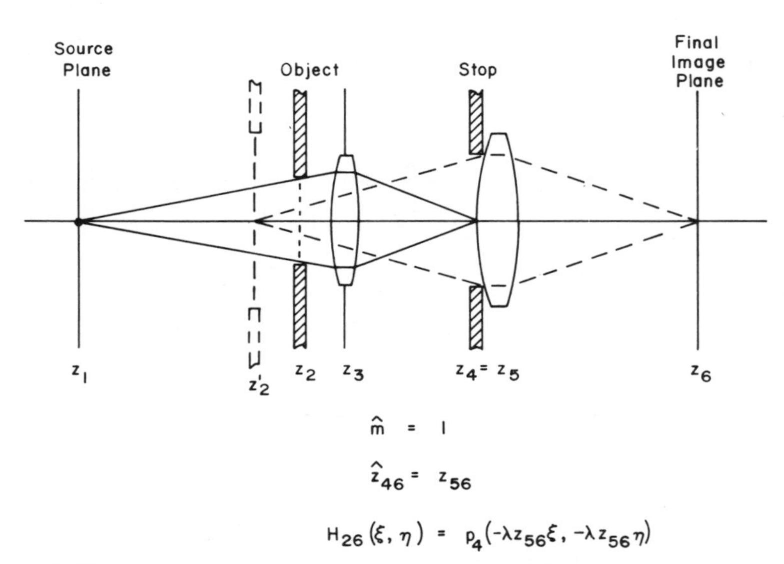

Figure 11-14 Second lens element located at the frequency plane.

As may be seen in Fig. 11-14, the exit pupil is coplanar with the aperture stop.

Separation of Lenses Equal to Focal Length of the Second

We now set $z_{35}=f_2$, which results in an overall magnification of

$$m''=\frac{-f_2}{z_{23}}. \tag{11.77}$$

Thus, when this system is used (see Fig. 11-15), *the magnification is inversely proportional to the object distance z_{23}.*

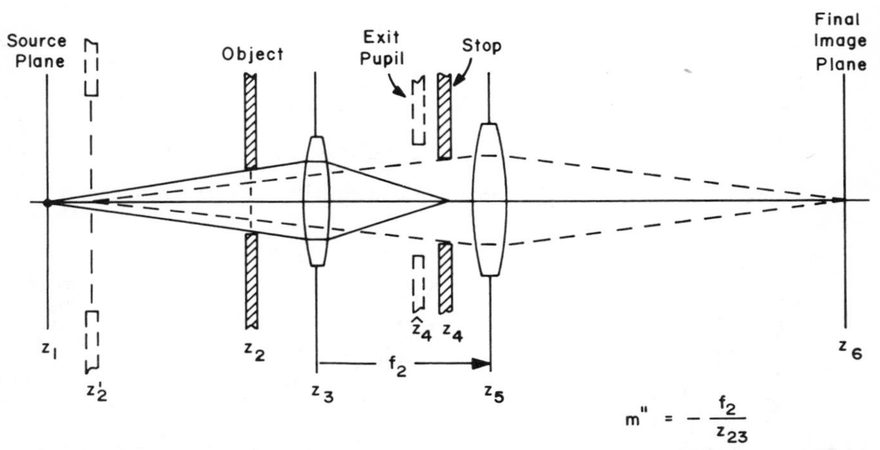

Figure 11-15 Lens elements separated by focal length of second.

Single-Lens Configuration

We can obtain various single-lens configurations simply by setting the focal length of either lens equal to infinity, and such configurations are also commonly seen in textbooks. However, such systems are shift invariant only for a few special cases, which we now investigate.

If we choose $f_2=\infty$, the second lens effectively vanishes and we have the system shown in Fig. 11-16. For diverging spherical wave illumination, as we have indicated in this figure, the frequency plane will still lie at the conjugate plane $z=z_4$, and that is where the stop must be located. However, substitution of $f_2=\infty$ into the various expressions of Eq. (11.48)

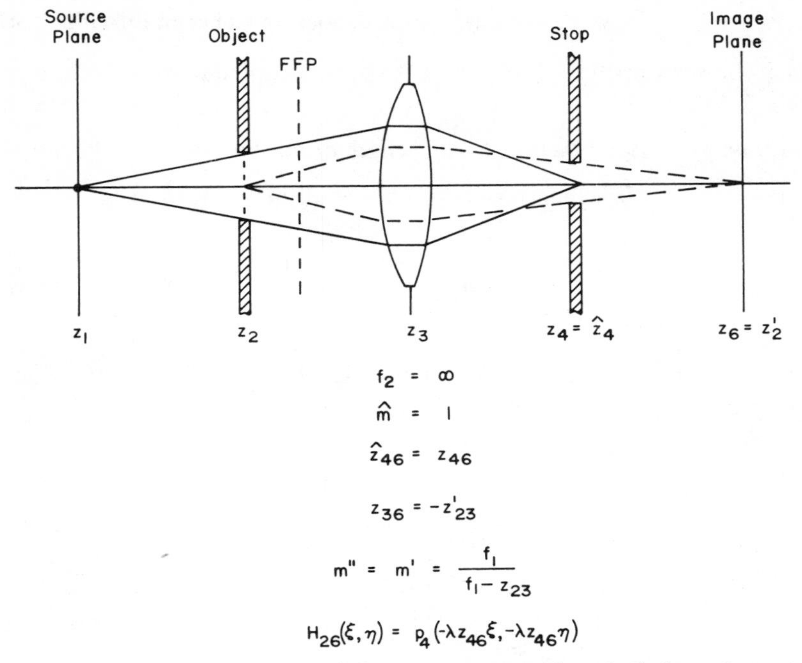

Figure 11-16 Shift-invariant imaging system with a single lens element.

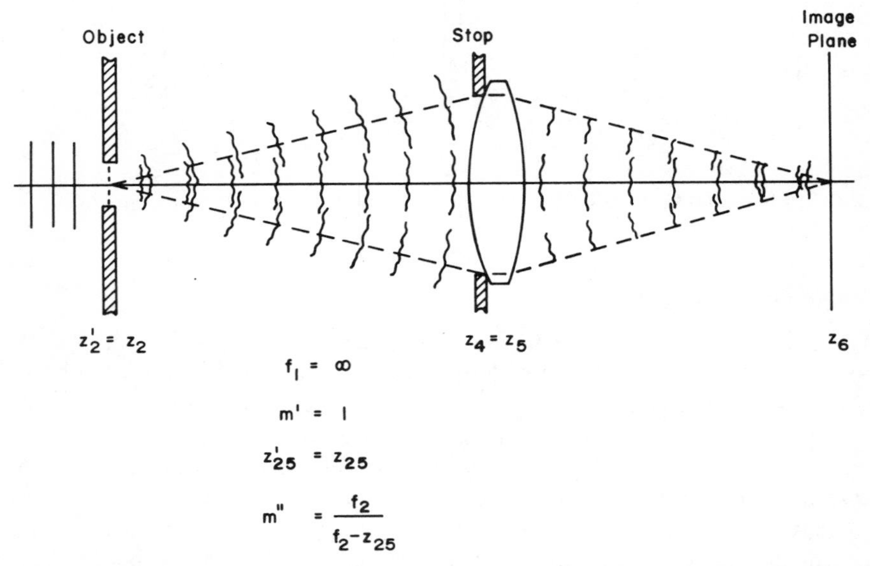

Figure 11-17 Single-element imaging system that is shift invariant only for small objects with limited spatial-frequency content.

yields

$$\hat{m}=1, \qquad \hat{z}_{46}=z_{46},$$

$$z_{36}=-z'_{23}, \quad m''=m'=\frac{f_1}{f_1-z_{23}},$$

$$H_{26}(\xi,\eta)=p_4(-\lambda z_{46}\xi, -\lambda z_{46}\eta), \tag{11.78}$$

and we see that the image plane is now conjugate with the object plane $z=z_2$. In other words, the planes $z=z_6$ and $z=z'_2$ are no longer conjugates, but are in fact the same plane. We note that the final image is virtual if the object is placed inside the FFP of the lens, and real if the object is placed outside the FFP. We have depicted this latter situation in Fig. 11-16.

The system of Fig. 11-16 will yield shift invariant imaging for either diverging or converging spherical wave illumination, or for plane-wave illumination, as long as the frequency plane is physically accessible and is colocated with the aperture stop. If the stop is not located at the frequency plane, as is the case for many textbook analyses, the system will be shift invariant only for small objects with no significant high-frequency structure. More precisely, the size and spectrum of the objects must be such that all of the diffracted light reaches the image plane, conditions that impose severe limitations on the use of such a system (see Fig. 11-17).

11-4. EXPERIMENTAL VERIFICATION OF THE FILTERING INTERPRETATION

As discussed previously, the imaging process causes an image to be a smoothed version of the object. This smoothing is a physical consequence of diffraction, but mathematically we regard it to be the result of a linear filtering operation. Hence, from the frequency-domain point of view, the image smoothing occurs because the object spectrum is altered as it passes through the system. In the preceding example involving the Ronchi ruling, the image did not exhibit sharp bar edges because the plane-wave components required to produce such sharp edges had been physically blocked by the aperture stop; the aperture stop acted as a *spatial filter*, allowing only the low-frequency spectral components to reach the image plane.

We now present experimentally derived verification of this spatial filtering action. The experimental setup used for the verification is described in Ref. 11-3. The object of interest was the two-dimensional wire mesh shown in Fig. 11-18(a), and its Fraunhofer diffraction pattern is pictured in Fig. 11-18(b). Note that the spaces between the wires are considerably larger

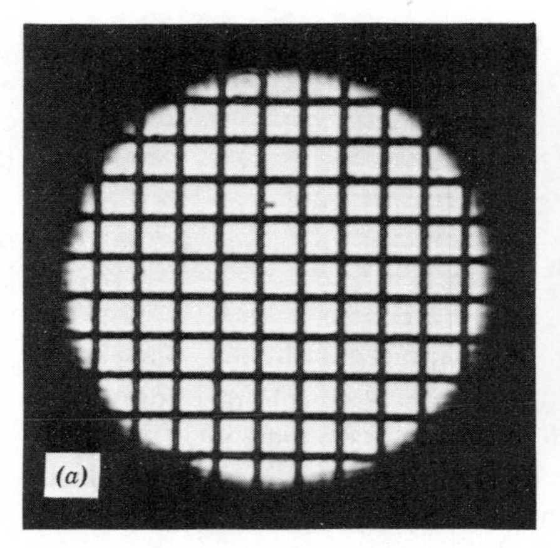

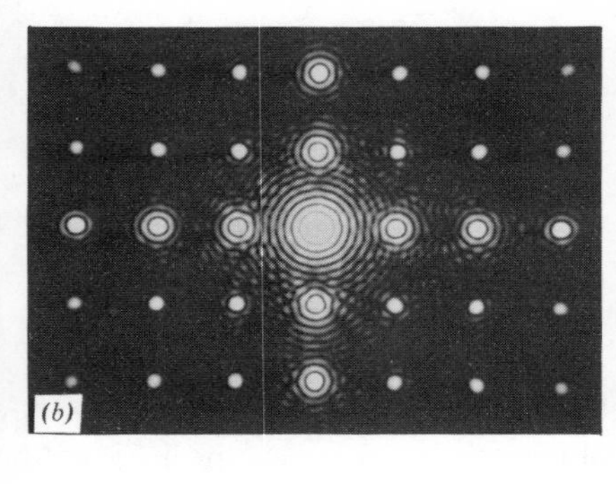

Figure 11-18 Filtering interpretation of coherent imaging. (*a*) 10-wire-per-millimeter mesh used for verification. (*b*) Fraunhofer diffraction pattern of mesh.

than the wires themselves; therefore, the transmittance function of this mesh contains both odd- and even-harmonic components, and its average value (zero-frequency component) is greater than 0.5. Each bright spot (diffraction order) of the Fraunhofer pattern is associated with a particular Fourier component of the transmittance function, and the size and shape of the aperture stop determines which of these components are allowed to reach the image plane, and which are not. In Figs. 11-19 through 11-22 we show photographic images obtained for different aperture stops; part (a) of each figure depicts the spectrum passed by the system, and part (b) displays the resulting image irradiance.

Figure 11-19 shows the situation for which the aperture stop is a narrow horizontal slit that behaves as a low-pass filter in the η-direction. The effective cutoff frequency is approximately one-half the fundamental frequency of the mesh pattern, and this causes the fundamental and all higher harmonic components in the η-direction to be blocked. Consequently, because the plane-wave components necessary for producing horizontal image detail are prevented from reaching the image plane, no horizontal wires appear in the image. The vertical wires, on the other hand, encounter very little smoothing because the cutoff frequency in the ξ-direction is high enough to pass most of the significant harmonics.

In Fig. 11-20 the aperture stop is rectangular in shape, and it acts as a low-pass filter in both ξ and η. The cutoff frequency in the η-direction is the same as in the previous case, which again eliminates horizontal wires from the image, but the cutoff frequency in the ξ-direction lies between the fundamental and second-harmonic frequencies. Thus, all ξ-components of

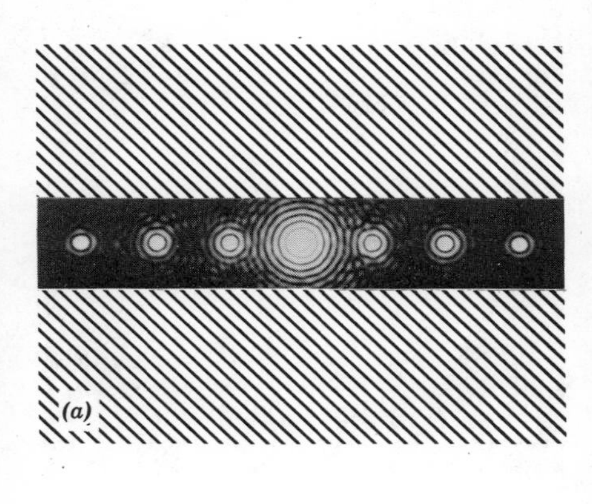

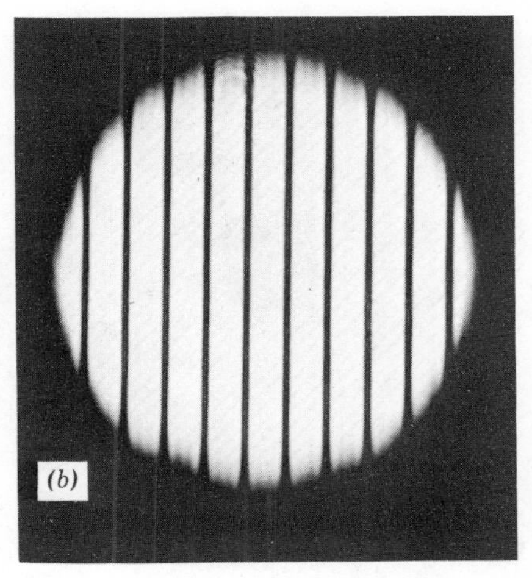

Figure 11-19 Filtering of all nonzero Fourier components in η-direction. (*a*) Spectrum passed by the system. (*b*) Filtered image, exhibiting no horizontal wires and sharply imaged vertical wires.

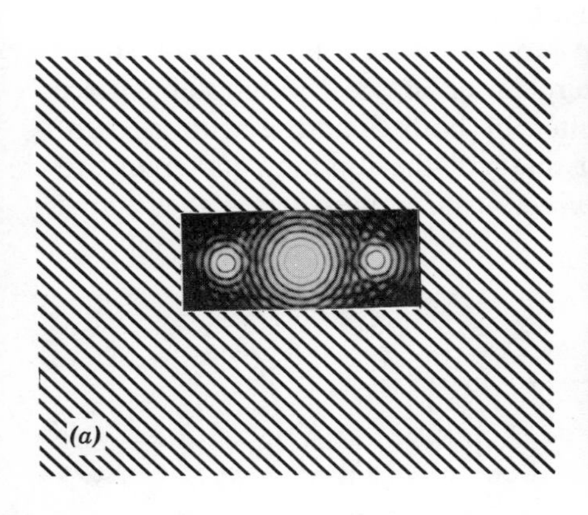

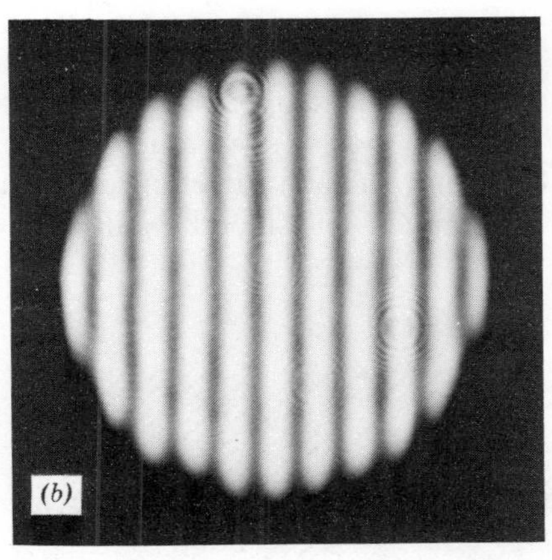

Figure 11-20 Filtering of all nonzero components in η-direction and all components higher than fundamental in ξ-direction. (*a*) Spectrum passed by the system. (*b*) Filtered image, exhibiting no horizontal wires and smoothed vertical wires.

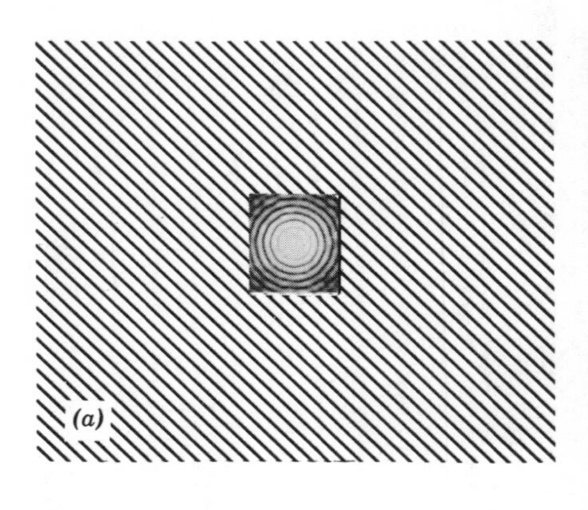

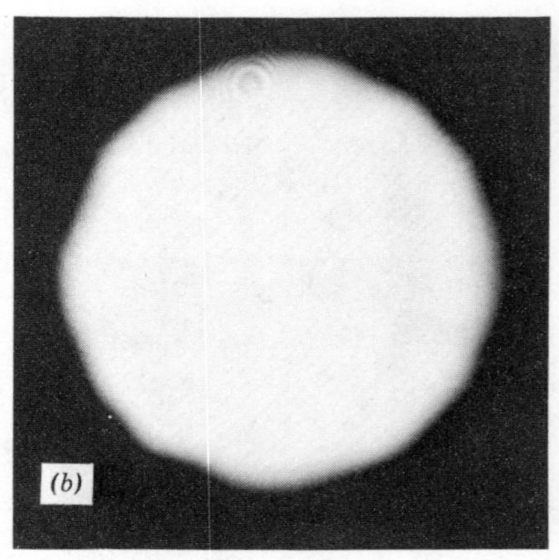

Figure 11-21 Filtering of all nonzero Fourier components in both ξ- and η-directions. (*a*) Only zero-frequency spectral component passed by system. (*b*) Filtered image, exhibiting neither horizontal nor vertical wires.

frequency higher than the fundamental are blocked by the stop, and the vertical wires are now a great deal smoother than in the previous image.

Figure 11-21 illustrates the smoothing produced by a small square aperture stop that passes only the zero order of the Fraunhofer pattern. This stop behaves as a low-pass filter in both ξ and η with a cutoff frequency in each direction of about one-half the fundamental frequency of the mesh. The smoothing in both directions is now so severe that neither

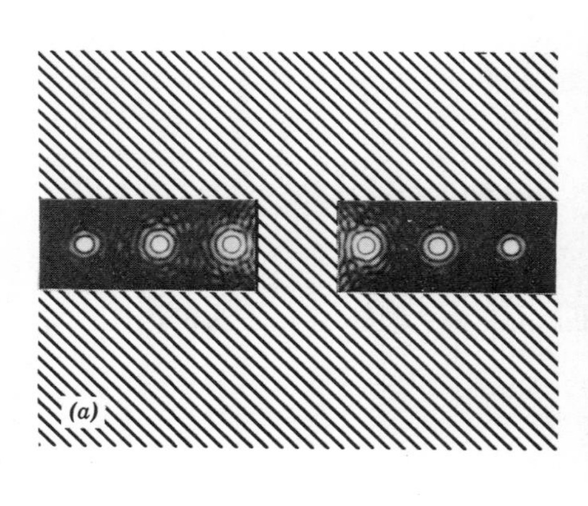

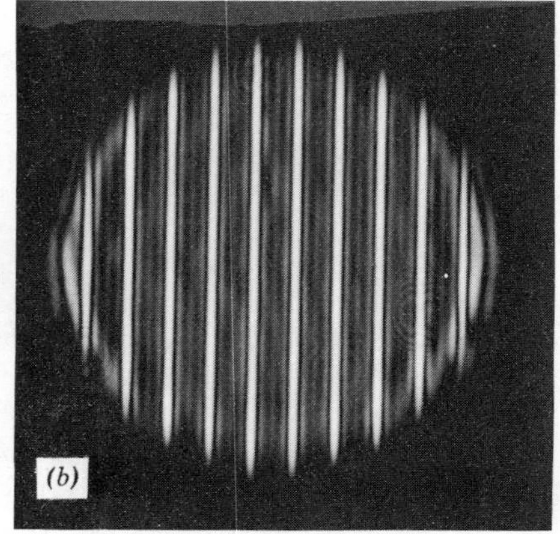

Figure 11-22 Contrast reversal by filtering out the zero-frequency component. (*a*) Spectrum passed by system. (*b*) Filtered image, exhibiting contrast reversal.

the vertical nor the horizontal wires appear in the image. We note that the edge of the pupil surrounding the mesh has also undergone the same sort of smoothing.

The aperture stop for the situation illustrated by Fig. 11-22 is a narrow horizontal slit with an obscuration just large enough to block the zero order of the Fraunhofer pattern added at the center. This system acts as a low-pass filter in the η-direction and a high-pass filter in the ξ-direction. The removal of the zero-frequency component, which corresponds to the average transmittance of the mesh, leads to a *contrast reversal* in the image: the wires now appear as narrow spaces and the spaces appear as wide wires. Can you explain this behavior?

11-5 IMAGE FORMATION WITH INCOHERENT LIGHT

In this section we derive image-irradiance expressions for the situation in which the light radiated by the object has a narrow temporal-frequency spectrum,† but is *spatially incoherent*. This differs from the previous situations in which the light was assumed to be monochromatic and therefore perfectly coherent. A coherent wave field is characterized by its ability to produce observable constructive and destructive interference (i.e., on a time-averaged basis) when different portions of it are combined at the same location. The fundamental reason for this behavior is that the relationship between the phase at any two points in a coherent wave field is *fixed in time*. A distinction is often made between *spatial coherence* and *temporal coherence*, and we now discuss the difference. Consider the amplitude of a wave field, $u(\mathbf{r};t)$, where $\mathbf{r}$ denotes a general observation point in space and t denotes the observation time. If the phase difference between $u(\mathbf{r}_1;t)$ and $u(\mathbf{r}_2;t)$ is independent of time for all points $\mathbf{r}_1$ and $\mathbf{r}_2$, the wave field is called spatially coherent. On the other hand, it is temporally coherent if the phase relationship between $u(\mathbf{r};t_1)$ and $u(\mathbf{r};t_2)$ depends only on the time difference t_2-t_1, and not on t_1 or t_2 separately. Spatial coherence may often be regarded as a characteristic of a wave field transverse to the direction of propagation, whereas temporal coherence may be thought of as a characteristic along the direction of propagation.

A monochromatic wave field would be both spatially and temporally coherent, but such a wave field cannot exist in the real world. However, no wave field in the real world is completely incoherent either. Thus, all real-world optical wave fields exhibit some degree of *partial coherence*, and the notions of coherent light and incoherent light are contrived to simplify

†This requirement may be specified as follows: if the light has a bandwidth of $\triangle\lambda$ centered about the wavelength λ, we require that $\triangle\lambda \ll \lambda$.

certain calculations. These notions are nevertheless important, and yield surprisingly accurate results for many problems of practical concern. For the interested reader, more details regarding the subject of coherence may be found in Refs. 11-4 and 11-5.

For our purposes here, the object may be either self-luminous or irradiated by some incident wave field; in either event, we shall consider the radiated light to be spatially incoherent. We may therefore regard the object to be a collection of independent point radiators, the strength of which varies from point to point according to some information-bearing characteristic of the object (e.g., its reflectance). The quantity of importance is the *radiance* of the object, which is a measure of the radiant flux emitted in a particular direction per unit of normally projected surface area per unit solid angle. The units of radiance are W/cm^2/steradian. As we shall see, an image-forming system maps the radiance distribution of the object into an irradiance distribution at the image plane.

In general, the radiance of a surface is a function of the angle at which the surface is viewed, but it can be nearly independent of viewing angle if the surface is highly diffuse.[§] We assume that our objects are not only highly diffuse, but also small in lateral extent relative to the object distance. In addition, we restrict the size of the aperture stop such that the cone of light accepted for each object point is of small angular extent. These size restrictions are also required to ensure that the Fresnel conditions are satisfied for all portions of the propagation path: from the object to the system, between elements of the system, and from the system to the image plane.

General Two-Element Lens System

Although we can now obtain shift-invariant imaging with a single-element lens when incoherent illumination is used, we begin with the two-element system shown in Fig. 11-23; the results for single-element systems are then obtained as special cases of the two-element expressions. We retain the configuration and notation used in the coherent-imaging case for the sake of consistency, but there are some important differences. We no longer specify a source plane, $z = z_1$, because the quantity of interest is now the radiance of the object rather than its transmittance. In addition, there is no longer a requirement to locate the aperture stop at a frequency plane. In fact, as long as the following conditions are satisfied, *the exact location of the aperture stop is of minimal importance in obtaining shift-invariant (or*

[§]Surfaces for which the radiance is independent of viewing angle are called *Lambertian surfaces*. They cannot exist physically, but are conceptually useful nevertheless.

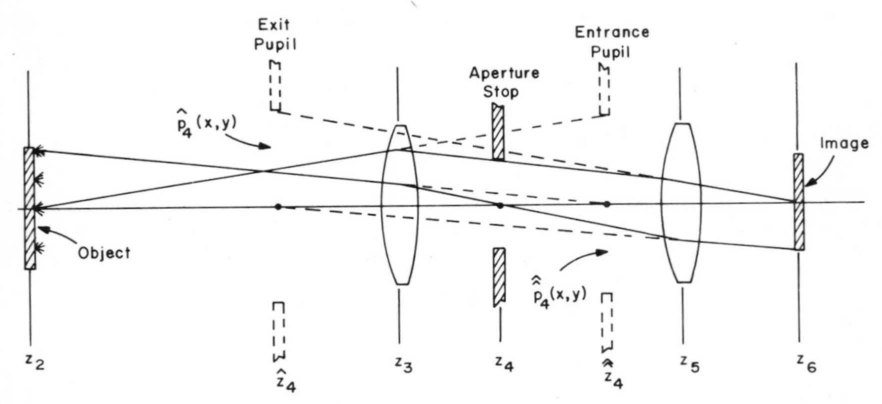

Figure 11-23 Two-element system for investigation of incoherent image formation.

isoplanatic) imaging: the lens elements must be relatively free of aberrations, the object and image field angles must not be too large, and the system must be free of vignetting. We have placed the stop between the lens elements only for convenience.

Each point radiator on the object will give rise to its own amplitude distribution at the image plane but, because all of the object radiators are mutually incoherent, these amplitude distributions do not add linearly. Rather, it is the associated irradiance distributions produced by each object point that add together to form the image. The irradiance distribution produced at the image plane due to a single object point radiator is referred to as the *incoherent impulse response*, or *point spread function*, of the system. It is a real-valued, nonnegative quantity, and *is proportional to the squared modulus of the system's coherent impulse response*. We denote the point spread function by $\mathscr{h}_{26}(x, y)$, and we consider the overall image of any object to be a superposition of appropriately weighted and shifted versions of $\mathscr{h}_{26}(x, y)$.

The first lens element forms an image at the plane $z = z_2'$, as before, and the second element relays this image to the final image plane $z = z_6$. However, we will find it convenient to approach the incoherent imaging problem from a different point of view. We recall that the exit pupil of the system is an image of the aperture stop, as seen from image space, and that it is located at the plane $z = \hat{z}_4$. We now define the *entrance pupil* $\hat{\hat{p}}_4(x, y)$ to be the image of the aperture stop as viewed from object space, and we denote its location by $z = \hat{\hat{z}}_4$. Then, with f_1 and f_2 the focal lengths of the first and second lens elements, respectively, and with $\hat{\hat{z}}_{i4} = \hat{\hat{z}}_4 - z_i$ and

$\hat{z}_{4i}=z_i-\hat{z}_4$, we obtain the following relationships:

$$\hat{\hat{z}}_{24}=z_{23}+\hat{\hat{z}}_{34}, \qquad \hat{z}_{46}=\hat{z}_{45}+z_{56},$$

$$\hat{\hat{z}}_{34}=\hat{\hat{m}}z_{34}, \qquad \hat{z}_{45}=\hat{m}z_{45},$$

$$\hat{\hat{m}}=\frac{f_1}{f_1-z_{34}}, \qquad \hat{m}=\frac{f_2}{f_2-z_{45}},$$

$$\hat{\hat{p}}_4(x,y)=\frac{1}{\hat{\hat{m}}}p_4\left(\frac{x}{\hat{\hat{m}}},\frac{y}{\hat{\hat{m}}}\right), \qquad \hat{p}_4(x,y)=\frac{1}{\hat{m}}p_4\left(\frac{x}{\hat{m}},\frac{y}{\hat{m}}\right),$$

$$z_{56}=-\hat{m}\left(\frac{m''\hat{\hat{z}}_{24}}{\hat{\hat{m}}}+z_{45}\right)=\frac{z_{35}\left(\frac{z_{23}}{f_1}-\frac{z_{25}}{z_{35}}\right)}{\frac{z_{23}}{f_1}\left(\frac{z_{35}}{f_2}-1\right)-\frac{z_{25}}{f_2}+1},$$

$$m''=\frac{-\hat{m}\hat{z}_{46}}{\hat{\hat{m}}\hat{\hat{z}}_{24}}=\left[\left(\frac{z_{35}}{f_2}-1\right)\left(\frac{z_{23}}{f_1}-1\right)-\frac{z_{23}}{f_2}\right]^{-1}. \tag{11.79}$$

In the absence of vignetting, the entrance pupil may be regarded as the limiting pupil of the system because it effectively determines how much of the light emitted by the object will reach the image plane. The exit pupil, on the other hand, once again governs the nature of the impulse response of the system, and thereby controls the degree to which the image is smoothed.[¶] When viewed in this fashion, the lens elements and aperture stop may effectively be discarded and replaced by the system shown in Fig. 11-24. Each object point emits a diverging spherical wave, a portion of which is collected by the entrance pupil, and the exit pupil then launches this light in the form of a spherical wave converging toward (or diverging from) the appropriate point in the image plane.

With the assumption that the object radiance distribution $N_2(x, y)$ is constant over the solid angle subtended by the entrance pupil, we define the irradiance of the *geometrical image* by**

$$I_2''(x,y)=\hat{\hat{K}}\left(\frac{1}{m''}\right)^2 N_2\left(\frac{x}{m''},\frac{y}{m''}\right), \tag{11.80}$$

¶This is not the only way to regard the effects of the entrance and exit pupils, but it is a convenient one. For another interpretation, see Goodman (Ref. 11-1).

**This will be valid only for small field angles in both object and image space. Even for Lambertian objects, the image irradiance falls off rapidly as the field angles become large (see Ref. 11-6 for details).

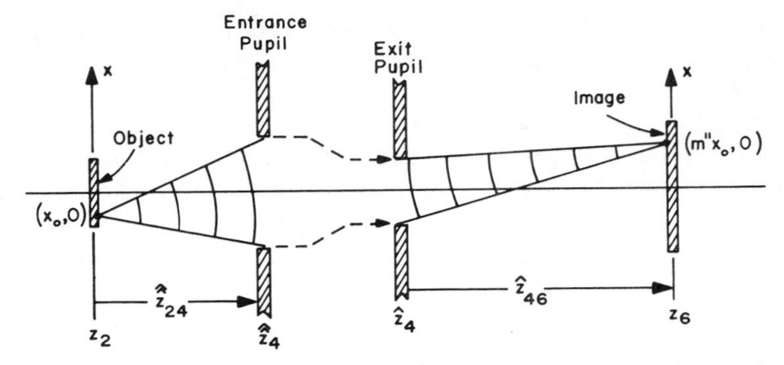

Figure 11-24 Simplified model of image-forming system of Fig. 11-23.

where

$$\hat{\hat{K}}=\left(\frac{1}{\hat{\hat{z}}_{24}}\right)^2\iint_{-\infty}^{\infty}\left|\hat{\hat{m}}\hat{\hat{p}}_4(\hat{\hat{\alpha}},\hat{\hat{\beta}})\right|^2 d\hat{\hat{\alpha}}\, d\hat{\hat{\beta}}$$

$$=\left(\frac{\hat{\hat{m}}}{\hat{\hat{z}}_{24}}\right)^2\iint_{-\infty}^{\infty}|p_4(\alpha,\beta)|^2 d\alpha\, d\beta. \tag{11.81}$$

Next we define the *diffraction image* to be

$$I_6''(x,y)=I_2''(x,y)**\mathscr{h}_{26}(x,y), \tag{11.82}$$

where $\mathscr{h}_{26}(x, y)$ is the point spread function of the system. As mentioned earlier, $\mathscr{h}_{26}(x, y)$ is proportional to the squared modulus of the coherent impulse response, i.e.,

$$\mathscr{h}_{26}(x,y)\propto|h_{26}(x,y)|^2, \tag{11.83}$$

and we shall determine the constant of proportionality in the next section. From Eq. (11.82) we see that the diffraction image is simply a smoothed version of the geometrical image.

Finally, with

$$K''=|B_{\ell1}B_{\ell2}|^2, \tag{11.84}$$

a constant that accounts for the total power losses of the lens elements, both absorption and reflection, the actual image irradiance is given by

$$I_6(x,y)=K''I_6''(x,y). \tag{11.85}$$

As long as our assumptions are valid, this expression accurately describes the irradiance of the image, both qualitatively and quantitatively.

In the event that the aperture stop is either clear or has only phase variations across it, such that $|p_4(x, y)|^2$ is a zero-one function, the constant $\hat{\hat{K}}$ is equal to the solid angle subtended by the entrance pupil at a distance of $\hat{\hat{z}}_{24}$. It may then be approximated by

$$\hat{\hat{K}} \cong \left(\frac{1}{\hat{\hat{z}}_{24}}\right)^2 \times \text{area of entrance pupil}$$

$$\cong \left(\frac{\hat{\hat{m}}}{\hat{\hat{z}}_{24}}\right)^2 \times \text{area of aperture stop.} \tag{11.86}$$

If we now consider the special case for a clear circular aperture of diameter d, we obtain

$$\hat{\hat{K}} \cong \frac{\pi}{4}\left(\frac{\hat{\hat{m}}d}{\hat{\hat{z}}_{24}}\right)^2. \tag{11.87}$$

Thus, the geometrical image becomes

$$I_2''(x, y) \cong \frac{\pi}{4}\left(\frac{\hat{\hat{m}}d}{m''\hat{\hat{z}}_{24}}\right)^2 N_2\left(\frac{x}{m''}, \frac{y}{m''}\right). \tag{11.88}$$

However, from Eq. (11.79) we know that

$$\frac{\hat{\hat{m}}d}{m''\hat{\hat{z}}_{24}} = \frac{-\hat{m}d}{\hat{z}_{46}}, \tag{11.89}$$

and from Eq. (10.138) we see that this quantity is related to the *effective F-number of the system in image space*, i.e.,

$$\left(\frac{\hat{\hat{m}}d}{m''\hat{\hat{z}}_{24}}\right)^2 = \left(\frac{\hat{m}d}{\hat{z}_{46}}\right)^2$$

$$= (F_{\text{eff}}^{\#})^{-2}. \tag{11.90}$$

Therefore, *the geometrical image irradiance depends inversely on the square of the effective image-space F-number*, i.e.,

$$I_2''(x, y) \cong \frac{\pi}{4}\left(\frac{1}{F_{\text{eff}}^{\#}}\right)^2 N_2\left(\frac{x}{m''}, \frac{y}{m''}\right). \tag{11.91}$$

An equivalent single-element system is illustrated in Fig. 11-25. In this system, the equivalent aperture stop is described by the original exit pupil

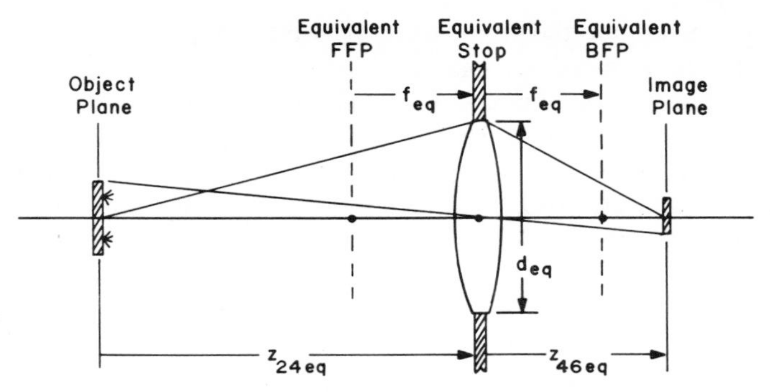

Figure 11-25 Equivalent single-element system for modeling two-element system of Fig. 11-23.

function and is located at the equivalent lens. With the stop at the lens, the new entrance and exit pupils are identical and they have diameters equal to $\hat{m}$ times the diameter of the original stop. These relationships are given by

$$p_{4\mathrm{eq}}(r)=\hat{p}_4(r)=\frac{1}{\hat{m}}\,p_4\left(\frac{r}{\hat{m}}\right),$$

$$\hat{p}_{4\mathrm{eq}}(r)=\hat{\hat{p}}_{4\mathrm{eq}}(r)=p_{4\mathrm{eq}}(r),$$

$$d_{\mathrm{eq}}=\hat{m}d. \tag{11.92}$$

We require that the image distance $z_{46\mathrm{eq}}$ and magnification m_{eq} of the equivalent system be equal to $\hat{z}_{46}$ and m'', respectively, and to satisfy this requirement the equivalent object distance $z_{24\mathrm{eq}}$ and focal length f_{eq} must obey the following:

$$z_{24\mathrm{eq}}=\frac{-\hat{z}_{46}}{m''}=\frac{\hat{m}\hat{\hat{z}}_{24}}{\hat{\hat{m}}},$$

$$f_{\mathrm{eq}}=\frac{(z_{24\mathrm{eq}})(z_{46\mathrm{eq}})}{(z_{24\mathrm{eq}})+(z_{46\mathrm{eq}})}=\frac{\hat{z}_{46}}{1-m''}. \tag{11.93}$$

Finally, the F-number of the equivalent lens is given by

$$F^{\#}_{\mathrm{eq}}=\frac{f_{\mathrm{eq}}}{d_{\mathrm{eq}}}=\frac{F^{\#}_{\mathrm{eff}}}{1-m''}, \tag{11.94}$$

such that, in terms of the equivalent single-element system, Eq. (11.91) becomes

$$I_2''(x,y)\cong\frac{\pi}{4}\left[\frac{1}{F^{\#}_{\mathrm{eq}}(1-m'')}\right]^2 N_2\left(\frac{x}{m''},\frac{y}{m''}\right). \tag{11.95}$$

We see, then, that the more general two-element lens system may be replaced by an equivalent single-element system and that the geometrical image irradiance varies inversely as the square of $F^{\#}_{\rm eq}(1-m'')$. This is an interesting conclusion, because we see that the object and image distances and the stop diameter are unimportant individually; the significant quantities are the various ratios of these parameters. We note that for very distant objects, such that

$$|m''|\ll 1, \tag{11.96}$$

the magnitude of the geometrical image irradiance is effectively independent of both object distance and magnification; i.e.,

$$\begin{aligned} I_2''(x,y) &\cong \frac{\pi}{4}\left(\frac{1}{F^{\#}_{\rm eq}}\right)^2 N_2\left(\frac{x}{m''},\frac{y}{m''}\right) \\ &\cong \frac{\pi}{4}\left(\frac{1}{F^{\#}_{\rm eff}}\right)^2 N_2\left(\frac{x}{m''},\frac{y}{m''}\right), \qquad |m''|\ll 1. \end{aligned} \tag{11.97}$$

11-6 LINEAR FILTER INTERPRETATION OF INCOHERENT IMAGING

The convolution operation of Eq. (11.82) again suggests that the diffraction image may be regarded as the output of a linear shift-invariant system. Thus, following the approach used for the coherent case, we devise a system comprised of a geometric scaler, an LSI system, and a multiplier (see Fig. 11-26). Because the constant K'' and the effect of the scaler are easily determined, we now concentrate on the filtering behavior of the LSI system.

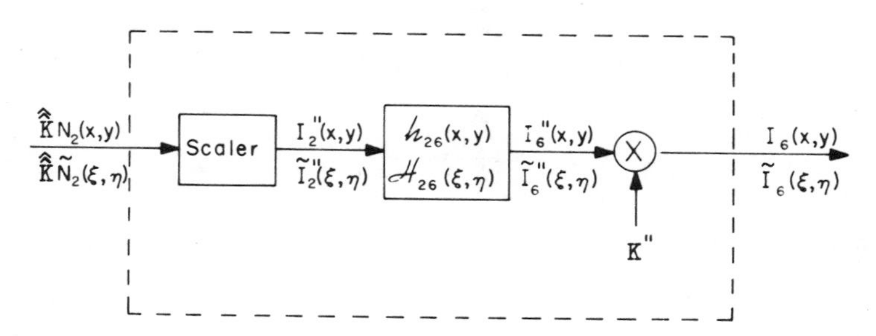

Figure 11-26 Operational representation of optical system used for incoherent image formation.

Point Spread Function

The relationship between the incoherent point spread function (PSF) and the coherent impulse response is given by Eq. (11.83), and we note that this relationship holds even when the location of the aperture stop does not yield shift-invariant coherent imaging. To specify the constant of proportionality, we recall that the coherent transfer function is given by

$$H_{26}(\xi,\eta)=\hat{m}\hat{p}_4(-\lambda\hat{z}_{46}\xi,\,-\lambda\hat{z}_{46}\eta), \tag{11.98}$$

and we define the complex autocorrelation of this function by

$$\gamma_H(\xi,\eta)=H_{26}(\xi,\eta)\star\star H^*_{26}(\zeta,\eta). \tag{11.99}$$

It can then be shown that the point spread function has the form

$$\mathscr{h}_{26}(x,y)=\frac{|h_{26}(x,y)|^2}{\gamma_H(0,0)}. \tag{11.100}$$

This expression may also be written in terms of the exit pupil and aperture stop as follows:

$$\begin{aligned}\mathscr{h}_{26}(x,y)&=\frac{\left|\hat{m}\left(\frac{1}{\lambda\hat{z}_{46}}\right)^2\hat{P}_4\left(\frac{x}{\lambda\hat{z}_{46}},\frac{y}{\lambda\hat{z}_{46}}\right)\right|^2}{\iint\limits_{-\infty}^{\infty}|\hat{m}\hat{p}_4(\lambda\hat{z}_{46}\xi,\lambda\hat{z}_{46}\eta)|^2\,d\xi\,d\eta}\\
&=\frac{\left(\frac{\hat{m}}{\lambda\hat{z}_{46}}\right)^2\left|\left(\frac{1}{\lambda\hat{z}_{46}}\right)\hat{P}_4\left(\frac{x}{\lambda\hat{z}_{46}},\frac{y}{\lambda\hat{z}_{46}}\right)\right|^2}{\left(\frac{\hat{m}}{\lambda\hat{z}_{46}}\right)^2\iint\limits_{-\infty}^{\infty}|\hat{p}_4(\hat{\alpha},\hat{\beta})|^2\,d\hat{\alpha}\,d\hat{\beta}}\\
&=\frac{\left(\frac{1}{\lambda\hat{z}_{46}}\right)^2\left|\hat{P}_4\left(\frac{x}{\lambda\hat{z}_{46}},\frac{y}{\lambda\hat{z}_{46}}\right)\right|^2}{\iint\limits_{-\infty}^{\infty}|\hat{p}_4(\hat{\alpha},\hat{\beta})|^2\,d\hat{\alpha}\,d\hat{\beta}}\\
&=\frac{\left(\frac{\hat{m}}{\lambda\hat{z}_{46}}\right)^2\left|P_4\left(\frac{\hat{m}x}{\lambda\hat{z}_{46}},\frac{\hat{m}y}{\lambda\hat{z}_{46}}\right)\right|^2}{\iint\limits_{-\infty}^{\infty}|p_4(\alpha,\beta)|^2\,d\alpha\,d\beta}.\end{aligned} \tag{11.101}$$

We note that the quantity

$$\iint_{-\infty}^{\infty} |\hat{p}_4(\hat{\alpha},\hat{\beta})|^2 \, d\hat{\alpha}\, d\hat{\beta} = \iint_{-\infty}^{\infty} \left| \frac{1}{\hat{m}} p_4\left(\frac{\hat{\alpha}}{\hat{m}}, \frac{\hat{\beta}}{\hat{m}}\right)\right|^2 d\hat{\alpha}\, d\hat{\beta}$$

$$= \iint_{-\infty}^{\infty} |p_4(\alpha,\beta)|^2 \, d\alpha\, d\beta \qquad (11.102)$$

is simply the "volume" of the squared modulus of either the exit pupil function or the aperture stop function and, for the case of a clear aperture, it has a value equal to the area of the aperture stop. In fact, this will be true whenever the modulus of the aperture function is a zero-one function, even if there are phase variations across the aperture. Thus, whenever $|p_4(x, y)|$ satisfies this condition, we have

$$\iint_{-\infty}^{\infty} |p_4(\alpha,\beta)|^2 \, d\alpha\, d\beta = \text{area of aperture stop.} \qquad (11.103)$$

Then, Eq. (11.101) becomes

$$\mathscr{h}_{26}(x,y) = \frac{\left(\dfrac{1}{\lambda\hat{z}_{46}}\right)^2 \left|\hat{P}_4\left(\dfrac{x}{\lambda\hat{z}_{46}}, \dfrac{y}{\lambda\hat{z}_{46}}\right)\right|^2}{\text{area of aperture stop}}$$

$$= \frac{\left(\dfrac{\hat{m}}{\lambda\hat{z}_{46}}\right)^2 \left|P_4\left(\dfrac{\hat{m}x}{\lambda\hat{z}_{46}}, \dfrac{\hat{m}y}{\lambda\hat{z}_{46}}\right)\right|^2}{\text{area of aperture stop}}. \qquad (11.104)$$

Once the point spread function has been obtained from Eq. (11.101) or Eq. (11.104), the irradiance of the diffraction image is determined from

$$I_6''(x,y) = I_2''(x,y) ** \mathscr{h}_{26}(x,y), \qquad (11.105)$$

where $I_2''(x, y)$ is the geometrical image irradiance.

Unlike the coherent impulse response, which in general is complex valued, the incoherent impulse response is a real-valued, nonnegative function. Consequently, the nature of incoherently formed images is significantly different from those formed with coherent light, as we now explain. Every image, whether coherent or incoherent, is a superposition of weighted and shifted impulse response functions. If the impulse response functions are complex valued, as in the coherent case, destructive interference can occur and the resulting image irradiance may exhibit zeros at points corresponding to object points for which the transmittance (or reflectance) is not zero. The so-called *speckle effect* is a consequence of this behavior (see Ref. 11-7). However, when the impulse response functions are real and nonnegative, as for the case of incoherent illumination, the situation is different: at any given point, the value of the sum of several such functions can never be less than the value of any single component at that point. This statement may be expressed mathematically as follows:

$$\iint_{-\infty}^{\infty} I_2''(\alpha,\beta)\,\mathscr{h}_{26}(x-\alpha,y-\beta)\,d\alpha\,d\beta \geqslant I_2''(x_0,y_0)\,\mathscr{h}_{26}(x-x_0,y-y_0), \tag{11.106}$$

for all (x,y) and for every (x_0,y_0). As a result, incoherently formed images do not suffer from the effects of speckle.

Optical Transfer Function

The incoherent transfer function of an imaging system is referred to as the *optical transfer function* (OTF) and is given by

$$\begin{aligned}\mathscr{H}_{26}(\xi,\eta) &= \mathscr{F}\mathscr{F}\{\mathscr{h}_{26}(x,y)\}\\ &= \mathscr{F}\mathscr{F}\left\{\frac{|h_{26}(x,y)|^2}{\gamma_H(0,0)}\right\}\\ &= \frac{\gamma_H(\xi,\eta)}{\gamma_H(0,0)}.\end{aligned} \tag{11.107}$$

In terms of the exit pupil and aperture stop function we have

$$\begin{aligned}\gamma_H(\xi,\eta) &= \hat{m}\hat{p}_4(-\lambda\hat{z}_{46}\xi,-\lambda\hat{z}_{46}\eta)\star\star\,\hat{m}\hat{p}_4^*(-\lambda\hat{z}_{46}\xi,-\lambda\hat{z}_{46}\eta)\\ &= \left(\frac{\hat{m}}{\lambda\hat{z}_{46}}\right)^2\gamma_{\hat{p}_4}(-\lambda\hat{z}_{46}\xi,-\lambda\hat{z}_{46}\eta)\\ &= \left(\frac{\hat{m}}{\lambda\hat{z}_{46}}\right)^2\gamma_{\hat{p}_4}^*(\lambda\hat{z}_{46}\xi,\lambda\hat{z}_{46}\xi)\\ &= \left(\frac{\hat{m}}{\lambda\hat{z}_{46}}\right)^2\gamma_{\hat{p}_4}^*\left(\frac{\lambda\hat{z}_{46}\xi}{\hat{m}},\frac{\lambda\hat{z}_{46}\eta}{\hat{m}}\right).\end{aligned} \tag{11.108}$$

Then, Eq. (11.107) becomes

$$\mathcal{H}_{26}(\xi,\eta)=\frac{\left(\dfrac{\hat{m}}{\lambda\hat{z}_{46}}\right)^2\gamma_{\hat{p}_4}^*(\lambda\hat{z}_{46}\xi,\lambda\hat{z}_{46}\eta)}{\left(\dfrac{\hat{m}}{\lambda\hat{z}_{46}}\right)^2\displaystyle\iint_{-\infty}^{\infty}|\hat{p}_4(\hat{\alpha},\hat{\beta})|^2\,d\hat{\alpha}\,d\hat{\beta}}$$

$$=\frac{\gamma_{\hat{p}_4}^*(\lambda\hat{z}_{46}\xi,\lambda\hat{z}_{46}\eta)}{\displaystyle\iint_{-\infty}^{\infty}|\hat{p}_4(\hat{\alpha},\hat{\beta})|^2\,d\hat{\alpha}\,d\hat{\beta}}$$

$$=\frac{\gamma_{p_4}^*\left(\dfrac{\lambda\hat{z}_{46}\xi}{\hat{m}},\dfrac{\lambda\hat{z}_{46}\eta}{\hat{m}}\right)}{\displaystyle\iint_{-\infty}^{\infty}|p_4(\alpha,\beta)|^2\,d\alpha\,d\beta}, \tag{11.109}$$

and we see that the optical transfer function is simply the normalized complex autocorrelation function of either the exit pupil or the aperture stop. The numerator of these expressions may be calculated easily by using either of the following expressions:

$$\gamma_{\hat{p}_4}^*(\lambda\hat{z}_{46}\xi,\lambda\hat{z}_{46}\eta)=\left[\hat{p}_4(x,y)\star\star\hat{p}_4^*(x,y)\right]^*\Big|_{x=\lambda\hat{z}_{46}\xi,\;y=\lambda\hat{z}_{46}\eta}$$

$$=\left[p_4(x,y)\star\star p_4^*(x,y)\right]^*\Big|_{x=\lambda\hat{z}_{46}\xi/\hat{m},\;y=\lambda\hat{z}_{46}\eta/\hat{m}}. \tag{11.110}$$

The autocorrelation function of the exit pupil (or aperture stop) is first obtained in the space domain, and then its conjugate is found. Finally, a change of variables is made to obtain the proper frequency-domain scaling.

If the modulus of $p_4(x, y)$ is a zero-one function, then the denominator of Eq. (11.109) is again equal to the area of the aperture stop. In this event, the OTF may be determined by

$$\mathcal{H}_{26}(\xi,\eta)=\frac{\left[\hat{p}_4(x,y)\star\star\hat{p}_4^*(x,y)\right]^*\Big|_{x=\lambda\hat{z}_{46}\xi,\;y=\lambda\hat{z}_{46}\eta}}{\text{area of aperture stop}}$$

$$=\frac{\left[p_4(x,y)\star\star p_4^*(x,y)\right]^*\Big|_{x=\lambda\hat{z}_{46}\xi/\hat{m},\;y=\lambda\hat{z}_{46}\eta/\hat{m}}}{\text{area of aperture stop}}. \tag{11.111}$$

We now investigate the general behavior of the OTF. Because the OTF is the Fourier transform of the real-valued function $h_{26}(x, y)$, we know that *it will be hermitian*. If we write it in terms of amplitude and phase transfer functions, i.e.,

$$\mathcal{H}_{26}(\xi,\eta)=A_{\mathcal{H}}(\xi,\eta)e^{-j\Phi_{\mathcal{H}}(\xi,\eta)}, \tag{11.112}$$

it follows that $A_{\mathcal{H}}(\xi,\eta)$ is an even function and $\Phi_{\mathcal{H}}(\xi, \eta)$ is odd. We recall from Chapter 8 that, unless $\Phi_{\mathcal{H}}(\xi,\eta)$ is a linear phase function, the point spread function will be asymmetrical and the image will suffer from phase distortion.[††]

The OTF is not only hermitian—*it has a maximum at the origin*. This latter conclusion is derived from the fact that its behavior is governed by an autocorrelation function. Furthermore, as a consequence of the normalization used in its definition [see Eq. (11.107)], the OTF has a value of unity at the origin. Thus, we have

$$\begin{aligned}|\mathcal{H}_{26}(\xi,\eta)| &= |A_{\mathcal{H}}(\xi,\eta)| \\ &\leqslant \mathcal{H}_{26}(0,0) \\ &\leqslant 1.\end{aligned} \tag{11.113}$$

The value of $A_{\mathcal{H}}(\xi,\eta)$ at any pair of frequencies (ξ,η) specifies the attenuation of the Fourier component having those frequencies relative to the attenuation of the zero-frequency component. Clearly, then, no component of nonzero frequency will ever undergo less attenuation than the zero-frequency component. The fact that the modulus of the amplitude transfer function falls off with increasing $|\xi|$ and $|\eta|$ leads to images that are smoother than those obtained with coherent illumination. For example, incoherent images of edges do not exhibit the ringing found in coherent images of such objects.

Once the OTF has been determined, the spectrum of the diffraction image may be written as

$$\tilde{I}_6''(\xi,\eta)=\tilde{I}_2''(\xi,\eta)\mathcal{H}_{26}(\xi,\eta), \tag{11.114}$$

where

$$\begin{aligned}\tilde{I}_6''(\xi,\eta) &= \mathcal{F}\mathcal{F}\{I_6''(x,y)\} \\ \tilde{I}_2''(\xi,\eta) &= \mathcal{F}\mathcal{F}\{I_2''(x,y)\}.\end{aligned} \tag{11.115}$$

[††]Of course, for an unaberrated system with a clear aperture, the phase transfer function will be zero and no phase distortion will occur.

The final irradiance is then determined from

$$I_6''(x,y) = K''\mathscr{F}^{-1}\mathscr{F}^{-1}\{\tilde{I}_2''(\xi,\eta)\mathcal{H}_{26}(\xi,\eta)\}, \tag{11.116}$$

where $K'' = |B_{\ell 1}B_{\ell 2}|^2$.

The *cutoff frequency* of a system used for incoherent imaging is defined in the same manner as the cutoff frequency for coherent imaging: it is the value of the frequency variable beyond which the optical transfer function is zero. This quantity is specified by a single number for circularly symmetric exit pupils, but its value depends on the signal orientation when the pupil is not circularly summetric. This aspect was covered in the discussion of coherent transfer functions and will not be repeated.

We now determine the OTF and cutoff frequency for an imaging system having a circular aperture stop of diameter d. We first calculate the autocorrelation function of the aperture stop function:

$$\begin{aligned} p_4(r) \star\star p_4^*(r) &= \mathrm{cyl}\left(\frac{r}{d}\right) \star\star \mathrm{cyl}^*\left(\frac{r}{d}\right) \\ &= \frac{\pi d^2}{4}\gamma_{\mathrm{cyl}}\left(\frac{r}{d};1\right), \end{aligned} \tag{11.117}$$

where $\gamma_{\mathrm{cyl}}(r;1)$ is defined by Eq. (9.62). Then, from Eq. (11.111) we obtain

$$\begin{aligned} \mathcal{H}_{26}(\rho) &= \frac{\left[\dfrac{\pi d^2}{4}\gamma_{\mathrm{cyl}}\left(\dfrac{\lambda\hat{z}_{46}\rho}{\hat{m}d};1\right)\right]^*}{\dfrac{\pi d^2}{4}} \\ &= \gamma_{\mathrm{cyl}}\left(\frac{\lambda\hat{z}_{46}\rho}{\hat{m}d};1\right). \end{aligned} \tag{11.118}$$

This function is zero for $\rho \geqslant |\hat{m}d/\lambda\hat{z}_{46}|$; thus, the cutoff frequency is

$$\rho_c = \left|\frac{\hat{m}d}{\lambda\hat{z}_{46}}\right|. \tag{11.119}$$

Note that the cutoff frequency in the present case is *twice* the cutoff frequency obtained for coherent imaging, an observation that might tempt one to conclude that incoherent imaging is "better" than coherent imaging. Such a conclusion, however, would be tenuous at best; the comparison of coherent and incoherent imaging is as unwarranted as the proverbial comparison of apples and oranges. (We shall have more to say on this subject later in the chapter.) The PSF of this system is given by

$$\begin{aligned} h_{26}(x,y) &= \mathcal{H}_0\left\{ \gamma_{\text{cyl}}\left(\frac{\lambda \hat{z}_{46}\rho}{\hat{m}d}\,;1 \right) \right\} \\ &= \frac{\pi}{4}\left(\frac{\hat{m}d}{\lambda \hat{z}_{46}} \right)^2 \text{somb}^2\left(\frac{\hat{m}dr}{\lambda \hat{z}_{46}} \right), \end{aligned} \tag{11.120}$$

where $\mathcal{H}_0\{\ \}$ denotes the zero-order Hankel transform operator here.

Modulation Transfer Function

The modulus of the OTF,

$$|\mathcal{H}_{26}(\xi,\eta)| = |A_{\mathcal{H}}(\xi,\eta)|, \tag{11.121}$$

is called the *modulation transfer function* (MTF), and it is used extensively to characterize the performance of imaging systems. This characterization is incomplete, of course, because the MTF specifies only the relative attenuation of the various spectral components and includes no information about the relative phase of these components. Loss of the phase information can be quite significant because the effects of phase distortion are frequently more severe than those of amplitude distortion; in fact, this will nearly always be the case when aberrations are present. Nevertheless, the MTF can be a useful system descriptor.

We now give a physical interpretation of the MTF. Let us specify the object radiance to be the sum of a constant and a cosinusoidal function, such that the geometrical image produced by the system has the form

$$I_2''(x,y) = B + D\cos\left[2\pi(\xi_0 x + \eta_0 y) - \theta_0\right]. \tag{11.122}$$

Here B and D are real-valued constants such that $B \geqslant 0$ and $B \geqslant |D|$. The *modulation* of this distribution is defined to be

$$M_2'' = \frac{|D|}{B}, \tag{11.123}$$

i.e., it is the ratio of the magnitude of the variable part to that of the constant part. As we learned in Chapter 5, a cosinusoidal function is an eigenfunction of any LSI system that has a real-valued impulse response. Consequently, when the geometrical image is applied as the input to the LSI system associated with the imaging device (as depicted in Fig. 11-26), the output will have the form [see Eq. (5.39)]

$$I_6''(x,y)=B+DA_{\mathcal{H}}(\xi_0,\eta_0)\cos\left[2\pi(\xi_0 x+\eta_0 y)-\theta_0-\Phi_{\mathcal{H}}(\xi_0,\eta_0)\right]. \quad (11.124)$$

The modulation of this distribution is then

$$M_6''(\xi_0,\eta_0)=\frac{|DA_{\mathcal{H}}(\xi_0,\eta_0)|}{B}, \quad (11.125)$$

and it depends explicitly on the frequencies (ξ_0,η_0). If we now form the ratio of the output modulation to the input modulation (or diffraction image modulation to geometrical image modulation), we obtain

$$\frac{M_6''(\xi_0,\eta_0)}{M_2''}=|A_{\mathcal{H}}(\xi_0,\eta_0)|, \quad (11.126)$$

which is simply the MTF of the system evaluated at the spatial frequencies (ξ_0,η_0). Thus, the quantity $|A_{\mathcal{H}}(\xi_0,\eta_0)|$ describes the reduction in modulation of the (ξ_0,η_0) spectral component as it passes through the system. If the above procedure is carried out for inputs of all frequencies (ξ,η), the entire MTF is obtained:

$$\frac{M_6''(\xi,\eta)}{M_2''}=|A_{\mathcal{H}}(\xi,\eta)|. \quad (11.127)$$

Review of Incoherent Image Formation

As we did previously for coherent image formation, we now briefly review the subject of incoherent image formation and collect the important equations for the convenience of the reader. We use the two-element lens system of Fig. 11-23, and we make the following assumptions: the object is a highly diffuse, narrowband, incoherent radiator of radiance $N_2(x,y)$; the lens elements are aberration free; the aperture stop is located between the lens elements and is the limiting aperture for all object points (no vignetting); the Fresnel conditions are satisfied for propagation between the object and the entrance pupil; and, finally, the Fresnel conditions are satisfied for propagation from the exit pupil to the image plane.

The focal lengths of the two lens elements may be either positive or negative, the entrance pupil is located at the plane $z=\hat{\hat{z}}_4$, and the exit pupil is located at $z=\hat{z}_4$. The important relationships for incoherent imaging are then:

$$I_6(x,y) = K'' I_6''(x,y),$$

$$K'' = |B_{\ell 1} B_{\ell 2}|^2,$$

$$I_6''(x,y) = I_2''(x,y) ** h_{26}(x,y),$$

$$I_2''(x,y) = \hat{\hat{K}} \left(\frac{1}{m''}\right)^2 N_2\left(\frac{x}{m''}, \frac{y}{m''}\right),$$

$$h_{26}(x,y) = \frac{\left(\frac{1}{\lambda \hat{z}_{46}}\right)^2 \left|\hat{P}_4\left(\frac{x}{\lambda \hat{z}_{46}}, \frac{y}{\lambda \hat{z}_{46}}\right)\right|^2}{\iint_{-\infty}^{\infty} |\hat{p}_4(\hat{\alpha}, \hat{\beta})|^2 \, d\hat{\alpha} \, d\hat{\beta}}$$

$$= \frac{\left(\frac{\hat{m}}{\lambda \hat{z}_{46}}\right)^2 \left|P_4\left(\frac{\hat{m}x}{\lambda \hat{z}_{46}}, \frac{\hat{m}y}{\lambda \hat{z}_{46}}\right)\right|^2}{\iint_{-\infty}^{\infty} |p_4(\alpha, \beta)|^2 \, d\alpha \, d\beta},$$

$$\hat{\hat{K}} = \left(\frac{1}{\hat{\hat{z}}_{24}}\right)^2 \iint_{-\infty}^{\infty} |\hat{\hat{m}} \hat{\hat{p}}_4(\hat{\hat{\alpha}}, \hat{\hat{\beta}})|^2 \, d\hat{\hat{\alpha}} \, d\hat{\hat{\beta}}$$

$$= \left(\frac{\hat{\hat{m}}}{\hat{\hat{z}}_{24}}\right)^2 \iint_{-\infty}^{\infty} |p_4(\alpha, \beta)|^2 \, d\alpha \, d\beta,$$

$$\tilde{I}_6''(\xi,\eta) = \tilde{I}_2''(\xi,\eta) \mathcal{H}_{26}(\xi,\eta),$$

$$\tilde{I}_2''(\xi,\eta) = \hat{\hat{K}} \tilde{N}_2(m''\xi, m''\eta),$$

$$\mathcal{H}_{26}(\xi,\eta) = \frac{\gamma_{\hat{p}_4}^*(\lambda \hat{z}_{46}\xi, \lambda \hat{z}_{46}\eta)}{\iint_{-\infty}^{\infty} |\hat{p}_4(\hat{\alpha}, \hat{\beta})|^2 \, d\hat{\alpha} \, d\hat{\beta}}$$

$$= \frac{\gamma_{p_4}^*\left(\frac{\lambda \hat{z}_{46}\xi}{\hat{m}}, \frac{\lambda \hat{z}_{46}\eta}{\hat{m}}\right)}{\iint_{-\infty}^{\infty} |p_4(\alpha, \beta)|^2 \, d\alpha \, d\beta}. \qquad (11.128)$$

Helpful auxiliary expressions are

$$\hat{\hat{z}}_{24} = z_{23} + \hat{\hat{z}}_{34}, \qquad \hat{z}_{46} = \hat{z}_{45} + z_{56},$$

$$\hat{\hat{z}}_{34} = \hat{\hat{m}} z_{34}, \qquad \hat{z}_{45} = \hat{m} z_{45},$$

$$\hat{\hat{m}} = \frac{f_1}{f_1 - z_{34}}, \qquad \hat{m} = \frac{f_2}{f_2 - z_{45}},$$

$$\hat{\hat{p}}_4(x,y) = \frac{1}{\hat{\hat{m}}} p_4\left(\frac{x}{\hat{\hat{m}}}, \frac{y}{\hat{\hat{m}}}\right), \qquad \hat{p}_4(x,y) = \frac{1}{\hat{m}} p_4\left(\frac{x}{\hat{m}}, \frac{y}{\hat{m}}\right),$$

$$z_{56} = -\hat{m}\left(\frac{m'' \hat{\hat{z}}_{24}}{\hat{\hat{m}}} + z_{45}\right) = \frac{z_{35}\left(\dfrac{z_{23}}{f_1} - \dfrac{z_{25}}{z_{35}}\right)}{\dfrac{z_{23}}{f_1}\left(\dfrac{z_{35}}{f_2} - 1\right) - \dfrac{z_{25}}{f_2} + 1},$$

$$m'' = \frac{-\hat{\hat{m}} \hat{z}_{46}}{\hat{m} \hat{\hat{z}}_{24}} = \left[\left(\frac{z_{35}}{f_2} - 1\right)\left(\frac{z_{23}}{f_1} - 1\right) - \frac{z_{23}}{f_2}\right]^{-1}. \tag{11.129}$$

Example

We now illustrate the imaging of an incoherently illuminated object having a rectangle-wave radiance distribution. Specifically, we assume the object radiance $N_1(x, y)$ to have the same form as the transmittance function of Eq. (11.49), which was used in the example of coherent imagery. In addition, we assume the imaging system to be the same as in that example (e.g., the same lens elements, element separation, stop location, etc.). The calculation of the various associated expressions is left as an exercise for the reader, but the important quantities are shown graphically in Fig. 11-27. Note the difference between the image irradiance obtained in the present case and that obtained in the coherent imaging example (compare with Fig. 11-6). ■

Line Spread and Edge Response Functions

The incoherent line response is commonly called the *line spread function* (LSF), and it is related to the optical transfer function by [see Eq. (11.62)]

$$\ell_x(x) = \mathcal{F}^{-1}\{\mathcal{H}_{26}(\xi, 0)\}; \tag{11.130}$$

i.e., it is the one-dimensional inverse Fourier transform of the ξ-axis profile of the OTF. Even though the coherent and incoherent line response

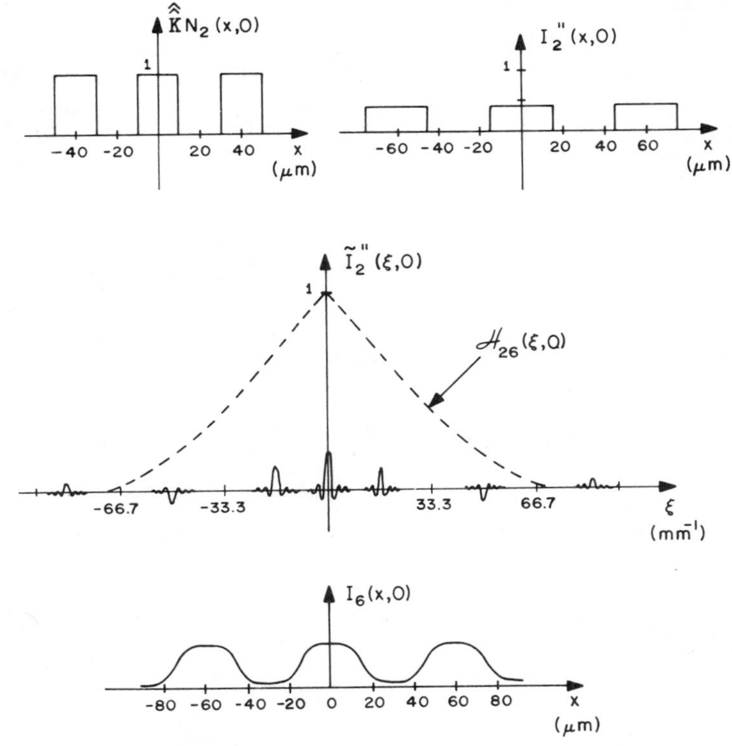

Figure 11-27 Effect of OTF on incoherent image formation.

functions are related to their respective transfer functions in the same manner, there is a significant difference in their behavior. You will recall that the coherent line response $l_x(x)$ is determined entirely by the x-axis profile of the aperture stop function, and that the exact shape of the aperture has no bearing on the form of $l_x(x)$. However, the incoherent LSF is highly dependent on the shape of the aperture due to the autocorrelation nature of the OTF.

We now investigate the LSF for a system having a circular aperture stop of diameter d. This function was first expressed mathematically by Struve (Ref. 11-8) without the aid of Fourier theory, and we show a method for calculating it with the use of Fourier theory. In doing so, we use the approximation of Eq. (9.63) to describe the OTF of the system, i.e.,

$$\gamma_{\text{cyl}}(r;1) \cong 0.25\left[5\,\text{tri}(r) - 1 + r^4\right]\text{cyl}\left(\frac{r}{2}\right). \tag{11.131}$$

First we set

$$F(\xi,\eta)=\gamma_{\text{cyl}}\left(\sqrt{\xi^2+\eta^2}\ ;1\right), \tag{11.132}$$

such that

$$\begin{aligned} F(\xi,0) &= \gamma_{\text{cyl}}(\xi;1) \\ &\cong 0.25\left[5\,\text{tri}(\xi)-1+\xi^4\right]\text{rect}\left(\frac{\xi}{2}\right). \end{aligned} \tag{11.133}$$

Then with $f(x)=\mathcal{F}^{-1}\{F(\xi,0)\}$, we obtain

$$\begin{aligned} f(x) &\cong \mathcal{F}^{-1}\left\{0.25\left[5\,\text{tri}(\xi)-1+\xi^4\right]\text{rect}\left(\frac{\xi}{2}\right)\right\} \\ &\cong 0.25\left[5\,\text{sinc}^2(x)-\delta(x)+\left(\frac{1}{2\pi}\right)^4\delta^{(4)}(x)\right]*2\,\text{sinc}(2x) \\ &\cong 0.25\left[5\,\text{sinc}^2(x)-2\,\text{sinc}(2x)+\left(\frac{1}{2\pi}\right)^4 2\,\text{sinc}^{(4)}(2x)\right], \end{aligned} \tag{11.134}$$

where

$$\text{sinc}^{(4)}(2x)=\frac{d^4}{dx^4}\,\text{sinc}(2x). \tag{11.135}$$

It can be shown that

$$\begin{aligned} \left(\frac{1}{2\pi}\right)^4 2\,\text{sinc}^{(4)}(2x) &= 2\left[\frac{1}{2\pi x}-\frac{12}{(2\pi x)^3}+\frac{24}{(2\pi x)^5}\right]\sin(2\pi x) \\ &\quad +2\left[\frac{4}{(2\pi x)^2}-\frac{24}{(2\pi x)^4}\right]\cos(2\pi x) \\ &= \left[1-\frac{12}{(2\pi x)^2}+\frac{24}{(2\pi x)^4}\right]2\,\text{sinc}(2x) \\ &\quad +\left[\frac{8}{(2\pi x)^2}-\frac{48}{(2\pi x)^4}\right]\cos(2\pi x). \end{aligned} \tag{11.136}$$

Thus, $f(x)$ becomes

$$f(x) \cong 1.25\,\mathrm{sinc}^2(x) + \left[\frac{12}{(2\pi x)^4} - \frac{6}{(2\pi x)^2}\right]\mathrm{sinc}(2x) + \left[\frac{2}{(2\pi x)^2} - \frac{12}{(2\pi x)^4}\right]\cos(2\pi x). \tag{11.137}$$

Finally, because

$$\mathcal{H}_{26}(\xi,0) \cong F\left(\frac{\lambda \hat{z}_{46}\xi}{\hat{m}d}, 0\right), \tag{11.138}$$

the LSF is simply given by

$$\ell_x(x) \cong \frac{\hat{m}d}{\lambda \hat{z}_{46}} f\left(\frac{\hat{m}dx}{\lambda \hat{z}_{46}}\right). \tag{11.139}$$

A graph of this function is shown in Fig. 11-28. Note that, unlike the coherent line response function of Fig. 11-7, the LSF has no zeros.

The *incoherent edge response* is again given by the integral of the line response function:

$$e_x(x) = \int_{-\infty}^{x} \ell_x(\alpha)\, d\alpha. \tag{11.140}$$

The edge response for a system having a circular pupil would be determined by integrating the expression of Eq. (11.137), and we leave that integration as an exercise for the ambitious reader. The resulting function, which is shown in Fig. 11-29, has been tabulated by Swantner and Hayslett

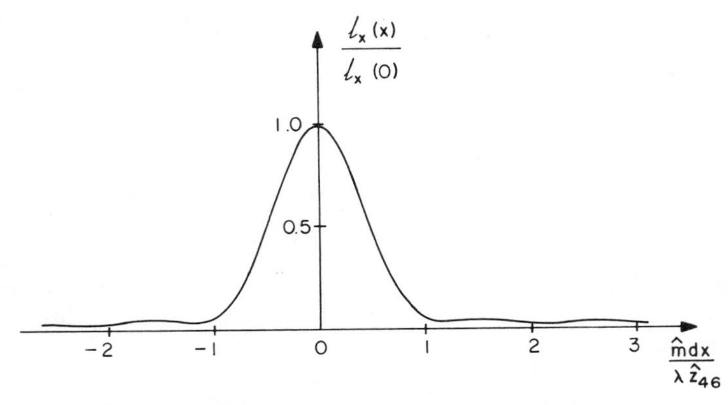

Figure 11-28 Normalized line spread function.

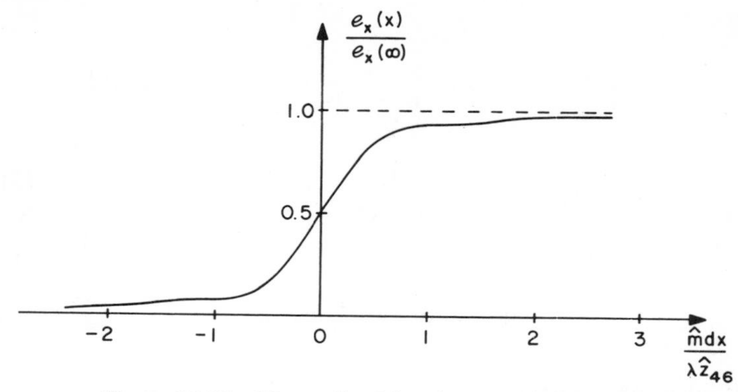

Figure 11-29 Normalized incoherent edge response.

(Ref. 11-9). If we again compare this edge response with that obtained for the coherent case (Figs. 11-7 and 11-8), we see that the previously observed ringing and apparent edge shifting are no longer present. In addition, $e_x(x)$ is now a monotonically increasing function of x.

11-7 SPECIAL CONFIGURATIONS FOR INCOHERENT IMAGING

As we did for coherent imaging, we now investigate a few special configurations of the general two-element lens systems shown in Fig. 11-23. Most of these configurations are similar to those discussed in Sec. 11-3.

Object in FFP of First Lens Element

With $z_{23}=f_1$ we find that

$$m''=\frac{-f_2}{f_1}, \qquad z_{56}=f_2. \tag{11.141}$$

Thus, *the image is always found in the BFP of the second lens element and the overall magnification is equal to the negative of the focal length ratio.*

Lens Elements Separated by the Sum of their Focal Lengths

This is the configuration once again for an *afocal system*, which was described in some detail in Sec. 11-3. As we found in that section,

$$m''=\frac{-f_2}{f_1}, \qquad z_{26}=\frac{(f_1+f_2)^2}{f_1}+z_{23}\left(1-\frac{f_2^2}{f_1^2}\right). \tag{11.142}$$

The *magnification is always independent of object position* and, in addition, the separation of object and image planes is independent of object position if $f_1 = f_2 = f$. In this case, $z_{26} = 4f$.

Separation of Lens Elements Equal to Focal Length of the Second

For this configuration we have $z_{35} = f_2$, which causes *the overall magnification to be inversely proportional to object distance*, i.e.,

$$m'' = \frac{-f_2}{z_{23}}. \tag{11.143}$$

Single Element Configuration: $f_1 = \infty$

By setting $f_1 = \infty$, we effectively eliminate the first lens element. In this event we obtain

$$\hat{\hat{m}} = 1, \qquad \hat{\hat{z}}_{24} = z_{24},$$

$$\hat{\hat{p}}_4(x, y) = p_4(x, y), \qquad z_{23} = 0,$$

$$z_{56} = \frac{z_{25} f_2}{z_{25} - f_2}, \qquad m'' = \frac{f_2}{f_2 - z_{25}}. \tag{11.144}$$

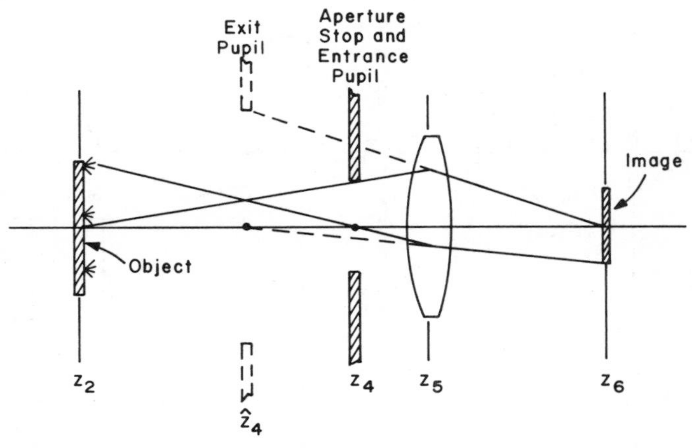

Figure 11-30 Single-element imaging system with stop between object and lens.

Thus, the aperture stop itself serves as the entrance pupil for this system, which is illustrated in Fig. 11-30.

Single Element Configuration: $f_2=\infty$

This configuration consists of a single lens element followed by the aperture stop as shown in Fig. 11-31. We now find

$$\hat{m}=1, \qquad \hat{z}_{46}=z_{46},$$

$$\hat{p}_4(x,y)=p_4(x,y), \qquad z_{56}=0,$$

$$z_{36}=\frac{z_{23}f_1}{z_{23}-f_1}, \qquad m''=\frac{f_1}{f_1-z_{23}},$$

$$\mathcal{H}_{26}(\xi,\eta)=\frac{\gamma_{p_4}^{*}(\lambda z_{46}\xi,\lambda z_{46}\eta)}{\int\int_{-\infty}^{\infty}|p_4(\alpha,\beta)|^2\,d\alpha\,d\beta}. \tag{11.145}$$

We now find that the aperture stop serves as the exit pupil.

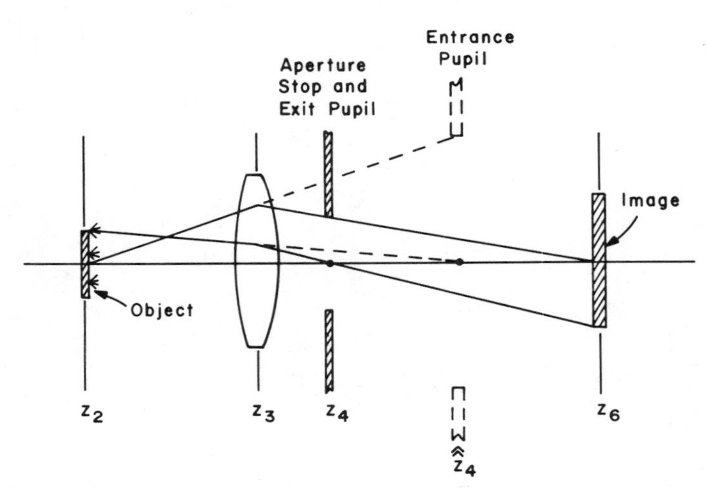

Figure 11-31 Single-element imaging system with stop between lens and image plane.

11-8 COHERENT AND INCOHERENT IMAGING: SIMILARITIES AND DIFFERENCES

We remarked earlier that comparisons between coherent and incoherent imaging may be improper and are likely to lead to incorrect conclusions; in this section we explore some of their similarities and differences.

Resolution

The *resolution* of an imaging system is generally regarded as a measure of the system's ability to distinguish between two closely spaced point sources, and it is therefore related to the width of the impulse response: the narrower the impulse response, the better the resolution (and vice versa). In discussing resolution, the assumption is usually made that the two point sources to be distinguished are of equal strength. This assumption is rather artificial, but the entire concept of two-point resolution is somewhat artificial except for a few well-defined imaging situations. Nevertheless, the general notion of system resolution can be useful in an imaging context.

There are a number of criteria used for specifying the resolution of a particular imaging system, with the *Rayleigh criterion* perhaps the most common. Based on this criterion, the resolution of a system having a circular aperture stop of diameter d is considered to be equal to the radius of the first zero of the impulse response, i.e.,

$$\text{resolution} = 1.22\left|\frac{\lambda \hat{z}_{46}}{\hat{m}d}\right|. \tag{11.146}$$

In other words, two point sources of equal strength are regarded as being "just resolved" if the separation of their ideal point images is equal to the distance specified by Eq. (11.146). Such a separation causes the peak of the diffraction image of each point to coincide with the first zero of the other. We point out, however, that the distance between the ideal image point and the first zero of the impulse response is not necessarily a good indicator of resolution for all imaging systems. To illustrate, consider the point spread functions depicted in Fig. 11-32; the distance to the first zero may be a good measure of resolution in Part (a), but it certainly is not in Part (b). Thus, we see that resolution depends on the overall behavior of the impulse response and not simply the radius of the first zero.

The resolution of an imaging system also depends on the coherence characteristics of the object to be imaged, as we now illustrate. Two incoherent point sources are indeed resolved by a system with a circular aperture when they are separated according to the Rayleigh criterion. However, if these two sources are mutually coherent, the ability of the

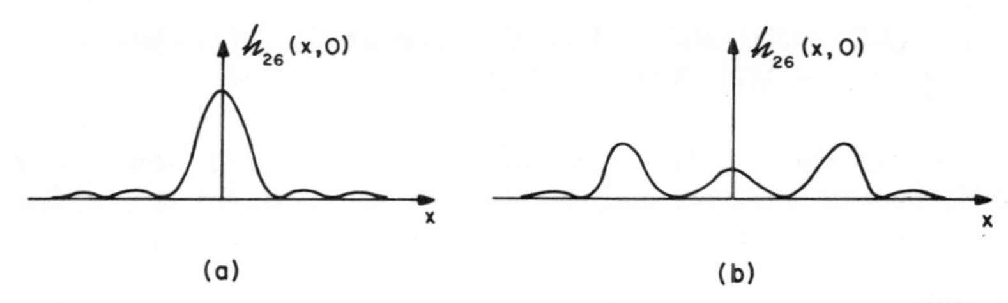

Figure 11-32 Effect of point spread function shape on resolution. (*a*) Radius of first zero a good measure of system resolution. (*b*) Radius of first zero not a good measure of system resolution.

system to resolve them is highly dependent on their relative phase. To show this we consider three situations; a phase difference of π, a phase difference of $\pi/2$, and no phase difference. For a phase difference of π, the two point sources are resolved more distinctly than for the incoherent situation, whereas, for a phase difference of $\pi/2$, the resolution is exactly the same as for the incoherent case. Finally, if the two point sources are in phase, they cannot be resolved.

The dependence of resolution on the degree of coherence and relative phase of the point sources may be understood with the following explanation. We assume that the coherent impulse response of the system is described by the function $h_{26}(x, y)$, and that the ideal geometrical images of the two sources are separated by the Rayleigh distance, which we denote by x_R. If the two sources are incoherent, the total image irradiance is simply the sum of the image irradiance distributions of the individual sources; thus, we might express the irradiance as

$$I_6(x,y) \propto |h_{26}(x,y)|^2 + |h_{26}(x-x_R,\, y)|^2. \tag{11.147}$$

If, on the other hand, the two sources are mutually coherent, the overall image irradiance is given by the squared modulus of the sum of the amplitude responses of the individual sources. Consequently, if we denote the phase difference between these two sources by Φ, the overall image irradiance may be expressed as

$$\begin{aligned} I_6(x,y) &\propto |h_{26}(x,y) + e^{j\Phi} h_{26}(x-x_R, y)|^2 \\ &\propto |h_{26}(x,y)|^2 + |h_{26}(x-x_R, y)|^2 \\ &\quad + e^{-j\Phi} h_{26}(x,y) h_{26}^*(x-x_R, y) \\ &\quad + e^{j\Phi} h_{26}^*(x,y) h_{26}(x-x_R, y). \end{aligned} \tag{11.148}$$

As can be seen, the first two terms of this expression are simply the image irradiance of two incoherent point sources; as a result, the overall irradiance of the image of two coherent point sources is simply a combination of the image irradiance obtained for two incoherent sources and a pair of interference terms. It is these interference terms that depend on the relative phase of the sources, and they in turn influence the behavior of the overall image irradiance. A detailed discussion of these effects may be found in Goodman (Ref. 11-1). In addition, an interesting discussion of the relationship between resolution and the degree of coherence of the sources is given by Grimes and Thompson (Ref. 11-10).

We add that a study of resolution is incomplete unless statistical considerations and measurement capabilities are included. However, the inclusion of these factors in a rigorous fashion is beyond the scope of this book. We merely remark that the resolution of an imaging system is dependent on the statistical behavior of the sources and the precision with which the image irradiance can be measured. For example, if the image irradiance can be measured with sufficiently high precision, it should be possible (theoretically) to resolve two point sources whose separation is much less than the Rayleigh distance and whose relative strengths are significantly different.

Cutoff Frequency

As mentioned previously, comparisons of system performance for coherent imaging and incoherent imaging are often inappropriate, and perhaps no comparison is more inappropriate than that of cutoff frequency. For a typical system, the incoherent cutoff frequency will nearly always be twice the coherent cutoff frequency. Nevertheless, because the system is linear in irradiance for incoherent objects and linear in complex amplitude for coherent objects, it is not fair to conclude that better images are obtained in the former case than in the latter. About the only thing that can be said with certainty is that the imaging process depends not only on the configuration of the image-forming elements and the coherence of the illumination source, but also on the spatial behavior of the object. In other words, the quality of an image is highly dependent on the object.

Although the cutoff frequency of an imaging system is an indicator of that system's performance, the behavior of the transfer function within the passband is perhaps more important. Consider the two OTF's shown in Fig. 11-33. Even though the cutoff frequency of the first is less than half that of the second, one should not necessarily conclude that it is a poorer system. Once again, the nature of the object is important in determining which system is better. To illustrate, let us consider a sinusoidal object with a frequency of $\xi_0 = 10$. Even though the image may suffer from amplitude

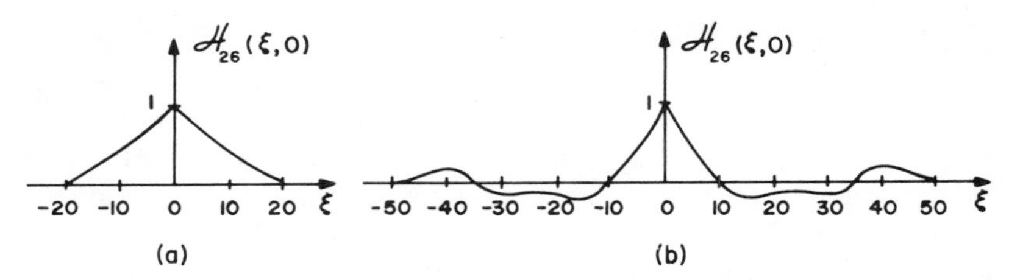

Figure 11-33 Optical transfer functions. (*a*) Unaberrated system with cutoff frequency of $\xi_c = 20$. (*b*) Aberrated system with cutoff frequency of $\xi_c = 50$.

distortion or a reduction of modulation, an image will be produced by the first system. The second system, on the other hand, will not form an image of the sinusoidal object because the OTF has a zero at $\xi = 10$.

If we next consider a periodic object with a fundamental frequency of $\xi_0 = 22$, we find the opposite situation. The first system will not image the periodic structure of this object because its fundamental frequency lies above the cutoff frequency of the system. On the other hand, the second system will form an image; this image may suffer from amplitude and phase distortion, but there will be an image. Consequently, we see that any statement regarding the imaging capabilities of a system may be entirely meaningless unless the conditions under which the system will be used are also specified.

Although the illustrations just presented were for incoherent imaging situations, the same conclusions may be drawn for coherent imaging.

Effects of Aberrations

All of the results derived in this chapter have been based on the assumption that field angles are small, optical elements are aberration free, etc. Consequently, these results must be altered whenever the assumed conditions are violated, and such violations are encountered frequently in the real world. If the various lens elements are not ideal, the wave field launched by the exit pupil (for each object point) will no longer be spherical and the impulse response will no longer be the Fraunhofer pattern of the aperture stop function. Even when the lens elements are ideal (or nearly so), a similar behavior is noted for object points lying far from the optical axis; this latter behavior may be attributed to a violation of the Fresnel diffraction conditions. At any rate, regardless of their origin, these effects may normally be associated with a *generalized exit pupil function* (see Goodman, Ref. 11-1).

Let us now consider the effect of a point source located at the point (x_0, y_0) in the object plane. We denote by $W(x, y; x_0, y_0)$ the difference in phase between an ideal spherical wave field and the effective aberrated wave field launched by the exit pupil. This function is referred to as an *aberration function*, and in general it is dependent on the location of the source. With $p_4(x, y)$ still representing the aperture stop, the generalized exit pupil function may be expressed as

$$\hat{p}_4(x,y) = \frac{1}{\hat{m}} p_4\left(\frac{x}{\hat{m}}, \frac{y}{\hat{m}}\right) e^{jk\,W(x,y;x_0,y_0)}. \tag{11.149}$$

As in the unaberrated case, the exit pupil function governs the performance of the system. The coherent transfer function of an aberrated system is given by

$$\begin{aligned} H_{26}(\xi,\eta) &= \hat{m}\hat{p}_4(-\lambda\hat{z}_{46}\xi, -\lambda\hat{z}_{46}\eta) \\ &= p_4\left(\frac{-\lambda\hat{z}_{46}\xi}{\hat{m}}, \frac{-\lambda\hat{z}_{46}\eta}{\hat{m}}\right) e^{jk\,W(-\lambda\hat{z}_{46}\xi,\, -\lambda\hat{z}_{46}\eta;\, x_0, y_0)}. \end{aligned} \tag{11.150}$$

In this form, we see that the coherent amplitude transfer function is governed by the aperture stop function, whereas the coherent phase transfer function is governed by the aberration function. As a result, the cutoff frequency of the system is unaffected by the presence of aberrations; their only effect is to alter the phase of the Fourier components whose frequencies lie within the passband of the system. This will lead to phase distortion, which can have quite severe consequences (see Chapter 8).

The situation is different for incoherent imaging because the incoherent transfer function is given by the normalized complex autocorrelation of the generalized exit pupil function; consequently, the presence of aberrations can alter the behavior of both the amplitude and phase transfer functions. Goodman (Ref. 11-1) shows that, for any specified pair of frequencies (ξ_0, η_0), the MTF of an aberrated system can never have a value greater than it would have were it free of aberrations and that in general it will have a smaller value at these frequencies. This result may be expressed mathematically by

$$|\mathcal{H}_{26}(\xi_0,\eta_0)|_{\substack{\text{with}\\\text{aberrations}}} \leqslant |\mathcal{H}_{26}(\xi_0,\eta_0)|_{\substack{\text{without}\\\text{aberrations}}}, \tag{11.151}$$

for all (ξ_0, η_0). Not only that, but aberrations can produce zeros at points where no zeros are found in the aberration-free MTF (see Fig. 11-34).

The exact behavior of the MTF (as well as of the phase transfer function) is dependent not only on the form of the aberration function W, but also on its magnitude. The effects of W will be entirely negligible—no

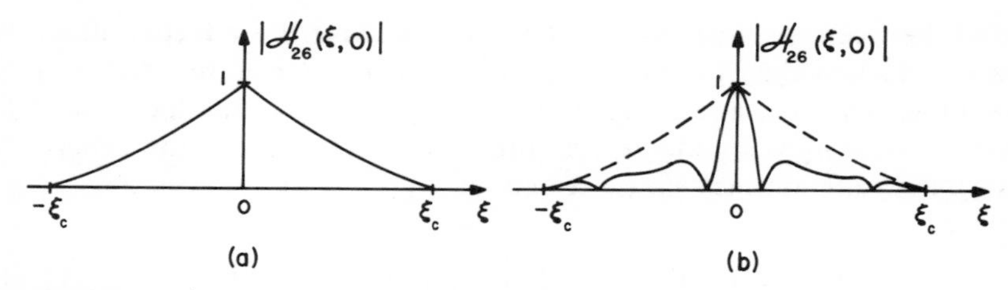

Figure 11-34 Effect of aberrations on MTF. (*a*) MTF of unaberrated system. (*b*) MTF of aberrated system.

matter what form it has—if its maximum value is small enough. On the other hand, even a well-behaved aberration function can produce undesired effects if its magnitude is great enough. To illustrate the former, we consider the function

$$e^{jkW(x,y;x_0,y_0)} = 1 + jkW(x,y;x_0,y_0) - \frac{k^2}{2}W^2(x,y;x_0,y_0) + \cdots. \tag{11.152}$$

If $W \ll \lambda/2\pi$, we have

$$e^{jkW(x,y;x_0,y_0)} \cong 1, \tag{11.153}$$

and the system behaves as an aberration-free system.

Even when W does not satisfy the above constraint, it has been found that many common aberrations do not seriously affect the imaging properties of a system as long as the maximum value of W does not exceed $\lambda/4$. This result has become known as *Rayleigh's quarter wavelength rule* (see Born and Wolf, Ref. 11-11), and is a useful criterion regarding the aberration tolerances of imaging systems. When a system suffers from aberrations, the peak value of its point spread function is generally less than the peak value of the point spread function of a similar, but unaberrated, system. The ratio of these peak values, known as the *Strehl ratio*, is useful as an indicator of the amount of aberration present. A system is often considered to be effectively unaberrated if it yields a Strehl ratio of 0.8 or higher, and for most of the commonly encountered aberrations (e.g. the Seidel aberrations discussed in Appendix 2), a Strehl ratio of 0.8 or higher is obtained as long as the aberration function satisfies the quarter wavelength rule. For less common aberrations, the quarter wavelength rule for W is not appropriate; Born and Wolf (Ref. 11-11) outline another rule, due to Maréchal, for these situations. This latter rule specifies that the root-mean-square value of the wavefront aberration function must not exceed $\lambda/14$ in order for a Strehl ratio greater than 0.8 to be obtained. When the aberrations exceed the limits discussed above, a Strehl ratio less than 0.8 generally results—in fact, it can be significantly less than 0.8. In addition, the shapes of the point spread function and optical transfer function may be altered substantially; this, in turn, can lead to a serious degradation of the imaging capabilities of the system.

A complete study of the effects of aberrations is not possible here, but the interested reader is encouraged to consult such references as Born and Wolf (Ref. 11-11) and Welford (Ref. 11-12) for more details. However, before leaving this subject, we must consider one final, and very important, aspect of aberrations. You will recall that we denoted the aberration function by $W(x, y; x_0, y_0)$, where (x_0, y_0) specified the location of the point source that produced the aberrated wavefield at the exit pupil. Note that if $W(x, y; x_0, y_0)$ indeed varies with (x_0, y_0), the impulse response will also vary with (x_0, y_0) and the system will no longer be shift invariant. Now the impulse response must be denoted by $h_{26}(x, y; x_0, y_0)$ [or $\mathscr{h}_{26}(x, y; x_0, y_0)$] to indicate that its form depends on the location of the source point. In addition, the transfer function must be denoted by $H_{26}(\xi, \eta; x_0, y_0)$ [or $\mathcal{H}_{26}(\xi, \eta; x_0, y_0)$] to show this dependence. In such a case, the system is no longer described by a single impulse response and a single transfer function; rather, it is described by families of these functions.

In some cases, the form of the impulse response remains nearly unchanged over some finite region of the image field, and it may be reasonable to describe the imaging characteristics over this region by a single impulse response and a single transfer function. The total number of different impulse response functions and transfer functions required to characterize the system over the entire image field is dependent on the number of these regions, known as *isoplanatic regions* (or isoplanatic patches), needed to fill the image field. Note that the various isoplanatic regions are not necessarily the same size; in fact, their size generally decreases with distance from the optical axis. As a result, one might expect the number of required isoplanatic regions to increase somewhat more rapidly than the square of the image field radius. Of course, the entire image field would be a single isoplanatic region for an ideal, aberration-free system.

Cascaded Systems

When two or more imaging systems are combined in cascade, it may be possible to describe the overall behavior of the combination in terms of a single equivalent shift-invariant system as discussed in Chapter 8. As you will recall, the impulse response of such a combination is given by the convolution of the individual impulse responses and the transfer function by the product of the individual transfer functions [see Eq. (8.59)]. However, for such a description to be valid, each of the component systems must itself be shift invariant—a condition that will be violated unless each of the component imaging systems has a magnification of unity. When the unity-magnification condition is not satisfied, the analysis must be approached in a different fashion. It is always possible to define an overall magnification as the product of the individual magnifications. In addition, an exit pupil and effective image distance can be determined for the overall system and, if no vignetting occurs as a result of the combination, the effective impulse response and transfer function can be obtained from the exit pupil in the usual fashion. Finally, we can model the cascade by combining a scaler, a multiplier, and an LSI system as we did previously (see Figs. 11-2 and 11-26).

One could always analyze each component system individually, determining the various intermediate images one at a time. However, as we now illustrate, caution is once again in order for incoherent imaging situations. Light emitted from an incoherent source becomes partially coherent as it propagates, and this may invalidate the assumption that each of the intermediate images behaves as an incoherent object for the following system. In addition, the assumption that the radiance of each intermediate image is independent of field angle may no longer be valid. Consequently, results obtained earlier in this chapter may be inappropriate in many situations involving cascaded systems. It is difficult to specify a universal set of results that apply to all imaging situations; however, there is one rule that should be applied universally: results should never be used unless it is first determined that the conditions for which they are valid have been satisfied.

REFERENCES

11-1 J. W. Goodman, *Introduction to Fourier Optics*, McGraw-Hill, New York, 1968.

11-2 D. A. Tichenor and J. W. Goodman, "Coherent Transfer Function," *J. Opt. Soc. Am.*, **62**(2):293 (1972).

11-3 J. D. Gaskill, "Demonstrations of Diffraction and Spatial Filtering," in *Novel Experiments in Physics: II*, American Association of Physics Teachers, Stony Brook, N.Y., 1975.

11-4 E. Hecht and Z. Zajac, *Optics*, Addison-Wesley, Reading, Mass., 1974.

11-5 M. J. Beran and G. B. Parrent, Jr., *Theory of Partial Coherence*, Prentice-Hall, Englewood Cliffs, N.J., 1964.

11-6 R. Kingslake, "Illumination in Optical Images," in *Applied Optics and Engineering*, Vol. II, R. Kingslake, Ed., Academic, New York, 1965.

11-7 P. S. Considine, "Effects of Coherence on Imaging Systems," *J. Opt. Soc. Am.*, **56**(8):1001 (1966).

11-8 H. Struve, *Ann. Physik*, **17**:1008 (1882).

11-9 W. H. Swantner and C. R. Hayslett, "Point Spread Functions, Edge Responses and Modulation Transfer Functions of Obscured Aperture Optical Systems," Research Projects Office Technical Memorandum 75-5, Instrumentation Directorate, WSMR, New Mexico, July 1975.

11-10 D. N. Grimes and B. J. Thompson, "Two Point Resolution with Partially Coherent Light," *J. Opt. Soc. Am.*, **57**(11):1330 (1967).

11-11 M. Born and E. Wolf, *Principles of Optics*, 3rd edition, Pergamon, New York, 1965.

11-12 W. T. Welford, *Aberrations of the Symmetrical Optical System*, Academic, London, 1974.

PROBLEMS

11-1. Use the configuration of Fig. 11-5 for this coherent imaging problem. A transparency with amplitude transmittance

$$t_2(x,y)=\left[\left(\frac{10}{d}\right)^2 \operatorname{comb}\left(\frac{10x}{d},\frac{10y}{d}\right) ** \operatorname{rect}\left(\frac{12.5x}{d},\frac{12.5y}{d}\right)\right] \times \operatorname{cyl}\left(\frac{\sqrt{x^2+y^2}}{d}\right)$$

is placed in the plane $z=z_2$ and illuminated with a monochromatic wavefield of wavelength λ. Additional information is as follows: $|A|^2=0.1$ mW/cm^2, $\lambda=514.5$ nm, $z_{13}=25$ cm, $z_{23}=4$ cm, $f_1=12$ cm, $z_{35}=29$ cm, $f_2=28$ cm, $d=1$ mm, $|B_{\ell 1}|=|B_{\ell 2}|=0.96$, and both lenses are 4 cm in diameter.

a. Find the locations of the frequency plane and the final image plane, and determine the overall magnification m''.
b. Find the coherent impulse response and coherent transfer function of the system when

$$p_4(r)=\operatorname{cyl}\left(\frac{r}{20d}\right).$$

Sketch profiles of these functions.

c. Find a reasonably accurate expression for the image irradiance $I_6(x, y)$, and sketch the profiles $I_6(x,0)$ and $I_6(0, y)$.

d. Compare the above results with Fig. 11-18.

11-2. A thin liquid gate‡ is now placed at the frequency plane of the system of Problem 11-1. Owing to a fabrication problem the gate exhibits a small wedge, and the effective aperture stop function is given by

$$p_4(x,y)=B_g\,e^{j2\pi(ax+by)}\,\mathrm{cyl}\left(\frac{\sqrt{x^2+y^2}}{20d}\right),$$

where $|B_g|=0.90$, $a=4\ \mathrm{mm}^{-1}$, and $b=2\ \mathrm{mm}^{-1}$. With the remaining parameter values as given in Problem 11-1:

a. Find the coherent impulse response and coherent transfer function, and sketch profiles of these functions. [Note: normalize such that $H_{26}(0,0)=1$.]

b. Find a reasonably accurate expression for the image irradiance $I_6(x, y)$, and sketch the profiles $I_6(x,0)$ and $I_6(0, y)$.

11-3. The liquid gate of Problem 11-2 is now removed from the frequency plane and placed at the object plane $z=z_2$.

a. Describe the behavior of the system in this configuration.

b. How would this behavior change for $a=40\ \mathrm{mm}^{-1}$ and $b=20\ \mathrm{mm}^{-1}$?

11-4. Given the same information as in Problem 11-1, find reasonably accurate expressions for the image irradiance $I_6(x, y)$, and sketch the profiles $I_6(x,0)$ and $I_6(0, y)$ when:

a. $p_4(x,y)=\mathrm{rect}\left(\frac{x}{20d},\frac{y}{d}\right)$.

b. $p_4(x,y)=\mathrm{rect}\left(\frac{x}{3d},\frac{y}{d}\right)$.

c. $p_4(x,y)=\mathrm{rect}\left(\frac{x}{d},\frac{y}{d}\right)$.

d. $p_4(x,y)=\mathrm{rect}\left(\frac{x}{20d},\frac{y}{d}\right)-\mathrm{rect}\left(\frac{x}{d},\frac{y}{d}\right)$.

e. Compare the above results with Figs. 11-19 through 11-22.

11-5. Design an afocal imaging system, for use with coherent light of wavelength 632.8 nm, to the following specifications: 16×24 mm object format, overall magnification $|m''|=2.5$, circular aperture stop with cutoff frequency in image space of $\rho_c \geqslant 20\ \mathrm{mm}^{-1}$, 10-cm maximum lens diameter.

‡Liquid gates are used to prevent undesirable effects caused by thickness variations in transparencies. We shall ignore any misfocusing produced by the axial thickness of the gate.

11-6. Use the configuration of Fig. 11-13 for this coherent imaging problem. A phase object is placed at the plane $z=z_2=z_3$ and illuminated with a monochromatic wave field of wavelength λ. The amplitude transmittance of the transparency is described by

$$t_2(x,y)=e^{j0.1\cos(2\pi\xi_0 x)}\,\mathrm{cyl}\left(\frac{\sqrt{x^2+y^2}}{10d}\right),$$

and additional information is as follows: $|A|^2=0.1\ \mathrm{mW/cm^2}$, $\lambda=632.8$ nm, $z_{12}=z_{13}=60$ cm, $f_1=15$ cm, $z_{35}=30$ cm, $f_2=15$ cm, $\xi_0=10\ \mathrm{mm}^{-1}$, $d=1$ mm, and $|B_{\ell 1}|=|B_{\ell 2}|=0.96$.

a. Find the image irradiance $I_6(x,y)$ when

$$p_4(r)=\mathrm{cyl}\left(\frac{r}{5d}\right).$$

b. Find the image irradiance when

$$p_4(r)=\mathrm{cyl}\left(\frac{r}{5d}\right)+\left[0.2e^{j0.5\pi}-1\right]\mathrm{cyl}\left(\frac{r}{0.5d}\right).$$

This describes an aperture with a small quarter-wave absorbing disc at the center, as used in phase-contrast imaging.

11-7. Use the configuration of Fig. 11-13 and the parameter values given for Problem 11-6, except that now the transparency is a rectangle-wave phase grating described by

$$t_2(x,y)=e^{-j\Phi(x)}\,\mathrm{cyl}\left(\frac{\sqrt{x^2+y^2}}{10d}\right),$$

where $\Phi(x)=\pi-\pi[\xi_0\,\mathrm{comb}(\xi_0 x)*\mathrm{rect}(4\xi_0 x)]$.

a. Find the image irradiance $I_6(x,y)$ when

$$p_4(r)=\mathrm{cyl}\left(\frac{r}{5d}\right).$$

b. Find the image irradiance $I_6(x,y)$ when

$$p_4(r)=\mathrm{cyl}\left(\frac{r}{5d}\right)-2\,\mathrm{cyl}\left(\frac{r}{0.5d}\right).$$

This describes an aperture with a small half-wave plate at the center. Note the similarity between this problem and Problem 8-4.

11-8. Use the configuration of Fig. 11-23 for this incoherent imaging problem. A telephoto lens, consisting of a positive lens followed by a negative lens, is to be used for imaging a distant scene that includes brick buildings, trees, etc. The iris diaphragm, which behaves as the aperture stop, is to be adjusted to the smallest diameter that will still permit the bricks to be observed in the image. Additional information is as follows: smallest period of brick pattern is 8 cm, $z_{23}=1000$ m, $f_1=10$ cm, $z_{34}=3$ cm, $z_{45}=2$ cm, $z_{35}=5$ cm, $f_2=-10$ cm, $\lambda=550$ nm, $p_4(r)\cong\mathrm{cyl}(r/d)$, diameter of the first lens element is 4 cm, and diameter of the second is 2 cm.

a. Find the location of the image plane and the overall magnification m''.
b. Ignoring any aberrations or atmospheric effects, find the smallest value of d that will permit the bricks to be observed in the image.
c. Using the value obtained for d in part b, find the effective F-number, $F^{\#}_{\mathrm{eff}}$.
d. Describe the equivalent single-lens system.

11-9. An incoherently illuminated object is to be photographed with a camera, which is modeled by the equivalent single-element system shown in Fig. 11-25. The object consists of four sets of white bar (rectangle-wave) patterns on a black background, each having a different period. Each pattern has a total of 25 white bars, the width of each bar is equal to the spacing of the bars, and the length of the bars is 10 times the bar width. If we denote the bar width of the ith pattern by b_i, we have $b_1=4.0$ mm, $b_2=2.0$ mm, $b_3=1.0$ mm, and $b_4=0.5$ mm. Two lenses are available for the camera, one having an equivalent focal length of 35 mm and the other of 105 mm. Each lens has an equivalent aperture stop diameter of 5.25 mm. In addition, two types of film are available, one having a cutoff frequency of 50 mm^{-1} and the other of 150 mm^{-1}. [Note: the cutoff frequency of a film is the value of the frequency variable above which sinusoidal components will not be recorded.] The equivalent object distance is 10.5 m, and we assume the wavelength of the illumination to be 500 nm.

a. Select the lens/film combination that will yield the "best" photograph of the object.
b. Does the choice of lens have an effect on the quality of the image irradiance incident on the film?
c. Sketch the image irradiance of a few bars of each pattern when

the lens selected for part a is used. Normalize your sketches in any convenient fashion.

11-10. A Cassegrain telescope with a 1.5-m aperture and a 12-m focal length is used for astronomical imaging. The primary-mirror cell has not been properly adjusted, with the result that the mirror exhibits a slight tilt. To simplify things, the telescope is modeled by the refractive system shown in Fig. 11-30, and the effective aperture stop is described by

$$p_4(x,y)=\left[\operatorname{cyl}\left(\frac{\sqrt{x^2+y^2}}{d}\right)-\operatorname{cyl}\left(\frac{\sqrt{x^2+y^2}}{0.25d}\right)\right]e^{jk(\gamma_x x+\gamma_y y)},$$

where $k=2\pi/\lambda$. Additional information is as follows: $z_{25}=\infty$, $f_2=9$ m, $z_{45}=4$ m, $d=1.5$ m, $\gamma_x=10^{-4}$, $\gamma_y=4\times10^{-4}$, $\lambda=550$ nm.

a. Find an expression for the transfer function, and sketch appropriate profiles of the amplitude and phase transfer functions.
b. Ignoring atmospheric turbulence, find a normalized expression for the irradiance of the image of a distant star located on the geometric axis of the telescope barrel. Sketch appropriate profiles.
c. Assuming the tilt error is not large enough to introduce appreciable off-axis aberrations, what effect does it have on the resolution of the telescope?

11-11. A multi-aperture telescope is to be constructed from four off-axis parabolic mirror segments, each of which has a diameter of 2.0 m. These four segments are arranged in a square array such that the center-to-center spacing of adjacent segments is 3.0 m. Under the assumption that proper alignment and phasing of the mirror segments can be achieved, the telescope is modeled by the refractive system of Fig. 11-31, and the effective aperture stop is described by

$$p_4(x,y)=\operatorname{cyl}\left(\frac{\sqrt{x^2+y^2}}{2d}\right)**\left(\frac{1}{1.5d}\right)^2\delta\delta\left(\frac{x}{1.5d}\right)\delta\delta\left(\frac{y}{1.5d}\right).$$

Additional information is as follows: $z_{23}=\infty$, $z_{34}=0$, $f_1=10$ m, $d=1$ m, $\lambda=550$ nm.

a. Find the transfer function, and sketch profiles along each axis and along a 45° diagonal.

b. Find the point spread function, and sketch profiles along each axis and along a 45° diagonal.

c. Suppose one of the mirror segments were to slip in its cell by just the amount necessary to introduce a half-wave of piston error, i.e., a phase shift of π radians over the entire segment. Find the value of the point spread function at the origin, and compare with the value determined in part b.

APPENDIX 1

SPECIAL FUNCTIONS

In this appendix we tabulate a number of commonly encountered functions. Included are

$$\mathrm{sinc}(x) = \frac{\sin(\pi x)}{\pi x}, \tag{A1.1}$$

$$\mathrm{sinc}^2(x) = \frac{\sin^2(\pi x)}{(\pi x)^2}, \tag{A1.2}$$

$$\mathrm{Gaus}(x) = e^{-\pi x^2}, \tag{A1.3}$$

$$\mathrm{somb}(x) = \frac{2J_1(\pi x)}{\pi x}, \tag{A1.4}$$

$$\mathrm{somb}^2(x) = \frac{4J_1^2(\pi x)}{(\pi x)^2}, \tag{A1.5}$$

where $J_\nu(x)$ is the νth order Bessel function of the first kind. The $\mathrm{somb}(x)$ and $\mathrm{somb}^2(x)$ functions were taken from Ref. A1-1, with permission of the authors. Although tabulations of the functions $J_0(x)$ and $J_1(x)$ may be found in many handbooks, we list them here for convenience.

In addition, the normalized cylinder-function cross correlation

$$\gamma_{\mathrm{cyl}}(r;a) = \frac{4}{\pi}\,\mathrm{cyl}(r) \star\star\, \mathrm{cyl}\left(\frac{r}{a}\right) \tag{A1.6}$$

is tabulated for various values of the parameter a. The mathematical description of this function is given by Eqs. (9.59) and (9.60).

Table A1-1

Special Functions

x	sinc(x)	sinc^2(x)	$J_0(x)$	$J_1(x)$	somb(x)	somb^2(x)	Gaus(x)	x
0.00	1.0000	1.0000	1.0000	0.0000	1.0000	1.0000	1.0000	0.00
0.10	0.9836	0.9675	0.9975	0.0499	0.9877	0.9756	0.9691	0.10
0.20	0.9355	0.8751	0.9900	0.0995	0.9515	0.9053	0.8819	0.20
0.30	0.8584	0.7368	0.9776	0.1483	0.8930	0.7975	0.7537	0.30
0.40	0.7568	0.5727	0.9604	0.1960	0.8152	0.6645	0.6049	0.40
0.50	0.6366	0.4053	0.9385	0.2423	0.7217	0.5209	0.4559	0.50
0.60	0.5046	0.2546	0.9120	0.2867	0.6170	0.3806	0.3227	0.60
0.70	0.3679	0.1353	0.8812	0.3290	0.5057	0.2558	0.2145	0.70
0.80	0.2339	0.0547	0.8463	0.3688	0.3929	0.1544	0.1339	0.80
0.90	0.1093	0.0119	0.8057	0.4059	0.2833	0.0803	0.0785	0.90
1.00	0.0000	0.0000	0.7652	0.4401	0.1812	0.0328	0.0432	1.00
1.10	−0.0894	0.0080	0.7196	0.4709	0.0902	0.0081	0.0223	1.10
1.20	−0.1559	0.0243	0.6711	0.4983	0.0133	0.0002	0.0108	1.20
1.30	−0.1981	0.0392	0.6201	0.5220	−0.0477	0.0023	0.0049	1.30
1.40	−0.2162	0.0468	0.5669	0.5419	−0.0920	0.0085	0.0021	1.40
1.50	−0.2122	0.0450	0.5118	0.5579	−0.1195	0.0143	0.0009	1.50
1.60	−0.1892	0.0358	0.4554	0.5699	−0.1315	0.0173	0.0003	1.60
1.70	−0.1515	0.0229	0.3980	0.5778	−0.1296	0.0168	0.0001	1.70
1.80	−0.1039	0.0108	0.3400	0.5815	−0.1164	0.0136	0.0000	1.80
1.90	−0.0518	0.0027	0.2818	0.5812	−0.0947	0.0090	0.0000	1.90
2.00	0.0000	0.0000	0.2239	0.5767	−0.0676	0.0046		2.00
2.10	0.0468	0.0022	0.1666	0.5683	−0.0381	0.0015		2.10
2.20	0.0850	0.0072	0.1104	0.5560	−0.0091	0.0001		2.20
2.30	0.1120	0.0125	0.0555	0.5399	0.0172	0.0003		2.30
2.40	0.1261	0.0159	0.0025	0.5202	0.0385	0.0015		2.40
2.50	0.1273	0.0162	−0.0484	0.4971	0.0538	0.0029		2.50
2.60	0.1164	0.0136	−0.0968	0.4708	0.0624	0.0039		2.60
2.70	0.0954	0.0091	−0.1424	0.4416	0.0644	0.0041		2.70
2.80	0.0668	0.0045	−0.1850	0.4097	0.0601	0.0036		2.80
2.90	0.0339	0.0012	−0.2243	0.3754	0.0507	0.0026		2.90
3.00	0.0000	0.0000	−0.2601	0.3391	0.0375	0.0014		3.00
3.10	−0.0317	0.0010	−0.2921	0.3009	0.0221	0.0005		3.10
3.20	−0.0585	0.0034	−0.3202	0.2613	0.0060	0.0000		3.20
3.30	−0.0780	0.0061	−0.3443	0.2207	−0.0092	0.0001		3.30
3.40	−0.0890	0.0079	−0.3643	0.1792	−0.0222	0.0005		3.40
3.50	−0.0909	0.0083	−0.3801	0.1374	−0.0320	0.0010		3.50
3.60	−0.0841	0.0071	−0.3918	0.0955	−0.0381	0.0015		3.60
3.70	−0.0696	0.0048	−0.3992	0.0538	−0.0400	0.0016		3.70

Table A1-1 (continued)

x	sinc(x)	sinc^2(x)	$J_0(x)$	$J_1(x)$	somb(x)	somb^2(x)	Gaus(x)	x
3.80	−0.0492	0.0024	−0.4026	0.0128	−0.0381	0.0015		3.80
3.90	−0.0252	0.0006	−0.4018	−0.0272	−0.0327	0.0012		3.90
4.00	0.0000	0.0000	−0.3971	−0.0660	−0.0246	0.0006		4.00
4.10	0.0240	0.0006	−0.3887	−0.1033	−0.0148	0.0002		4.10
4.20	0.0445	0.0020	−0.3766	−0.1386	−0.0043	0.0000		4.20
4.30	0.0599	0.0036	−0.3610	−0.1719	0.0059	0.0000		4.30
4.40	0.0688	0.0047	−0.3423	−0.2028	0.0149	0.0002		4.40
4.50	0.0707	0.0050	−0.3205	−0.2311	0.0218	0.0005		4.50
4.60	0.0658	0.0043	−0.2961	−0.2566	0.0262	0.0007		4.60
4.70	0.0548	0.0030	−0.2693	−0.2791	0.0279	0.0008		4.70
4.80	0.0390	0.0015	−0.2404	−0.2985	0.0268	0.0007		4.80
4.90	0.0201	0.0004	−0.2097	−0.3147	0.0233	0.0005	0.0000	4.90

REFERENCES

A1-1. C. R. Hayslett and W. H. Swantner, "Tables of the Coherent and Incoherent Impulse Responses of an Aberration Free Optical System with an Unobscured Circular Pupil," Research Projects Office Technical Memorandum 75-3, Instrumentation Directorate, WSMR, New Mexico, March 1975.

Table A1-2

Values of $\gamma_{cyl}(r; a)$ for Various Values of r and a.

r	$a=1.0$	$a=1.2$	$a=1.4$	$a=1.6$	$a=1.8$	$a=2.0$	$a=3.0$	$a=4.0$	$a=5.0$	$a=6.0$	r
0.00	1.0000	1.0000	1.0000	1.0000	1.0000	1.0000	1.0000	1.0000	1.0000	1.0000	0.00
0.10	0.8729	1.0000									0.10
0.20	0.7471	0.9051	1.0000								0.20
0.30	0.6238	0.7806	0.9173	1.0000							0.30
0.40	0.5046	0.6536	0.7982	0.9241	1.0000						0.40
0.50	0.3910	0.5291	0.6716	0.8091	0.9283	1.0000					0.50
0.60	0.2848	0.4099	0.5453	0.6837	0.8166	0.9312					0.60
0.70	0.1881	0.2984	0.4232	0.5568	0.6925	0.8221					0.70
0.80	0.1041	0.1970	0.3083	0.4331	0.5654	0.6992					0.80
0.90	0.0374	0.1089	0.2037	0.3159	0.4406	0.5721					0.90
1.00	0.0000	0.0391	0.1127	0.2089	0.3220	0.4467	1.0000				1.00
1.10		0.0000	0.0404	0.1156	0.2131	0.3268	0.9383				1.10
1.20			0.0000	0.0416	0.1180	0.2166	0.8362				1.20
1.30				0.0000	0.0425	0.1200	0.7173				1.30
1.40					0.0000	0.0431	0.5914				1.40
1.50						0.0000	0.4645	1.0000			1.50
1.60							0.3418	0.9413			1.60

Table A1-2 (continued)

r	$a=1.0$	$a=1.2$	$a=1.4$	$a=1.6$	$a=1.8$	$a=2.0$	$a=3.0$	$a=4.0$	$a=5.0$	$a=6.0$	r
1.70							0.2276	0.8422			1.70
1.80							0.1265	0.7256			1.80
1.90							0.0457	0.6005			1.90
2.00							0.0000	0.4736	1.0000		2.00
2.10								0.3495	0.9427		2.10
2.20								0.2333	0.8456		2.20
2.30								0.1302	0.7303		2.30
2.40								0.0470	0.6058		2.40
2.50								0.0000	0.4787	1.0000	2.50
2.60									0.3542	0.9437	2.60
2.70									0.2369	0.8478	2.70
2.80									0.1324	0.7334	2.80
2.90									0.0480	0.6094	2.90
3.00									0.0000	0.4823	3.00
3.10										0.3573	3.10
3.20										0.2394	3.20
3.30										0.1339	3.30
3.40										0.0486	3.40
3.50	0.0000	0.0000	0.0000	0.0000	0.0000	0.0000	0.0000	0.0000	0.0000	0.0000	3.50

APPENDIX 2

ELEMENTARY GEOMETRICAL OPTICS

We present the basic notions of geometrical optics in this Appendix because valuable information about the size, location, and aberrations of images can be obtained from geometric considerations alone. Once the fundamentals of geometrical optics are understood, the study of diffraction and its effects are much more easily undertaken.

A2-1 SIMPLE LENSES

Although mirrors play important roles in many optical systems, we shall restrict our attention here to systems containing only lenses; the behavior of mirrors is readily obtained by extending the studies of lenses. Contrary to popular opinion, a lens is not necessarily a single piece of glass with curved surfaces. In the most general case, a lens includes not only several pieces of glass, but baffles, mounting rings, and diaphragms as well, all held together by some sort of lens cell. For example, a high-quality 35-mm camera lens contains many glass elements, a focusing mechanism, and a variable iris diaphragm. However, we shall direct most of our attention to single-element lenses and their properties.

It would often be desirable for a lens to convert an incident spherical wave field into another spherical wave field. In practice, lenses are very seldom capable of this type of performance because of various difficulties encountered in the fabrication process; for example, there are problems in grinding and polishing the lens surfaces to the desired figure, the glass is inhomogeneous, it is difficult to center the lens elements in the cell, etc. Nevertheless, if such a lens could be manufactured, it would be theoretically possible to design ideal imaging systems for which point objects are mapped into ideal point images over some limited field of view (ignoring diffraction, of course). Thus, we now assume an ideal lens to be one that behaves as described above: it converts an incident spherical wave into another spherical wave.

There are several configurations of single-element lenses that exhibit this ideal behavior to some extent, and the most frequently encountered are: biconvex, plano-convex, meniscus convex, biconcave, plano-concave, and meniscus concave (see Fig. A2-1). The first three types are called *positive lenses*, and the last three are called *negative lenses*. A positive lens is one that causes a diverging spherical wave either to diverge less rapidly or to

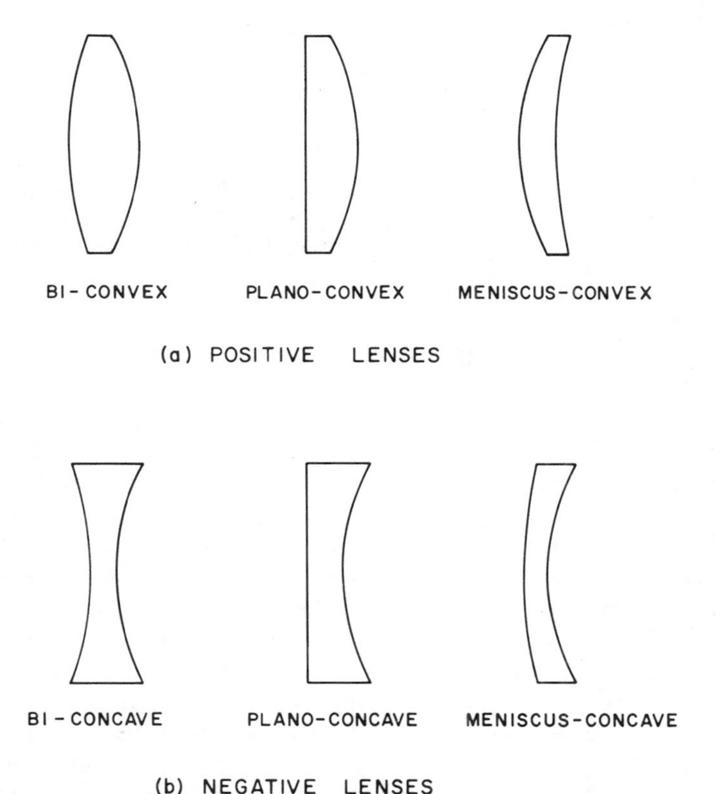

Figure A2-1 Single-element lenses of various shapes. (*a*) Positive lenses. (*b*) Negative lenses.

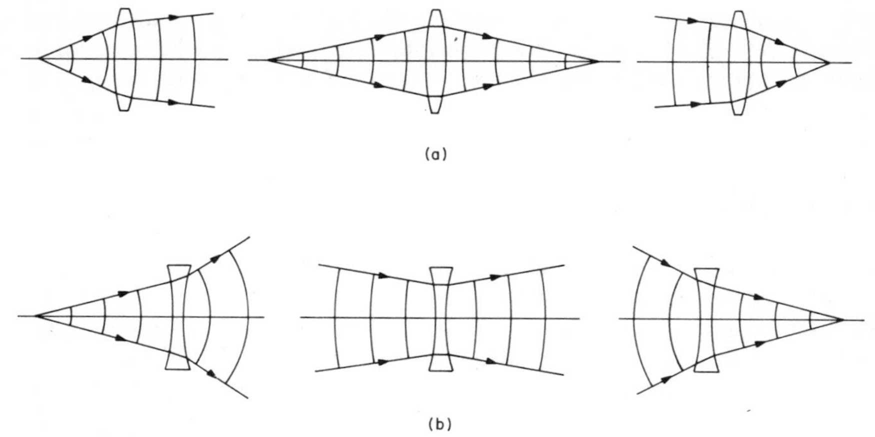

Figure A2-2 Effect of lens on incident spherical wave field. (*a*) Positive lens. (*b*) Negative lens.

converge, and causes a converging spherical wave to converge more rapidly. On the other hand, a negative lens causes a diverging spherical wave to diverge more rapidly, and causes a converging spherical wave either to converge less rapidly or to diverge. This behavior is illustrated in Fig. A2-2.

A2-2 CARDINAL POINTS OF A LENS

We now discuss the cardinal points for a simple lens. Let us assume that the lens surfaces are spherical, possibly of different radius, and construct a line through the centers of curvature of these surfaces. We shall refer to this line as the *optical axis* of the lens. The cardinal points of the lens are located on the optical axis, and there are six of them: the *first focal point* F_1, the *second focal point* F_2, the *first principal point* P_1, the *second principal point* P_2, the *first nodal point* N_1, and the *second nodal point* N_2. These points are illustrated in Fig. A2-3. This figure also shows the *first and second principal surfaces*; these two surfaces lie normal to the optical axis and contain the first and second principal points, respectively.

Spherical waves emanating from F_1 emerge from the lens as plane waves, or, in the terminology of ray optics, a ray passing through the point F_1 and striking the left side of the lens will emerge from the right side parallel to the optical axis. In addition, a ray traveling parallel to the optical axis and striking the left side of the lens will emerge from the right

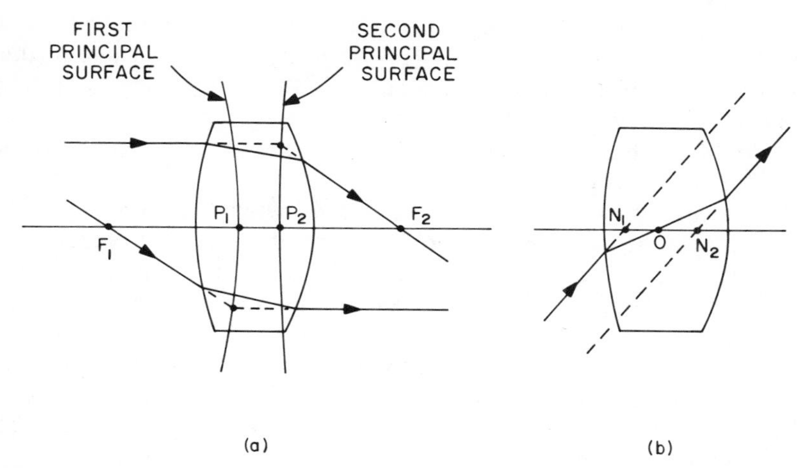

Figure A2-3 Cardinal points of a lens. (*a*) Focal points F_1 and F_2 and the principal points P_1 and P_2. (*b*) Nodal points N_1 and N_2.

side such that it passes through F_2. In the former case, the ray emanating from the point F_1 is refracted at each surface of the lens, but, if extensions of the incident ray and exiting ray are constructed, it will be found that they intersect at the first principal surface. In the latter case, extensions of the entering and exiting rays will intersect at the second principal surface. Because the principal surfaces are often found to be very nearly plane surfaces in the vicinity of the optical axis, they are commonly referred to as *principal planes*. These surfaces are depicted as planes in Fig. A2-4.

As indicated in Fig. A2-3(b), a ray passing through the *optical center* of the lens, point O, emerges parallel to the incident ray. If extensions of these

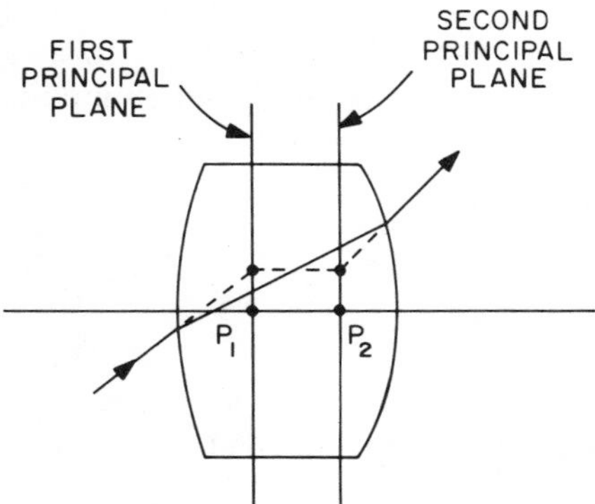

Figure A2-4. Method for determining ray paths through a lens.

incident and exiting rays are constructed, they will intersect the optical axis at the first and second nodal points, respectively. When the same medium is present on both sides of the lens, as for a lens in air, the nodal points and principal points coincide. This is the case for all of our discussions here. We note that for a general lens element, either or both of the principal points may lie outside the glass.

If the locations of the cardinal points are known, the behavior of the lens can be determined without worrying about the details of the refraction process at each surface. Thus, the lens can be effectively modeled by its two focal points and its two principal planes. It should be pointed out that all rays passing through the lens behave as if they travel directly to the first principal plane without refraction at the first glass surface, then parallel to the optical axis until they reach the second principal plane, and finally exit directly from that point without refraction at the second glass surface. This is indicated in Fig. A2-4.

A2-3 FOCAL LENGTH OF A LENS

Let us consider a general lens, the first and second surfaces of which have radii of curvature of R_1 and R_2, respectively. These radii may be positive or negative, and we choose the following convention for determining their polarity: R_1 is positive if its center of curvature lies to the right of the vertex V_1, and negative otherwise; R_2 is positive if its center of curvature lies to the left of the vertex V_2, and negative otherwise. For the biconvex lens illustrated in Fig. A2-5, both radii are positive.*

The distance from either principal point to the corresponding focal point is known as the *effective focal length* of the lens (or simply *focal length*), and we denote it by f. The focal length may be either positive or negative; when f is positive, F_1 lies to the left of P_1 and F_2 lies to the right of P_2; when f is negative, the converse is true. A lens with a positive focal length is depicted in Fig. A2-6. If the lens has a refractive index of n, an axial thickness of T, and is surrounded by air, the focal length is determined by the expression

$$\frac{1}{f}=(n-1)\left[\frac{1}{R_1}+\frac{1}{R_2}-\frac{(n-1)T}{nR_1R_2}\right], \tag{A2.1}$$

which is known as the lens maker's formula.

The *front focal length* f_1 of the lens is the distance from the vertex V_1 to the first focal point F_1, and the *back focal length* f_2 is the distance from V_2

*Note that this convention differs from that found in many optics texts.

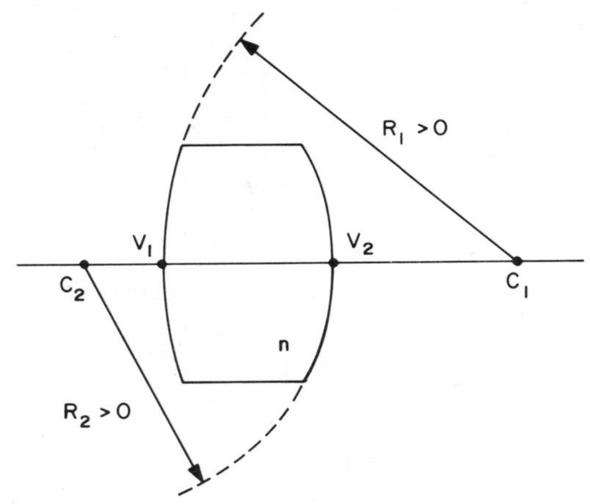

Figure A2-5 Centers of curvature C_1 and C_2, vertices V_1 and V_2, and radii of curvature R_1 and R_2 for a positive lens.

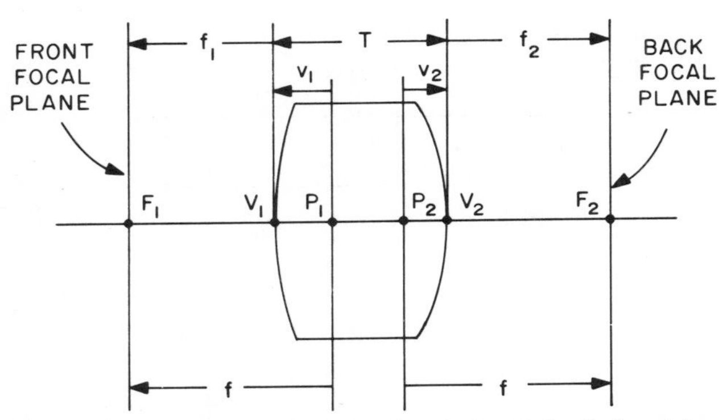

Figure A2-6 Front focal length f_1, front focal plane, back focal length f_2, back focal plane, and effective focal length f for a positive lens.

to F_2. The distances from P_1 to V_1 and from P_2 to V_2 are denoted by v_1 and v_2, respectively, as shown in Fig. A2-6, and are given by (see Ref. A2-1)

$$v_1 = \frac{f(n-1)T}{nR_2}, \tag{A2.2}$$

$$v_2 = \frac{f(n-1)T}{nR_1}. \tag{A2.3}$$

As a result we have

$$f_1 = f - v_1 = f\left[1 - \frac{(n-1)T}{nR_2}\right], \tag{A2.4}$$

$$f_2 = f - v_2 = f\left[1 - \frac{(n-1)T}{nR_1}\right]. \tag{A2.5}$$

For *thin lenses* T is small and the above expressions reduce to the following:

$$\frac{1}{f} \cong (n-1)\left[\frac{1}{R_1} + \frac{1}{R_2}\right], \tag{A2.6}$$

$$v_1 \cong v_2 \cong 0, \tag{A2.7}$$

$$f_1 \cong f_2 \cong f. \tag{A2.8}$$

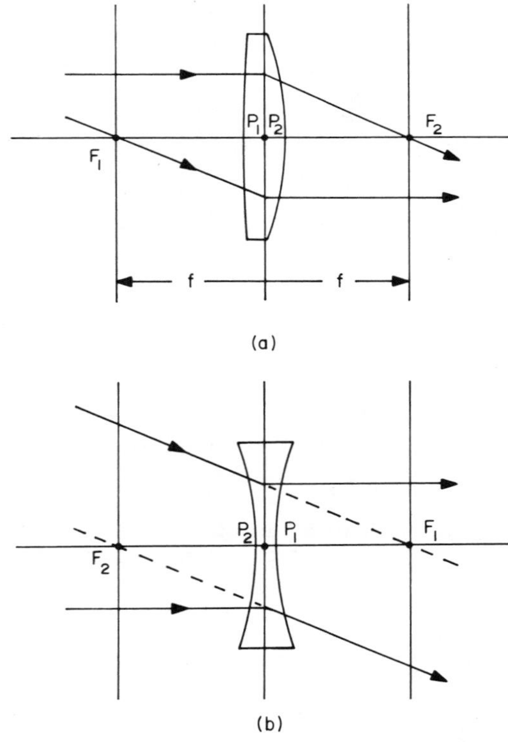

Figure A2-7 Determination of ray paths. (*a*) Positive thin lens. (*b*) Negative thin lens.

Thus, for thin lenses, it is usually assumed that the principal planes coincide as shown in Fig. A2-7.

A2-4 ELEMENTARY IMAGING SYSTEMS

We now investigate the geometrical imaging properties of the single-element biconvex lens modeled in Fig. A2-8. We assume that the cardinal points are known and that the object to be imaged is the arrow shown to the left of the lens. Each point on the object will emit spherical waves, some of which will be intercepted by the lens and brought to a focus at the image plane.

The location of the image may be determined as follows. We already know that a ray emanating from the object parallel to the optical axis will exit from the lens so as to pass through the second focal point F_2. We also know that a ray passing through the first focal point F_1 will exit from the lens parallel to the optical axis. If two such rays emanate from the same point on the object, the point at which they intersect in image space defines the location of the image plane. It can be shown that all other rays emanating from the same point on the object will also pass through this point in the inage plane, and thus each point on the object is mapped into a point in the image. We now define the additional symbols shown in Fig. A2-8, with the distances positive in the directions of the corresponding arrows and negative otherwise:

h_1—object height (positive upward),

d_1—object distance (distance from first principal plane to object; positive for object to left of P_1),

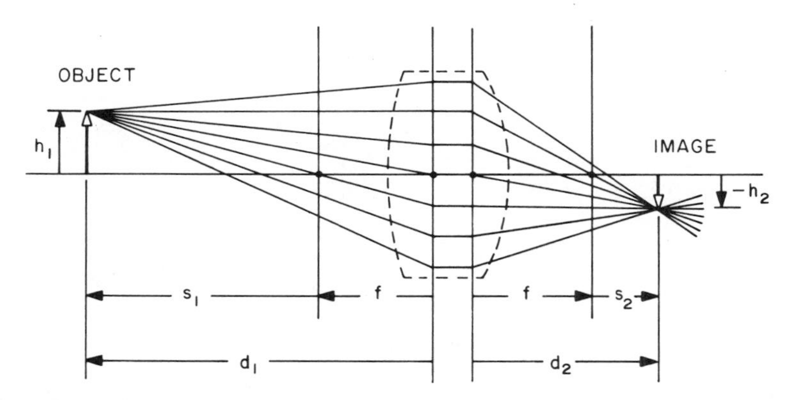

Figure A2-8 Determination of image location and magnification by ray tracing.

s_1—distance from F_1 to object (positive for object to left of F_1),
h_2—image height (positive upward),
d_2—image distance (distance from second principal plane to image; positive for image to right of P_2),
s_2—distance from F_2 to image (positive for image to right of F_2).

The relationships between the various parameters, distances, and heights can be shown to be as follows:

1. *Positive focal length* ($f > 0$).

$$\frac{1}{d_1}+\frac{1}{d_2}=\frac{1}{f}, \tag{A2.9}$$

$$d_2=\frac{d_1 f}{d_1-f}, \tag{A2.10}$$

$$s_1 s_2 = f^2. \tag{A2.11}$$

The *lateral magnification*, which is defined to be the ratio of the image height to object height, is related to these quantities in the following way:

$$m=\frac{h_2}{h_1}=\frac{-d_2}{d_1}=\frac{-f}{d_1-f}=\frac{f-d_2}{f}=\frac{-f}{s_1}=\frac{-s_2}{f}. \tag{A2.12}$$

Equation (A2.9) is often referred to as the *lens law*, and the object and image planes for which this expression is satisfied are called *conjugate planes*.

2. *Negative focal length* ($f < 0$).

$$\frac{1}{d_1}+\frac{1}{d_2}=-\frac{1}{|f|} \tag{A2.13}$$

$$d_2=\frac{-d_1|f|}{d_1+|f|} \tag{A2.14}$$

$$s_2=-\frac{3|f|+2s_1}{s_1/|f|+2} \tag{A2.15}$$

$$m=\frac{h_2}{h_1}=\frac{-d_2}{d_1}=\frac{d_2+|f|}{|f|}=\frac{|f|}{d_1+|f|}. \tag{A2.16}$$

A2-5 IMAGE CHARACTERISTICS

We have assumed that the light in our imaging system propagates from left to right, and if this light emanates from an object to the left of the lens, the object is a *real object*. If the resulting image lies to the right of the lens it is a *real image*, and if it lies to the left of the lens it is a *virtual image*. A real image is formed when the light is physically brought to a focus at the image location, whereas a virtual image is formed when the light only appears to be brought to a focus at the image position. To illustrate, when a magnifying glass is used to start a fire, the effect may be regarded as one of simply concentrating the incident radiant flux into a small area on the material to be burned. However, in an imaging context, the lens merely forms a real image of the sun on the material. Conversely, there is no physical concentration of radiant flux at the location of a virtual image. A jeweler's eye loupe may be used to form an enlarged virtual image of a small object at a comfortable viewing distance, but the light only appears to come from this image.

Often two or more lenses are arranged in a cascade. For such a configuration the image formed by the first lens acts as the object for the second lens, the image formed by the second lens acts as the object for the third, and so on. Let us now consider a two-lens imaging cascade. If the image formed by the first lens—whether real or virtual—lies to the left of the second lens, it behaves as a real object for that lens. If, on the other hand, the second lens intercepts the rays coming from the first lens before an image can be formed, then this would-be image must be regarded as a *virtual object* for the seond lens. In other words, if the image produced by the first lens lies to the right of the second lens when the second lens is removed, we have a virtual object situation. This implies that d_1 is negative.

If the lateral magnification m is positive, the image is erect; if m is negative, the image is inverted. If $|m|>1$, the image is larger than the object and, if $|m|<1$, the image is smaller than the object.

Although the image distance and magnification are easily determined by using Eqs. (A2.9)–(A2.16) above, it would often be convenient to have a graphic method for determining these quantities. Figure A2-9 was designed for just this purpose, and although its accuracy is limited, it can be quite helpful in visualizing how image location and magnification vary with changes of object position. We now explain its use:

1. *Positive lens.* We first locate the normalized object distance d_1/f on the object line of the chart, noting that this distance is positive when the object is to the left of the lens and negative when it is to the right; e.g., object A is

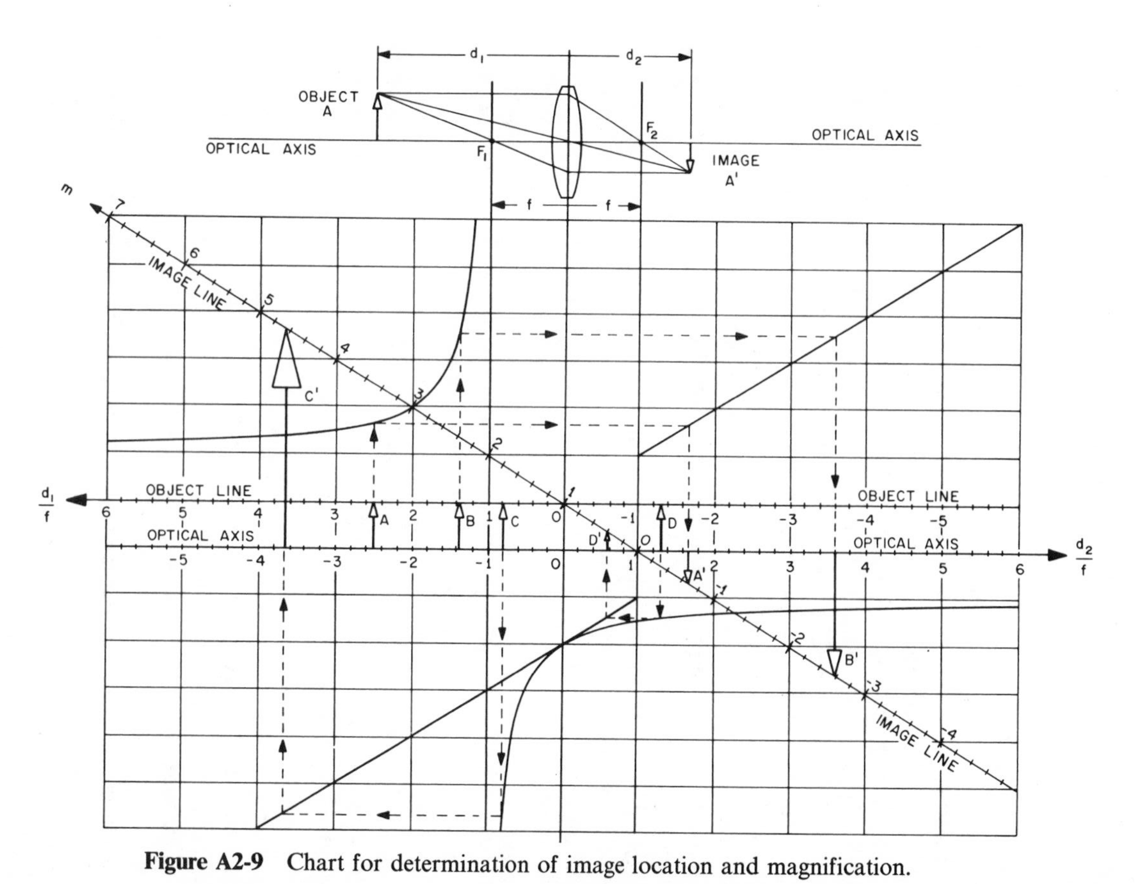

Figure A2-9 Chart for determination of image location and magnification.

located at a distance of $d_1/f = +2.5$ from the lens. From this point, proceed up (or down) to the hyperbola, then right (or left) to the heavy sloping line, and finally down (or up) to the optical axis. This point on the optical axis corresponds to the normalized image distance d_2/f, which is positive when the image is to the right of the lens and negative otherwise; e.g., image A' is located at a distance of $d_2/f \cong 1.67$. The head of the image arrow is located on the *image line* directly below (or above) the image distance point on the optical axis, and the magnification may be read directly from the scale on the image line; e.g., image A' displays a magnification of $m \cong -0.67$. Other examples are illustrated in Fig. A2-9 to facilitate the use of the chart, and all objects are erect and normalized in height for easy comparison.

2. *Negative lens*. The use of the chart is effectively the same for negative lenses; however, because f is now negative, the positive and negative directions for normalized object and image distance must be interchanged. To illustrate, object A is located at a normalized distance of $d_1/f = 2.5$, but since f is negative we have $d_1/|f| = -2.5$. Thus, A now represents a virtual object physically located to the right of the negative lens. This leads to a normalized image distance of $d_2/|f| = -d_2/f = -1.67$, which means that the image A' is a virtual image located to the left of the lens. The magnification is again found to be $m = -0.67$, and A' is again inverted.

A2-6 STOPS AND PUPILS

The *aperture stop* is the element in an imaging system that physically limits the angular size of the cone of light accepted by the system, and it therefore governs the total radiant flux reaching the image plane. It may be simply the edge of one of the lenses in the system, or it may be an opaque screen with a hole in it specifically introduced for the purpose. In a camera, the iris diaphragm acts as an aperture stop with a variable diameter. The behavior of an aperture stop is illustrated in Fig. A2-10.

The *field stop* is the element that physically restricts the size of the image (or field of view) as shown in Fig. A2-11. It may be an opaque screen with a hole in it specifically introduced for that purpose, or, as in a camera, the film may effectively serve as a field stop.

The *entrance pupil* is the image of the aperture stop, as viewed from object space, formed by all of the optical elements preceding it. Frequently it is a virtual image, as shown, in Fig. A2-12, and thus is the “apparent” limiting element for determining the angular size of the cone of light accepted by the system.

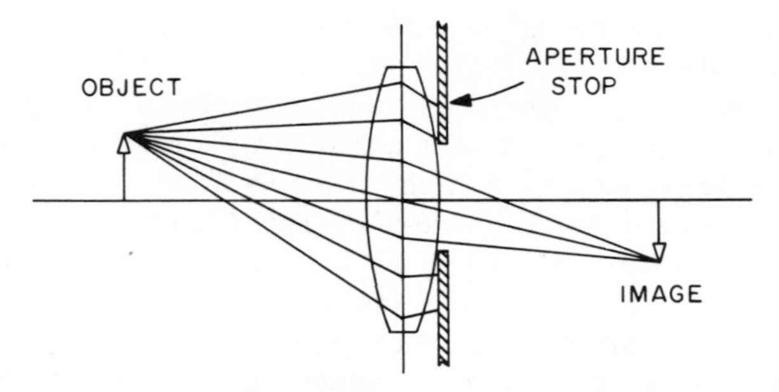

Figure A2-10 Effect of an aperture stop.

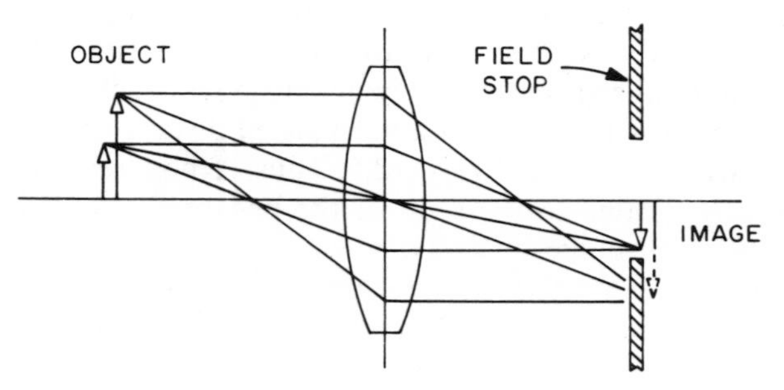

Figure A2-11 Effect of a field stop.

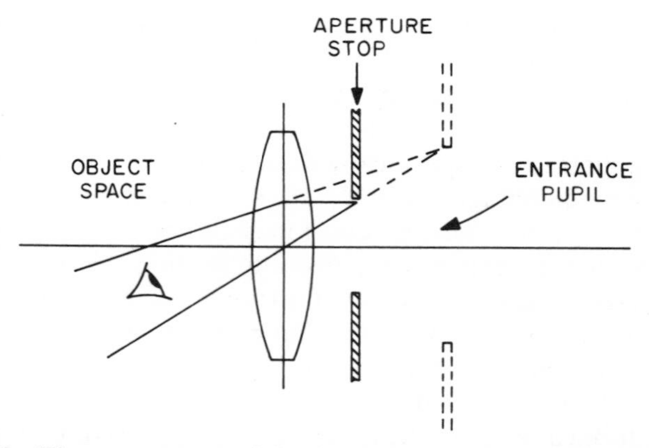

Figure A2-12 The entrance pupil is the image of the aperture stop seen from object space.

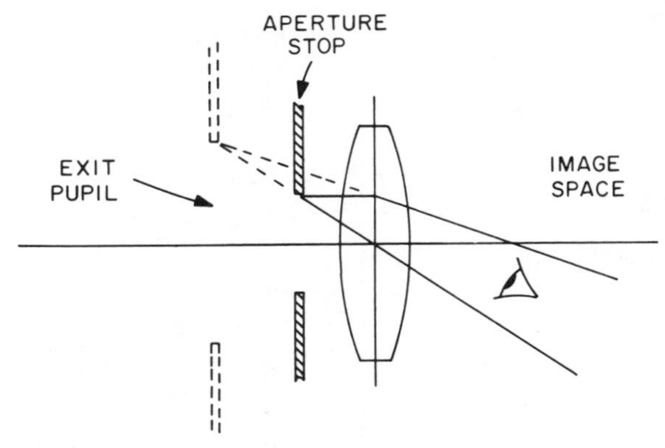

Figure A2-13 The exit pupil is the image of the aperture stop seen from image space.

The *exit pupil* is the image of the aperture stop, as seen looking back from image space, formed by all of the optical elements following it (see Fig. A2-13). The aberrations of a system, as well as its resolution, are often associated with the exit pupil. Ideally, for a point object, a spherical wave is launched by the exit pupil and converges to (or diverges from) an indeal point image.

A2-7 CHIEF AND MARGINAL RAYS

Any ray that emanates from an off-axis object point and physically passes through the center of the aperture stop is called a *chief ray*. A chief ray is directed toward the center of the entrance pupil as it enters the system and appears to emanate from the center of the exit pupil as it leaves the system.

Any ray emanating from an on-axis object point that physically grazes the rim of the aperture stop is called a *marginal ray*. A marginal ray appears to be directed toward the edge of the entrance pupil as it enters the system and appears to emanate from the edge of the exit pupil as it leaves the system. We note that an image plane is located at every axial position where the marginal ray crosses the optical axis, and that the height of the chief ray at such a point determines the height and magnification of the corresponding image.

In Fig. A2-14 we have illustrated the chief and marginal rays for a two-element lens with the aperture stop located between the elements. For the arrangement shown, both the entrance and exit pupils are virtual.

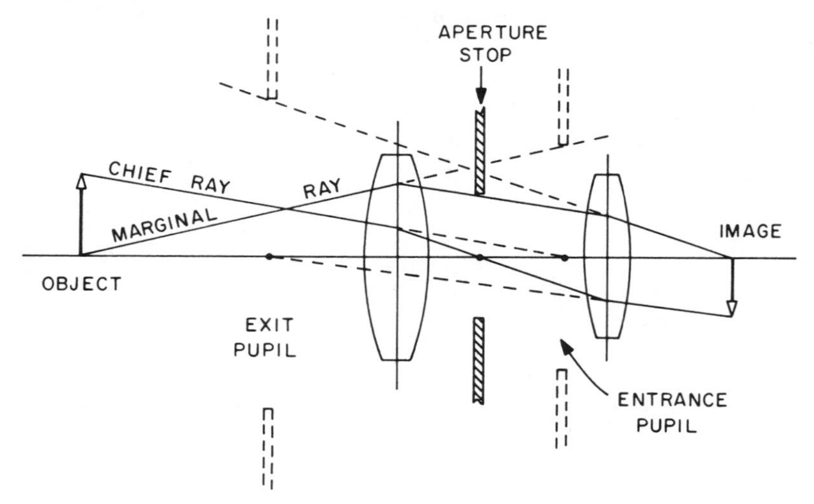

Figure A2-14 Chief and marginal rays.

A2-8 ABERRATIONS AND THEIR EFFECTS

The discussion so far has been based on the assumption that all of our imaging systems exhibit ideal behavior, i.e., they cause a point object to be mapped into a point image at the proper location. In practice, however, we may encounter *aberrations* that cause nonideal images to be formed, and these aberrations may be of such magnitude that they seriously degrade the image (even after diffraction effects are accounted for). The subject of optical aberrations is quite complex, and here we only mention the causes and effects of a few of the most commonly encountered aberrations. For those who are interested, a reasonably thorough introduction may be found in Ref. A2-1, and a more advanced development may be found in Ref. A2-2.

We may divide aberrations into two general categories: those that are wavelength dependent and those that are wavelength independent. The former are called *chromatic aberrations* and the latter *monochromatic aberrations*.

1. *Monochromatic aberrations*. The monochromatic aberrations we discuss here are the five *primary aberrations* of third-order theory (also known as Seidel aberrations): *spherical, coma, astigmatism, field curvature*, and *distortion*. The first three are aberrations that tend to blur or smear the image, and they preclude the formation of ideal point images. On the other

hand, the last two tend to deform the image in some fashion even though they permit ideal point image formation. For each of these primary aberrations, some of the rays emanating from an object point are prevented from passing through the associated ideal image point, and this results in some form of image degradation.

(a) *Spherical aberration.* Spherical aberration may be encountered for object points both on and off the optical axis, and its nature is illustrated in Fig. A2-15 for an on-axis object point. Those rays passing through the lens near the optical axis are brought to a focus in a different location than those passing through the lens near its edge. The point at which the ideal image is formed is often referred to as the *paraxial focus*; the transverse deviation of a ray from this point is called a *transverse (or lateral) aberration*, whereas the longitudinal deviation of a ray from this point is called a *longitudinal (or axial) aberration*. Both are known as *ray aberrations*, as opposed to *wavefront aberrations*.

The locus of the various refracted rays is called the *caustic*. The intersection of the caustic with the postfocus cone of marginal rays locates the *circle of least confusion*, which is often regarded as the "best" image of a point source when spherical aberration is present. As a result, the "best" image plane is no longer at the paraxial focal plane.

The severity of spherical aberration depends on the *third power of the aperture-stop diameter*, and it can therefore be reduced by decreasing the size of the aperture stop, i.e., by stopping the system down. In addition, if the object and image locations (conjugate planes) are fixed, the effects of spherical aberration can be minimized by the proper selection of lens shape. If one of the conjugate planes is farther from the lens than the other, as shown in Fig. A2-16, a single-element lens should generally

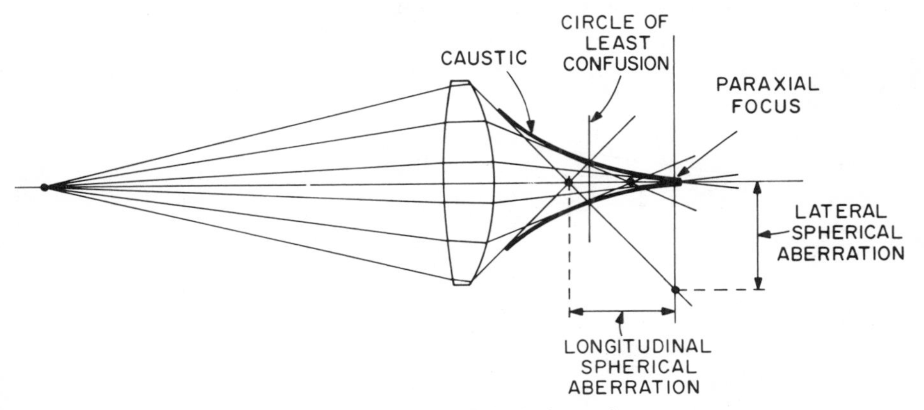

Figure A2-15 Spherical aberration.

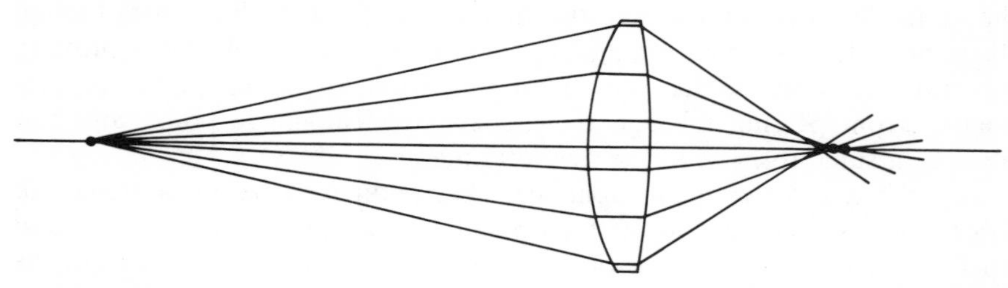

Figure A2-16 Proper orientation of lens to minimize spherical aberration.

"bulge" toward this more distant point (this is a good way to remember the proper orientation for a lens when minimum spherical aberration is desired).

(b) *Comatic aberration (coma).* Coma produces an asymmetric blurring of an image that is cometlike in shape (hence the name). It is encountered only for off-axis points, and its severity increases as the *square of the aperture-stop diameter and linearly with the off-axis distance of the object point*. As a result, coma may be reduced by stopping the system down and by limiting the size of the field stop. The effects of coma may also be minimized by shifting the position of the aperture stop appropriately (called stop shifting) or by selecting an appropriate lens shape. In fact, for fixed conjugates a lens shape can be found that will completely eliminate coma, and this shape will also be very nearly the shape required to minimize spherical aberration.

Portions of the comatic image are not found at the predicted location of the ideal image because, in making that prediction, the principal surfaces of the lens are considered to be planes when in fact they are not. The effects of coma on various rays are shown in Fig. A2-17.

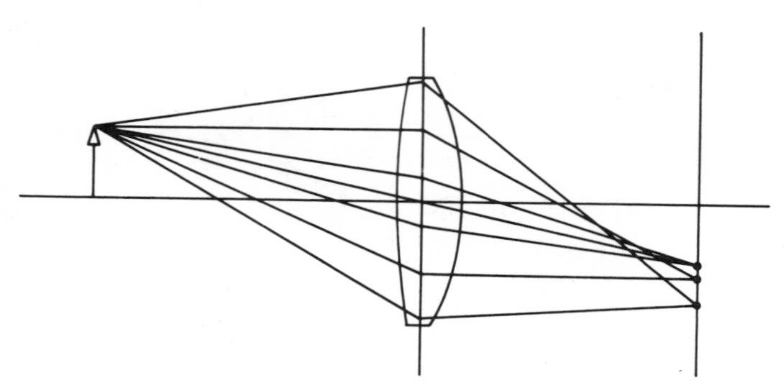

Figure A2-17 Coma.

(c) *Astigmatism.* Astigmatism is another off-axis aberration of the image-blurring type, and its severity varies as the *square of the off-axis distance but only linearly with the diameter of the aperture stop*—just the opposite of coma. As a result, it is less sensitive to a change in stop diameter, but it may be reduced by proper choice of lens shape.

We shall not discuss astigmatism in detail, but it arises basically because object rays lying in different planes are brought to focus at different distances from the lens. To illustrate, we define the *tangential plane* to be the plane containing both the chief ray and the optical axis, and the *sagittal plane* to be the plane that is both perpendicular to the tangential plane and contains the chief ray. Rays lying in the tangential plane are brought to focus at the *tangential image surface*, whereas rays lying in the sagittal plane are brought to focus at the *sagittal image surface*. As a result, the circle of least confusion will lie on an intermediate surface called the *medial image surface*. These three surfaces, designated T, S, and M, respectively, are illustrated in Fig. A2-18, which also shows their relationship to the *Petzval surface* (P) and the paraxial focal plane (see Ref. A2-1 for more details).

(d) *Field curvature.* When field curvature is present, object points lying in a plane are imaged onto the Petzval surface (see Fig. A2-18), which is a parabola of revolution. Even though an ideal point image is formed for each object point, the image points do not lie in a plane; as a result, it is not possible to obtain an image that is sharply in focus on a plane surface. Field curvature is sometimes corrected for by using field flattening lenses,

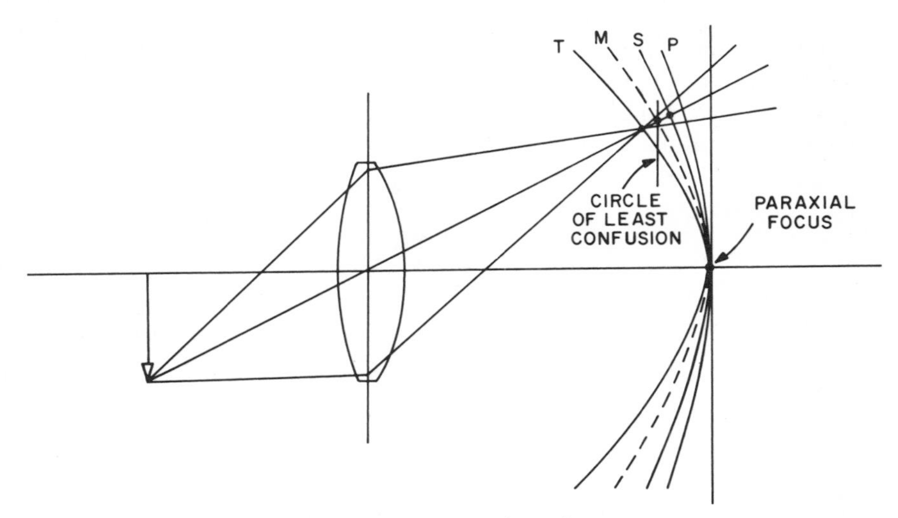

Figure A2-18 Astigmatism.

but since it is closely related to astigmatism, these two aberrations must generally be considered together. (In fact, an attempt to correct any of the primary aberrations generally has an effect on the magnitudes of one or more of the others.)

(e) *Distortion.* Distortion is an aberration produced when the lateral magnification varies with off-axis distance. If the magnification increases with off-axis distance, *pin-cushion distortion* results. On the other hand, a decrease of magnification with off-axis distance produces *barrel distortion*. The type of distortion, as well as its severity, may be associated with the location of the aperture stop. For example, a stop located in front of a thin positive lens will produce barrel distortion, whereas a stop placed behind this type of lens will cause pin-cushion distortion. If the stop is located at the lens, the distortion will vanish.

2. *Chromatic aberrations.* We now briefly discuss chromatic aberrations, which arise because of dispersion in the elements of the imaging system. Such aberrations are of little interest as long as light of a single wavelength is used (e.g., as in many laser applications), but they have significant effects when polychromatic light is used. Chromatic aberrations cause light of different wavelengths to be brought to focus at different locations even though this light emanates from the same object point. As a result, an image suffering from such aberrations can be in sharp focus for only one wavelength and will exhibit a wavelength-dependent magnification. Consequently, the overall image will be a superposition of largely out-of-focus images of different colors and sizes. Chromatic aberrations can be reduced by combining positive and negative elements, and by using elements of different refractive indices. They can also be minimized by restricting the spectral width of the light passing through the system.

We shall leave a more detailed investigation of chromatic aberrations to the reader (see Refs. A2-1 and A2-2).

REFERENCES

A2-1 Eugene Hecht and Alfred Zajac, *Optics*, Addison-Wesley, Reading, Mass., 1974.

A2-2 W. T. Welford, *Aberrations of the Symmetrical Optical System*, Academic, London, 1974.

Index